Basic Mathematics
through Applications

Annotated Instructor's Edition

Geoffrey Akst • **Sadie Bragg**
Borough of Manhattan Community College,
City University of New York

Addison
Wesley

Boston   San Francisco   New York
London   Toronto   Sydney   Tokyo   Singapore   Madrid
Mexico City   Munich   Paris   Cape Town   Hong Kong   Montreal

Publisher	Jason Jordan
Acquisitions Editor	Jennifer Crum
Project Manager	Kari Heen
Assistant Editor	Lauren Morse
Developmental Editor	Elka Block, Twin Prime Editorial
Managing Editor	Ron Hampton
Production Supervisor	Kathleen A. Manley
Production Services	Kathy Diamond
Copy Editor	Jerrold Moore
Proofreader	Jennifer Bagdigian
Art Editor	Janet Theurer, Theurer Briggs Design
Illustrator	Scientific Illustrators
Design Direction	Susan Carsten Raymond
Text Design	Delgado Design, Inc.
Cover Design and Illustration	Leslie Haimes
Composition	The Beacon Group, Inc.
Indexer	Bernice Eisen
Marketing Manager	Dona Kenly
Marketing Coordinator	Elan Hanson
Prepress Supervisor	Caroline Fell
Print Buyer	Evelyn Beaton

Photo Credits: p. 1: Jack Star/PhotoLink; p. 11: PhotoLink; p. 40: Reprinted by permission of the publisher from Ginsburg and Smith, *Numbers and Numerals*, (New York: Teachers College Press, © 1937 by Teachers College, Columbia University. All rights reserved, p. 22); p. 79: Ryan McVay; p. 107: PhotoLink; p. 149: Reprinted by permission of the publisher from Ginsburg and Smith, *Numbers and Numerals*, (New York: Teachers College Press, © 1937 by Teachers College, Columbia University. All rights reserved, p. 36); p. 161: PhotoLink; p. 162: PhotoDisc; p. 178: By permission of the British Library. C.112.c.8; p. 208: Sexto Sol; p. 221: Nick Rowe; p. 227: © the British Museum; p. 247: Hisham F. Ibrahim; p. 257L: PhotoLink; p. 257R: Stocktrek; p. 267: © Bob Krist/CORBIS; p. 279: PhotoLink; p. 289: Keith Brofsky; p. 293: S. Solum/PhotoLink; p. 300: PhotoLink; p. 302: © Archivo Iconografico, S.A./CORBIS; p. 337: Ryan McVay; p. 346: Nick Koudis; p. 348: PhotoDisc; p. 387: PhotoLink; p. 390: Reprinted by permission of the publisher from David, F.N., *Games, Gods and Gambling*, © 1962, Arnold Publishers, London; p. 427: Steven Peters/Stone; p. 428: Jules Frazier; p. 449: R. Morley/PhotoLink; p. 452L: The Metropolitan Museum of Art, Arthur Hoppock Hearn Fund, 1916. (16.53) Photograph © 1997 the Metropolitan Museum of Art; p. 452R: © Bettmann/CORBIS; p. 459: C. Borland/PhotoLink; p. 470: Reprinted by permission of the publisher from Struik, Dirk J., *A Concise History of Mathematics*, © 1967, Dover Publications Inc., NY; p. 479: PhotoLink; p. 489: Andy Sotiriou; p. 514: PhotoLink; p. 516: © Charles O'Rear/CORBIS

Library of Congress Cataloging-in-Publication Data
Akst, Geoffrey.
  Basic mathematics through applications / Geoffrey Akst and Sadie Bragg. — 1st ed.
    p. cm.
  ISBN 0-201-31222-0 (se pbk.) — ISBN 0-201-66224-8 (aie: pbk.)
    1. Mathematics.   I. Bragg, Sadie II. Title.
  QA39.2.A37 2000
  510 — dc21

                            00-022809

Printed in the U.S.A.
23456789-VH-04 03 02 01

Table of Contents

This book is dedicated to our mothers,
Mag Dora Chavis and Anne Akst.

Preface

From the Authors

Our goal in writing *Basic Mathematics through Applications* was to help motivate students and to establish a strong foundation for their success in a developmental mathematics program. Our text provides the appropriate coverage and review of whole numbers, fractions, decimals, ratio and proportion, percents, signed numbers, a preview of statistics, measurement and units, and an introduction to geometry. Compared to other texts on the market, we have introduced algebra earlier in our text to stress the importance of algebraic concepts and skills and to allow us to adopt an algebraic approach in successive chapters.

For all topics covered in this text, we have carefully selected applications that we believe are relevant, interesting, and motivating. This thoroughly integrated emphasis on applications reflects our view that college students need to master basic mathematics not so much for its own sake but rather to be able to apply this understanding to their everyday lives and to the demands of subsequent college courses.

Our goal throughout the text has been to address many of the issues raised by the American Mathematical Association of Two-Year Colleges and the National Council of Teachers of Mathematics by writing a flexible, approachable, and readable text that reflects

- an emphasis on applications that model real-world situations,
- explanations that foster conceptual understanding,
- exercises with connections to other disciplines,
- appropriate use of technology,
- an emphasis on estimation,
- the integration of geometric visualization with concepts and applications,
- exercises in student writing and groupwork that encourage interactive and collaborative learning, and
- the use of real data in charts, tables, and graphs.

The following key content and key features stem from our strong belief that mathematics is logical, useful, and fun.

Key Content

Applications   One of the main reasons to study mathematics is its applications to a wide range of disciplines, to a variety of occupations, and to everyday situations. Each chapter begins with a real-world application to show the usefulness of the topic under discussion and to motivate student interest. These opening applications vary widely from whole numbers in the census to fractions and the stock market crash of 1929, and decimals and blood tests (see pages 1, 79, and 161). When appropriate to the chapter content, applications are highlighted in section exercise sets with an Applications heading (see pages 50, 230, and 279).

Concepts   Explanations in each section foster intuition by promoting student understanding of underlying concepts. To stress these concepts, we included discovery-type exercises on reasoning and pattern recognition that encourage students to be logical in their problem-solving techniques, promote self-confidence, and allow students with varying learning styles to be successful (see pages 29, 291, and 321).

Skills   Practice is necessary to reinforce and maintain skills. In addition to comprehensive chapter problem sets, chapter review exercises include mixed applications, requiring students to use skills learned in previous sections and chapters (see pages 155, 256, and 333).

Writing   Writing both enhances and demonstrates students' understanding of concepts and skills. In addition to the user-friendly worktext format, open-ended questions throughout the text give students the opportunity to explain their answers in full sentences. Students can build on these questions by keeping individual journals. Suggestions for optional journal entries are included only in the Annotated Instructor's Edition (see pages 6, 43, and 203).

Estimation   Students need to develop estimation skills to distinguish between reasonable and unreasonable solutions, as well as to check their solutions. The chapters on whole numbers (Chapter 1), fractions (Chapter 2), decimals (Chapter 3), and percents (Chapter 6) cover these skills (see pages 1, 79, 161, and 289).

Use of Geometry   Students need to develop their abilities to visualize and compare objects. Throughout this text, students have opportunities to use geometric concepts and drawings to solve problems (see pages 16, 38, and 230). In addition, Chapter 11 is dedicated to geometry topics (see page 489).

Use of Technology   Each student should be familiar with a range of problem-solving techniques — mental, paper-and-pencil, and calculator arithmetic — depending on the problem and the student's level of mathematical preparation. This text includes optional calculator inserts (see pages 57, 193, and 311), which provide explanations for calculator techniques. These inserts also feature paired side-by-side examples and practice exercises, as well as a variety of optional calculator exercises in the section exercise sets that use the power of scientific calculators to perform arithmetic operations (see pages 197, 217, and 285). Both calculator inserts and calculator exercises are indicated by a calculator icon. At the end of the text is an Introduction to Calculators that combines and extends the calculator inserts.

Key Features

Side-by-Side Example/Practice Format   A distinctive side-by-side format pairs each numbered example with a corresponding practice exercise, encouraging students to get actively involved in the mathematical content from the start. Examples are immediately followed by solutions so that students can have a ready guide to follow as they work (see pages 5, 135, and 203).

Historical Notes   To show how mathematics has evolved over the centuries — in many cultures and throughout the world — each chapter features a compelling historical note that investigates and illustrates the origins of mathematical concepts. Historical notes give students further evidence that mathematics grew out of a universal need to find efficient solutions to everyday problems. Diverse topics include the evolution of digit notation, the popularization of decimals, and the ancient practice of using a scale to find an unknown weight (see pages 40, 178, and 227). Each historical note is indicated by an icon.

Mindstretchers   For every appropriate section in the text, related investigation, critical thinking, mathematical reasoning, pattern recognition, and writing exercises — along with corresponding groupwork and historical connections — are incorporated into one broad-ranged problem set called a mindstretcher. Mindstretchers target different levels and types of student understanding and can be used for enrichment, homework, or extra credit (see pages 130, 188, and 272). Each mindstretcher is indicated by an icon.

Tips   Throughout the text, students find helpful suggestions for understanding certain concepts, skills, or rules, and advice on avoiding common mistakes (see pages 16, 56, and 137).

Pretests and Posttests   To promote individualized learning—particularly in a self-paced or lab environment—pretests and posttests help students gauge their level of understanding of chapter topics both at the beginning and at the end of each chapter. The pretests and posttests also allow students to target topics for which they may need to do extra work to achieve mastery. All answers to pretests and posttests are given in the answer section of the Student Edition.

Section Objectives   At the beginning of each section, clearly stated learning objectives help students and instructors identify and organize individual competencies covered in the upcoming content.

For Extra Help   Several valuable study aids accompany this text. At the beginning of every exercise set are references to the appropriate videotape, the tutorial software, the AWL Tutor Center, the *Student's Solutions Manual,* and the www.awlonline.com/akstbragg Web site, to make it easy for students to find the correct support materials.

Key Concepts and Skills   At the end of each chapter, a comprehensive chart organized by section relates the key concepts and skills to a corresponding description and example, giving students a unique tool to help them review and translate the main points of the chapter.

Chapter Review Exercises   Following the Key Concepts and Skills at the end of each chapter, a variety of relevant exercises organized by section help students test their comprehension of the chapter content. As mentioned earlier, included in these exercises are mixed applications, which give students an opportunity to practice their reasoning skills by requiring them to choose and apply an appropriate problem-solving method from several previously presented (see pages 256, 285, and 333).

Cumulative Review Exercises   At the end of Chapter 2 and for every chapter thereafter, Cumulative Review Exercises help students maintain and build on the skills learned in previous chapters.

Supplements for the Student

Student's Solutions Manual (ISBN 0-201-66228-0)
This useful manual contains solutions to all odd-numbered exercises in each exercise set and solutions to all chapter pretests and posttests, practice exercises, review exercises, and cumulative review exercises.

Videotapes (ISBN 0-201-70290-8)
This new video series includes a section-by-section correlation with many odd-numbered exercises taken from the corresponding text exercise set. Each video segment includes a stop-tape feature that encourages students to stop the tape, work an example, and resume playing the tape to go over the solution with the video instructor.

Digital Video Tutor (ISBN 0-201-70981-3)
AWL mathematics videos are available on CD-ROM, making it easy and convenient for students to watch video segments from a computer either at home or on campus. This complete video set, affordable and portable for student use, is ideal for distance learning or extra instruction.

For Extra Help

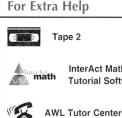

 Tape 2

InterAct Math Tutorial Software

 AWL Tutor Center

Student's Solutions Manual

www.awlonline.com/ akstbragg

InterAct Math® Tutorial Software (ISBN 0-201-70295-9)

This tutorial software correlates directly to the odd-numbered exercises in the text. The program is highly interactive, with sample problems and interactive guided solutions accompanying these exercises. The program recognizes common student errors and provides step-by-step customized feedback with sophisticated answer recognition capabilities. The management system (InterAct Math® Plus) allows instructors to create, administer, and track tests, and to monitor student performance during practice sessions.

Addison Wesley Longman Tutor Center

This center is staffed by qualified mathematics instructors who provide students with tutoring on text examples, exercises, and problems. Tutoring is provided by toll-free telephone, fax, and e-mail, and is available five days a week, seven hours a day.

InterAct MathXL: www.mathxl.com (ISBN 0-201-71630-5)

This software helps students prepare for tests by allowing them to take practice tests that are similar to the chapter tests in their text. The software automatically scores their tests and each student gets a personalized study guide that identifies individual student strengths and pinpoints topics that need more review. Links from the study guide take the student directly to the appropriate sections in the InterAct Math tutorial software for more practice and review. Through the course-management feature professors can track student results online.

World Wide Web Supplement: www.awlonline.com/akstbragg

The Web site accompanying this textbook provides additional tutorials, activities, practice materials, and learning resources to help students be more successful in their mathematics courses as well as a syllabus manager for instructor use.

For more information on these and other helpful supplements published by Addison Wesley Longman, please contact your bookstore.

Supplements for the Instructor (available free to qualifying adopters)

Annotated Instructor's Edition (ISBN 0-201-66224-8)

The annotated version of the student text includes answers to all exercises printed in blue on the same page as those exercises. Teaching tips in each chapter provide instructors with alternative explanations and approaches, typical student misconceptions, and connections to previous and future topics to consider conveying to their students, including annotations for suggested student journal entries (see pages 7, 83, and 203).

Instructor's Solutions Manual (ISBN 0-201-70291-6)

This manual contains worked-out solutions to every even-numbered text exercise as well as solutions to mindstretcher exercises.

Printed Test Bank/Instructor's Resource Guide (ISBN 0-201-70292-4)

This supplement includes four different versions of chapter tests for every chapter, and two final exams.

TestGen-EQ with QuizMaster-EQ dual-platform CD-ROM (ISBN 0-201-70293-2)

TestGen-EQ has a friendly graphical interface that enables instructors to easily view, edit, and add questions, transfer questions to tests, and print tests in a variety of fonts and forms. Search and sort features let the instructor quickly locate questions and arrange them in a preferred order. Six question formats are available, including short-answer, true/false, multiple-choice, essay, matching, and bi-modal. A built-in question editor gives the user power to create graphs, import graphics, insert mathematical symbols and templates, and insert variable numbers or text. The computerized testbank includes algorithmically defined problems organized according to the textbook. An export to HTML feature lets instructors create practice tests for the Web.

QuizMaster-EQ enables instructors to create and save tests by using TestGen-EQ so that students can take them for practice or for a grade on a computer network. Instructors can set preferences for how and when tests are administered. QuizMaster-EQ automatically grades the exams, stores the results on disk, and allows the instructor to view or print a variety of reports for individual students, classes, or courses.

MathPass, Version 2.2 for Windows (ISBN 0-201-71942-8)

MathPass helps students succeed in their developmental mathematics courses by creating customized study plans based on diagnostic test results from ACT, Inc.'s Computer-Adaptive Placement Assessment and Support System (COMPASS®). MathPass pinpoints topics where the student needs in-depth study or targeted review and correlates these topics with the student's textbook and related supplements (such as videos, student's solutions manuals, Web sites, and tutorial software). The MathPass Learning System provides diagnostic assessment, focused instruction, and exit placement all in one package.

For more information on these and other helpful supplements published by Addison Wesley Longman, please contact your AWL sales representative.

Acknowledgments

We are grateful to everyone who has helped to shape this textbook by responding to questionnaires, participating in telephone surveys and focus groups, reviewing the manuscript, and using the text in their classes. We wish to thank all of you, and in particular, the following:

Michele Bach, *Kansas City Kansas Community College*

Irma Bakenhus, *San Antonio College*

Tim Bremer, *Prestonburg Community College*

Robert Denitti, *Westmoreland County Community College*

Eunice Everett, *Seminole Community College*

Alan Greenhalgh, *Borough of Manhattan Community College*

Barbara Gardner, *Carroll Community College*

Janet C. Guynn, *Blue Ridge Community College*

Judith M. Jones, *Valencia Community College—East Campus*

Joanne Kendall, *College of the Mainland*

Roberta Lacefield, *Waycross College*

Lider-Manuel Lamar, *Seminole Community College*

Christopher McNally, *Tallahassee Community College*

Pat Roux, *Delgado Community College*

Joyce Saxon, *Morehead State University*

Radha Shrinivas, *Forest Park Community College*

Sharon A. Testone, *Onondaga Community College*

James VanArk, *University of Detroit Mercy*

Betty Vix Weinberger, *Delgado Community College*

Harvey S. Wiener, *Marymount Manhattan College*

J.W. Wing, *Angelina College*

James C. Woodall, *Salt Lake Community College*

Writing a textbook requires the contributions of many individuals. Special thanks go to Jason Jordan, our publisher at Addison Wesley Longman, for encouraging and supporting us throughout the entire process. We are very grateful to Elka Block, our Developmental Editor, who assisted us in more ways than one could imagine and whose unwavering support made our work more manageable. We thank Kari Heen for her patience and tact in gently reminding us of deadlines, Jennifer Crum for keeping us abreast of market trends, Lauren Morse for attending to the endless details connected with the project, Jerrold Moore for his editorial assistance, Kathy Diamond and Kathy Manley for their support throughout the production process, Susan Raymond and Janet Theurer for producing an appealing design and art program for the book, and the entire Addison Wesley Longman Developmental Mathematics team for helping to make this text one of which we are very proud.

Geoffrey Akst   Sadie Bragg
gakst@bmcc.cuny.edu   sbragg@bmcc.cuny.edu

Walkthrough

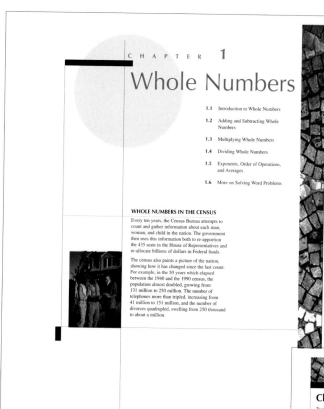

CHAPTER OPENERS:

The focus of this textbook is applications, and you will find them everywhere—in the chapter openers, in the explanation of the material, in the examples, and in the exercise sets. Each chapter opener introduces students to the material that lies ahead through an interesting and real career application, in an effort to grab students' attention and help them understand the relevance of mathematics to their lives and different careers.

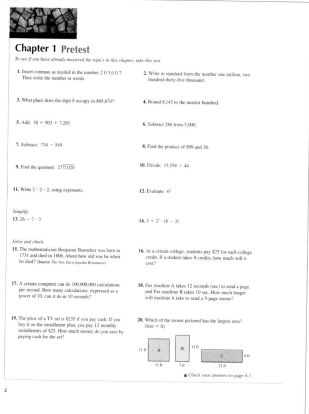

PRETESTS:

Pretests, found at the beginning of each chapter, help students to gauge their understanding of the chapter ahead. Answers can be found in the back of the book.

TEACHING TIPS:

These tips, found only in the Annotated Instructor's Edition, help instructors with explanations, reminders of previously covered material, and tips on encouraging students to write in a journal.

GEOMETRY:

The authors integrate this important topic where appropriate throughout the text so students can see its relevance to their surroundings.

ESTIMATION:

In order to help students develop their reasoning skills, the authors integrate the topic of estimation throughout the text.

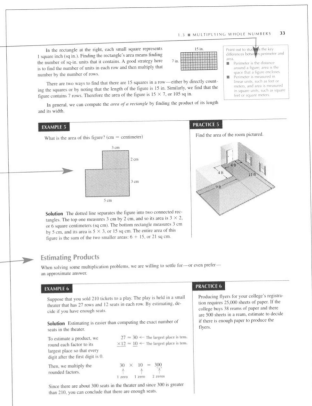

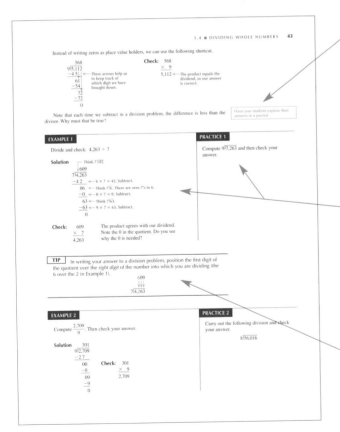

WRITING EXERCISES:

Students understand a concept better if they have to explain it in their own words. Journal assignments (provided in the Teaching Tips) give students an opportunity to improve their mathematical vocabulary and communication skills, thus improving their understanding of mathematical concepts.

SIDE-BY-SIDE FORMAT:

A unique side-by-side format pairs examples with corresponding practice exercises, encouraging active learning from the start. Students use this format for solving skill exercises, application problems, and technology exercises throughout the text.

STUDENT TIPS:

These insightful tips help students avoid common errors and provide other helpful suggestions to foster understanding of the material at hand.

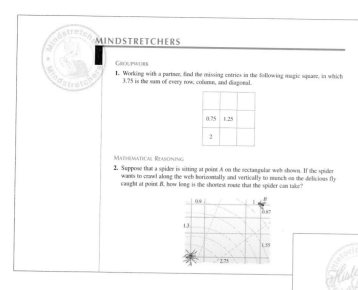

MINDSTRETCHERS

GROUPWORK

1. Working with a partner, find the missing entries in the following magic square, in which 3.75 is the sum of every row, column, and diagonal.

	0.75	1.25
	2	

MATHEMATICAL REASONING

2. Suppose that a spider is sitting at point *A* on the rectangular web shown. If the spider wants to crawl along the web horizontally and vertically to munch on the delicious fly caught at point *B*, how long is the shortest route that the spider can take?

MINDSTRETCHERS:

At the end of almost every section, students find these engaging activities that incorporate related investigation, critical thinking, mathematical reasoning, pattern recognition, and writing exercises along with corresponding groupwork and historical connections in one comprehensive problem set. These problem sets target different levels and types of student understanding.

HISTORICAL NOTES:

In an effort to get students to realize that mathematics was created out of a need to solve problems in everyday life, Historical Notes investigate the origins of mathematical concepts, discussing and illustrating the evolution of mathematics over the centuries, in many cultures, and throughout the world.

HISTORICAL NOTE

Throughout history, the concepts of percent and taxation have been interrelated. At the peak of the Roman Empire, the Emperor Augustus instituted an inheritance tax of 5% to provide retirement funds for the military. Another emperor, Julius Caesar, imposed a 1% sales

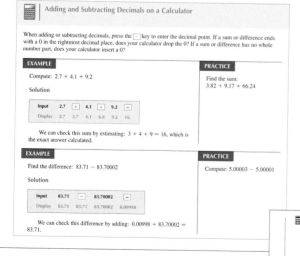

Adding and Subtracting Decimals on a Calculator

When adding or subtracting decimals, press the ⊡ key to enter the decimal point. If a sum or difference ends with a 0 in the rightmost decimal place, does your calculator drop the 0? If a sum or difference has no whole number part, does your calculator insert a 0?

EXAMPLE

Compute: 2.7 + 4.1 + 9.2

Solution

Input	2.7	+	4.1	+	9.2	=
Display	2.7	2.7	4.1	6.8	9.2	16.

We can check this sum by estimating: 3 + 4 + 9 = 16, which is the exact answer calculated.

EXAMPLE

Find the difference: 83.71 − 83.70002

Solution

Input	83.71	−	83.70002	=
Display	83.71	83.71	83.70002	0.00998

We can check this difference by adding: 0.00998 + 83.70002 = 83.71.

PRACTICE

Find the sum:
3.82 + 9.17 + 66.24

PRACTICE

Compute: 5.00003 − 5.00001

TECHNOLOGY INSERTS:

In order to familiarize students with a range of problem-solving methods—mental, paper-and-pencil, and calculator arithmetic—the authors include optional technology inserts that instruct students on how to use the scientific calculator to perform arithmetic operations. Note that even in these inserts, the authors use their side-by-side format to provide consistency in the students' learning environment.

CALCULATOR EXERCISES:

These optional exercises can be found in the exercise sets, giving students the opportunity to use a calculator to solve a variety of real-life applications.

⊞ *Use a calculator to solve the following problems, giving (a) the operation(s) carried out in your solution, (b) the exact answer, and (c) an estimate of the answer.*

65. To prevent anemia, your doctor advises you to take at least 18 mg of iron each day. Suppose that the table shows the amount of iron in the food that you ate yesterday. Did you get enough iron? If not, how much more do you need?

Food	Iron (mg)
Tomato juice	1.1
Oat flakes	2.4
Milk	0.1
Peanut butter sandwich	1.8

END-OF-CHAPTER MATERIAL:

In order to reinforce the concepts presented in current and previous chapters, the authors provide a wealth of end-of-chapter material designed to help students retain the concepts they have learned.

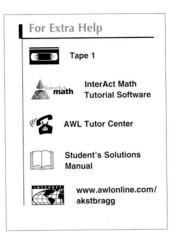

For Extra Help

▆▆	Tape 1
InterAct math	InterAct Math Tutorial Software
☎	AWL Tutor Center
📖	Student's Solutions Manual
🌐	www.awlonline.com/ akstbragg

FOR EXTRA HELP:

These boxes, found at the beginning of each exercise set, direct students to helpful resources to aid in their study of the material.

KEY CONCEPTS AND SKILLS:

These give students a quick overview of what they have learned in the chapter. Each concept/skill is keyed to the section in which it was introduced, and a brief description and example are provided for a one-stop quick review of the chapter material.

CHAPTER REVIEW EXERCISES:

These exercises are keyed to the corresponding sections for easy student reference. Numerous mixed application problems complete each of these exercise sets, always keeping the focus on the applicability of what students are learning.

CHAPTER POSTTEST:

Just as every chapter begins with a Pretest to test student understanding BEFORE attempting the material, every chapter ends with a Posttest to measure student understanding AFTER completing the chapter material. Answers to these tests are provided in the back of the book.

CUMULATIVE REVIEW EXERCISES:

Beginning at the end of Chapter 2, students have the opportunity to maintain their skills by completing the Cumulative Review Exercises. These exercises are invaluable, especially when students need to recall a previously learned concept or skill before beginning the next chapter, or when studying for mid-term and final examinations.

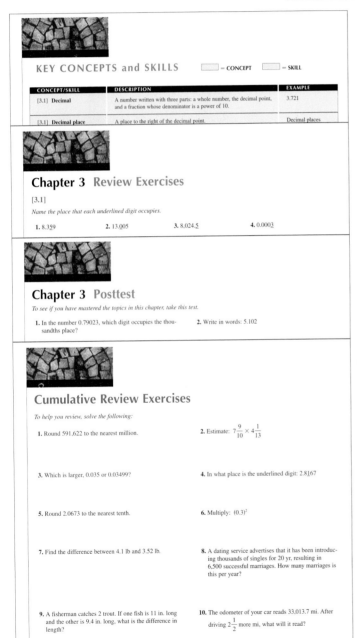

KEY CONCEPTS and SKILLS   ▢ = CONCEPT   ▢ = SKILL

CONCEPT/SKILL	DESCRIPTION	EXAMPLE
[3.1] Decimal	A number written with three parts: a whole number, the decimal point, and a fraction whose denominator is a power of 10.	3.721
[3.1] Decimal place	A place to the right of the decimal point.	Decimal places

Chapter 3 Review Exercises

[3.1]

Name the place that each underlined digit occupies.

1. 8.3<u>5</u>9   2. 13.0<u>0</u>5   3. 8,024.<u>5</u>   4. 0.000<u>3</u>

Chapter 3 Posttest

To see if you have mastered the topics in this chapter, take this test.

1. In the number 0.79023, which digit occupies the thousandths place?   2. Write in words: 5.102

Cumulative Review Exercises

To help you review, solve the following:

1. Round 591,622 to the nearest million.   2. Estimate: $7\frac{9}{10} \times 4\frac{1}{13}$

3. Which is larger, 0.035 or 0.03499?   4. In what place is the underlined digit: 2.8<u>1</u>67

5. Round 2.0673 to the nearest tenth.   6. Multiply: $(0.3)^2$

7. Find the difference between 4.1 lb and 3.52 lb.   8. A dating service advertises that it has been introducing thousands of singles for 20 yr, resulting in 6,500 successful marriages. How many marriages is this per year?

9. A fisherman catches 2 trout. If one fish is 11 in. long and the other is 9.4 in. long, what is the difference in length?   10. The odometer of your car reads 33,013.7 mi. After driving $2\frac{1}{2}$ more mi, what will it read?

■ *Check your answers on page A-6.*

Basic Mathematics

through Applications

Annotated Instructor's Edition

CHAPTER 1

Whole Numbers

WHOLE NUMBERS IN THE CENSUS

Every ten years, the Census Bureau attempts to count and gather information about each man, woman, and child in the nation. The government then uses this information both to re-apportion the 435 seats in the House of Representatives and re-allocate billions of dollars in Federal funds.

The census also paints a picture of the nation, showing how it has changed since the last count. For example, in the 50 years which elapsed between the 1940 and the 1990 census, the population almost doubled, growing from 131 million to 250 million. The number of telephones more than tripled, increasing from 41 million to 151 million, and the number of divorces quadrupled, swelling from 250 thousand to about a million.

Chapter 1 Pretest

To see if you have already mastered the topics in this chapter, take this test.

1. Insert commas as needed in the number 2 0 5,0 0 7. Then write the number in words. Two hundred five thousand seven

2. Write in standard form the number one million, two hundred thirty-five thousand. 1,235,000

3. What place does the digit 8 occupy in 805,674? Hundred thousands

4. Round 8,143 to the nearest hundred. 8,100

5. Add: 38 + 903 + 7,285 8,226

6. Subtract 286 from 5,000. 4,714

7. Subtract: 734 − 549 185

8. Find the product of 809 and 36. 29,124

9. Find the quotient: $27\overline{)7,020}$ 260

10. Divide: 13,558 ÷ 44 308 R6 or $308\frac{3}{22}$

11. Write 2 · 2 · 2, using exponents. 2^3

12. Evaluate: 6^2 36

Simplify.

13. 26 − 7 · 3 5

14. 3 + 2^3 · (8 − 3) 43

Solve and check.

15. The mathematician Benjamin Banneker was born in 1731 and died in 1806. About how old was he when he died? (*Source: The New Encyclopedia Britannica*) 75 years old

16. At a certain college, students pay $75 for each college credit. If a student takes 9 credits, how much will it cost? $675

17. A certain computer can do 100,000,000 calculations per second. How many calculations, expressed as a power of 10, can it do in 10 seconds? 10^9

18. Fax machine A takes 12 seconds (sec) to send a page, and Fax machine B takes 10 sec. How much longer will machine A take to send a 5-page memo? 10 sec

19. The price of a TV set is $235 if you pay cash. If you buy it on the installment plan, you pay 12 monthly installments of $25. How much money do you save by paying cash for the set? $65

20. Which of the rooms pictured has the largest area? (feet = ft) Room C which measures 126 ft^2

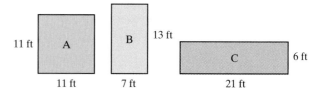

■ *Check your answers on page A-1.*

1.1 Introduction to Whole Numbers

OBJECTIVES

■ *To read and write whole numbers*
■ *To write whole numbers in expanded form*
■ *To round whole numbers*

What the Whole Numbers Are and Why They Are Important

We use whole numbers for counting, whether it is the number of *e*'s on this page or the number of stars in the sky.

The whole numbers are 0, 1, 2, 3, 4, 5, 6, 7, 8, 9, 10, 11, 12, 13, ... An important property of whole numbers is that there is always a next whole number. This property means that they go on without end, as the three dots above indicate.

Every whole number is either *even* or *odd*. The even whole numbers are 0, 2, 4, 6, 8, 10, 12, ... The odd whole numbers are 1, 3, 5, 7, 9, 11, 13, ...

We can represent the whole numbers on a number line. Similar to a ruler, the number line starts with 0 and extends without end to the right, as the arrow indicates.

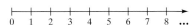

Reading and Writing Whole Numbers

Generally speaking, we *read* whole numbers in words, but we use the **digits** 0, 1, 2, 3, 4, 5, 6, 7, 8, and 9 to *write* them. For instance, we read the whole number *fifty-one* but write it *51*, which we call **standard form**.

Each of the digits in a whole number in standard form has a **place value**. Our place value system is very important because it underlies both the way we write and the way we compute with numbers.

The following chart shows the place values in whole numbers up to twelve digits long. For instance in the number 1,234,056 the digit 2 occupies the hundred thousands place. Study the place values in the chart now.

BILLIONS			MILLIONS			THOUSANDS			ONES			Period
Hundreds	Tens	Ones	Hundreds	Tens	Ones	Hundreds	Tens	Ones	Hundreds	Tens	Ones	Place value
					1	2	3	4	0	5	6	
		8	1	6	8	9	3	1	0	4	7	

TIP   We read whole numbers from left to right, but it is easier in the place-value chart to learn the names of the places *from right to left*.

When we write a large whole number in standard form, we insert *commas* to separate its digits into groups of three, called **periods**. For instance the number 8,168,931,047 has four periods: *ones, thousands, millions,* and *billions.*

EXAMPLE 1

In each number, identify the place that the digit 7 occupies.

a. 207

b. 7,654,000

c. 5,700,000,001

Solution

a. The ones place

b. The millions place

c. The hundred millions place

PRACTICE 1

What place does the digit 8 occupy in each number?

a. 278,056   Thousands

b. 803,746   Hundred thousands

c. 3,080,700,059   Ten millions

The following rule provides a short cut for *reading a whole number*.

To Read Whole Numbers

Working from left to right,

■ read the number in each period and then
■ name the period in place of the comma.

For instance, 1,234,056 is read "one million, two hundred thirty-four thousand, fifty-six."

EXAMPLE 2

How do you read the number 422,000,085?

Solution Beginning at the left in the millions period, we read this number as "four hundred twenty-two million, eighty-five." Note that, because there are all zeros in the thousands period, we do not read "thousands."

PRACTICE 2

Write 8,000,376,052 in words.
Eight billion, three hundred seventy-six thousand, fifty-two

EXAMPLE 3

The display on your calculator shows the answer 3578002105. Insert commas in this answer and then read it.

Solution The number with commas is 3,578,002,105, which is read "three billion, five hundred seventy-eight million, two thousand, one hundred five."

PRACTICE 3

A company is worth $7372050. Read this amount. $7,372,050
Seven million, three hundred seventy-two thousand, fifty dollars

Until now, we have discussed how to *read* whole numbers in standard form. Now let's turn to the question of how they are *written* in standard form. We simply reverse the process just described. For instance, the number eight billion, one hundred sixty-eight million, nine hundred thirty-one thousand, forty-seven in standard form is 8,168,931,047. Here we use the 0 as a **placeholder** in the hundreds place because there are no hundreds.

To Write Whole Numbers

Working from left to right,

■ write the number named in each period and
■ replace each period name with a comma.

When writing large whole numbers in standard form, we must remember that the number of commas is always one less than the number of periods. For instance, the number one million, two hundred thirty-four thousand, fifty-six—1,234,056—has three periods and two commas. Similarly, the number 8,168,931,047 has four periods and three commas.

> Show students that a good way to avoid a mistake in writing large whole numbers is first to write down the number's structure in terms of commas and blanks:
>
> _ , _ _ _ , _ _ _ , _ _ _

EXAMPLE 4

Write the number eight billion, seven in standard form.

Solution This number involves billions, so there are four periods— billions, millions, thousands, and ones—and three commas. Writing the number named in each period and replacing each period name with a comma, we get 8,000,000,007. Note that we write three 0's when no number is named in a period.

PRACTICE 4

Use digits and commas to write the amount ninety-five million, three dollars.
$95,000,003

EXAMPLE 5

The treasurer of a company writes a check in the amount of four hundred thousand seven hundred dollars. Using digits, how would you write this amount on the check?

Solution We write this amount with one comma because its largest period is thousands. We write $400,700 as shown on the check below.

UNITED INDUSTRIES
Atlanta, Georgia   2066

DATE *November 1, 1998*

PAY TO THE ORDER OF *American Vendors, Inc.* $ 400,700—

Four Hundred Thousand Seven Hundred and ⁰⁰⁄₁₀₀   DOLLARS

SB Southern Bank

MEMO _____   *Jack Jones*

⑆721107560⑆ 022000658711ⁱ 2066

PRACTICE 5

A rich alumna donates three hundred seventy-five thousand dollars to her college's scholarship fund.

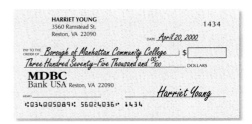

Using digits, how would she write this amount on the check?
$375,000

When writing checks, we write the amount in both digits and words. Why do we do this?

> Have your students start a journal in which they can record their answers to questions like this one.

Writing Whole Numbers in Expanded Form

We have just described how to write whole numbers in standard form. Now let's turn to how we write these numbers in **expanded form**.

Let's consider the whole number 4,325 and examine the place value of its digits.

$$4{,}325 = 4 \text{ thousands} + 3 \text{ hundreds} + 2 \text{ tens} + 5 \text{ ones}$$

This last expression is called the expanded form of the number, and it can be written as follows.

$$4{,}000 + 300 + 20 + 5$$

The expanded form of a number spells out its value in terms of place value, helping us understand what the number really means. For instance, think of the numbers 92 and 29. By representing them in *expanded* form, can you explain why they differ in value even though their *standard* form consists of the same digits?

EXAMPLE 6

Express the number 906 in expanded form.

Solution The 6 is in the ones place, the 0 is in the tens place, and the 9 is in the hundreds place.

ONES		
Hundreds	**Tens**	**Ones**
9	0	6

So 906 is 9 hundreds + 0 tens + 6 ones = 900 + 0 + 6 in expanded form.

PRACTICE 6

Write 27,013 in expanded form.
2 ten thousands + 7 thousands + 0 hundreds + 1 ten + 3 ones = 20,000 + 7,000 + 10 + 3

EXAMPLE 7

Write 3,203,000 in expanded form.

Solution Using the place-value chart, we see that 3,203,000 = 3 millions + 2 hundred thousands + 3 thousands = 3,000,000 + 200,000 + 3,000.

PRACTICE 7

Express 1,270,093 in expanded form.
1 million + 2 hundred thousands + 7 ten thousands + 0 thousands + 0 hundreds + 9 tens + 3 ones = 1,000,000 + 200,000 + 70,000 + 90 + 3

Rounding Whole Numbers

Most people equate mathematics with precision, but some problems require sacrificing precision for simplicity. In this case, we use the technique called **rounding** to approximate the exact answer by a number that ends in a given number of zeros. Rounded numbers have special advantages: They seem clearer to us than other numbers, and they make computation easier—especially when we are trying to compute in our heads.

Of these two headlines, which do you prefer? Why?

Study the following chart to see the connection between place value and rounding.

Rounding to the nearest	Means that the rounded number ends in at least
10	One 0
100	Two 0's
1,000	Three 0's
10,000	Four 0's
100,000	Five 0's
1,000,000	Six 0's

Note in the chart that the place value tells us how many 0's the rounded number must have at the end. Having more 0's than indicated is possible. Can you think of an example?

Have your students explain their answers in a journal.

When rounding, we use an underlined digit to indicate the place to which we are rounding.

Now let's consider the following rule for rounding whole numbers.

To Round Whole Numbers

■ Underline the place to which you are rounding.
■ Look at the digit to the right of the underlined digit, called the **critical digit**. If this digit is 5 or more, add 1 to the underlined digit; if it is less than 5, leave the underlined digit unchanged.
■ Replace all the digits to the right of the underlined digit with zeros.

EXAMPLE 8

Round 79,680 to

a. the nearest hundred and

b. the nearest thousand.

Solution

a. First, we underline the 6 because that digit occupies the hundreds place: 79,680. The critical digit is 8: 79,680. As 8 is greater than 5, we add 1 to the underlined digit. Then, we replace all digits to the right with 0's, getting 79,700. We write 79,680 ≈ 79,700, meaning that 79,680 when rounded to the nearest hundred is 79,700.

b. 79,680 = 79,680 ← Underline the digit in the thousands place.

    = 79,680 ← The critical digit 6 is greater than 5; add 1 to the underlined digit.

    = 80,000 ← Change the digits to the right of the underlined digit to 0's.

Note that adding 1 to the underlined digit gave us 10 and forced us to write 0 and carry 1 to the next column, changing the 6 to 7.

PRACTICE 8

Round 58,760 to

a. the nearest thousand and   59,000

b. the nearest ten thousand.   60,000

For Example 8, consider this number line.

79,680

78,000   79,000   80,000   81,000

The number line shows that 79,860 lies between 79,000 and 80,000 and that it is closer to 80,000, as the rule indicates.

EXAMPLE 9	PRACTICE 9

In your Anatomy and Physiology class, you learned that the adult human skeleton contains 206 bones. How many bones is this to the nearest hundred bones?

Solution We first write 2̲06. The critical digit 0 is less than 5, so we do *not* add 1 to the underlined digit. However, we do change both the digits to the right of the 2 to 0's. So 2̲06 ≈ 200, and there are approximately 200 bones in the human body.

A college president earned an annual salary of $99,839. What was this salary to the nearest thousand dollars? $100,000

Exercises 1.1

Insert commas as needed, and then write the number in words.

1. 1,0 0 0,0 0 0,0 0 0   One billion

2. 3 7 9,0 5 2   Three hundred seventy-nine thousand fifty-two

3. 2,3 5 0,0 0 0   Two million, three hundred fifty thousand

4. 1,3 5 0,1 3 2   One million, three hundred fifty thousand, one hundred thirty-two

5. 9 7 5,1 3 5,0 0 0   Nine hundred seventy-five million, one hundred thirty-five thousand

6. 2 1,0 0 0,1 3 2   Twenty-one million, one hundred thirty-two

7. 4 8 7,5 0 0   Four hundred eighty-seven thousand, five hundred

8. 5 2 8,0 5 0   Five hundred twenty-eight thousand, fifty

9. 2,0 0 0,0 0 0,3 5 2   Two billion, three hundred fifty-two

10. 4,1 0 0,0 0 0,0 0 7   Four billion, one hundred million seven

Write each number in standard form.

11. Ten thousand, one hundred twenty
10,120

12. Three billion, seven hundred million
3,700,000,000

13. One hundred fifty thousand, eight hundred fifty-six   150,856

14. Twenty million, five thousand
20,005,000

15. Six million, fifty-five
6,000,055

16. Two million, one hundred twenty-two
2,000,122

17. Fifty million, six hundred thousand, one hundred ninety-five   50,600,195

18. Nine hundred thousand, eight hundred eleven   900,811

19. Four hundred thousand, seventy-two
400,072

20. Nine hundred billion
900,000,000,000

Underline the digit that occupies the given place.

21. 4,867   Thousands place

22. 075   Hundreds place

23. 316   Tens place

24. 41,722   Ten thousands place

25. 28,461,013   Millions place

26. 762,800   Hundred thousands place

Identify the place occupied by the underlined digit.

27. 691,400   Hundred thousands

28. 72,109   Ten thousands

29. 7,380   Thousands

30. 351   Hundreds

31. 8,450,000,000   Billions

32. 35,832,775   Ten millions

Write each number in expanded form.

33. 3   3 ones = 3

34. 6,300   6 thousands + 3 hundreds = 6,000 + 300

35. 858   8 hundreds + 5 tens + 8 ones = 800 + 50 + 8

36. 9,000,000   9 millions = 9,000,000

37. 2,500,004   2 millions + 5 hundred thousands + 4 ones = 2,000,000 + 500,000 + 4

38. 7,251,380   7 millions + 2 hundred thousands + 5 ten thousands + 1 thousand + 3 hundreds + 8 tens = 7,000,000 + 200,000 + 50,000 + 1,000 + 300 + 80

Round to the indicated place.

39. 671 to the nearest ten   670

40. 838 to the nearest hundred   800

41. 7,103 to the nearest hundred   7,100

42. 46,099 to the nearest thousand 46,000

43. 28,241 to the nearest ten thousand 30,000

44. 7,802,555 to the nearest million 8,000,000

45. 705,418 to its largest place 700,000

46. 81 to its largest place 80

47. 31,972 to its largest place 30,000

48. 4,913,440 to its largest place 5,000,000

Remind students that when rounding numbers to different places they should start with the original number each time.

Round each number as indicated.

49.

To the nearest	135,800	816,533
Hundred	135,800	816,500
Thousand	136,000	817,000
Ten thousand	140,000	820,000
Hundred thousand	100,000	800,000

50.

To the nearest	972,055	3,189,602
Thousand	972,000	3,190,000
Ten thousand	970,000	3,190,000
Hundred thousand	1,000,000	3,200,000
Million	1,000,000	3,000,000

Applications

Write each whole number in words.

51. Biologists have classified more than 700,000 species of insects.   Seven hundred thousand

52. Each pair of human lungs contains some 300,000,000 tiny air sacs. Three hundred million

53. More than 300,000,000,000 Oreo cookies have been sold, making this cookie the most popular in the world. (*Source:* www.nabisco.com)   Three hundred billion

54. Mercury is the closest planet to the Sun, a distance of approximately 36,000,000 miles (mi). Thirty-six million

55. The satellite flew a total of 1,400,085 mi before burning up in the atmosphere. One million, four hundred thousand, eighty-five

56. The Pyramid of Khufu in Egypt has a base of approximately 2,315,000 blocks. (*Source:* The New Encyclopedia Britannica)   Two million, three hundred fifteen thousand

Write each whole number in standard form.

57. Some one hundred billion nerve cells are part of the human brain. 100,000,000,000

58. The number of people employed rose to four million, seven hundred thousand. 4,700,000

59. The bank invested eighty-five million, sixty-one thousand, fifty-eight dollars in mortgages. $85,061,058

60. Twenty-three million five thousand people voted for the winning candidate.   23,005,000

61. The hydroelectric plant produces one billion, four hundred million watts of electricity annually.   1,400,000,000

62. Two million, three hundred forty-six thousand, seven hundred sixty-one passengers flew on this airline last year.   2,346,761

Round to the indicated place.

63. The Statue of Liberty is 152 ft high. What is its height to the nearest 10 ft? 150 ft

64. The Nile, with a length of 4,180 mi, is the longest river in the world. Round to the nearest thousand miles.   4,000 mi

65. In 1949, Air Force Captain James Gallagher led the first team to make an around-the-world flight. The team flew 23,452 mi. Round to the nearest ten thousand miles. (*Source:* Taylor and Mondey, *Milestones of Flight*)   20,000 mi

66. The element copper changes from a liquid to a gas at the temperature 2,567 degrees Celsius (°C). Round this temperature to the nearest hundred degrees Celsius.   2,600°C

67. A weight of 373 grams (g) is equivalent to 1 pound. How many grams is this to the nearest hundred?   400 g

68. Suppose that last year you earned $27,502. What was your income to the nearest ten thousand dollars? $30,000

■ *Check your answers on page A-1.*

MINDSTRETCHERS

MATHEMATICAL REASONING

1. I am thinking of a certain whole number. My number, rounded to the nearest hundred, is 700. When it is rounded to the nearest ten, it is 750. What numbers could I be thinking of?   745, 746, 747, 748, 749

WRITING

2. How does the number 10 play a special role in the way that we write whole numbers? Do you think that it would be possible to have the number 2 play this role? Explain.

In decimal notation, there are ten digits (0, 1, 2, … , 9), and place values are powers of

ten. In binary notation, there are two digits (0 and 1), and place values are powers of

two.

GROUPWORK

3. Here are several ways of writing the number seven:

Working with a partner, express each of the numbers 1, 2, … , 9 in these three ways.

1, I, | 2, II, || 3, III, ||| 4, IV, |||| 5, V, ⊬

6, VI, ⊬| 7, VII, ⊬|| 8, VIII, ⊬||| 9, IX, ⊬||||

1.2 Adding and Subtracting Whole Numbers

OBJECTIVES

- *To add and subtract whole numbers*
- *To estimate the sum and difference of whole numbers*
- *To solve word problems involving the addition or subtraction of whole numbers*

The Meaning and Properties of Addition and Subtraction

Addition is perhaps the most fundamental of all operations. One way to think about this operation is as *combining sets*. For example, suppose that we have two distinct sets of pens, with 5 pens in one set and 3 in the other. If we put the two sets together, we get a single set that has 8 pens.

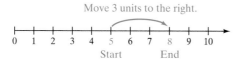

So we can say that 5 added to 3 is 8, or here, 5 pens plus 3 pens equals 8 pens. Note that we are adding quantities of the same thing, or *like quantities*.

Another good way to think about addition is as *moving to the right on a number line*. In this way, we start at the point on the line corresponding to the first number, 5. Then to add 3, we move 3 units to the right, ending on the point that corresponds to the answer, 8.

Now let's look at subtraction. One way to look at this operation is as *taking away*. For instance, when we subtract 5 pens from 8 pens, we take 5 pens away from 8 pens, leaving 3 pens.

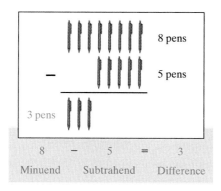

We say that the *difference* between 8 and 5 is 3.

As in the preceding example, we can only subtract *like quantities*: we cannot subtract 5 pens from 8 scissors.

We can also think of subtraction as the *opposite of addition*.

$$8 - 5 = 3 \quad \text{because} \quad 5 + 3 = 8$$

Subtraction   Related Addition

In a subtraction problem, remind students to be careful of the order in which they read the numbers.

Note in this example that, if we add the 5 pens to the 3 pens, we get 8 pens.

Addition and subtraction problems can be written either horizontally or vertically.

$$5 + 3 = 8 \qquad 8 - 5 = 3$$
Horizontal

$$\begin{array}{r} 5 \\ +3 \\ \hline 8 \end{array} \qquad \begin{array}{r} 8 \\ -5 \\ \hline 3 \end{array}$$

Vertical

Either format gives the correct answer. But it is generally easier to figure out the sum and difference of large numbers if the problems are written vertically.

Now let's briefly consider several special properties of addition that we use frequently. Examples appear to the right of each property.

The Identity Property of Addition

The sum of a number and zero is the original number.

$$3 + 0 = 3$$
$$0 + 5 = 5$$

The Commutative Property of Addition

Changing the order in which two numbers are added does not affect the sum.

$$3 + 2 = 2 + 3$$

The Associative Property of Addition

When adding three numbers, regrouping addends gives the same sum. Note that the parentheses tell us which numbers to add first.

$$(4 + 7) + 2 = 4 + (7 + 2)$$
$$11 + 2 = 4 + 9$$

Adding Whole Numbers

We add whole numbers by arranging the numbers vertically, keeping the digits with the same place value in the same column. Then we add the digits in each column.

Consider the sum $32 + 65$. In the vertical format at the right, the sum of the digits in each column is 9 or less. The sum is 97. When the sum of the digits in a column is greater than 9, we must **carry** because only a single digit can occupy a single place. Example 1 illustrates carrying.

$$\begin{array}{r} 32 \\ +65 \\ \hline 97 \end{array}$$

EXAMPLE 1

Add 47 and 28.

Solution First, we write the addends in expanded form. Then, we add down the ones column.

We carry the 1 ten to the tens place.

1 ten

$$47 = 4 \text{ tens} + 7 \text{ ones} = 4 \text{ tens} + 7 \text{ ones}$$
$$+28 = 2 \text{ tens} + 8 \text{ ones} = 2 \text{ tens} + 8 \text{ ones}$$
$$\overline{\qquad 15 \text{ ones} \qquad\qquad 5 \text{ ones}}$$

Next, we add down the tens column.

1 ten

$$4 \text{ tens} + 7 \text{ ones}$$
$$\underline{2 \text{ tens} + 8 \text{ ones}}$$
$$7 \text{ tens} + 5 \text{ ones} = 75$$

PRACTICE 1

Add: $198 + 37$ 235

The following rule tells how to add whole numbers without using expanded form.

To Add Whole Numbers

- Write the addends vertically, lining up the place values.
- Add the digits in the ones column, writing the sum on the bottom. If the sum has two digits, carry the left digit to the top of the next column on the left.
- Add the digits in the tens column, as in the preceding step.
- Repeat this process until you reach the last column on the left, writing down the entire sum of that column on the bottom.

EXAMPLE 2	**PRACTICE 2**
Add: $9{,}824 + 356 + 2{,}976$	Find the total: $838 + 96 + 1{,}002$ 1,936

Solution We write the problem vertically, with the addends lined up on the right.

$$
\begin{array}{r}
\overset{1}{9},\,8\;2\;4 \\
3\;5\;6 \\
+2,\,9\;7\;6 \\
\hline
6
\end{array}
$$

⟵ The sum of the ones digits is 16 ones. We write the 6 and carry the 1 to the tens column.

$$
\begin{array}{r}
\overset{1}{9},\,\overset{1}{8}\;2\;4 \\
3\;5\;6 \\
+2,\,9\;7\;6 \\
\hline
5\;6
\end{array}
$$

The sum of the tens digits is 15 tens. We write the 5 and carry the 1 to the hundreds column.

$$
\begin{array}{r}
\overset{2}{9},\,\overset{1}{8}\;\overset{1}{2}\;4 \\
3\;5\;6 \\
+2,\,9\;7\;6 \\
\hline
1\;5\;6
\end{array}
$$

The sum of the hundreds digits is 21 hundreds. We write the 1 and carry the 2 to the thousands column.

$$
\begin{array}{r}
\overset{2}{9},\,\overset{1}{8}\;\overset{1}{2}\;4 \\
3\;5\;6 \\
+2,\,9\;7\;6 \\
\hline
13,\,1\;5\;6
\end{array}
$$

The sum of the digits in the thousands column is 13, which we write completely—no need to carry here.

The sum is 13,156.

In Example 3, let's apply the operation of addition to finding the geometric perimeter of a figure. The **perimeter** is the distance around a figure, which we can find by adding the lengths of its sides.

EXAMPLE 3	PRACTICE 3

Find the perimeter of this figure.

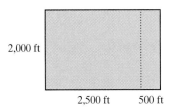

How long a fence would you need to enclose the piece of land sketched?   16 mi

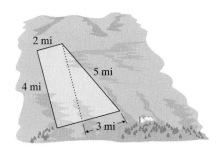

Solution This figure consists of two rectangles placed side by side. We note that the opposite sides of each rectangle are equal in length.

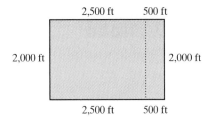

To compute the figure's perimeter, we need to add the lengths of all its sides.

$$\begin{array}{r} 2,000 \\ 2,500 \\ 500 \\ 2,000 \\ 500 \\ +2,500 \\ \hline 10,000 \end{array}$$

The figure's perimeter is 10,000 ft.

Subtracting Whole Numbers

Consider the subtraction $59 - 36$, written vertically at the right. We write the whole numbers underneath one another, lined up on the right, so each column contains digits with the same place value. Then we subtract the digits within each column, the bottom digit from the top, getting a difference of 23.

$$\begin{array}{r} 59 \\ -36 \\ \hline 23 \end{array}$$

Keep in mind two useful properties of subtraction.

■ When you subtract a number from itself, the result is 0. $6 - 6 = 0$

■ When you subtract 0 from a number, you get the original number. $25 - 0 = 25$

TIP   When writing a subtraction problem vertically, be sure that

■ the minuend—the number from which you are subtracting—goes on the *top* and that
■ the subtrahend—the number that you are taking away—goes on the *bottom*.

Now we consider subtraction problems that involve *borrowing*. In these problems a digit on the bottom is too large to subtract from the corresponding digit on top.

EXAMPLE 4

Subtract: $329 - 87$

Solution We first write these numbers vertically in expanded form.

$$329 = \quad \text{3 hundreds} + \text{2 tens} + \text{9 ones}$$
$$-\ 87 = -\ \underline{\hspace{3cm} \text{8 tens} + \text{7 ones}}$$

We then subtract the digits in the ones column: 7 ones from 9 ones gives 2 ones.

$$\begin{array}{r} \text{3 hundreds} + \text{2 tens} + \text{9 ones} \\ -\ \underline{\hspace{2cm} \text{8 tens} + \text{7 ones}} \\ \text{2 ones} \end{array}$$

10 tens + 2 tens = 12 tens

We next go to the tens column. We cannot take 8 tens from 2 tens. But we can *borrow* 1 hundred from the 3 hundreds, leaving 2 in the hundreds place. We *exchange* this hundred for 10 tens (1 hundred = 10 tens). Then combining the 10 tens with the 2 tens gives 12 tens.

$$\begin{array}{r} \overset{2}{\cancel{3}}\ \text{hundreds} + \overset{1}{2}\ \text{tens} + \text{9 ones} \\ -\ \underline{\hspace{2cm} \text{8 tens} + \text{7 ones}} \\ \text{2 ones} \end{array}$$

We next take 8 from 12 in the tens column, giving 4 tens. Finally, we bring down the 2 hundreds. The difference is 242 in standard form.

$$\begin{array}{r} \overset{2}{\cancel{3}}\ \text{hundreds} + \overset{1}{2}\ \text{tens} + \text{9 ones} \\ -\ \underline{\hspace{2cm} \text{8 tens} + \text{7 ones}} \\ \text{2 hundreds} + \text{4 tens} + \text{2 ones} = 242 \end{array}$$

PRACTICE 4

Subtract: $748 - 97$ 651

Although we can always rewrite whole numbers in expanded form so as to subtract them, the following rule provides a shortcut.

To Subtract Whole Numbers

- On top, write the number *from which* you are subtracting. On the bottom, write the number that you are *taking away*, lining up the place values. Subtract in each column separately.
- Start with the ones column.
  a. If the digit on top is *larger* than or *equal* to the digit on the bottom, subtract and write the difference below the bottom digit.
  b. If the digit on top is *smaller* than the digit on the bottom, borrow from the digit to the left on top. Then subtract and write the difference below the bottom digit.
- Repeat this process until you have finished the last column on the left.

EXAMPLE 5	PRACTICE 5

Find the difference between 293 and 500.

Solution We rewrite the problem vertically.

$$\begin{array}{r} 500 \\ -293 \end{array}$$

We cannot subtract 3 ones from 0 ones, and we cannot borrow from 0 tens. So we borrow from the 5 hundreds.

$$\begin{array}{r} \overset{4}{\cancel{5}}\overset{1}{} 0\ 0 \\ -2\ 9\ 3 \end{array} \quad \longleftarrow 5\ \text{hundreds} = 4\ \text{hundreds} + 10\ \text{tens}$$

We now borrow from the tens column.

$$\begin{array}{r} \overset{9}{} \longleftarrow 10\ \text{tens} = 9\ \text{tens} + 10\ \text{ones} \\ \overset{4}{\cancel{5}}\ \overset{\cancel{10}}{\cancel{0}}\ \overset{1}{0} \\ -2\ 9\ 3 \\ \hline 2\ 0\ 7 \end{array}$$

As usual, we check by adding back what we subtracted. The sum turns out to be the original minuend, so our answer is correct.

$$\begin{array}{r} 207 \\ +293 \\ \hline 500 \end{array}$$

Subtract 3,253 from 8,000.   4,747

EXAMPLE 6	PRACTICE 6

A large factory has 1,230 employees. Yesterday, 195 of these employees were absent. How many were at work?

Solution The total number of employees equals the number of employees at work plus the number of employees absent.

Employees At Work (?)	Employees Absent (195)

All employees (1,230)

To compute the number of employees at work, we subtract the number of employees absent from the total number of employees.

Total number of employees	1,230
− Number of employees absent	− 195
Number of employees at work	?

$$\begin{array}{r} 1,230 \\ -\ \ \ 195 \\ \hline 1,035 \end{array}$$

So 1,035 employees were at work.

In a theater, 372 of the 750 seats were occupied. How many seats were empty?
378

Estimating Sums and Differences

Because everyone occasionally makes a mistake, we need to know how to check an answer so that we can correct it if it is wrong.

One method of checking addition and subtraction is by *estimation*. In this approach, we first compute and then estimate the answer. Then we compare the estimate and our "exact answer" to see if they are close. If they are, we can be confident that our answer is reasonable. If they are not close, we should redo the computation.

EXAMPLE 7

Add 58,776 + 7,107 + 3,022. Check by estimation.

Solution To check by estimation, we round the addends so that they will be easy to add mentally.

Solution		Check	
58,776	≈	60,000	
7,107	≈	7,000	
+3,022	≈	+3,000	
68,905	Exact sum	70,000	Estimated sum

Our estimated sum is reasonably close to—and therefore supports—our exact sum.

PRACTICE 7

Find the total of 196 + 5,056 + 22,097. Check by estimation.   27,349

Note that we can get different estimates for an answer, depending on how we round. For instance, think how our estimated sum in Example 7 would have changed had all the addends been rounded to the nearest thousand instead of the nearest ten thousand.

EXAMPLE 8

Compute the sum of 1,923 + 898 + 754 + 2,873. Check by estimation.

Solution 1,923
898
754
+2,873
6,448   Exact sum

Check   We round each addend to the nearest thousand.

1,923 ≈   2,000
898 ≈   1,000
754 ≈   1,000
+2,873 ≈ +3,000
7,000   Estimated sum

Our exact answer, 6,448, is reasonably close to the estimate, so we are done.

PRACTICE 8

Add 838 + 962 + 1,002. Check by estimating the sum.   2,802

| **EXAMPLE 9** | **PRACTICE 9** |

Subtract 1,994 from 8,253. Check by estimating.

Find the difference between 17,836 and 15,045. Then estimate this difference to check.   2,791

Solution First, we compute the exact answer.

$$\begin{array}{r} 8,253 \\ -1,994 \\ \hline 6,259 \end{array}$$

Now, we check this answer by estimation. We first round both the minuend and subtrahend to the nearest thousand.

$$\begin{array}{r} 8,253 \approx 8,000 \\ -1,994 \approx -2,000 \\ \hline \end{array}$$

Then, we compute the difference between the rounded numbers.

$$\begin{array}{r} 8,000 \\ -2,000 \\ \hline 6,000 \end{array}$$ **Estimated difference**

Our exact answer (6,259) and estimated difference (6,000) are fairly close.

With practice, we can mentally estimate and check differences like this one, both quickly and easily.

> Calculator inserts like the one below are based on the use of either four-function or scientific calculators.

 ## Adding and Subtracting Whole Numbers on a Calculator

Calculators are handy and powerful tools for carrying out complex computations. But it is easy to press a wrong key, so be sure to estimate your answer and compare it to the displayed answer to see if it is reasonable.

| **EXAMPLE** | **PRACTICE** |

On a calculator, compute the sum of 3,125 and 9,391.

Use a calculator to add:
39,822 + 9,710   49,532

Solution

Input	3125	$+$	9391	$=$
Display	3125.	3125.	9391.	12516.

To check this answer, we mentally round the addends and then add.

$$\begin{array}{r} 3125 \approx 3,000 \\ 9391 \approx 9,000 \\ \hline 12,000 \end{array}$$

This estimate is reasonably close to our answer 12,516.

When you enter a whole number on a calculator, note that in the display

- no commas appear,
- the calculator inserts a decimal point to the right of the number, and
- pressing the clear key (usually [AC] or [C]) cancels the number. Press this key whenever you complete a computation to be sure that no number remains to affect the next problem.

EXAMPLE

Calculate: 39 + 48 + 277

Solution

Input	39	+	48	+	277	=
Display	39.	39.	48.	87.	277.	364

A reasonable estimate is the sum of 40, 50, and 300, or 390—close to our calculated answer 364.

Note in this example that, after entering the second addend (48), it was not necessary to press the = key.

PRACTICE

Find the sum on a calculator:
23,801 + 7,116 + 982
31,899

When using a calculator to subtract,

■ enter the numbers in the correct order—first enter the number **from which** you are subtracting and then the number **being** subtracted; and
■ be sure not to confuse the *negative sign key* (−) that some calculators have with the *subtraction key* − .

EXAMPLE

Subtract on a calculator: 3,000 − 973

Solution

Input	3000	−	973	=
Display	3000.	3000.	973.	2027.

A good estimate is 3,000 − 1,000, or 2,000, which is close to 2,027.

PRACTICE

Use a calculator to find the difference between 5,280 ft and 2,781 ft.   2,499 ft

Exercises 1.2

Add and then check.

1. 100,250
+ 77,528
177,778

2. 3,505
+ 11
3,516

3. 9,261
+ 412
9,673

4. 46,319
+10,520
56,839

5. 4,663
+ 371
5,034

6. 81,452
+25,199
106,651

7. 8,132
+6,578
14,710

8. 60,725
+38,928
99,653

9. 3,750
1,725
+4,992
10,467

10. 3,505
5,700
+2,888
12,093

11. 3,227
2,806
+5,481
11,514

12. 90,316
10,882
+ 5,281
106,479

13. 7,481
702
+5,819
14,002

14. 99,103
33,450
+ 6,627
139,180

15. 49,002
1,999
+ 5,187
56,188

16. 55,998
40,003
+17,827
113,828

17. 1,903 + 5,075   6,978

18. 7,406 + 12,381   19,787

19. 800 + 20 + 4,000   4,820

20. 40,000 + 800 + 60   40,860

21. 31 + 93 + 277 + 12   413

22. 418 + 47 + 365 + 95   925

23. 6,482 meters + 9,027 meters
15,509 m

24. 17,812 mi + 4,283 mi   22,095 mi

25. 35 hours + 47 hours   82 hr

26. 225 square feet + 896 square feet
1,121 sq ft

27. $845 + $39 + $1,871 $2,755

28. $5,233 + $481 + $82 $5,796

29. $92,258 + $7,447 + $5,126
$104,831

30. $55,709 + $2,822 + $30,819
$89,350

31. 281 + 758 + 104 + 533   1,676

32. 3,911 + 2,947 + 8,007   14,865

33. 5,374 + 4,055 + 20,173   29,602

34. 7,900,471 + 201,387   8,101,858

35. $1,863 + $1,089 + $9,772 $12,724

36. 5,009 ft + 7,993 ft   13,002 ft

37. 8,300 tons + 22,900 tons   31,200 tons

38. 420,057 pounds + 900,808 pounds
1,320,865 lb

In each addition table, fill in the empty spaces. Compute all sums two ways, first working downward and then across. Check that you get the same grand total both ways.

39.

+	400	200	1,200	300	Total
300	700	500	1,500	600	3,300
800	1,200	1,000	2,000	1,100	5,300
Total	1,900	1,500	3,500	1,700	8,600

40.

+	4,000	300	3,000	2,000	Total
100	4,100	400	3,100	2,100	9,700
900	4,900	1,200	3,900	2,900	12,900
Total	9,000	1,600	7,000	5,000	22,600

41.

+	389	172	1,155	324	Total
255	644	427	1,410	579	3,060
799	1,188	971	1,954	1,123	5,236
Total	1,832	1,398	3,364	1,702	8,296

42.

+	3,749	279	2,880	1,998	Total
134	3,883	431	3,014	2,132	9,460
896	4,445	1,193	3,776	2,894	12,308
Total	8,328	1,624	6,790	5,026	21,768

43.
```
  3,088,281
  5,658,137
+ 4,550,239
 13,296,657
```
44.
```
   638,719
    40,003
 + 984,035
 1,662,757
```
45.
```
   2,008,490
   8,948,227
+ 11,956,174
  22,912,891
```
46.
```
 1,938,722
   325,411
 + 517,827
 2,781,960
```

In each group of three sums, one is wrong. Use estimation to explain which is incorrect.

47. a.
```
    814
  9,106
+ 2,811
 15,731
```
b.
```
 30,812
 47,045
+ 9,338
 87,195
```
c.
```
 183,066
  78,911
+ 96,527
 358,504
```

48. a.
```
  1,035
  5,210
+ 7,992
 14,237
```
b.
```
  5,801
  3,882
+12,644
 32,327
```
c.
```
  801,716
   78,001
+5,009,635
 5,889,352
```

49. a. $711,488
          a   102,663
          +   95,003
          $809,154

b. $62,933
        51,858
        + 49,612
        $164,403

c. $106,729
        99,821
        + 103,277
        $309,827

50. a. $9,512,622
          c   8,038,517
          + 2,615,334
          $20,166,473

b. $4,277,020
        915,611
        + 3,688,402
        $8,881,033

c. $200,312
        102,683
        + 504,113
        $707,108

Subtract. Then check either by addition or by estimation.

51.   379
         −162
         217

52.   362
         −110
         252

53.   200
         −110
          90

54.   210
         −100
         110

55.   513
         − 92
         421

56.   741
         −350
         391

57.   20,005
        −13,002
          7,003

58.   982,111
        −613,101
         369,010

59.   401
         − 39
         362

60.   728
         −539
         189

61.   70,000
        − 1,759
         68,241

62.   8,000
        −1,691
         6,309

63.   304
         −129
         175

64.   5,062
         −2,777
          2,285

65.   3,005
         −1,666
          1,339

66.   72,000
         −19,001
          52,999

67.   2,001
         −   2
          1,999

68.   3,000
         −   57
          2,943

69.   52,947
         −27,997
          24,950

70.   729,888
         −192,889
          536,999

71. $8,286 - 3,100$ 5,186

72. $12,799 - 2,357$ 10,442

73. $5,900 - 1,500$ 4,400

74. $14,400 - 12,300$ 2,100

75. $350,840 - 230,530$ 120,310

76. $410,700 - 280,900$ 129,800

77. $6,922 - 3,002$ 3,920

78. $27,733 - 22,500$ 5,233

79. $550 - 182$ 368

80. $1,448 - 962$ 486

81. $6,000 - 1,004$ 4,996

82. $8,602 - 907$ 7,695

83. $3,570 - 2,588$ 982

84. $2,182 - 899$ 1,283

85. $5,000 \text{ mi} - 3,005 \text{ mi}$ 1,995 mi

86. 701 square feet − 206 square feet   495 sq ft

87. $800 − $131 $669

88. 622 hr − 137 hr   485 hr

89. $4,812 − $1,203 $3,609

90. 402 mi − 57 mi   345 mi

91. 500 books − 227 books   273 books

92. $537 − $196 $341

93. 527 m − 318 m   209 m

94. 1,266 tons − 597 tons   669 tons

⊞ **95.**   30,000,000   ⊞ **96.**   1,973,000   ⊞ **97.**   3,402,331   ⊞ **98.**   14,500,007
    −27,999,000   −   997,001   −2,588,902   −13,972,008
     2,001,000   975,999   813,429   527,999

In each group of three differences, one is wrong. Use estimation to explain which is incorrect.

99. a.   817,770   **b.**   11,172,055   **c.**   120,426,811
c   −502,966   − 7,892,106   − 98,155,772
     314,804   3,279,949   32,271,039

100. a.   67,812   **b.**   3,997,401   **c.**   316,134
b   −12,180   −1,125,166   − 89,164
     55,632   1,872,235   226,970

101. a. $381,882   **b.** $479,116   **c.** $200,072,639
a   − 173,552   − 102,663   − 150,038,270
    $108,330 $376,453 $ 50,034,369

102. a. $3,810,662   **b.** $4,718,287   **c.** $381,975
b   −   299,137   − 1,002,875   − 117,263
    $3,511,525 $5,721,162 $264,712

Applications

Solve each problem and check your answer.

103. In 1800, the population of the United States was 5,000,000. During the next 100 years, the population grew by 70,000,000 people. What was the population in 1900?   75,000,000 people

104. You were driving along a road at 38 miles per hour (mph). To pass another car, you increased your speed by 19 mph. If the speed limit on the road was 55 mph, were you in violation of the law?   Yes (your actual speed was 57 mph).

105. You ran up a dental bill of $800. Your dental insurance reimbursed you for $300 of this amount. How much of this bill was not reimbursed? $500

106. The pressure in a tank of oxygen dropped from 1,400 pounds per square inch (psi) to 600 psi. By how much did the pressure drop?   800 psi

107. In a baseball game, the Huskies played the Ravens. The scoreboard shows how many runs each team scored during the nine innings of the game.

Inning	1	2	3	4	5	6	7	8	9
Huskies	0	1	1	0	1	1	1	0	3
Ravens	1	2	0	0	1	1	2	0	2

a. How many runs did Huskies score? Ravens?   Huskies: 8; Ravens: 9

b. Which team won the game?   Ravens

108. Suppose that the deposit slip shown is yours.

a. Estimate how much money you are depositing.   Possible answer: $3,800

b. Then fill in the exact total. $3,725

109. Blues singer Bessie Smith was born in 1894 and died in 1937. Estimate how old she was when she died. (*Source: Encyclopedia of World Biography*)   Possible answer: 43 years old

110. The United States entered the First World War in 1917 and the Second World War in 1941. Estimate how many years apart these two events were.   Possible answer: 20 yr

111. A sign in an elevator reads: MAXIMUM CAPACITY—1,000 POUNDS. The passengers in the elevator weigh 187 lb, 147 lb, 213 lb, 162 lb, 103 lb, and 151 lb. Will the elevator be overloaded?   No, the elevator is not overloaded. The total weight of passengers is 963 lb.

112. You would like to use a computer program that requires at least 600 kilobytes (K) of memory. Unfortunately, the memory of your old computer is only 512K. If you increase your computer's memory by 128K, will there be enough memory to run the program?   Yes; new memory = 640K

113. The thermometer at the right shows the boiling point and the freezing point of water in degrees Fahrenheit (°F). What is the difference between these two temperatures?
180°F

114. You bought a car after seeing this ad in a local newspaper. How much below the MSRP (manufacturer's suggested retail price) is the selling price?

$2,950

115. Find the perimeter of the figure shown, which is made up of a triangle and a square.   31 m

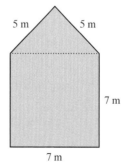

5 m   5 m

7 m

7 m

116. What is the length of the molding along the perimeter of the room pictured (yards = yd)?   22 yd

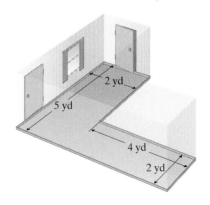

2 yd

5 yd

4 yd

2 yd

117. Your annual income is $16,998. How much more money must you make to have an annual income of $20,000? $3,002

118. Among the 9,572 students at a college, 938 worked full time. How many students at the college did not work full time?   8,634

119. The Earth has four oceans. The areas of these oceans, in round numbers, are the Pacific, 64,000,000 square miles (sq mi); the Atlantic, 32,000,000 sq mi; the Indian, 25,000,000 sq mi; and the Arctic, 5,000,000 sq mi. What is their total area? 126,000,000 sq mi

120. An oil tanker broke apart at sea. It spilled 150,000 gallons (gal) of crude oil the first day, 400,000 gal the second day, and 1,000,000 gal the third day. How much oil was spilled in all?   1,550,000 gal

121. You promise to give the editor of your magazine a bonus if its circulation rises by more than 10,000 readers. What will you do if the circulation increases from 92,570 to 103,871?   Give the bonus; rise = 11,301

122. You have $2,358 in a bank account. After withdrawing $755, will there be enough money in the account to pay a $1,500 bill?   Yes; remaining balance is $1,603.

▦ *Use a calculator to solve each problem, giving (a) the operation(s) carried out in your solution, (b) the exact answer, and (c) an estimate of the answer.*

123. Do you agree with the total amount of the deposits shown on the following bank statement?

MBU Bank & Trust Co.

```
Your Account

   Deposits    Date
   $    83     2/13
   $    59     2/14
   $   727     2/16
   $   183     2/17
   $   511     2/21

TOTAL $1,563
```

(a) Addition; (b) yes, $1,563; (c) possible estimate: $1,600

124. At its first eight games this season, a professional baseball team had the following paid attendance:

Game	Attendance
1	11,862
2	18,722
3	14,072
4	9,713
5	25,913
6	28,699
7	19,302
8	18,780

What was the combined attendance for these games?   (a) Addition; (b) 147,063; (c) possible estimate: 148,000

■ *Check your answers on page A-1.*

MINDSTRETCHERS

WRITING

1. There are many different ways of putting expressions into words.

a. For example, 3 + 2 can be expressed as

the sum of 3 and 2,   2 more than 3,   or   3 increased by 2.

Can you think of any other ways of reading this expression?   Possible answers: 2 added to 3; the total of 3 and 2; etc.

b. For example, 5 − 2 can be expressed as

the difference between 5 and 2,   5 take away 2,   or   5 decreased by 2.

Write two other ways.   Possible answers: 5 minus 2; 2 subtracted from 5; etc

CRITICAL THINKING

2. In a **magic square**, the sum of every row, column, and diagonal is the same number. Using the given information, complete the square at the right, which contains the whole numbers from 1 to 16. (*Hint*: The sum of every row, column, and diagonal is 34.)

16	3	2	13
5	10	11	8
9	6	7	12
4	15	14	1

GROUPWORK

3. Two methods for borrowing in a subtraction problem are illustrated as follows. In method (a)—the method that we have already discussed—we borrow by taking 1 from the top and in method (b) by adding 1 to the bottom.

$$
\textbf{a.} \quad
\begin{array}{r}
{}^{7}\!\!{}^{1} \\
8\;5\;9 \\
-3\;7\;6 \\
\hline
4\;8\;3
\end{array}
\qquad
\textbf{b.} \quad
\begin{array}{r}
{}^{1} \\
8\;5\;9 \\
{}^{4}\!\!-3\;7\;6 \\
\hline
4\;8\;3
\end{array}
$$

Note that we get the same answer with both methods. Discuss with a partner the advantages of each method.   Answers may vary.

1.3 Multiplying Whole Numbers

OBJECTIVES

■ *To multiply whole numbers*
■ *To estimate the product of whole numbers*
■ *To solve word problems involving the multiplication of whole numbers*

The Meaning and Properties of Multiplication

What does it mean to multiply whole numbers? A good answer to this question is *repeated addition*.

For instance, suppose that you buy 4 packages of pens and that each package contains 3 pens. How many pens are there altogether?

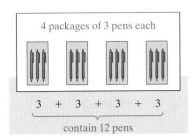

That is, $4 \times 3 = 3 + 3 + 3 + 3 = 12$. Generally, *multiplication means adding the same number repeatedly*.

We can also picture multiplication in terms of a rectangular figure, like this one, that represents 4×3.

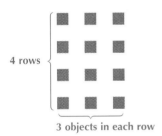

There are several ways to write a multiplication problem.

		Factor	Factor	Product
$\times$	the times sign	4	$\times$ 3	= 12
$\cdot$	a multiplication dot	4	$\cdot$ 3	= 12
( )( )	parentheses		(4)(3)	= 12

Like addition and subtraction, multiplication problems can be written either horizontally or vertically.

$$8 \times 5 = 40$$
Horizontal

$$\begin{array}{r} 8 \\ \times\ 5 \\ \hline 40 \end{array}$$
Vertical

The operation of multiplication has several important properties that we use frequently, often without thinking.

The Identity Property of Multiplication

The product of any number and 1 is that number.
$$1 \times 12 = 12$$
$$5 \times 1 = 5$$

The Multiplication Property of 0

The product of any number and 0 is 0.
$$49 \times 0 = 0$$
$$0 \times 8 = 0$$

The Commutative Property of Multiplication

Changing the order in which we multiply two numbers gives the same product.
$$2 \times 9 = 9 \times 2$$
$$\downarrow \qquad \downarrow$$
$$18 \quad = \quad 18$$

The Associative Property of Multiplication

When we multiply three numbers, regrouping the factors gives the same product.

We multiply inside the parentheses first.
$$\downarrow \qquad\qquad \downarrow$$
$$(3 \times 4) \times 5 = 3 \times (4 \times 5)$$
$$\downarrow \qquad\qquad\qquad \downarrow$$
$$12 \times 5 = 3 \times 20$$
$$\downarrow \qquad\qquad \downarrow$$
$$60 = 60$$

The next—and last—property of multiplication also involves addition.

The Distributive Property

Multiplying a factor by the sum of two numbers gives us the same result as multiplying the factor by each of the two numbers and then adding.

$$2 \times (5 + 3) = (2 \times 5) + (2 \times 3)$$
$$\downarrow \qquad\qquad \downarrow \qquad\quad \downarrow$$
$$2 \times 8 = 10 + 6$$
$$\downarrow \qquad\qquad\qquad \downarrow$$
$$16 \qquad = \qquad 16$$

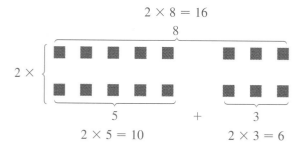

Before going on to the next section, study these properties of multiplication.

Multiplying Whole Numbers

Now let's consider problems in which we multiply any whole number by a single-digit whole number.

Note that, to multiply whole numbers with reasonable speed, you must commit to memory the products of all single-digit whole numbers.

EXAMPLE 1

Multiply: $98 \cdot 4$

Solution We recall that the dot means multiplication. We first write the problem vertically.

$$\begin{array}{r} \overset{3}{9}\,8 \\ \times\ 4 \\ \hline 2 \end{array}$$

← The product of 4 and 8 ones is 32 ones. We write the 2 and carry the 3 to the tens column.

We recall that the 9 in 98 means 9 tens. The product of $98 \cdot 4$ is 392.

$$\begin{array}{r} \overset{3}{9}\,8 \\ \times\ \ 4 \\ \hline 3\,9\,2 \end{array}$$

← The product of 4 and 9 tens is 36 tens. We add the 3 tens to the 36 tens to get 39 tens.

PRACTICE 1

Find the product of 76 and 8.   608

EXAMPLE 2

Calculate: $(806)(7)$

Solution We recall that parentheses side by side means to multiply. We write this problem vertically.

$$\begin{array}{r} 8\,\overset{4}{0}\,6 \\ \times\ \ \ 7 \\ \hline 5,\,6\,4\,2 \end{array}$$

└── Here 7×0 tens = 0 tens. Add the carried 4 tens to the 0 tens to get 4 tens.

The product of 806 and 7 is 5,642.

Now let's look at multiplying any two whole numbers.

Consider multiplying 32 by 48. We can write 32×48 as follows.

$$32 \times 48 = 32 \times (40 + 8)$$

We then use the distributive property to get the answer.

$$32 \times (40 + 8) = (32 \times 40) + (32 \times 8)$$
$$= 1,280 + 256$$
$$= 1,536$$

Generally, we solve this problem vertically.

$$\begin{array}{r} \overset{1}{3}\,2 \\ \times\ 4\,8 \\ \hline 2\,5\,6 \\ 1\,2\,8\,0 \\ \hline 1,\,5\,3\,6 \end{array}$$

← Partial product (8×32)
← Partial product (40×32)
← Add the partial products.

PRACTICE 2

Find the product: $(705)(6)$ 4,230

Remind students about the multiplication property of 0.

Shortcut

$$\begin{array}{r} 32 \\ \times\ 48 \\ \hline 256 \\ 1\,28 \\ \hline 1,536 \end{array}$$

← (8×32)
← (4×32)

If we use just the tens digit 4, we must write the product 128 leftward, starting at the tens column.

Example 2 suggests the following rule for multiplying whole numbers.

> **To Multiply Whole Numbers**
> ■ multiply the top factor by the ones digit in the bottom factor, and write down this product,
> ■ multiply the top factor by the tens digit in the bottom factor, and write this product leftward, beginning with the tens column,
> ■ repeat this process until you use all the digits in the bottom factor, then
> ■ add the partial products, writing down this sum.

EXAMPLE 3

Multiply: 300×50

Solution

$$
\begin{array}{r}
300 \\
\times\ 50 \\
\hline
000 \leftarrow 0 \times 300 = 0 \\
15\ 00 \leftarrow 5 \times 300 = 1{,}500 \\
\hline
15{,}000
\end{array}
$$

Note that the number of zeros in the product equals the total number of zeros in the factors. This result suggests a shortcut for multiplying factors that end in zeros.

$$
\begin{array}{r}
300 \leftarrow \text{2 zeros} \\
\times\ 50 \leftarrow \text{1 zero} \\
\hline
15{,}000 \leftarrow 2 + 1 = \text{3 zeros}
\end{array}
$$

PRACTICE 3

Find the product of 1,200 and 400.
480,000

TIP   When multiplying two whole numbers that end in zeros, multiply the nonzero parts of the factors and then attach the total number of zeros to the product.

EXAMPLE 4

Simplify: $739 \cdot 305$

Solution

$$
\begin{array}{r}
739 \\
\times\ 305 \\
\hline
3\ 695 \leftarrow 5 \times 739 \\
0\ 00 \leftarrow 0 \times 739 = 0 \\
221\ 7\ \ \leftarrow 3 \times 739 \\
\hline
225{,}395
\end{array}
$$

We don't have to write the row 000. Here is a shortcut.

$$
\begin{array}{r}
739 \\
\times\ 305 \\
\hline
3\ 695 \\
221\ 70 \leftarrow \\
\hline
225{,}395
\end{array}
$$
This one 0 represents the product of the tens digit 0 and 739. This 0 lines up the products correctly.

PRACTICE 4

Find the product of 987 and 208.
205,296

Now let's apply the operation of multiplication to geometric area. Area means the number of square units that a figure contains.

In the rectangle at the right, each small square represents 1 square inch (sq in.). Finding the rectangle's area means finding the number of sq-in. units that it contains. A good strategy here is to find the number of units in each row and then multiply that number by the number of rows.

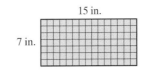
15 in.

7 in.

Point out to students the key differences between perimeter and area.
■ Perimeter is the distance around a figure; area is the space that a figure encloses.
■ Perimeter is measured in linear units, such as feet or meters, and area is measured in square units, such as square feet or square meters.

There are two ways to find that there are 15 squares in a row—either by directly counting the squares or by noting that the length of the figure is 15 in. Similarly, we find that the figure contains 7 rows. Therefore the area of the figure is 15×7, or 105 sq in.

In general, we can compute the *area of a rectangle* by finding the product of its length and its width.

EXAMPLE 5

What is the area of this figure? (cm = centimeter)

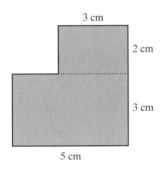
3 cm
2 cm
3 cm
5 cm

Solution The dotted line separates the figure into two connected rectangles. The top one measures 3 cm by 2 cm, and so its area is 3×2, or 6 square centimeters (sq cm). The bottom rectangle measures 3 cm by 5 cm, and its area is 5×3, or 15 sq cm. The entire area of this figure is the sum of the two smaller areas: $6 + 15$, or 21 sq cm.

PRACTICE 5

Find the area of the room pictured. 107 sq ft

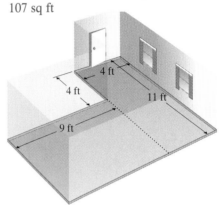
4 ft
4 ft
11 ft
9 ft

Estimating Products

When solving some multiplication problems, we are willing to settle for—or even prefer—an approximate answer.

EXAMPLE 6

Suppose that you sold 210 tickets to a play. The play is held in a small theater that has 27 rows and 12 seats in each row. By estimating, decide if you have enough seats.

Solution Estimating is easier than computing the exact number of seats in the theater.

To estimate a product, we round each factor to its largest place so that every digit after the first digit is 0.

$27 \approx 30 \leftarrow$ The largest place is tens.
$\times 12 \approx 10 \leftarrow$ The largest place is tens.

Then, we multiply the rounded factors.

$30 \times 10 = 300$
1 zero   1 zero   2 zeros

Since there are about 300 seats in the theater and since 300 is greater than 210, you can conclude that there are enough seats.

PRACTICE 6

Producing flyers for your college's registration requires 25,000 sheets of paper. If the college buys 38 reams of paper and there are 500 sheets in a ream, estimate to decide if there is enough paper to produce the flyers.   No; possible estimate = 20,000

TIP   When checking by estimation, round the factors to a large enough place so that you can multiply them *in your head*.

As we mentioned before, estimation is also a valuable technique for checking an exact answer.

EXAMPLE 7

Multiply 328 by 179. Check the answer by estimation.

Solution We first find the exact product.

$$
\begin{array}{r}
328 \\
\times\ 179 \\
\hline
2\ 952 \\
22\ 96 \\
32\ 8 \\
\hline
58{,}712
\end{array}
$$

We now check by estimating.

$$
\begin{array}{r}
328 \approx\quad 300 \\
\times\ 179 \approx \times\ 200 \\
\hline
58{,}712\quad 60{,}000
\end{array}
$$

328 ≈ 300 ← The largest place is hundreds.
179 ≈ 200 ← The largest place is hundreds.

Exact product ⌞ Fairly close ⌟ Estimated product

So we can say that our answer, 58,712, is close to the estimate.

PRACTICE 7

Find the product of 455 and 248. Use estimation to check your answer.   112,840

Multiplying Whole Numbers on a Calculator

When you are using a calculator to multiply large whole numbers, the answer may be too big to fit in the display. If this overflow occurs, some calculators will display an error sign (often an E) to indicate that the answer shown is wrong, and the system freezes. To unlock the system, try pressing a clear key (C or AC).

EXAMPLE

Use a calculator to multiply: $3{,}192 \times 41$

Solution

Input	3192	×	41	=
Display	3192.	3192.	41.	130872.

A reasonable estimate for this product is $3{,}000 \times 40$, or 120,000, which supports our answer, 130,872.

PRACTICE

Find the product:
$\$2{,}811 \times 365$ $\$1{,}026{,}015$

EXAMPLE

Calculate: 61 · 24 · 19

Solution

Input	61	$\times$	24	$\times$	19	$=$
Display	61.	61.	24.	1464.	19.	27816

A good estimate is 60 · 20 · 20, or 24,000—in the ballpark of 27,816.

PRACTICE

Multiply: 2,133 · 18 · 9
345,546

EXAMPLE

Examine your calculator's answer to 999,999,999 × 888,888,888.

Solution

Input	999999999	$\times$	888888888	$=$
Display	999999999.	999999999.	888888888.	?

For most calculators, this product is too large to fit in the display. If that is the case for your calculator model, note how it deals with an overflow.

PRACTICE

Try to compute the product

1,234,567,890 · 987,654,321

on your calculator.
Possible answer:
$1.219326311 \times 10^{18}$

Exercises 1.3

Compute.

1. 4 × 100
400

2. 100 × 12
1,200

3. 71 × 100
7,100

4. 27 × 10
270

5. 85 × 10
850

6. 68 × 1,000
68,000

7. 10,000 × 700
7,000,000

8. 1,000 × 8,000
8,000,000

Multiply and check by estimation.

9. 398
× 3
1,194

10. 550
× 9
4,950

11. 964
× 2
1,928

12. 8,381
× 5
41,905

13. 6,350
× 2
12,700

14. 8,864
× 7
62,048

15. 209
× 2
418

16. 703
× 9
6,327

17. 5,420,000
× 2
10,840,000

18. 357,000
× 3
1,071,000

19. 812,000
× 4
3,248,000

20. 19,250
× 8
154,000

21. 892
× 35
31,220

22. 882
× 74
65,268

23. 992
× 68
67,456

24. 881
× 28
24,668

25. 43 · 19
817

26. 85 · 72
6,120

27. 709 · 48
34,032

28. 602 · 34
20,468

29. 273
× 11
3,003

30. 607
× 65
39,455

31. 301
× 12
3,612

32. 513
× 34
17,442

33. 3,001
× 19
57,019

34. 4,005
× 72
288,360

35. 5,072
× 48
243,456

36. 301
× 34
10,234

37. 8,801
× 25
220,025

38. 5,003
× 40
200,120

39. 2,881
× 70
201,670

40. 719
× 55
39,545

41. (302)(403)
121,706

42. (699)(101)
70,599

43. 8,500 × 17
144,500

44. 700 × 207
144,900

45. 406 × 305
123,830

46. 702 × 59
41,418

47. 46 · 8 · 9
3,312

48. 13 · 11 · 5
715

49. $81 \times 2 \times 13$
2,106

50. $3 \times 5 \times 88$
1,320

51. $(10)(10)(400)$
40,000

52. $(20)(80)(30)$
48,000

53. $57 \times 81 \times 5$
23,085

54. $73 \times 4 \times 33$
9,636

55. 8,972
  $\times\ \ 365$
3,274,780

56. 7,552
  $\times\ \ \ 36$
271,872

57. 18,650
  $\times\ \ 2,949$
54,998,850

58. 8,783
  $\times 7,159$
62,877,497

In each group of three products, one is wrong. Use estimation to explain which product is incorrect.

59. a. $802 \times 755 = 605,510$ **b.** $39 \times 4,722 = 184,158$ **c.** $77 \times 6,005 = 46,385$
c; Possible estimate: 480,000

60. a. $618 \times 555 = 342,990$ **b.** $86,331 \times 21 = 18,129,511$ **c.** $380 \times 772 = 293,360$
b; Possible estimate: 1,800,000

61. a. $9 \times 37,118 = 334,062$ **b.** $82 \times 961 = 7,882$ **c.** $13 \times 986 = 12,818$
b; Possible estimate: 80,000

62. a. $3,002 \times 9 = 2,718$ **b.** $58 \times 891 = 51,678$ **c.** $106 \times 68 = 7,208$
a; Possible estimate: 27,000

Applications

Solve. Then check your answer by estimating.

63. Underwater explorers in the eastern Mediterranean Sea found the wreck of an Egyptian ship that had sunk 33 centuries earlier. How long ago in years did the ship sink? (*Hint*: 1 century = 100 yr)   3,300 yr

64. Suppose that every day you take 4 vitamin tablets and that each tablet contains 10,000 International Units (I.U.) of Vitamin A. How much Vitamin A do you take daily?   40,000 I.U.

65. The walls of a human heart are made of muscles that contract about 100,000 times a day. How many contractions are there in 30 days? (*Source: American Heart Association's Your Heart: An Owner's Manual*)
3,000,000

66. A computer disk stores 640K bytes of information. How much information does it have? (*Hint*: The letter K stands for "kilo," which means 1,000).
640,000 bytes

67. Your car gets 32 miles per gallon (mpg) of gasoline. If a full tank contains 18 gal, can you drive to a town 520 mi away without refilling your car's gas tank?   Yes; you could go 576 mi without refilling.

68. Long Lasting light bulbs last 2,560 hours (hr), whereas other bulbs last 745 hr. Is Long Lasting's claim true that its light bulbs last 3 times as long as the others?   Yes

69. The monthly rent for an apartment is $725. What does this rent amount to for a year? $8,700

70. You type about 35 words a minute (min). Approximately how many words can you type in 45 min?   1,575 words

71. On the following map, 1 in. corresponds to 250 mi in the real world. How many miles actually separate towns A and B?   1,750 mi

72. Angles are measured in either degrees (°) or radians (rad). A radian is about 57°. How big in degrees is the angle shown?   About 171°

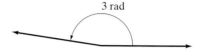

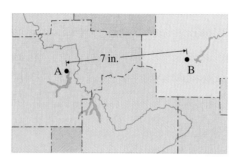

73. You hired 7 moving men to work 19 hr at $21/hr each. How much money did you have to pay them in all? $2,793

74. A dishwasher takes about 15 gal of water per wash. How much water does it use in 7 days with 2 washes per day?   210 gal

🖩 *Use a calculator to solve each problem, giving (a) the operation(s) carried out in your solution, (b) the exact answer, and (c) an estimate of the answer.*

75. The state of Colorado is rectangular in shape, as shown. If the area of Kansas is about 82,000 sq mi, which state is larger? (**Source:** *The Columbia Gazeteer of the World*)   (a) Multiplication   (b) Colorado; area = 106,700 sq mi   (c) Possible estimate: 120,000 sq mi

76. Tuition at a certain college is $2,125/yr for every full-time student. If there are 10,975 full-time students at the college, how much revenue is generated from student tuition?   (a) Multiplication   (b) $23,321,875   (c) Possible estimate: $22,000,000

■ *Check your answers on page A-2.*

MINDSTRETCHERS

WRITING

1. Study the following diagram. Explain how it justifies the distributive property.

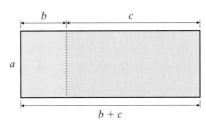

Using areas: Area of the big rectangle = sum of the areas of the two smaller rectangles.

$a(b + c) = ab + ac$

MATHEMATICAL REASONING

> Suggest to students that they may want to use a calculator to solve this problem.

2. Consider the six digits 1, 3, 5, 7, 8, and 9. Fill in the blanks with these digits, using each digit only once, so as to maximize the product:

$$\underline{9} \ \underline{5} \ \underline{1} \times \underline{8} \ \underline{7} \ \underline{3}$$

HISTORICAL

3. Centuries ago in India and Persia, the **lattice method** of multiplication was popular. The following example, in which we multiply 57 by 43, illustrates this method. Explain how it works.

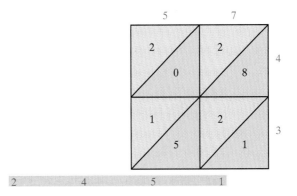

$57 \times 43 = (5 \text{ tens} + 7 \text{ ones}) \times (4 \text{ tens} + 3 \text{ ones}) = 20 \text{ hundreds} + 15 \text{ tens} + 28 \text{ tens}$
$+ \ 21 \text{ ones} = 2 \text{ thousands} + 43 \text{ tens} + 21 \text{ ones} = 2 \text{ thousands} + 4 \text{ hundreds} + 3 \text{ tens}$
$+ \ 2 \text{ tens} + 1 \text{ one} = 2 \text{ thousands} + 4 \text{ hundreds} + 5 \text{ tens} + 1 \text{ one} = 2,451$

HISTORICAL NOTE

1	2	3	4	5	6	7	8	9	0	
										Twelfth century
										1197 A.D.
										1275 A.D.
										c. 1294 A.D.
										c. 1303 A.D.
										c. 1360 A.D.
										c. 1442 A.D.

The way the ten digits are written has evolved over time. Early Hindu symbols found in a cave in India date from more than two thousand years ago. About twelve hundred years ago, an Indian manuscript on arithmetic which had been translated into Arabic was carried by merchants to Europe where it was later rendered into Latin.

This table shows European examples of digit notation from the twelfth to the fifteenth century, when the printing press led to today's standardized notation. Through international trade, these symbols became known throughout the world.

Source: David Eugene Smith and Jekuthiel Ginsburg, *Numbers and Numerals, A Story Book for Young and Old* (New York: Bureau of Publications, Teachers College, Columbia University, 1937).

1.4 Dividing Whole Numbers

OBJECTIVES

■ *To divide whole numbers*
■ *To estimate the quotient of whole numbers*
■ *To solve word problems involving the division of whole numbers*

The Meaning and Properties of Division

What does it mean to divide? One good answer is to think of division as *breaking up a set of objects* into a given number of equal smaller sets.

For instance, suppose that we want to divide 15 by 3. We can look at this problem as splitting a set of 15 objects, say pens, evenly among 3 boxes.

From the diagram we see that each box ends up with 5 pens. We therefore say that 15 divided by 3 is 5, which we can write as follows.

$$\overset{5}{3)\overline{15}}$$

Divisor · · · Quotient · · · Dividend

We can also think of division as the *opposite* (*inverse*) of multiplication. Consider the following pair of problems that illustrate this point.

$$\overset{5}{3)\overline{15}} \qquad \text{because} \qquad 5 \times 3 = 15$$

Division · · · Related Multiplication

In general, the following relationship is true.

> Quotient × Divisor = Dividend

Do you see that this relationship allows us to check our answer to a division problem by multiplying?

There are several common ways to write a division problem.

$$\overset{5}{3)\overline{15}}, \qquad \frac{15}{3} = 5, \qquad \text{or} \qquad 15 \div 3 = 5$$

Tell students that some calculators will show $\frac{15}{3}$ as 15/3.

Usually, we use the first of these to compute the answer. However, no matter which way we write this problem, 3 is the divisor, 15 is the dividend, and 5 is the quotient.

TIP When reading a division problem, we say that we are dividing either the divisor *into* the dividend or the dividend *by* the divisor. For instance, $3)\overline{15}$ is read either "3 divided into 15" or "15 divided by 3."

When calculating a quotient, we frequently use the following properties of division.

	Division	**Related Multiplication**
■ Any whole number (except 0) divided by itself is 1.	$\dfrac{1}{6\overline{)6}}$	$1 \times 6 = 6$
■ Any whole number divided by 1 is the number itself.	$\dfrac{12}{1\overline{)12}}$	$12 \times 1 = 12$
■ Zero divided by any whole number (other than 0) is 0.	$\dfrac{0}{8\overline{)0}}$	$0 \times 8 = 0$
■ Division by 0 is not permitted.	$\dfrac{?}{0\overline{)5}}$	$? \times 0 = 5$

There is no number that when multiplied by 0 equals 5.

> Ask your students to consider the problem $0 \div 0$. Let them decide if this problem makes sense, and if it does, to identify the solutions.

Dividing Whole Numbers

Multiplication is the opposite of division. So in a simple division problem, such as $3\overline{)15}$, we know that the answer is 5 because we have memorized that $5 \cdot 3$ is 15. But what should we do when the dividend is a larger number?

Consider the following problem: Divide 9 into 5,112 and check the answer.

■ We start with the greatest place (thousands) in the dividend. We consider the dividend to be 5 thousands and think $9\overline{)5}$. Since $9 \cdot 1 = 9$ and 9 is larger than 5, there are no thousands in the quotient.

$$\begin{array}{r} 0 \longleftarrow \text{Thousands} \\ 9\overline{)5{,}112} \end{array}$$

■ So we go to the hundreds place in the dividend. We consider the dividend to be 51 hundreds and think $9\overline{)51}$. Since $9 \cdot 5 = 45$, we position the 5 in the hundreds place of the quotient.

$$\begin{array}{r} 5 \longleftarrow \text{Hundreds} \\ 9\overline{)5{,}112} \\ -4{,}500 \longleftarrow 500 \cdot 9 = 4500 \\ \hline 612 \longleftarrow \text{Difference} \end{array}$$

■ Next, we move to the tens place of the difference, 612. We consider the new dividend to be 61 tens and think $9\overline{)61}$. Since $9 \cdot 6 = 54$, we position the 6 in the tens place of the quotient.

$$\begin{array}{r} \text{Tens} \\ 56 \\ 9\overline{)5{,}112} \\ -4{,}500 \\ \hline 612 \\ -540 \longleftarrow 60 \cdot 9 = 540 \\ \hline 72 \longleftarrow \text{Difference} \end{array}$$

■ Finally, we go to the ones place of the difference, 72. We consider the new dividend to be 72 ones. So we think $9\overline{)72}$. Since $9 \cdot 8 = 72$, we position the 8 in the ones place of the quotient.

So 568 is our answer.

$$\begin{array}{r} \text{Ones} \\ 568 \\ 9\overline{)5{,}112} \\ -4{,}500 \\ \hline 612 \\ -540 \\ \hline 72 \\ -72 \longleftarrow 8 \cdot 9 = 72 \\ \hline 0 \longleftarrow \text{Difference} \end{array}$$

Instead of writing zeros as place value holders, we can use the following shortcut.

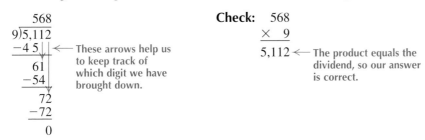

```
      568
   9)5,112
    -4 5↓↓   ←  These arrows help us
    ─────         to keep track of
       61↓         which digit we have
      -54↓         brought down.
      ────
        72
       -72
       ────
         0
```

Check: 568
            × 9
           ─────
           5,112 ← The product equals the
                       dividend, so our answer
                       is correct.

Note that each time we subtract in a division problem, the difference is less than the divisor. Why must that be true?

> Have your students explain their answers in a journal.

EXAMPLE 1	PRACTICE 1

EXAMPLE 1

Divide and check: 4,263 ÷ 7

Solution

```
            ┌─ Think 7)‾42‾.
           ↓609
        7)4,263
         -4 2      ← 6 × 7 = 42. Subtract.
         ─────
            06      ← Think 7)‾6‾. There are zero 7's in 6.
           -0       ← 0 × 7 = 0. Subtract.
          ─────
            63      ← Think 7)‾63‾.
           -63      ← 9 × 7 = 63. Subtract.
          ─────
             0
```

Check: 609   The product agrees with our dividend.
            ×   7   Note the 0 in the quotient. Do you see
            ─────   why the 0 is needed?
            4,263

PRACTICE 1

Compute 9)‾7,263‾ and then check your answer.   807

TIP

TIP   In writing your answer to a division problem, position the first digit of the quotient over the *right digit* of the number into which you are dividing (the 6 over the 2 in Example 1).

```
        609
        ↓↓↓
    7)4,263
```

> For students who are having difficulty lining up quotients, have them do division problems on grid or graph paper, or on lined paper turned sideways.

EXAMPLE 2	PRACTICE 2

EXAMPLE 2

Compute $\dfrac{2,709}{9}$. Then check your answer.

Solution

```
         301
      9)2,709
       -2 7
       ─────
          00      Check:   301
         -0              ×   9
        ─────           ─────
          09            2,709
         -9
        ─────
          0
```

PRACTICE 2

Carry out the following division and check your answer.

8)‾56,016‾

7,002

In Examples 1 and 2, note that the remainder is 0; that is, the divisor goes evenly into the dividend. However, in some division problems, that is not the case. Consider, for instance, the problem of dividing 16 pens *equally* among 3 boxes.

From the diagram, we see that each box contains 5 pens *but* that 1 pen—the *remainder*—is left over.

$$\begin{array}{r} 5 \quad \longleftarrow \text{Number of pens in each box} \\ \text{Number of boxes} \longrightarrow 3\overline{)16} \quad \longleftarrow \text{Total number of pens} \\ -15 \quad \longleftarrow \text{Total number of pens in the boxes} \\ \hline 1 \quad \longleftarrow \text{Number of pens remaining} \end{array}$$

We write the answer to this problem as 5 R1 (read "5 Remainder 1.") Note that $(3 \times 5) + 1 = 16$. The following relationship is always true.

$$\boxed{(\text{Quotient} \times \text{Divisor}) + \text{Remainder} = \text{Dividend}}$$

When a division problem results in a remainder as well as a quotient, we use this relationship for checking.

EXAMPLE 3	**PRACTICE 3**
Find the quotient of 55,811 and 6. Then check.	Compute $8\overline{)42,329}$ and check. 5,291 R1

Solution
$$\begin{array}{r} 9{,}301 \text{ R5} \\ 6\overline{)55{,}811} \\ -54 \phantom{{,}811} \\ \hline 1\,8 \\ -1\,8 \\ \hline 0\,1 \\ -0 \\ \hline 11 \\ -6 \\ \hline 5 \end{array}$$

Our answer is therefore 9,301 R5.

Check: $(9{,}301 \times 6) + 5 =$
$$55{,}806 + 5 = 55{,}811$$

The dividend is 55,811, so our answer is correct.

Now let's consider division problems in which a divisor has more than one digit. Notice that such problems involve rounding.

EXAMPLE 4

Compute $\dfrac{2,574}{34}$ and check.

Solution In order to estimate the first digit of the quotient, we round 34 to 30 and 257 to 260.

$$
\begin{array}{r}
8 \\
34\overline{)2,574} \\
-2\,72 \\
\end{array}
$$
← Think 260 ÷ 30, or 26 ÷ 3. The quotient 8 goes over the 7 because we are dividing 34 into 257.
← 8 × 34 = 272. Try to subtract.

Because 272 is too large, we reduce our estimate in the quotient by 1 and try 7.

$$
\begin{array}{r}
76 \\
34\overline{)2,574} \\
-2\,38 \\
\hline
194 \\
-204 \\
\end{array}
$$
← 7 × 34 = 238. Subtract.
← Think 190 ÷ 30, or 19 ÷ 3.
← 6 × 34 = 204. Try to subtract.

Because 204 is too large, we reduce our estimate in the quotient by 1 and try 5.

$$
\begin{array}{r}
75 \\
34\overline{)2,574} \\
-2\,38 \\
\hline
194 \\
-170 \\
\hline
24 \\
\end{array}
$$

Our answer is 75 R24.

Check: $(75 \times 34) + 24 = 2,574$

PRACTICE 4

Divide 23 into 1,817. Then check.   79

EXAMPLE 5

Divide $26\overline{)1,849}$ and then check.

Solution First we round 26 to 30 and 184 to 180.

$$
\begin{array}{r}
6 \\
26\overline{)1,849} \\
-1\,56 \\
\hline
28 \\
\end{array}
$$
← This difference is larger than the divisor, so we increase the 6 in the quotient by 1.

$$
\begin{array}{r}
71 \\
26\overline{)1,849} \\
-182 \\
\hline
29 \\
-26 \\
\hline
3 \\
\end{array}
$$

Our answer is therefore 71 R3.

Check: $(71 \times 26) + 3 = 1,849$

PRACTICE 5

Compute and check: $15\overline{)1,420}$ 94 R10

TIP If the divisor has more than one digit, estimate each digit in the quotient by rounding and then dividing. If the product is too large or too small, adjust it up or down by 1 and then try again.

EXAMPLE 6

Find the quotient of 13,559 and 44. Then check.

Solution

```
        308 R7
    44)13,559
      −13 2
          35  ←── This number is smaller than the divisor, so the next
          −0       digit in the quotient is 0.
         359
        −352
           7
```

Check: $(308 \times 44) + 7 = 13,559$

PRACTICE 6

Divide 16,999 by 28. Then check your answer.   607 R3

EXAMPLE 7

Divide and check: $6,000 \div 20$

Solution We set up the problem as before.

```
        300
    20)6,000        Check:   300
      −60                  × 20
       00                 6,000
      −00
       00
      −00
        0
```

Because the divisor and dividend both end in zero, a quicker way to do Example 7 is by dropping zeros.

```
    20)6,000  ←── Drop one 0 from
                  both the divisor
                  and the dividend.

        300
     2)600  ←── Then divide.
```

PRACTICE 7

Compute 40)8,000 and then check.   200

```
You may want to preview
simplifying fractions here:

  6,000     600 × 10̸
  ─────  =  ──────────
   20        2 × 10̸

          600
       =  ───
           2

       = 300
```

TIP If you drop the same number of zeros at the right end of the divisor and the dividend, you do not change the quotient.

Estimating Quotients

As for other operations, estimating is an important skill for division. Checking a quotient by estimation is faster than checking by multiplication, although less exact. And in some division problems, we need only an approximate answer.

How do we estimate a quotient? A good way is to round the divisor to its greatest place. The new divisor then contains only one nonzero digit and so is relatively easy to divide by mentally. Then we round the dividend to the place of our choice. Finally we compute the estimated quotient by calculating its first digit and then attaching the appropriate number of zeros.

> Remind students that when they estimate, there is no single correct answer. However, the closer the estimate comes to the exact answer, the better the estimate is.

EXAMPLE 8

Calculate $\dfrac{7{,}004}{34}$ and then check by estimation.

Solution

$$
\begin{array}{r}
206 \\
34\overline{)7{,}004} \\
-6\,8 \\
\hline
204 \\
-204 \\
\hline
\end{array}
$$
 Exact quotient

Check: $34\overline{)7{,}004}$ Round 34 to 30 and round 7,004 to 7,000.

$\downarrow\quad\downarrow$

$30\overline{)7{,}000}$ Think 70 ÷ 30, or 7 ÷ 3.

$\begin{array}{r} 200 \\ 30\overline{)7{,}000} \end{array}$ Estimated Quotient

Note that, to the right of the 2 in the estimated quotient, we added a 0 over each of the digits in the dividend. Our answer (206) is close to our estimate (200), and so our answer is reasonable.

PRACTICE 8

Compute 100,568 ÷ 104 and use estimation to check.   967

Dividing Whole Numbers on a Calculator

When using a calculator to divide, we must enter the numbers in the correct order to get the correct answer. We first enter the number *into* which we are dividing (the dividend) and then the number *by* which we are dividing (the divisor).

Most calculators express remainders not as whole numbers but rather as decimals—for instance, 3.5 instead of 3 R1 for the answer to $2\overline{)7}$. (We discuss these remainders in Chapter 3.)

EXAMPLE

Use a calculator to divide $18\overline{)11{,}718}$.

Solution

Input	11718	÷	18	=
Display	11718.	11718.	18.	651.

A reasonable estimate is 10,000 ÷ 20, or 500, which is fairly close to 651.

PRACTICE

Find the following quotient with a calculator:

$$\frac{47{,}034}{78}$$
603

> In a division problem have students enter the numbers in the wrong order to see that they get the wrong answer.

Exercises 1.4

Divide and then check.

1. $5\overline{)2{,}000}$
400

2. $4\overline{)10{,}000}$
2,500

3. $3\overline{)9{,}000{,}000}$
3,000,000

4. $6\overline{)12{,}000}$
2,000

5. $3\overline{)150}$
50

6. $2\overline{)482}$
241

7. $6\overline{)726}$
121

8. $4\overline{)904}$
226

9. $5\overline{)2{,}800}$
560

10. $8\overline{)12{,}504}$
1,563

11. $9\overline{)2{,}709}$
301

12. $2\overline{)5{,}780}$
2,890

13. $7\overline{)21{,}021}$
3,003

14. $8\overline{)41{,}008}$
5,126

15. $5\overline{)27{,}450}$
5,490

16. $4\overline{)3{,}500}$
875

17. $3\overline{)606}$
202

18. $7\overline{)497}$
71

19. $9\overline{)1{,}971}$
219

20. $2\overline{)30{,}534}$
15,267

21. $9\overline{)4{,}500}$
500

22. $6\overline{)834}$
139

23. $5\overline{)4{,}205}$
841

24. $3\overline{)4{,}512}$
1,504

25. $300 \div 10$
30

26. $400 \div 20$
20

27. $700 \div 50$
14

28. $6{,}000 \div 20$
300

29. $\dfrac{8{,}000}{50}$
160

30. $\dfrac{700}{10}$
70

31. $\dfrac{2{,}000}{40}$
50

32. $\dfrac{50{,}000}{20}$
2,500

33. 6,512 ÷ 10
651 R2

34. 8,922 ÷ 25
356 R22

35. 304 ÷ 27
11 R7

36. 206 ÷ 45
4 R26

37. 10,179 ÷ 87
117

38. 99,980 ÷ 66
1,514 R56

39. 79,992 ÷ 99
808

40. 12,948 ÷ 13
996

41. 6,996 ÷ 44
159

42. 9,660 ÷ 92
105

43. 80,295 ÷ 15
5,353

44. 936 ÷ 72
13

45. 39,078 ÷ 39
1,002

46. 49,497 ÷ 21
2,357

47. 249,984 ÷ 36
6,944

48. 499,992 ÷ 24
20,833

49. 47)‾34,000
723 R19

50. 25)‾21,000
840

51. 14)‾6,000
428 R8

52. 32)‾3,007
93 R31

53. 52)‾52,052
1,001

54. 24)‾48,072
2,003

55. 12)‾36,600
3,050

56. 36)‾25,560
710

57. 537)‾387,177
721

58. 265)‾197,160
744

59. 638)‾98,890
155

60. 152)‾34,048
224

In each group of three quotients, one is wrong. Use estimation to explain which quotient is incorrect.

61. **a.** 455,260 ÷ 65 = 704   **b.** 11,457 ÷ 57 = 201   **c.** 10,044 ÷ 93 = 108
a; Possible estimate: 7,000

62. a. 18,473 ÷ 91 = 203   **b.** 43,364 ÷ 74 = 586   **c.** 14,562 ÷ 18 = 8,009
     c; Possible estimate: 800

63. a. 43,710 ÷ 93 = 47   **b.** 71,048 ÷ 107 = 664   **c.** 11,501 ÷ 31 = 371
     a; Possible estimate: 400

64. a. 178,267 ÷ 89 = 2,003   **b.** 350,007 ÷ 21 = 1,667   **c.** 37,185 ÷ 37 = 1,005
     b; Possible estimate: 20,000

Applications

Solve and check.

65. A family spent $445 for 5 nights in a motel. How much money did they spend per night? $89

66. Your new laser printer prints 6 pages per minute. How long will it take to print a 42-page report?   7 min

67. In round numbers, the area of the Pacific Ocean is 64 million sq mi, and the area of the Atlantic Ocean is 32 million sq mi. The Pacific is how many times as large as the Atlantic? (*Source: The New Encyclopedia Britannica*)   2 times

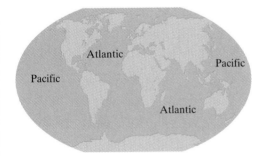

68. The diameter of the Earth is about 8 thousand mi, whereas the diameter of the Moon is about 2 thousand mi. How many times the Moon's diameter is the Earth's? (*Source: The New Encyclopedia Britannica*)   4 times

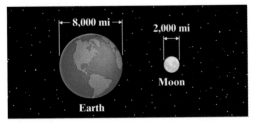

69. A car used 15 gal of gas on a 300-mi trip. How many miles per gallon (mpg) of gas did the car get?   20 mpg

70. You save $25 a week for a $625 stereo system. In how many weeks will you have enough money to buy the stereo system?   25 weeks

71. You want to buy enough 120-min blank video tapes to record a TV special that will last 250 min. How many tapes should you buy?   3

72. You type approximately 45 words/min. How long will it take you to type a 900-word essay for your English class?   20 min

73. What is the monthly income for someone with an annual salary of $29,952? (*Reminder*: There are 12 months (mo) in a year.) $2,496

74. Three friends split the driving evenly going from Chicago to Los Angeles. How many miles did each drive if the total distance is 2,835 mi?   945 mi

▦ *Use a calculator to solve each problem, giving (a) the operation(s) carried out in your solution, (b) the exact answer, and (c) an estimate of the answer.*

75. Suppose that you can read 275 words/min. How long will it take you to read an article containing 15,950 words?   (a) Division, (b) exact answer: 58 min, (c) possible estimate: 50 min

76. Light travels at a speed of 186,000 miles per second (mps). How long will it take for light to reach the Earth from a star 6,882,000 mi away?   (a) Division, (b) exact answer: 37 sec, (c) possible estimate: 35 sec

■ *Check your answers on page A-2.*

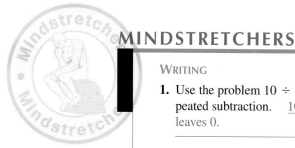

MINDSTRETCHERS

WRITING

1. Use the problem $10 \div 2 = 5$ to help you explain why you can think of division as repeated subtraction. _$10 - 2 - 2 - 2 - 2 - 2 = 0$; taking away 2 from 10 five times_ leaves 0.

___

Consider using this problem as a springboard for discussing fractions.

MATHEMATICAL REASONING

2. Consider the following pair of problems.

a. $2\overline{)7}$ **b.** $4\overline{)13}$

Do you think the answers are the same? Explain. _They both appear to give the same_ result: 3 R1, but the values are different. a. $3\frac{1}{2}$, b. $3\frac{1}{4}$

___

GROUPWORK

3. In the following division problem, A, B, and C each stand for a different digit. Working with a partner, identify all the digits. (*Hint*: There are two answers.)

$$
\begin{array}{r}
ABA \\
AB\overline{)CACAB} \\
-CAB \\
\hline
CA \\
-B \\
\hline
CAB \\
-CAB \\
\hline
\end{array}
$$

If students are stuck, consider giving them the hint that B must take on the value 0.

Solution 1:
$$
\begin{array}{r}
505 \\
50\overline{)25{,}250} \\
25\,0 \\
\hline
250 \\
-250 \\
\hline
\end{array}
$$
A = 5, B = 0, C = 2

Solution 2:
$$
\begin{array}{r}
606 \\
60\overline{)36{,}360} \\
36\,0 \\
\hline
360 \\
-360 \\
\hline
\end{array}
$$
A = 6, B = 0, C = 3

1.5 Exponents, Order of Operations, and Averages

OBJECTIVES

- *To evaluate expressions involving exponents*
- *To evaluate expressions, using the rule for order of operations*
- *To compute averages*

Exponents

There are many mathematical situations in which we multiply a number by itself repeatedly. Writing such expressions in **exponential form** provides a shorthand method for representing this repeated multiplication of the same factor.

For instance, we can write $5 \cdot 5 \cdot 5 \cdot 5$ in exponential form as

$$5^4 \longleftarrow \text{Exponent}$$
$$\text{Base}$$

This expression is read "5 to the fourth *power*" or simply "5 to the fourth."

Definition

An **exponent** (or **power**) is a number that indicates how many times another number (called the **base**) is multiplied by itself.

We read the power 2 or the power 3 in a special way. For instance, 5^2 is usually read "5 *squared*" rather than "5 to the second power." Similarly, we usually read 5^3 as "5 *cubed*" instead of "5 to the third power."

> Explain to your students that we use the terms *squared* and *cubed* because of the formulas for the area of a geometric square and the volume of a geometric cube.

Let's look at a number written in exponential form—namely, 2^4. To evaluate this expression, we multiply the factor 2 by itself 4 times.

$$2^4 = 2 \cdot 2 \cdot 2 \cdot 2$$
$$= 4 \cdot 2 \cdot 2$$
$$= 8 \cdot 2$$
$$= 16$$

In short, $2^4 = 16$. Do you see the difference between 2^4 and $2 \cdot 4$?

> Have your students explain their answers in a journal.

Sometimes we prefer to shorten expressions by writing them in exponential form. For instance, we can write $3 \cdot 3 \cdot 4 \cdot 4 \cdot 4$ in terms of powers of 3 and 4.

$$\underbrace{3 \cdot 3}_{\substack{\text{2 factors} \\ \text{of 3}}} \cdot \underbrace{4 \cdot 4 \cdot 4}_{\substack{\text{3 factors} \\ \text{of 4}}} = 3^2 \cdot 4^3$$

EXAMPLE 1

Rewrite
$6 \cdot 6 \cdot 6 \cdot 10 \cdot 10 \cdot 10 \cdot 10$
in exponential form.

Solution $\underbrace{6 \cdot 6 \cdot 6}_{\text{3 factors of 6}} \cdot \underbrace{10 \cdot 10 \cdot 10 \cdot 10}_{\text{4 factors of 10}} = 6^3 \cdot 10^4$

PRACTICE 1

Write
$5 \cdot 5 \cdot 5 \cdot 5 \cdot 5$
in terms of a power. 5^5

EXAMPLE 2

Expand: 1^5

Solution $1^5 = \underbrace{1 \cdot 1 \cdot 1 \cdot 1 \cdot 1}$

$\qquad = \qquad 1$

Note that 1 raised to any power is 1.

PRACTICE 2

Compute: 1^8 1

EXAMPLE 3

Evaluate: 15^2

Solution $15^2 = 15 \cdot 15$

$\qquad = 225$

After considering this example, can you explain the difference between squaring and doubling a number?

PRACTICE 3

Find the value of 11^3.   1,331

Have your students explain their answers in a journal.

EXAMPLE 4

Write $4^3 \cdot 5^3$ in standard form.

Solution $4^3 \cdot 5^3 = (4 \cdot 4 \cdot 4) \cdot (5 \cdot 5 \cdot 5)$

$\qquad = 64 \cdot 125$

$\qquad = 8{,}000$

From this example, do you see the difference between cubing and tripling a number?

PRACTICE 4

Express $7^2 \cdot 2^4$ in standard form.   784

Have your students write their answers in a journal.

It is especially easy to compute powers of ten.

$$10^2 = 10 \cdot 10 = 1\underbrace{00}, \qquad 10^3 = 10 \cdot 10 \cdot 10 = 1\underbrace{,000}$$
$$\qquad\qquad\quad \text{2 zeros} \qquad\qquad\qquad\qquad\qquad\qquad\quad \text{3 zeros}$$

$$10^4 = 10 \cdot 10 \cdot 10 \cdot 10 = 1\underbrace{0{,}000}$$
$$\qquad\qquad\qquad\qquad\qquad \text{4 zeros}$$

and so on.

Do you see the pattern?

EXAMPLE 5

The distance from the Sun to the star Alpha-one Crucis is about 1,000,000,000,000,000 mi. Express this number in terms of a power of 10.

Solution $\underbrace{1{,}000{,}000{,}000{,}000{,}000} = 10^{15}$

$\qquad\qquad \text{15 zeros}$

PRACTICE 5

In 1850, the world population was approximately 1,000,000,000. Represent this number as a power of 10. (*Source: World Almanac and Book of Facts 2000*) 10^9

Order of Operations

Some mathematical expressions involve more than one mathematical operation. For instance, consider $5 + 3 \cdot 2$. This expression seems to have two different values, depending on the order in which we perform the given operations.

Adding first	**Multiplying first**
$5 + 3 \cdot 2$	$5 + 3 \cdot 2$
$= \quad 8 \cdot 2$	$= 5 + 6$
$= \quad 16$	$= \quad 11$

How are we to know which operation to carry out first? By consensus we agree to follow the rule called the **order of operations** so that everyone always gets the same value for an answer.

ORDER OF OPERATIONS RULE

To evaluate mathematical expressions, carry out the operations *in the following order.*

1. First, perform the operations within any grouping symbols, such as parentheses ( ) or brackets [ ].
2. Then, raise any number to its power 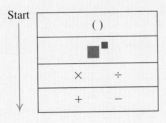.
3. Next, perform all multiplications and divisions as they appear from left to right.
4. Finally, do all additions and subtractions as they appear from left to right.

Applying this rule to the preceding example gives us the following result.

$$5 + 3 \cdot 2 \quad \text{Multiply first.}$$
$$= 5 + 6 \quad \text{Then add.}$$
$$= \quad 11$$

So 11 is the correct answer.

Let's consider more examples that depend on the order of operations rule.

EXAMPLE 6	**PRACTICE 6**
Simplify: $18 - 7 \cdot 2$	Evaluate: $2 \cdot 8 + 4 \cdot 3$ 28
Solution Applying the rule, we multiply first, and then subtract.	$18 - 7 \cdot 2 =$ $18 - 14 = 4$

EXAMPLE 7

Find the value of $3 + 2 \cdot (8 + 3^2)$.

Solution

$\begin{aligned}
&3 + 2 \cdot (8 + \underbrace{3^2}) && \text{First, perform the operations in} \\
&&& \text{parentheses: square and} \\
&= 3 + 2 \cdot (\underbrace{8 + 9}) && \text{then add.} \\
&= 3 + \underbrace{2 \cdot \quad 17} && \text{Next, multiply.} \\
&= \underbrace{3 + \quad 34} && \text{Finally, add.} \\
&= \quad\quad 37
\end{aligned}$

PRACTICE 7

Simplify: $(4 + 1)^2 \times 6 - 4$ 146

TIP   When a division problem is written in the format $\dfrac{\square}{\square}$, parentheses are understood to be around both the dividend and the divisor. For instance,

$$\frac{10 - 2}{3 + 1} \quad \text{means} \quad \frac{(10 - 2)}{(3 + 1)}.$$

EXAMPLE 8

Evaluate: $6 \cdot 2^3 - \dfrac{21 - 11}{2}$

Solution

$\begin{aligned}
&6 \cdot 2^3 - \frac{21 - 11}{2} && \text{First, simplify the dividend by subtracting.} \\
&= 6 \cdot 2^3 - \frac{10}{2} && \text{Then, cube.} \\
&= 6 \cdot 8 - \frac{10}{2} && \text{Next, multiply and divide.} \\
&= 48 - 5 && \text{Finally, subtract.} \\
&= 43
\end{aligned}$

PRACTICE 8

Simplify: $10 + \dfrac{24}{12 - 8} - 3 \times 4$ 4

Averages

We use an **average** to represent a set of numbers. Averages allow us to compare two or more sets. (For example, do the men or women in your class spend more time studying?) Averages also allow us to compare an individual with a set. (For example, is the amount of time you spend studying above or below the class average?) The following definition shows how to compute an average.

Tell students that other types of averages are discussed in Chapter 8.

Definition

The **average** or **mean** of a list of numbers is the sum of those numbers divided by however many numbers are on the list.

EXAMPLE 9

What is the average of 100, 94, and 100?

Solution The average equals the sum of these three numbers divided by 3.

$$\frac{100 + 94 + 100}{3} = \frac{294}{3} = 98$$

PRACTICE 9

Find the average of $30, $0 and $90. $40

EXAMPLE 10

Your test scores in a class were 85, 94, 93, 86, and 92. If the average of these scores was 90 or above, you earned an A. Did you earn an A?

Solution The average is

$$\frac{\text{the sum of the scores}}{\text{the number of scores}}$$

$$\frac{85 + 94 + 93 + 86 + 92}{5}$$

$$= \frac{450}{5}$$

$$= 90$$

Therefore you did earn an A.

PRACTICE 10

Last year, your average monthly electricity bill was $110. This year, these bills amounted to $84, $85, $88, $92, $80, $96, $150, $175, $100, $95, $75, and $80. On average, were your bills higher this year?
No; the average this year is $100.

 Powers and Order of Operations on a Calculator

Let's use a calculator to carry out computations that involve either powers or the order of operations rule.

EXAMPLE

Calculate 11^3.

Solution On many calculators, you can raise a number to a power *without having to enter the base more than once.*

Input	11	$\times$	$=$	$=$
Display	11.	11.	121.	1331.

Note that the number in the display is multiplied by 11 each time that you press the $=$ key.

PRACTICE

Use a calculator to compute 375^2.   140,625

Check to see whether any student has a calculator with a special key (x^y) for raising numbers to powers.

Point out to students that the number of times they hit the $=$ key is *one less* than the power to which they are raising the base.

Now let's discuss how to use a calculator to evaluate an expression involving the order of operations rule.

continued

EXAMPLE

Combine: $2 + 3 \times 4$

Solution Try entering this problem on your calculator as it is written, from left to right.

■ *If you get the correct answer* (14),

Input	2	+	3	×	4	=
Display	2.	2.	3.	3.	4.	14.

your calculator obeys the order of operations rule, and you can do similar problems this way.

■ *If you get the incorrect answer* (20),

Input	2	+	3	×	4	=
Display	2.	2.	3.	5.	4.	20.

your calculator does not obey the order of operations rule. In this case, you need to enter the operations in the order specified by the order of operations rule.

Input	3	×	4	+	2	=
Display	3.	3.	4.	12.	2.	14.

PRACTICE

On a calculator compute $135 - 44 \div 11$.   131

Show students how to do this problem, using their calculators' memory keys (MC , M+ , etc.).

Exercises 1.5

Complete each table by squaring the numbers given.

1.

n	0	2	4	6	8	10	12	14	16	18	20
n^2	0	4	16	36	64	100	144	196	256	324	400

2.

n	1	3	5	7	9	11	13	15	17	19	21
n^2	1	9	25	49	81	121	169	225	289	361	441

Complete each table by cubing the numbers given.

3.

n	0	2	4	6	8	10
n^3	0	8	64	216	512	1,000

4.

n	1	3	5	7	9	11
n^3	1	27	125	343	729	1,331

Express each number as a power of 10.

5. $100 = 10^{}$ 10^2 **6.** $1,000 = 10^{}$ 10^3

7. $10,000 = 10^{}$ 10^4 **8.** $100,000 = 10^{}$ 10^5

9. $1,000,000 = 10^{}$ 10^6 **10.** $10,000,000 = 10^{}$ 10^7

Express each number in terms of powers of 2 and 3.

11. $2 \cdot 2 \cdot 3 \cdot 3 = 2^{} \cdot 3^{}$ $2^2 \cdot 3^2$ **12.** $2 \cdot 2 \cdot 2 \cdot 3 = 2^{} \cdot 3^{}$ $2^3 \cdot 3^1$

13. $2 \cdot 3 \cdot 3 \cdot 3 = 2^{} \cdot 3^{}$ $2^1 \cdot 3^3$ **14.** $3 \cdot 2 \cdot 3 \cdot 2 \cdot 3 = 2^{} \cdot 3^{}$ $2^2 \cdot 3^3$

15. $2 \cdot 3 \cdot 2 = 2^{} \cdot 3^{}$ $2^2 \cdot 3^1$ **16.** $3 \cdot 2 \cdot 3 \cdot 2 = 2^{} \cdot 3^{}$ $2^2 \cdot 3^2$

Evaluate.

17. $8 + 5 \cdot 2$
18

18. $9 + 10 \div 2$
14

19. $8 - 2 \times 3$
2

20. $12 - 6 \div 2$
9

21. $10 + 5^2$
35

22. $9 - 2^3$
1

23. $(9 - 2)^3$
343

24. $(10 + 5)^2$
225

25. 10×5^2
250

26. $12 \div 2^2$
3

27. $(12 \div 2)^2$
36

28. $(10 \times 5)^2$
2,500

29. $(24 \div 4) + 2$
8

30. $(15 \cdot 6) - 2$
88

31. $15 \cdot 6 + 2$
92

32. $24 \div 4 - 2$
4

33. $15 \cdot (6 - 2)$
60

34. $24 \div (4 + 2)$
4

35. $2^6 - 6^2$
28

36. $3^5 - 5^3$
118

37. $8 + 5 - 2 \times 2$
9

38. $7 - 1 + 3 \cdot 2$
12

39. $\dfrac{8 + 2}{7 - 2}$
2

40. $\dfrac{9 - 1}{3 + 5}$
1

41. $(10 - 1)(10 + 1)$
99

42. $(8 + 1)(8 - 1)$
63

43. $10^2 - 1$
99

44. $8^2 - 1$
63

45. $\left(\dfrac{8 + 1}{5 - 2}\right)^2$
9

46. $\dfrac{3^2 + 5^2}{2}$
17

47. $32 + 9 \cdot 215 \div 5$
419

48. $84 \cdot 27 + 32 \cdot 27^2 \div 2$
13,932

49. $48(48 - 31)(48 - 24)(48 - 41)$
137,088

50. $137^2 - 4(36)(22)$
15,601

In each exercise, the three squares stand for the numbers 4, 6, and 8 in some order. Fill in the squares to make true statements.

51. $\boxed{4} \cdot 3 + \boxed{6} \cdot 5 + \boxed{8} \cdot 7 = 98$

52. $\boxed{6} + 10 \times \boxed{4} - \dfrac{\boxed{8}}{2} = 42$

53 . $(\boxed{8})(3 + \boxed{4}) - 2 \cdot \boxed{6} = 44$

54. $\boxed{8} \cdot 3 + \boxed{6} \cdot 5 + \boxed{4} \cdot 7 = 82$

55. $\boxed{8} + 10 \times \boxed{4} - \boxed{6} \div 2 = 45$

56. $\dfrac{48}{\boxed{6}} - \dfrac{\boxed{4}}{2} + (3 + \boxed{8})^2 = 127$

Insert parentheses if needed to make the expression on the left equal to the number on the right.

57. $5 + 2 \cdot 4^2 = 112$ $(5 + 2) \cdot 4^2 = 112$

58. $5 + 2 \cdot 4^2 = 69$ $5 + (2 \cdot 4)^2 = 69$

59. $5 + 2 \cdot 4^2 = 169$ $(5 + 2 \cdot 4)^2 = 169$

60. $5 + 2 \cdot 4^2 = 37$ none

61. $8 - 4 \div 2^2 = 1$ $(8 - 4) \div 2^2 = 1$

62. $8 - 4 \div 2^2 = 7$ none

Find the area of each shaded region.

63. 242 cm²

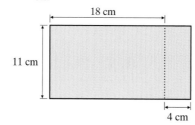

64. 44 cm²

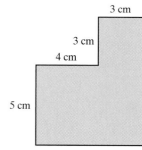

Point out to students that Exercises 63–66 are geometric applications of the order of operations rule. For example, in Exercise 63 to find the area, students could evaluate
$11(18 + 4)$ or
$11 \cdot 18 + 11 \cdot 4$.

65. 3,120 in.²

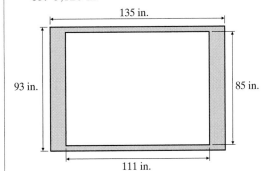

66. 1,600 cm²

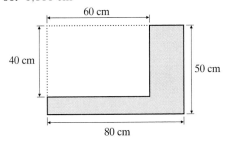

Complete each table.

67.

Input	Output
0	$21 + 3 \times \mathbf{0} = 21$
1	$21 + 3 \times \mathbf{1} = 24$
2	$21 + 3 \times \mathbf{2} = 27$

68.

Input	Output
0	$14 - 5 \times \mathbf{0} = 14$
1	$14 - 5 \times \mathbf{1} = 9$
2	$14 - 5 \times \mathbf{2} = 4$

Find the average of each set of numbers.

69. 20 and 30   25

70. 10 and 50   30

71. 30, 60, and 30   40

72. 17, 17, and 26   20

73. 10, 0, 3, and 3   4

74. 5, 7, 7, and 17   9

75. 3,527 mi, 1,788 mi, and 1,921 mi   2,412 mi

76. 7 hr, 6 hr, 10 hr, 9 hr, and 8 hr   8 hr

77. Six 10's and four 5's   8

78. Sixteen 5's and four 0's   4

Applications

Solve and check.

79. In 7 games, you bowled 163, 185, 154, 127, 156, 153, and 140. What was your average score for these games?   154

80. For the last 5 days, a traffic officer gave out 11, 22, 12, 3, and 17 tickets. On how many days was the number of tickets below average?   3 days; average = 13

81. According to the following table, how much greater was your average salary for all 3 years than for the first 2?

Year	1	2	3
Salary	$19,400	$21,400	$23,700

$1,100

82. The following grade book shows your last 3 math grades.

Test No. 1	Test No. 2	Test No. 3
85	63	98

If you were to get a 90 on the next test, by how much would your average test score increase?   2 points

83. The following table shows the results of a survey on the number of hours various men and women had watched television the previous day.

Man or Woman	Hours of TV
W	4
M	2
W	0
W	2
M	4

On average, did the men or the women spend more time watching TV? Explain how you know.   Men spent more time; men's average was 3 hr, while women's was 2 hr.

84. Because of a storm, your flight was 45 minutes late in leaving the airport. Six other flights had the delays tabulated below.

Flight To	Delay (in min)
Washington	0
Dallas	5
Miami	95
Columbus	20
San Jose	30

Was your delay more than the average of these other delays? If so, show why. Yes; the average delay was 30 min.

85. A 40-story office building has 25,000 sq ft of space to rent. What is the average rental space on a floor? 625 sq ft

86. The total area of the 50 states in the United States is about 3,700,000 sq mi. If the state of Georgia's area is about 60,000 sq mi, is its size above the average of all the states? Explain.
(**Source:** *The New Encyclopedia Britannica*)
No; the average size is 74,000 sq mi.

87. In a branch of mathematics called number theory, the numbers 3, 4, and 5 are called a *Pythagorean triple* because $3^2 + 4^2 = 5^2$ (that is, $9 + 16 = 25$). Show that 5, 12, and 13 are a Pythagorean triple. $5^2 + 12^2 = 13^2$
$25 + 144 = 169$

88. If an object is dropped off a cliff, after 10 sec it will have fallen $\frac{32 \cdot 10^2}{2}$ ft, ignoring air resistance. Express this distance in standard form, without exponents.   1,600 ft

▦ *Use a calculator to solve each problem giving (a) the operation(s) carred out in your solution, (b) the exact answer, and (c) an estimate of the answer.*

To help students compute the average in Exercise 90, point out the advantage of tabulating and multiplying

89. Last week, Newspaper A had an average daily circulation of 82,073. The daily circulation for Newspaper B was as follows.

Day	B's Circulation
M	85,774
Tu	72,503
W	68,513
Th	74,812
F	89,002
Sa	92,331
Su	102,447

Which newspaper was more popular? By how many people?   (a) Addition, division, and subtraction. (b) Newspaper B was more popular by 1,553. (c) The numbers are too close to estimate which is larger.

90. The hospital chart shown is a record of a patient's temperature.

Time	Temp.	Time	Temp.
6 A.M.	98°	6 A.M.	101°
10 A.M.	100°	10 A.M.	102°
2 P.M.	98°	2 P.M.	101°
6 P.M.	100°	6 P.M.	102°
10 P.M.	98°	10 P.M.	100°
2 A.M.	100°	2 A.M.	100°

What was her average temperature for this period of time?   (a) Addition and division (b) 100° (c) 100°

■ *Check your answers on page A-2.*

MINDSTRETCHERS

WRITING

1. Evaluate the expressions in parts (a) and (b).

a. $7^2 + 4^2$ _65_

b. $(7 + 4)^2$ _121_

c. Are your answers to parts (a) and (b) the same? _No_ If not, explain why not.

In part (a), we add the squares of the individual numbers, but in part (b) we add the individual numbers and then square the sum.

MATHEMATICAL REASONING

2. The square of any whole number (called a **perfect square**) can be represented as a geometric square, as follows:

$$\begin{matrix} \bullet & \bullet \\ \bullet & \bullet \end{matrix} \quad 4$$

$$\begin{matrix} \bullet & \bullet & \bullet \\ \bullet & \bullet & \bullet \\ \bullet & \bullet & \bullet \end{matrix} \quad 9$$

Try to represent the numbers 16, 25, 5, and 8 the same way.

16	25	5	8
· · · · · · · · · · · · · · · ·	· ·	Cannot be represented this way.	Cannot be represented this way.

CRITICAL THINKING

3. What is the average of the whole numbers from 1 through 999? Explain how you got your answer.   <u>Average is 500.</u>

$$\frac{1 + 2 + \cdots + 499 + 500 + 501 + \cdots + 998 + 999}{999}$$

$$= \frac{1{,}000 \cdot 499 + 500}{999} = \frac{499{,}500}{999} = 500$$

1.6 More on Solving Word Problems

OBJECTIVES

■ *To solve word problems involving the addition, subtraction, multiplication, or division of whole numbers using various problem-solving tips*

What Word Problems Are and Why They Are Important

In this section, we consider some general tips to help solve word problems.

Word problems can deal with any subject—from shopping to physics and geography to business. Each problem is a brief story that describes a particular situation and ends with a question. Our job, after reading and thinking about the problem, is to answer that question by using the given information.

Although there is no magic formula for solving word problems, you should keep the following problem-solving steps in mind.

To Solve Word Problems

■ Read the problem carefully.
■ Choose a strategy (such as drawing a picture, breaking up the question, substituting simpler numbers, or making a table.)
■ Decide which basic operation(s) are relevant and then translate the words into mathematical symbols.
■ Perform the operations.
■ Check the solution to see if the answer is reasonable. If it is not, start again by rereading the problem.

Reading the Problem

In a math problem, each word counts. So it is important to read the problem slowly and carefully, and not to scan it as you would a magazine or newspaper article.

When reading a problem, you need to understand the problem's key points: *What information is given* and *what question is posed*. Once these points are clear to you, jot them down in your own words to help you keep them in mind.

After taking notes, you should decide on a plan of action that will lead to the answer. For many problems, a solution will come to mind if you just think back to the meaning of the four basic operations.

Operation	Meaning
+	Combining
−	Taking away
×	Adding repeatedly
÷	Splitting up

Many word problems contain *clue words* that suggest performing particular operations. If you spot a clue word in a problem, you should consider whether the operation indicated in the following table will lead you to a solution.

+	**−**	**×**	**÷**
■ add	■ subtract	■ multiply	■ divide
■ sum	■ difference	■ product	■ quotient
■ total	■ take away	■ times	■ over
■ plus	■ minus	■ double	■ split up
■ more	■ less	■ twice	■ fit into
■ increase	■ decrease	■ triple	■ per
■ gain	■ loss	■ of	■ goes into

However be on guard—a clue word can be misleading. For instance, in the problem *What number increased by 2 is 6?*, you have to solve by subtracting, not adding.

See if you agree with the following "translations" of these clues.

The patient's fever increased by 5°.	+ 5
The number of unemployed people tripled.	× 3
The length of the bedroom is 8 ft less than the kitchen's.	− 8
The company's earnings were split among the four partners.	÷ 4

Choosing a Strategy

If no method of solution comes to mind after you read a problem, there are a number of problem-solving strategies that may help. Here we discuss four of these strategies: drawing a picture, breaking up the question, substituting simpler numbers, and making a table.

Drawing a Picture

Sketching even a rough representation of a problem—say, a diagram or a map—can give you insight into its solution, provided that the sketch accurately reflects the given information.

EXAMPLE 1

In an election, everyone voted for one of three candidates. The winner received 188,000 votes, and the second-place candidate got 177,000 votes. If 380,000 people voted in the election, how many people voted for the third candidate?

Solution To help us understand the given information, let's draw a diagram to represent the situation.

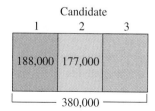

We see from this diagram that to find the answer we need to do two things.

PRACTICE 1

A company slashed its work force by laying off 1,150 employees during one month and laying off 2,235 employees during another month. Afterward, 7,285 employees remained. How many employees worked for the company before the layoffs began?   10,670

■ First, we need to add 188,000 to 177,000.

$$
\begin{array}{r}
188{,}000 \\
+\,177{,}000 \\
\hline
365{,}000
\end{array}
$$

■ Then, we need to subtract this sum from 380,000.

$$
\begin{array}{r}
380{,}000 \\
-\,365{,}000 \\
\hline
15{,}000
\end{array}
$$

A good way to check our answer here is by adding.

$$188{,}000 + 177{,}000 + 15{,}000 = 380{,}000$$

Our answer checks, so 15,000 people voted for the third candidate.

Breaking Up the Question

Another effective problem-solving strategy is to break up the given question into a chain of simpler questions.

EXAMPLE 2	PRACTICE 2

Suppose that you took a math test consisting of 20 questions, and that you answered 3 questions wrong. How many more questions did you get right than wrong?

Solution Say that you do not know how to answer this question directly. Try to split it into several easier questions that lead to a solution.

■ How many questions did you get *right*? $20 - 3 = 17$

■ How many questions did you get *wrong*?   3

■ How many *more* questions did you get right than wrong? $17 - 3 = 14$

So you had 14 more questions right than wrong.

   This answer seems reasonable, because it must be less than 17, the number of questions that you answered correctly.

Teddy and Franklin Roosevelt were both U.S. presidents. Teddy was born in 1858 and died in 1919. Franklin was born in 1882 and died in 1945. How much longer did Franklin live than Teddy? (*Source:* Foner and Garraty, *The Reader's Companion to American History*)   2 yr

Substituting Simpler Numbers

A word problem involving large numbers often seems difficult just because of these numbers. A good problem-solving strategy here is to consider first the identical problem but with simpler numbers. Solve the revised problem and then return to the original problem.

EXAMPLE 3	PRACTICE 3

Raffle tickets cost $4 each. How many tickets must be sold for the raffle to break even if the prizes total $4,736?

Solution Suppose that we are not sure which operation to perform to solve this problem. Let's try substituting a simpler amount (say, $8) for the breakeven amount of $4,736 and see if we can solve the resulting problem.

A college has 47 sections of Math 110. If 33 students are enrolled in each section, how many students are taking Math 110?   1,551

continued

The question would then become: How many $4 tickets must be sold to make back $8? Because it is a "fit in" question, we must *divide* the $8 by the $4. Going back to the original problem, we see that we must divide $4,736 by 4.

$$\$4,736 \div 4 = 1,184 \text{ tickets}$$

Is this answer reasonable? We can check either by estimating ($5,000 \div 4 = 1,250$, which is close to our answer) or by multiplying ($1,184 \times 4 = 4,736$, which also checks).

Making a Table

Finally, let's consider a strategy for solving word problems that involve many numbers. Organizing these numbers into a table often leads to a solution.

EXAMPLE 4

A borrower promises to pay back $50/mo until a $1,000 loan is settled. What is the remaining loan balance at the end of 5 mo?

Solution We can solve by organizing the information in a table.

After month	Remaining balance
1	$1,000 - 50 = 950$
2	$950 - 50 = 900$
3	$900 - 50 = 850$
4	$850 - 50 = 800$
5	$800 - 50 = 750$

From the table, we see that the remaining balance after 5 mo is $750.

We can also solve this problem by breaking up the question into simpler questions.

■ How much money did the borrower pay after 5 mo? $5 \cdot 50 = \$250$

■ How much money did the borrower still owe after 5 mo? $1000 - 250 = \$750$

Again, the remaining balance after 5 mo is $750.

PRACTICE 4

You weigh 210 lb and decide to go on a diet. If you lose 2 lb a week while on the diet, how much will you weigh after 15 weeks?   180 lb

---

Key Points in Problem Solving

■ **Read** the problem **carefully**.
■ Focus on the **meaning of the four basic operations** and think how they apply to the problem.
■ Identify any **clue words** that suggest a basic operation.
■ See if you can **draw a picture** that reflects the given information.
■ See if you can **break up the question** into a chain of simpler questions.
■ If the problem involves large numbers, try **substituting small numbers** and first solving the new problem.
■ See if you can **organize** the given information in **a table**.

Exercises 1.6

Choose a strategy. Solve and check.

1. In retailing, the difference between the *gross sales* and *customer returns and allowances* is called the *net sales*. If a store's gross sales were $2,538 and customer returns and allowances amounted to $388, what was the store's net sales? $2,150

2. The population of the United States in 1800 was 5,308,483. Ten years later, the population had grown to 7,239,881. During this period of time, did the country's population double? Justify your answer. (*Source: The Time Almanac 2000*)   No, if it had doubled, the population would have been at least 10,616,966.

3. Suppose that you drive 27 mi north, 31 mi east, 45 mi west, and 14 mi east. How far are you from your starting point?   27 mi

4. In your office, there are 19 reams of paper. If you need 7,280 sheets of paper for a printing job, do you have enough reams? (*Hint*: A ream is 500 sheets of paper.)   Yes, we have 9,500 sheets of paper available.

5. You are installing shelves for your collection of 400 compact disks. If 36 disks fit on each shelf, how many shelves will you need to house your entire collection?   12 shelves

6. Sound travels through air at a speed of about 1,000 feet per second (fps), whereas light travels at about 1,000,000,000 fps. How many times as fast as sound is light?   1,000,000 times

7. Last year, your food expenses amounted to $7,228. This year, they were $9,055. How much of an increase was this? $1,827

8. You flew from Los Angeles to Miami (2,339 mi), then to New York (1,092 mi), and finally back to LA (2,451 mi). How many total miles did you fly?   5,882 mi

9. Two of the great naval disasters of the century involved the sinking of British ships—the Titanic and the Lusitania. The Titanic, which weighed about 93,000,000 lb, was the most luxurious liner of its time; it struck an iceberg on its maiden voyage in 1912. The Lusitania, which weighed about 63,000,000 lb, was sunk by a German submarine in 1915. How much heavier was the Titanic than the Lusitania? (*Source: The Oxford Companion to Ships and the Sea*)   30,000,00 lb

10. Immigrants from all over the world came to the United States between 1931 and 1940 in the following numbers: 348,289 (Europe), 15,872 (Asia), 160,037 (Americas, outside the United States), 1,750 (Africa), and 2,231 (Australia/New Zealand). What was the total number of immigrants? (*Source:* George Thomas Kurian, *Datepedia of the United States*)   528,179

11. A photograph store sells 3 rolls of film for $15 and 2 rolls for $12. How much less expensive per roll is it if you buy 3 rolls? $1 less expensive

12. If you type 55 words/min, how long will it take you to type an 11-page report with about 400 words on each page?   80 min

13. Your boss asks you to order 1,000 pens from an office supply catalog. If the catalog sells pens by the gross (that is, in sets of 144) and you order 7 gross, how many extra pens did you order?   8 extra pens

14. While shopping, you decide to buy 3 shirts costing $39 apiece and 2 pairs of shoes at $62 per pair. If you have $300 with you, is that enough money to pay for these gifts? Explain.   Yes, the cost of gifts is $241.

15. Eisenhower beat Stevenson in the 1952 and 1956 presidential elections. In 1952, Eisenhower received 442 electoral votes and Stevenson 89. In 1956, Eisenhower got 457 electoral votes and Stevenson 73. Which election was closer? By how many electoral votes? (***Source:** World Almanac*)   1952 was closer by 31 votes.

16. Your garden is rectangular in shape— 26 ft in length and 14 ft in width. If your favorite fencing costs $13 a foot, how much will you have to pay to enclose the garden with this fencing? $1,040

🖩 *Use a calculator to solve each problem, giving (a) the operation(s) carried out in your solution, (b) the exact answer, and (c) an estimate of the answer.*

17. You agree to pay the seller of the house of your dreams $165,000. You put down $23,448 and promise to pay the balance in 144 equal installments. How much money will each installment be?
(a) Subtraction and division, (b) $983, (c) Possible estimate: $1,000

18. The Earth revolves around the Sun in 365 days, but the planet Mercury does so in only 88 days. Compared to the Earth, how many more complete revolutions will Mercury make in 1,000 days?

(a) Division and subtraction, (b) 9 more, (c) Possible estimate: 9 more

■ *Check your answers on page A-2.*

KEY CONCEPTS and SKILLS

▢ = CONCEPT   ▢ = SKILL

CONCEPT/SKILL	DESCRIPTION	EXAMPLE
[1.1] Place value	<table><tr><td colspan="3">**Thousands**</td><td colspan="3">**Ones**</td></tr><tr><td>Hundreds</td><td>Tens</td><td>Ones</td><td>Hundreds</td><td>Tens</td><td>Ones</td></tr></table>	846,120 ↑ 4 is in the ten thousands place.
[1.1] To round a whole number	■ Underline the place to which you are rounding. ■ Look at the digit to the right of the underlined digit, called the *critical digit*. If this digit is 5 or more, add 1 to the underlined digit; if it is less than 5, leave the underlined digit unchanged. ■ Replace all the digits to the right of the underlined digit with zeros.	$3\underline{8}6 \approx 390$
[1.2] Addend, Sum	In an addition problem, the numbers being added are called *addends*. The result is called their *sum*.	$\underset{\text{Addend}}{6} + \underset{\text{Addend}}{4} = \underset{\text{Sum}}{10}$
[1.2] The identity property of addition	The sum of a number and zero is the original number.	$4 + 0 = 4$ $0 + 7 = 7$
[1.2] The commutative property of addition	Changing the order in which two numbers are added does not affect the sum	$7 + 8 = 8 + 7$
[1.2] The associative property of addition	When adding three numbers, regrouping addends gives the same sum.	$(5 + 4) + 1 =$ $5 + (4 + 1)$
[1.2] To add whole numbers	■ Write the addends vertically, lining up the place values. ■ Add the digits in the ones column, writing the sum on the bottom. If the sum has two digits, carry the left digit to the top of the next column on the left. ■ Add the digits in the tens column as in the preceding step. ■ Repeat this process until you reach the last column on the left, writing the entire sum of that column on the bottom.	$\begin{array}{r} {}^{1}\quad{}^{1}\;{}^{1} \\ 7,3\,8\,5 \\ 9\,2,5\,5\,1 \\ +\quad 2,0\,0\,7 \\ \hline 1\,0\,1,9\,4\,3 \end{array}$
[1.2] Minuend, Subtrahend, Difference	In a subtraction problem, the number that is being subtracted from is called the *minuend*. The number that is being subtracted is called the *subtrahend*. The answer is called the *difference*.	$\underset{\text{Minuend}}{10} - \underset{\text{Subtrahend}}{6} = \overset{\text{Difference}}{\underset{\downarrow}{4}}$

[1.2] To subtract whole numbers	■ On top, write the number *from which* you are subtracting. On the bottom, write the number that you are *taking away*, lining up the place values. Subtract in each column separately.
	■ Start with the ones column.
	   a. If the digit on top is *larger* than or *equal* to the digit on the bottom, subtract and write the difference below.
	   b. If the digit on top is *smaller* than the digit on the bottom, borrow from the digit to the left on top. Exchanging and then subtracting, write the difference below.
	■ Repeat this process until you reach the last column on the left, subtracting and writing its difference below.
		$\begin{array}{r} {}^{8}\;{}^{1}4\;{}^{1} \\ 7,9\,5\,2 \\ -1,8\,8\,3 \\ \hline 6,0\,6\,9 \end{array}$

continued

CONCEPT/SKILL	DESCRIPTION	EXAMPLE
[1.3] Factors, Product	In a multiplication problem, the numbers being multiplied are called *factors*. The result is called their *product*.	Factor Product $4 \times 5 = 20$
[1.3] The identity property of multiplication	The product of any number and 1 is that number.	$1 \times 6 = 6$ $7 \times 1 = 7$
[1.3] The multiplication property of 0	The product of any number and 0 is 0.	$51 \times 0 = 0$ $0 \times 9 = 0$
[1.3] The commutative property of multiplication	Changing the order in which two numbers are multiplied does not affect their product.	$3 \times 2 = 2 \times 3$
[1.3] The associative property of multiplication	Regrouping three factors gives the same product.	$(4 \times 5) \times 6 =$ $4 \times (5 \times 6)$
[1.3] The distributive property	Multiplying a factor by the sum of two numbers gives the same product as multiplying the factor by each of the two numbers and then adding.	$2 \times (4 + 3) =$ $(2 \times 4) + (2 \times 3)$
[1.3] To multiply whole numbers	■ Multiply the top factor by the ones digit in the bottom factor and write this product. ■ Multiply the top factor by the tens digit in the bottom factor and write this product leftward, beginning with the tens column. ■ Repeat this process until you use all the digits in the bottom factor. ■ Add the partial products, writing this sum.	$\begin{array}{r} 693 \\ \times\ \ 71 \\ \hline 693 \\ 48\,51 \\ \hline 49{,}203 \end{array}$
[1.4] Divisor, Dividend, Quotient	In a division problem, the number that is being used to divide another number is called the *divisor*. The number into which it is being divided is called the *dividend*. The result is called the *quotient*.	Quotient 3 $4\overline{)12}$ Divisor ⌐ ⌐ Dividend
[1.4] To divide whole numbers	■ Divide 17 into 39, which gives 2. Multiply the 17 by 2 and subtract the result (34) from 39. Beside the difference (5), bring down the next digit (3) of the dividend. ■ Repeat this process, dividing the divisor (17) into 53. ■ At the end, there is a remainder of 2. Write it beside the quotient on top.	$\begin{array}{r} 23\ \text{R2} \\ 17\overline{)393} \\ \underline{34}\ \ \ \\ 53 \\ \underline{51} \\ 2 \end{array}$
[1.5] Exponent (or Power)	An *exponent* (or *power*) is a number that indicates how many times another number (called the *base*) is multiplied by itself.	Exponent $5^3 = 5 \times 5 \times 5$ Base
[1.5] Average (or Mean)	An *average* (or *mean*) is the sum of the numbers on a list divided by however many numbers are on the list.	The average of 3, 4, 10, and 3 is 5 because $\dfrac{3 + 4 + 10 + 3}{4}$ $= \dfrac{20}{4} = 5$

Chapter 1 Review Exercises

To help you review this chapter, solve these problems.

[1.1]

In each whole number, identify the place that the digit 3 occupies.

1. 23
Ones

2. 30,802
Ten thousands

3. 385,000,000
Hundred millions

4. 30,000,000,000
Ten billions

Write each number in words.

5. 497
Four hundred ninety-seven

6. 2,050
Two thousand fifty

7. 3,000,007
Three million, seven

8. 85,000,000,000
Eighty-five billion

Write each number in standard form.

9. Two hundred fifty-one
251

10. Nine thousand, two
9,002

11. Fourteen million, twenty-five
14,000,025

12. Three billion, three thousand
3,000,003,000

Express each number in expanded form.

13. 308   3 hundreds + 8 ones = 300 + 8

14. 2,500,000   2 millions + 5 hundred thousands =
2,000,000 + 500,000

15. 42,770   4 ten thousands + 2 thousands +
7 hundreds + 7 tens = 40,000 + 2,000 + 700 + 70

16. 30,000,012   3 ten millions + 1 ten + 2 ones =
30,000,000 + 10 + 2

Round each number to the place indicated.

17. 571 to the nearest hundred
600

18. 938 to the nearest thousand
1,000

19. 384,056 to the nearest ten thousand
380,000

20. 68,332 to its largest place
70,000

[1.2]

Find the sum.

21.
```
    102
  4,251
 +5,133
  9,486
```

22.
```
  53,569
  10,000
 + 2,123
  65,692
```

23.
```
  48,758
  37,226
 +87,559
 173,543
```

24.
```
  95,000
  25,895
 +30,000
 150,895
```

25. 972,558 + 87,055
1,059,613

26. $138,865 + $729 + $8,002 + $75,471
$223,067

Find the difference.

27.
```
   876
  −431
   445
```

28.
```
  500,000
 −200,000
  300,000
```

29.
```
  98,118
 −87,009
  11,109
```

30.
```
   7,100
  −1,500
   5,600
```

31. 60,000,000 − 48,957,777
11,042,223

32 $5,000,000 − $2,937,148
$2,062,852

[1.3]

Find the product.

33. 72
  × 6
  432

34. 400
  × 3
  1,200

35. 2,923
  × 51
  149,073

36. 6,000
  ×2,000
  12,000,000

37. 2,751 · 508
1,397,508

38. (681)(498)(555)
188,221,590

[1.4]

Find the quotient and remainder.

39. $4\overline{)376}$
94

40. $\dfrac{975}{25}$
39

41. $13\overline{)491}$
37 R10

42. 16,000 ÷ 20
800

43. $8\overline{)205,000}$
25,625

44. $347\overline{)332,079}$
957

[1.5]

Compute:

45. 7^3
343

46. 1^{10}
1

47. $2^3 \cdot 3^2$
72

48. $5^2 \cdot 5^3$
3,125

49. $20 - 3 \times 5$
5

50. $(9 + 4)^2$
169

51. $10 - \dfrac{6 + 4}{2}$
5

52. $3 + (5 - 1)^2$
19

53. 98(50 − 1)(50 − 2)(50 − 3)
10,833,312

54. $\dfrac{28^3 + 29^3 + 37^3 - 10}{(7 - 1)^2}$
2,694

Rewrite each expression, using exponents.

55. $7 \cdot 7 \cdot 5 \cdot 5 = 7^{\square} \cdot 5^{\square}$
$7^2 \cdot 5^2$

56. $5 \cdot 2 \cdot 5 \cdot 2 \cdot 5 = 2^{\square} \cdot 5^{\square}$
$2^2 \cdot 5^3$

Find the average.

57. 34 and 44
39

58. 20, 0, and 1
7

59. 5, 8, and 5
6

60. 4, 6, 3, and 7
5

Mixed Applications

Solve.

61. Beetles about the size of a pinhead destroyed 2,400,000 acres of forest. Express this number in words.   Two million, four hundred thousand

62. Scientists in Utah found a dinosaur egg one hundred fifty million years old. Write this number in standard form.   150,000,000

63. In a part-time job, you earned $15,964 a year. How much money did you earn per week? (*Hint*: 1 yr equals 52 weeks.) $307 per week

64. Halley's Comet was sighted in 1682. If it appeared 76 yr later, in what year was this sighting? (*Source: The Time Almanac 2000*)   1758

65. The largest of the U.S. national parks is located in Alaska and has an area of 8,223,618 acres. What is this area rounded to the nearest million acres? (*Source: The Time Almanac 2000*)   8,000,000 acres

66. It took 27 weeks for the work crew to build a road 8,316 ft in length. What length road did the crew build per week?   308 ft/week

67. The Empire State Building is 1,250 ft high, and the Statue of Liberty is 152 ft in height. What is the minimum number of Statues of Liberty that would have to be stacked to be taller than the Empire State Building? 9

68. A millipede—a little insect with 68 body segments—has 4 legs per segment. How many legs does a millipede have?   272 legs

69. The walls of your heart are composed of muscles that contract about 100,000 times a day. About how many contractions are there in a month?   Possible answers: For a 28-day month: 2,800,000; for a 30-day month: 3,000,000; for a 31-day month: 3,100,000

70. A college library contains 84,601 books. How many books is this to the nearest ten thousand? 80,000 books

71. The approximate areas of the five Great Lakes in square kilometers (sq km) are shown in the following table.

Lake Huron	60,000
Lake Ontario	20,000
Lake Michigan	58,000
Lake Erie	26,000
Lake Superior	82,000

What is the total area of the lakes? (*Source: The Time Almanac 2000*)   246,000

72. Complete the following business *skeletal profit and loss statement*.

Net sales	$430,000
−Cost of merchandise sold	− 175,000
Gross margin	$255,000
−Operating expenses	− 135,000
Net profit	$120,000

73. You buy a 12-ft by 12-ft rug for your living room, which measures 15-ft by 15-ft. How much of the floor will be exposed after the rug is laid?   81 sq ft

74. Both a singles tennis court and a football field are rectangular in shape. The tennis court measures 78 ft by 27 ft, whereas the football field measures 300 ft by 160 ft. About how many times the area of the tennis court is the football field?   20

75. Richard Nixon ran for the U.S. presidency three times. According to the table below, which was greater—the increase from 1960 to 1972 in the number of votes he got or the increase from 1968 to 1972? (*Source: World Almanac & Book of Facts 2000*)   1968 to 1972 (15,379,754 votes)

Year	Number of Votes for Nixon
1960	34,108,546
1968	31,785,480
1972	47,165,234

76. On a business trip, you flew from Chicago to Los Angeles to Boston. The chart below shows the air distances in miles between these cities.

Air Distance	Chicago	Los Angeles	Boston
Chicago	—	1,745	1,042
Los Angeles	1,745	—	2,596
Boston	1,042	2,596	—

If you earn a frequent flier point for each mile flown, how many points did you earn?   4,341 points

77. You are saving to buy a $175 portable compact disk player. During the past three months you have saved $55, $27, and $34. How much more money do you need? $59

78. In an astronomy course, you learn that light from a star $24 \cdot 10^{13}$ mi away takes 4 yr to reach you. Express this distance in standard form, without exponents. 240,000,000,000,000 mi

79. Find the area of the figure.   29 mi^2

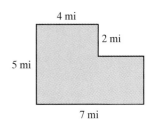

80. Find the perimeter of the figure.   162 cm

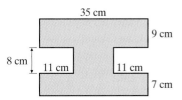

■ *Check your answers on page A-2.*

Chapter 1 Posttest

To see if you have already mastered the topics in this chapter, take this test.

1. Write two hundred twenty-five thousand, sixty-seven in standard form.   225,067

2. Underline the digit that occupies the ten thousands place in 1,7<u>6</u>8,405.

3. Write 1,205,007 in words.   One million, two hundred five thousand, seven

4. Round 196,593 to the nearest hundred thousand.   200,000

5. Find the sum of 398 and 1,496.   1,894

6. Subtract 398 from 1,005.   607

7. Subtract: $2,000 - 1,853$ 147

8. Multiply: 328×907 297,496

9. Find the quotient: $300\overline{)21,000}$ 70

10. Compute: $\dfrac{23,923}{47}$ 509

11. Evaluate: 5^3 125

12. Write $2 \cdot 2 \cdot 3 \cdot 3 \cdot 3$ using exponents. $2^2 \cdot 3^3$

Simplify.

13. $4 \cdot 9 + 3 \cdot 4^2$ 84

14. $29 - 3^3 \cdot (10 - 9)$ 2

Solve and check.

15. The two largest continents in the world are Asia and Africa. To the nearest thousand square miles, Asia's area is 17,400 and Africa's is 11,700. How much larger is Asia than Africa? (*Source: The Time Almanac 2000*)   5,700 sq mi

16. A club at your college spent $175 per person to send 17 of its members to a regional meeting. How much did the club pay for the trip? $2,975

Africa   Asia

17. A stock exchange's computer records 100,000 transactions a day. How many transactions can it record in 10 days, expressed as a power of 10? 10^6 transactions

18. In a math class, you earned test scores of 82, 85, 87, 92, and 89. What was your average score?   87

19. Yellow roses cost $2 each, and pink roses cost $4 each. For $40, how many more yellow roses than pink roses can you buy?   10 more yellow roses than pink roses.

20. For breakfast you have an 8-ounce (oz) serving of yogurt, a cup of black coffee, and 2 cups of pineapple juice. How much more Vitamin C do you need to reach the recommended 60 milligrams (mg)?

Food	Quantity	Vitamin C Content
Pineapple juice	1 cup	23 mg
Yogurt	8 oz	2 mg
Black coffee	1 cup	0 mg

12 mg

■ *Check your answers on page A-3.*

Fractions

FRACTIONS AND THE STOCK MARKET CRASH OF 1929

In September 1929, the New York Stock Exchange reached its highest level in history. People were so eager to play the market that $\frac{1}{3}$ of the nation's more than 3 million stockholders had bought shares on margin, borrowing to invest.

In the next few months, the crash gathered momentum as investor confidence sank. On October 24, shares of Montgomery Ward sold for $\$83\frac{3}{8}$. This was only $\frac{1}{2}$ their 1929 high. Stock of the gigantic Electric Bond and Share Company which had recently reached a high of $189\frac{5}{16}$ went for $91\frac{1}{8}$; a few days later, these shares sold at $49\frac{7}{8}$. The president of Union Cigar, overwhelmed when his company's shares dropped from $113\frac{1}{2}$ to 4, tumbled from the ledge of a New York hotel. On October 28, shares of White Sewing Machine Co., which had recently been $47\frac{7}{8}$ and were $11\frac{1}{4}$ only the day before, went for $\frac{15}{16}$, reportedly because a bright messenger boy made the only offer. In September, the market's stocks had been worth $80 billion. By November, $\frac{3}{8}$ of this had vanished into thin air.

Chapter 2 Pretest

To see if you have already mastered the topics in this chapter, take this test.

1. Find all the factors of 20. $1, 2, 4, 5, 10, 20$

2. Express 72 as the product of prime factors.
$2 \times 2 \times 2 \times 3 \times 3$

3. What fraction does the diagram represent? $\frac{2}{5}$

4. Write $20\frac{1}{3}$ as an improper fraction. $\frac{61}{3}$

5. Express $\frac{31}{30}$ as a mixed number. $1\frac{1}{30}$

6. Write $\frac{9}{12}$ in simplest form. $\frac{3}{4}$

7. What is the least common multiple of 10 and 4? 20

8. Which is greater, $\frac{1}{8}$ or $\frac{1}{9}$? $\frac{1}{8}$

Add and simplify.

9. $\frac{1}{2} + \frac{7}{10}$ $1\frac{1}{5}$

10. $7\frac{1}{3} + 5\frac{1}{2}$ $12\frac{5}{6}$

Subtract and simplify.

11. $8\frac{1}{4} - 6$ $2\frac{1}{4}$

12. $12\frac{1}{2} - 7\frac{7}{8}$ $4\frac{5}{8}$

Multiply and simplify.

13. $2\frac{1}{3} \times 1\frac{1}{2}$ $3\frac{1}{2}$

14. $\frac{5}{8} \times 96$ 60

Divide and simplify.

15. $3\frac{1}{3} \div 5$ $\frac{2}{3}$

16. $2\frac{2}{5} \div 1\frac{1}{6}$ $2\frac{2}{35}$

Solve. Write your answer in simplest form.

17. You pay your doctor $90. If your insurance company reimburses you $40 toward this bill, what fraction of the expense was reimbursed? $\frac{4}{9}$

18. In your biology class, three-fourths of the students received a passing grade. If there are 24 students in the class, how many students received failing grades?
6 students

19. Find the perimeter of the triangle shown. $20\frac{7}{8}$ ft

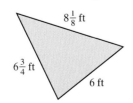

$8\frac{1}{8}$ ft

$6\frac{3}{4}$ ft

6 ft

20. A builder has $13\frac{1}{3}$ acres of land on which he plans to build 40 houses. If all the houses will be built on plots of the same size, how large will each plot be? $\frac{1}{3}$ acre

■ *Check your answers on page A-3.*

2.1 Factors and Prime Numbers

OBJECTIVES

■ *To identify prime and composite numbers*
■ *To find the factors and prime factorization of a whole number*
■ *To find the least common multiple of two or more numbers*
■ *To solve word problems using factoring or the LCM*

What Factors Mean and Why They Are Important

Recall that in a multiplication problem, the whole numbers that we are multiplying are called **factors**. For instance, 2 is said to be a factor of 8 because $2 \cdot 4 = 8$. Likewise, 4 is a factor of 8.

Another way of expressing the same idea is in terms of division: We say that 8 is *divisible* by 2, meaning that there is a remainder 0 when we divide 8 by 2.

$$\frac{8}{2} = 4 \text{ R} 0$$

Note that 1, 2, 4, and 8 are all factors of 8.

Although we factor whole numbers, a major application of factoring involves working with fractions, as we demonstrate in the next section.

Finding Factors

To identify the factors of a whole number, we divide the whole number by the numbers 1, 2, 3, 4, 5, 6, and so on, looking for remainders of 0.

EXAMPLE 1

Find all the factors of 6.

Solution Starting with 1, we divide each whole number into 6.

$$\frac{6}{1} = 6 \text{ R} 0 \qquad \frac{6}{2} = 3 \text{ R} 0 \qquad \frac{6}{3} = 2 \text{ R} 0 \qquad \frac{6}{4} = 1 \text{ R} 2 \qquad \frac{6}{5} = 1 \text{ R} 1 \qquad \frac{6}{6} = 1 \text{ R} 0$$

| A factor | A factor | A factor | Not a factor | Not a factor | A factor |

So the factors of 6 are 1, 2, 3, and 6. Note that

■ 1 is a factor of 6 and that
■ 6 is a factor of 6.
■ We did not need to divide 6 by the numbers 7 or greater. The reason is that no number larger than 6 could divide evenly into 6, that is, divide into 6 with no remainder.

PRACTICE 1

What are the factors of 7?   1, 7

TIP   For any whole number, both *the number itself* and *1* are always factors. Therefore all whole numbers (except 1) have at least two factors.

When checking to see if one number is a factor of another, it is generally faster to use the following **divisibility tests** than to divide.

The number is divisible by	if
2	the ones digit is 0, 2, 4, 6, or 8, that is, if the number is even.
3	the sum of the digits is divisible by 3.
4	the number named by the last two digits is divisible by 4.
5	the ones digit is either 0 or 5.
6	the number is even and the sum of the digits is divisible by 3.
9	the sum of the digits is divisible by 9.
10	the ones digit is 0.

> Point out that divisibility by 6 is equivalent to divisibility by both 2 and 3.

EXAMPLE 2

What are the factors of 45?

Solution Let's see if 45 is divisible by 1, 2, 3, and so on, using the divisibility tests wherever they apply.

Is 45 divisible by	Answer
1?	Yes, because 1 is a factor of any number; $\frac{45}{1} = 45$, so 45 is also a factor.
2?	No, because the ones digit is not even.
3?	Yes, because the sum of the digits, $4 + 5 = 9$, is divisible by 3; $\frac{45}{3} = 15$, so 15 is also a factor.
4?	No, because 4 will not divide into 45 evenly.
5?	Yes, because the ones digit is 5; $\frac{45}{5} = 9$, so 9 is also a factor.
6?	No, because 45 is not even.
7?	No, because $45 \div 7$ has remainder 3.
8?	No, because $45 \div 8$ has remainder 5.
9?	We already know that 9 is a factor.

The factors of 45 are therefore 1, 3, 5, 9, 15, and 45.

Note that we really didn't have to check to see if 9 was a factor—we learned that it was when we checked for divisibility by 5. Also, because the factors were beginning to repeat with 9, there was no need to check numbers greater than 9.

PRACTICE 2

Find all the factors of 75.   1, 3, 5, 15, 25, 75

EXAMPLE 3

Identify all the factors of 60.

Solution Let's check to see if 60 is divisible by 1, 2, 3, 4, and so on.

Is 60 divisible by	Answer
1?	Yes, because 1 is a factor of all numbers; $\frac{60}{1} = 60$, so 60 is also a factor.
2?	Yes, because the ones digit is even; $\frac{60}{2} = 30$, so 30 is also a factor.
3?	Yes, because the sum of the digits, $6 + 0 = 6$, is divisible by 3; $\frac{60}{3} = 20$, so 20 is also a factor.
4?	Yes, because 4 will divide into 60 evenly; $\frac{60}{4} = 15$, so 15 is also a factor.
5?	Yes, because the ones digit is 0; $\frac{60}{5} = 12$, so 12 is also a factor.
6?	Yes, because the ones digit is even and the sum of the digits is divisible by 3; $\frac{60}{6} = 10$, so 10 is also a factor.
7?	No, because $60 \div 7$ has remainder 4.
8?	No, because $60 \div 8$ has remainder 4.
9?	No, because the sum of the digits, $6 + 0 = 6$, is not divisible by 9.
10?	We already know that 10 is a factor.

The factors of 60 are therefore 1, 2, 3, 4, 5, 6, 10, 12, 15, 20, 30, and 60. Can you explain how we knew that 10 was a factor of 60 when we checked for divisibility by 6?

Have students write their answers in a journal.

PRACTICE 3

What are the factors of 90?   1, 2, 3, 5, 6, 9, 10, 15, 18, 30, 45, 90

EXAMPLE 4

A presidential election takes place in the United States every year that is a multiple of 4. Was there a presidential election in 1866?

Solution The question is: Does 4 divide into 1866 evenly? Using the divisibility test for 4, we check whether 66 is a multiple of 4.

$$\frac{66}{4} = 16 \text{ R2}$$

Because $\frac{66}{4}$ has remainder 2, 4 is not a factor of 1866. So there was no presidential election in 1866.

PRACTICE 4

The doctor instructs you to take a pill every 3 hr. If you took a pill at 8:00 this morning, should you take one tomorrow at the same time? (*Reminder*: There are 24 hr in a day.)
Yes (24 is a multiple of 3.)

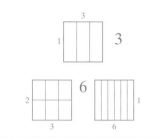

Identifying Prime and Composite Numbers

Now let's discuss the difference between prime numbers and composite numbers.

> **Definitions**
>
> A **prime number** is a whole number that has exactly two factors: itself and 1.
>
> A **composite number** is a whole number that has more than two factors.

Note that the numbers 0 and 1 are neither prime nor composite. But every whole number greater than 1 is either prime or composite, depending on its factors.

For instance, 5 is prime because its only factors are 1 and 5. But 8 is composite because it has more than two factors (it has four factors: 1, 2, 4, and 8).

Let's practice distinguishing between primes and composites.

EXAMPLE 5

Indicate whether each number is prime or composite.

a. 2   **b.** 78   **c.** 51   **d.** 19   **e.** 31

Solution

a. The only factors of 2 are 1 and 2. Therefore 2 is prime.

b. Because 78 is even, it is divisible by 2. Having 2 as an "extra" factor—in addition to 1 and 78—means that 78 is composite. Do you see why all even numbers, except for 2, are composite?

c. Using the divisibility test for 3, we see that 51 is divisible by 3 because the sum of the digits 5 and 1 is divisible by 3. Because 51 has more than two factors, it is composite.

d. The only factors of 19 are itself and 1. Therefore 19 is prime.

e. Because 31 has no factors other than itself and 1, it is prime.

PRACTICE 5

Decide whether each number is prime or composite.

a. 3   **b.** 57   **c.** 29
Prime   Composite   Prime

d. 34   **e.** 17
Composite   Prime

Finding the Prime Factorization of a Number

Every composite number can be written as the product of prime factors. This product is called its **prime factorization**. For instance, the prime factorization of 12 is $2 \cdot 2 \cdot 3$.

> **Definition**
>
> The **prime factorization** of a whole number is the number written as the product of its prime factors.

Being able to find the prime factorization of a number is an important skill to have for working with fractions, as we show later in this chapter. A good way to find the prime factorization of a number is by making a **factor tree**, as illustrated in Example 6.

Tell students that there is exactly one prime factorization for any composite number. This statement is called the *fundamental theorem of arithmetic.*

EXAMPLE 6

Write the prime factorization of 72.

Solution We start building a factor tree for 72 by dividing 72 by the smallest prime, 2. Because 72 is 2 · 36, we write both 2 and 36 underneath the 72. Then we circle the 2 because it is prime.

72
2   36 ←———————— Continue to divide by 2.
  2   18 ←———————— Continue to divide by 2.
    2   9 ←———————— 9 cannot be divided by 2, so we try 3.
      3   3 ← 3 is prime, so we stop dividing.

Next we divide 36 by 2, writing both 2 and 18, and circling 2 because it is prime. Below the 18, we write 2 and 9, again circling the 2. Because 9 is not divisible by 2, we divide it by the next smallest prime, 3. We continue this process until all the factors in the bottom row are prime. The prime factorization of 72 is the product of the circled factors.

$$72 = 2 \times 2 \times 2 \times 3 \times 3$$

We can also write this prime factorization as $2^3 \times 3^2$.

An alternative approach is the following.

$$\begin{array}{r} 2\ \underline{|72} \\ 2\ \underline{|36} \\ 2\ \underline{|18} \\ 3\ \underline{|9} \\ 3 \end{array}$$
$$72 = 2 \times 2 \times 2 \times 3 \times 3$$
$$= 2^3 \times 3^2$$

PRACTICE 6

Write the prime factorization of 56, using exponents. $2^3 \times 7$

EXAMPLE 7

Express 60 as the product of prime factors.

Solution The factor tree method for 60 is as shown.

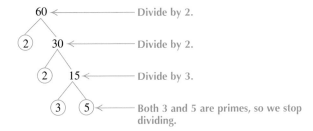

60 ←———————— Divide by 2.
2   30 ←———————— Divide by 2.
  2   15 ←———————— Divide by 3.
    3   5 ←———————— Both 3 and 5 are primes, so we stop dividing.

The prime factorization of 60 is $2 \times 2 \times 3 \times 5$, or $2^2 \times 3 \times 5$.

PRACTICE 7

What is the prime factorization of 45? $3^2 \times 5$

Finding the Least Common Multiples

The *multiples* of a number are the products of that number and the whole numbers. For instance the multiples of 5 are the following.

Point out that "15 is a multiple of 3" means the same as "3 is a factor of 15."

$$\underbrace{0}_{0 \times 5} \qquad \underbrace{5}_{1 \times 5} \qquad \underbrace{10}_{2 \times 5} \qquad \underbrace{15}_{3 \times 5}$$

A number that is a multiple of two or more numbers is called a *common multiple* of these numbers. To find the common multiples of 6 and 8, we first list the multiples of 6 and the multiples of 8 separately.

- The multiples of 6 are 0, 6, 12, 18, 24, 30, 36, 42, 48, 54, 60, ...
- The multiples of 8 are 0, 8, 16, 24, 32, 40, 48, 56, 64, ...

So the common multiples of 6 and 8 are 0, 24, 48, ... Of the nonzero common multiples, the *least* common multiple of 6 and 8 is 24.

Definition

The **least common multiple** (LCM) of two or more numbers is the smallest nonzero number that is a multiple of each number.

A shortcut for finding the LCM—faster than listing multiples—involves prime factorization.

To Compute the Least Common Multiple (LCM)

- find the prime factorization of each number,
- identify the prime factors that appear in each factorization, and
- multiply these prime factors, using each factor the greatest number of times that it occurs in any of the factorizations.

EXAMPLE 8

Find the LCM of 8 and 12.

Solution We first find the prime factorization of each number.

$$8 = 2 \times 2 \times 2 = 2^3 \qquad 12 = 2 \times 2 \times 3 = 2^2 \times 3$$

The factor 2 appears *three times* in the factorization of 8 and *twice* in the factorization of 12, so it must be included three times in forming the least common multiple.

$$\overset{\text{The highest power of 2}}{\text{LCM} = 2^3 \times 3 = 8 \times 3 = 24}$$

As always, it is a good idea to check that our answer makes sense. We do so by verifying that 8 and 12 really are factors of 24.

PRACTICE 8

What is the LCM of 9 and 6?   18

If the question arises, give students an example to show that this LCM technique may not work if any of the factors are composite:

$$30 = 5 \times 6 \qquad 15 = 3 \times 5$$

The LCM appears to be $3 \times 5 \times 6$ or 90, but in fact is 30.

EXAMPLE 9	**PRACTICE 9**

Find the LCM of 5 and 9.

Solution First we write each number as the product of primes.

$$5 = 5 \qquad 9 = 3 \times 3 = 3^2$$

To find the LCM we multiply the highest power of each prime.

$$LCM = 5 \times 3^2 = 5 \times 9 = 45$$

So the LCM of 5 and 9 is 45. Note that 45 is also the product of 5 and 9. Checking our answer, we see that 45 is a multiple of both 5 and 9.

Find the LCM for 3 and 22.   66

TIP   If two or more numbers have no common factor (other than 1), the LCM is their product.

Now let's find the LCM of three numbers.

EXAMPLE 10	**PRACTICE 10**

Find the LCM of 3, 5, and 6.

Solution First we find the prime factorizations of these three numbers.

$$3 = 3 \qquad 5 = 5 \qquad 6 = 2 \times 3$$

The LCM is therefore the product $2 \times 3 \times 5$, which is 30. Note that 30 is a multiple of 3, 5, and 6, which supports our answer.

Find the LCM of 2, 3, and 4.   12

EXAMPLE 11	**PRACTICE 11**

A gym that is open every day of the week offers aerobic classes every third day and gymnastic classes every fourth day. You took both classes this morning. In how many days will the gym offer both classes on the same day?

Solution To answer this question, we ask: What is the LCM of 3 and 4? As usual, we begin by finding prime factorizations.

$$3 = 3 \qquad 4 = 2 \times 2 = 2^2$$

To find the LCM, we multiply 3 by 2^2.

$$LCM = 2^2 \times 3 = 12$$

Both classes will be offered again on the same day in 12 days.

Suppose that a senator and a representative were both elected this year. If the senator runs for reelection every 6 yr and the representative every 2 yr, in how many years will they both be up for reelection?   6 yr

Exercises 2.1

List all the factors of each number.

1. 21
1, 3, 7, 21

2. 6
1, 2, 3, 6

3. 17
1, 17

4. 9
1, 3, 9

5. 12
1, 2, 3, 4, 6, 12

6. 15
1, 3, 5, 15

7. 31
1, 31

8. 47
1, 47

9. 36
1, 2, 3, 4, 6, 9, 12, 18, 36

10. 45
1, 3, 5, 9, 15, 45

11. 29
1, 29

12. 73
1, 73

13. 100
1, 2, 4, 5, 10, 20, 25, 50, 100

14. 98
1, 2, 7, 14, 49, 98

15. 28
1, 2, 4, 7, 14, 28

16. 35
1, 5, 7, 35

Indicate whether each number is prime or composite. If it is composite, identify a factor other than the number itself and 1.

17. 5
Prime

18. 7
Prime

19. 16
Composite
(2, 4, 8)

20. 24
Composite
(2, 3, 4, 6, 8, 12)

21. 49
Composite
(7)

22. 75
Composite
(3, 5, 15, 25)

23. 11
Prime

24. 19
Prime

25. 81
Composite
(3, 9, 27)

26. 57
Composite
(3, 19)

Write the prime factorization of each number.

27. 8
2^3

28. 10
2×5

29. 12
$2^2 \times 3$

30. 14
2×7

31. 24
$2^3 \times 3$

32. 18
2×3^2

33. 50
2×5^2

34. 40
$2^3 \times 5$

35. 77
7×11

36. 45
$3^2 \times 5$

37. 51
3×17

38. 57
3×19

39. 25
5^2

40. 49
7^2

41. 32
2^5

42. 64
2^6

43. 21
3×7

44. 22
2×11

45. 104
$2^3 \times 13$

46. 105
$3 \times 5 \times 7$

47. 121
11^2

48. 169
13^2

49. 142
2×71

50. 62
2×31

51. 100
$2^2 \times 5^2$

52. 200
$2^3 \times 5^2$

53. 125
5^3

54. 90
$2 \times 3^2 \times 5$

55. 135
$3^3 \times 5$

56. 400
$2^4 \times 5^2$

2.2 |

What F

A fraction
question, F
day we use
of a class (
(indicating
three of the

A frac

sense, the f

3 by the wh

Definit

A **fracti**
whole r

From t

When

■ T
w

■ T
c

■ T
s

Alterna
decimals an

Fraction

Diagrams h
$\frac{3}{4}$.

Note that in
shaded.

The nu

its denomin

Find the LCM in each case.

57. 3 and 15
15

58. 9 and 12
36

59. 8 and 10
40

60. 4 and 6
12

61. 9 and 30
90

62. 20 and 21
420

63. 10 and 11
110

64. 15 and 60
60

65. 18 and 24
72

66. 30 and 150
150

67. 40 and 180
360

68. 100 and 90
900

69. 12, 5, and 50
300

70. 2, 8, and 10
40

71. 3, 5, and 6
30

72. 2, 3, and 4
12

73. 3, 5, and 7
105

74. 6, 8, and 12
24

75. 5, 15, and 20
60

76. 8, 24, and 56
168

Applications

Solve.

77. The government conducts a national census every year that is a multiple of 10. Was there a census in 1985? In 1990? Explain.   There was not a census in 1985 because 1985 is not a multiple of 10; there was a census in 1990 because 1990 is a multiple of 10.

78. Because of production considerations, the number of pages in a book you are writing must be a multiple of 4. Can the book be 196 pages long? 198 pages?   The book can be 196 pages long because 196 is a multiple of 4; the book cannot be 198 pages long because 198 is not a multiple of 4.

79. What are the dimensions of the smallest square that you can make using 6-in. by 8-in. rectangular tiles?   24 in. × 24 in.

80. There are 9 players on a baseball team and 11 players on a soccer team. What is the smallest number of students in a college that can be split evenly into either baseball or soccer teams?   99 students

■ *Check your answers on page A-3.*

EXAMPLE 11

Suppose that your annual income is $39,000. If you pay $9,000 for rent and $3,000 for food per year, rent and food account for what fraction of your income? Simplify your answer.

Solution You must first find the total part of the income that you pay for rent and food per year.

$$\$9,000 + \$3,000 = \$12,000$$

Rent   Food   Total part

The total part is $12,000 and the whole is $39,000, so the fraction is $\dfrac{12,000}{39,000}$. You can simplify this fraction in the following way.

$$\frac{12,000}{39,000} = \frac{12,\cancel{000}}{39,\cancel{000}} = \frac{12}{39} \qquad \text{Note that canceling a 0 is the same as dividing by 10.}$$

$$= \frac{3 \cdot 4}{3 \cdot 13} = \frac{\overset{1}{\cancel{3}} \cdot 4}{\underset{1}{\cancel{3}} \cdot 13} = \frac{4}{13}$$

Therefore $\dfrac{4}{13}$ of your income goes for rent and food.

PRACTICE 11

An acre is a unit of area approximately equal to 4,900 sq yd. You own a parcel of land 50 yd by 30 yd. What fraction of an acre is this? $\dfrac{15}{49}$

Comparing Fractions

Some situations require us to *compare* fractions, that is, to rank them in order of size.

For instance, suppose that $\dfrac{5}{8}$ of one airline's flights arrive on time, in contrast to $\dfrac{3}{5}$ of another airline's flights. To decide which airline has a better record for on-time arrivals, we need to compare the fractions.

Or to take another example, suppose that the drinking water in your home, according to a lab report, has 2 parts per million (ppm) of lead. Is the water safe to drink? If the federal limit on lead in drinking water is 15 parts per billion (ppb), again you need to compare fractions.

One way to handle such problems is to draw diagrams corresponding to the fractions in question. The larger fraction corresponds to the larger shaded region.

For instance, the diagrams to the right show that $\dfrac{3}{4}$ is greater than $\dfrac{1}{4}$. The symbol $>$ stands for "greater than."

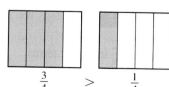

Both $\dfrac{3}{4}$ and $\dfrac{1}{4}$ have the same denominator, so we can rank them simply by comparing their numerators.

Have students note that the symbols $<$ and $>$ always point to the smaller number.

$$\frac{3}{4} > \frac{1}{4} \qquad \text{because} \qquad 3 > 1$$

For **like fractions**, the fraction with the larger numerator is the larger fraction.

Definitions

Like fractions are fractions with the same denominator.

Unlike fractions are fractions with different denominators.

To Compare Fractions

- compare the numerators of like fractions, and
- express unlike fractions as equivalent fractions having the same denominator and then compare their numerators.

EXAMPLE 12	PRACTICE 12

Compare $\dfrac{7}{15}$ and $\dfrac{4}{9}$.

Solution These fractions are unlike because they have different denominators. Therefore we need to express them as equivalent fractions having the same denominator. But what should that denominator be?

One common denominator that we can use is the *product of the denominators:* $15 \cdot 9 = 135$.

$$\frac{7}{15} = \frac{63}{135} \qquad 135 = 15 \cdot 9, \text{ so the new numerator is } 7 \cdot 9, \text{ or } 63.$$

$$\frac{4}{9} = \frac{60}{135} \qquad 135 = 9 \cdot 15, \text{ so the new numerator is } 4 \cdot 15, \text{ or } 60.$$

Next, we compare the numerators of the like fractions that we just found.

Because $63 > 60$, $\dfrac{63}{135} > \dfrac{60}{135}$. Therefore $\dfrac{7}{15} > \dfrac{4}{9}$.

Another common denominator that we can use is the least common multiple of the denominators.

$$15 = 3 \times 5 \qquad 9 = 3 \times 3 = 3^2$$

The LCM is $3^2 \times 5 = 9 \times 5 = 45$. We then compute the equivalent fractions.

$$\frac{7}{15} \qquad \frac{4}{9}$$
$$\downarrow \qquad \downarrow \qquad \text{is equivalent to}$$
$$\frac{21}{45} \qquad \frac{20}{45}$$

Because $\dfrac{21}{45} > \dfrac{20}{45}$, we know that $\dfrac{7}{15} > \dfrac{4}{9}$.

Practice 12: Which is larger, $\dfrac{13}{24}$ or $\dfrac{11}{16}$? $\dfrac{11}{16}$

Point out that another approach here is to say that $135 \div 15$ is 9 and that $9 \times 7 = 63$.

Note that in Example 12 we computed the LCM of the two denominators. This type of computation is used frequently in working with fractions.

> **Definition**
>
> For any set of fractions, their **least common denominator** (LCD) is the least common multiple of their denominators.

In Example 13, pay particular attention to how we use the LCD.

EXAMPLE 13

Order from smallest to largest: $\dfrac{3}{4}, \dfrac{7}{10}$, and $\dfrac{29}{40}$.

Solution Because these fractions are unlike, we need to find equivalent fractions with a common denominator. Let's use their LCD as that denominator.

$$4 = 2 \times 2 = 2^2$$
$$10 = 2 \times 5$$
$$40 = 2 \times 2 \times 2 \times 5 = 2^3 \times 5$$

The LCD $= 2^3 \times 5 = 8 \times 5 = 40$. Check: 4 and 10 are both factors of 40.

We write each fraction with a denominator of 40.

$$\frac{3}{4} = \frac{3 \cdot 10}{4 \cdot 10} = \frac{30}{40} \qquad \frac{7}{10} = \frac{7 \cdot 4}{10 \cdot 4} = \frac{28}{40} \qquad \frac{29}{40} = \frac{29}{40}$$

Then we order the fractions from smallest to largest. (The symbol $<$ stands for "less than.")

$$\frac{28}{40} < \frac{29}{40} < \frac{30}{40}, \qquad \text{or} \qquad \frac{7}{10} < \frac{29}{40} < \frac{3}{4}$$

PRACTICE 13

Arrange $\dfrac{9}{10}, \dfrac{23}{30}$, and $\dfrac{8}{15}$ from smallest to largest.

$$\frac{8}{15}, \frac{23}{30}, \frac{9}{10}$$

EXAMPLE 14

About $\dfrac{7}{10}$ of the Earth's surface is covered by water and $\dfrac{1}{20}$ is covered by desert. Does water or desert cover more of the Earth?

Solution We need to compare $\dfrac{7}{10}$ with $\dfrac{1}{20}$.

$$\frac{7}{10} = \frac{14}{20}$$
$$\frac{1}{20} = \frac{1}{20}$$

Since $\dfrac{14}{20} > \dfrac{1}{20}, \dfrac{7}{10} > \dfrac{1}{20}$. Therefore water covers more of the Earth than desert does.

PRACTICE 14

You work $\dfrac{1}{3}$ of a day and sleep $\dfrac{7}{24}$ of a day. Do you spend more time working or sleeping? Working; $\dfrac{1}{3} > \dfrac{7}{24}$

Exercises 2.2

Identify a fraction or mixed number that represents the shaded part of each figure.

1.

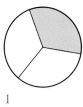

$\dfrac{1}{3}$

2.

$\dfrac{1}{4}$

3.

$\dfrac{3}{6}$

4.

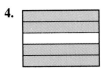

$\dfrac{4}{5}$

5.

$1\dfrac{1}{4}$

6.

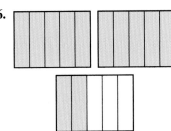

$2\dfrac{2}{5}$

7.

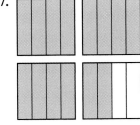

$3\dfrac{2}{4}$

8.

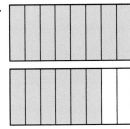

$1\dfrac{6}{8}$

Draw a diagram to represent each fraction or mixed number.

9. $\dfrac{5}{8}$

10. $\dfrac{6}{11}$

11. $\dfrac{2}{9}$

12. $\dfrac{4}{10}$

13. $\dfrac{6}{6}$

14. $\dfrac{11}{11}$

15. $\dfrac{6}{5}$
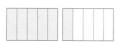

16. $\dfrac{7}{3}$

17. $2\dfrac{1}{2}$

18. $4\dfrac{1}{5}$

19. $1\dfrac{2}{3}$

20. $3\dfrac{4}{9}$

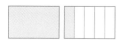

Indicate whether each number is a proper fraction, improper fraction, or mixed number.

21. $\dfrac{3}{4}$

Proper

22. $\dfrac{7}{12}$

Proper

23. $\dfrac{10}{9}$

Improper

24. $\dfrac{11}{10}$

Improper

25. $16\frac{2}{3}$

Mixed

26. $12\frac{1}{2}$

Mixed

27. $\frac{5}{5}$

Improper

28. $\frac{4}{4}$

Improper

29. $\frac{5}{8}$

Proper

30. $\frac{5}{6}$

Proper

31. $66\frac{2}{3}$

Mixed

32. $10\frac{3}{4}$

Mixed

Write each number as an improper fraction.

33. $2\frac{3}{5}$

$\frac{13}{5}$

34. $1\frac{1}{3}$

$\frac{4}{3}$

35. $6\frac{1}{9}$

$\frac{55}{9}$

36. $10\frac{2}{3}$

$\frac{32}{3}$

37. $11\frac{2}{5}$

$\frac{57}{5}$

38. $12\frac{3}{4}$

$\frac{51}{4}$

39. 5

$\frac{5}{1}$

40. 8

$\frac{8}{1}$

41. $7\frac{3}{8}$

$\frac{59}{8}$

42. $6\frac{5}{6}$

$\frac{41}{6}$

43. $9\frac{7}{9}$

$\frac{88}{9}$

44. $10\frac{1}{2}$

$\frac{21}{2}$

45. $12\frac{2}{3}$

$\frac{38}{3}$

46. $20\frac{1}{8}$

$\frac{161}{8}$

47. $19\frac{3}{5}$

$\frac{98}{5}$

48. $11\frac{5}{7}$

$\frac{82}{7}$

49. 14

$\frac{14}{1}$

50. 10

$\frac{10}{1}$

51. $4\frac{10}{11}$

$\frac{54}{11}$

52. $2\frac{7}{13}$

$\frac{33}{13}$

53. $8\frac{3}{14}$

$\frac{115}{14}$

54. $4\frac{1}{6}$

$\frac{25}{6}$

55. $8\frac{2}{25}$

$\frac{202}{25}$

56. $14\frac{1}{10}$

$\frac{141}{10}$

Express each fraction as a mixed or whole number.

57. $\frac{4}{3}$

$1\frac{1}{3}$

58. $\frac{6}{5}$

$1\frac{1}{5}$

59. $\frac{10}{9}$

$1\frac{1}{9}$

60. $\frac{12}{5}$

$2\frac{2}{5}$

61. $\dfrac{9}{3}$ **62.** $\dfrac{12}{12}$ **63.** $\dfrac{15}{15}$ **64.** $\dfrac{62}{3}$

3   1   1 $20\dfrac{2}{3}$

65. $\dfrac{99}{5}$ **66.** $\dfrac{31}{2}$ **67.** $\dfrac{82}{9}$ **68.** $\dfrac{100}{100}$

$19\dfrac{4}{5}$ $15\dfrac{1}{2}$ $9\dfrac{1}{9}$ 1

69. $\dfrac{45}{45}$ **70.** $\dfrac{40}{3}$ **71.** $\dfrac{74}{9}$ **72.** $\dfrac{41}{8}$

1     $13\dfrac{1}{3}$ $8\dfrac{2}{9}$ $5\dfrac{1}{8}$

73. $\dfrac{27}{2}$ **74.** $\dfrac{58}{11}$ **75.** $\dfrac{100}{9}$ **76.** $\dfrac{19}{1}$

$13\dfrac{1}{2}$ $5\dfrac{3}{11}$ $11\dfrac{1}{9}$ 19

77. $\dfrac{27}{1}$ **78.** $\dfrac{72}{9}$ **79.** $\dfrac{56}{7}$ **80.** $\dfrac{38}{3}$

27   8   8 $12\dfrac{2}{3}$

Find two fractions equivalent to each fraction.

81. $\dfrac{1}{8}$ **82.** $\dfrac{2}{5}$ **83.** $\dfrac{2}{11}$ **84.** $\dfrac{1}{10}$

Possible answer: $\dfrac{2}{16},\dfrac{3}{24}$ $\dfrac{4}{10},\dfrac{6}{15}$ $\dfrac{4}{22},\dfrac{6}{33}$ $\dfrac{2}{20},\dfrac{3}{30}$

85. $\dfrac{3}{4}$ **86.** $\dfrac{5}{6}$ **87.** $\dfrac{1}{7}$ **88.** $\dfrac{3}{5}$

Possible answer: $\dfrac{6}{8},\dfrac{9}{12}$ $\dfrac{10}{12},\dfrac{15}{18}$ $\dfrac{2}{14},\dfrac{3}{21}$ $\dfrac{6}{10},\dfrac{9}{15}$

Find n and check.

89. $\dfrac{3}{4}=\dfrac{n}{12}$ **90.** $\dfrac{2}{9}=\dfrac{n}{18}$ **91.** $\dfrac{5}{8}=\dfrac{n}{24}$ **92.** $\dfrac{7}{10}=\dfrac{n}{20}$

9   4   15   14

93. $4=\dfrac{n}{10}$ **94.** $5=\dfrac{n}{15}$ **95.** $\dfrac{3}{5}=\dfrac{n}{60}$ **96.** $\dfrac{4}{9}=\dfrac{n}{63}$

40   75   36   28

97. $\dfrac{5}{8}=\dfrac{n}{64}$ **98.** $\dfrac{3}{10}=\dfrac{n}{40}$ **99.** $3=\dfrac{n}{18}$ **100.** $2=\dfrac{n}{21}$

40   12   54   42

101. $\dfrac{4}{9} = \dfrac{n}{81}$

36

102. $\dfrac{7}{8} = \dfrac{n}{24}$

21

103. $\dfrac{6}{7} = \dfrac{n}{49}$

42

104. $\dfrac{5}{6} = \dfrac{n}{48}$

40

105. $\dfrac{2}{17} = \dfrac{n}{51}$

6

106. $\dfrac{1}{3} = \dfrac{n}{90}$

30

107. $\dfrac{7}{12} = \dfrac{n}{84}$

49

108. $\dfrac{1}{4} = \dfrac{n}{100}$

25

109. $\dfrac{2}{3} = \dfrac{n}{48}$

32

110. $\dfrac{7}{8} = \dfrac{n}{56}$

49

111. $\dfrac{3}{10} = \dfrac{n}{100}$

30

112. $\dfrac{5}{6} = \dfrac{n}{144}$

120

Simplify each fraction.

113. $\dfrac{2}{4}$

$\dfrac{1}{2}$

114. $\dfrac{6}{8}$

$\dfrac{3}{4}$

115. $\dfrac{6}{9}$

$\dfrac{2}{3}$

116. $\dfrac{9}{12}$

$\dfrac{3}{4}$

117. $\dfrac{10}{10}$

1

118. $\dfrac{21}{21}$

1

119. $\dfrac{5}{15}$

$\dfrac{1}{3}$

120. $\dfrac{4}{24}$

$\dfrac{1}{6}$

121. $\dfrac{42}{10}$

$\dfrac{21}{5}$ or $4\dfrac{1}{5}$

122. $\dfrac{40}{25}$

$\dfrac{8}{5}$ or $1\dfrac{3}{5}$

123. $\dfrac{66}{99}$

$\dfrac{2}{3}$

124. $\dfrac{25}{49}$

$\dfrac{25}{49}$

125. $\dfrac{25}{100}$

$\dfrac{1}{4}$

126. $\dfrac{75}{100}$

$\dfrac{3}{4}$

127. $\dfrac{125}{1,000}$

$\dfrac{1}{8}$

128. $\dfrac{875}{1,000}$

$\dfrac{7}{8}$

129. $\dfrac{20}{16}$

$\dfrac{5}{4}$ or $1\dfrac{1}{4}$

130. $\dfrac{15}{9}$

$\dfrac{5}{3}$ or $1\dfrac{2}{3}$

131. $\dfrac{66}{32}$

$\dfrac{33}{16}$ or $2\dfrac{1}{16}$

132. $\dfrac{30}{18}$

$\dfrac{5}{3}$ or $1\dfrac{2}{3}$

133. $\dfrac{18}{32}$

$\dfrac{9}{16}$

134. $\dfrac{36}{45}$

$\dfrac{4}{5}$

135. $\dfrac{50}{1,000}$

$\dfrac{1}{20}$

136. $\dfrac{25}{1,000}$

$\dfrac{1}{40}$

137. $\dfrac{36}{28}$

$\dfrac{9}{7}$ or $1\dfrac{2}{7}$

138. $\dfrac{7}{24}$

$\dfrac{7}{24}$

139. $\dfrac{19}{51}$

$\dfrac{19}{51}$

140. $\dfrac{10}{2}$

5

141. $\dfrac{27}{9}$

3

142. $\dfrac{36}{144}$

$\dfrac{1}{4}$

143. $\dfrac{12}{84}$

$\dfrac{1}{7}$

144. $\dfrac{21}{36}$

$\dfrac{7}{12}$

145. $\dfrac{36}{48}$

$\dfrac{3}{4}$

146. $\dfrac{75}{20}$

$\dfrac{15}{4}$ or $3\dfrac{3}{4}$

147. $\dfrac{375}{1,000}$

$\dfrac{3}{8}$

148. $\dfrac{168}{1,000}$

$\dfrac{21}{125}$

149. $4\dfrac{71}{142}$

$4\dfrac{1}{2}$

150. $3\dfrac{38}{57}$

$3\dfrac{2}{3}$

151. $5\dfrac{200}{300}$

$5\dfrac{2}{3}$

152. $10\dfrac{400}{1,600}$

$10\dfrac{1}{4}$

153. $7\dfrac{6}{15}$

$7\dfrac{2}{5}$

154. $11\dfrac{51}{102}$

$11\dfrac{1}{2}$

155. $2\dfrac{100}{100}$

3

156. $1\dfrac{144}{144}$

2

Between each pair of numbers, insert the appropriate sign: <, =, or >.

157. $\dfrac{7}{20} < \dfrac{11}{20}$ **158.** $\dfrac{5}{10} > \dfrac{3}{10}$ **159.** $\dfrac{1}{8} > \dfrac{1}{9}$ **160.** $\dfrac{5}{6} < \dfrac{7}{8}$

161. $\dfrac{2}{3} = \dfrac{6}{9}$ **162.** $\dfrac{9}{12} = \dfrac{3}{4}$ **163.** $2\dfrac{1}{3} < 2\dfrac{9}{15}$ **164.** $2\dfrac{3}{7} > 1\dfrac{1}{2}$

For each group of three numbers, choose the number whose value is between the other two.

165. $\dfrac{1}{2}, \dfrac{1}{3}, \dfrac{1}{4}$ $\dfrac{1}{3}$ **166.** $\dfrac{3}{2}, \dfrac{3}{3}, \dfrac{3}{4}$ $\dfrac{3}{3}$

167. $\dfrac{2}{3}, \dfrac{7}{12}, \dfrac{5}{6}$ $\dfrac{2}{3}$ **168.** $\dfrac{3}{4}, \dfrac{5}{6}, \dfrac{7}{8}$ $\dfrac{5}{6}$

169. $\dfrac{3}{5}, \dfrac{2}{3}, \dfrac{8}{9}$ $\dfrac{2}{3}$ **170.** $\dfrac{5}{8}, \dfrac{1}{2}, \dfrac{4}{11}$ $\dfrac{1}{2}$

Applications

Solve. Write your answer in simplest form.

171. In a session of Congress, there were 98 male senators and 2 female senators.

a. What fraction of the senators were women? $\frac{1}{50}$

b. What fraction of the senators were men? $\frac{49}{50}$

172. At a party, 6 of the guests are men and 8 are women.

a. What fraction of the guests are men? $\frac{3}{7}$

b. If 1 man leaves, what fraction of the guests are men? $\frac{5}{13}$

173. Suppose that your college newspaper accepts 30 of every 50 articles submitted. What fraction of the articles are rejected? $\frac{2}{5}$

174. A baseball player got 16 hits in 40 times at bat. What fraction of his times at bat did he not get a hit? $\frac{3}{5}$

175. The gutter on your roof overflows whenever more than $\frac{1}{4}$ in. of rain falls. Yesterday $\frac{23}{100}$ in. of rain fell. Did the gutter overflow? Explain. No; $\frac{1}{4} = \frac{25}{100}$ which is greater than $\frac{23}{100}$.

176. You learn in a course on probability and statistics that, when you roll a pair of dice, the probability of rolling a 5 is $\frac{1}{9}$ and that the probability of rolling a 6 is $\frac{5}{36}$. Does rolling a 5 or a 6 have a greater probability? Explain.

Rolling a 6, because $\frac{5}{36}$ is greater than $\frac{1}{9}$.

177. You drilled three holes in a piece of wood. The diameters of the three holes are $\frac{1}{8}$ in., $\frac{3}{16}$ in., and $\frac{3}{8}$ in.

a. Which hole is largest? $\frac{3}{8}$ in.

b. If you have a bolt with a $\frac{5}{32}$-in. diameter, will the bolt fit through all three holes?

No

178. When fog hit the New York City area, visibility was reduced to one-sixteenth mile at Kennedy Airport, one-eighth mile at LaGuardia Airport and one-half mile at Newark Airport.

a. Which of the three airports had the best visibility? Newark

b. Which of the three airports had the worst visibility? Kennedy

179. The following chart gives the rainfall for each of the past 4 days.

M	Tu	W	Th
2 in.	1 in.	2 in.	0 in.

What was the average rainfall for the 4 days? $1\frac{1}{4}$ in.

180. The following chart gives the age of the first six American presidents at the time of their inauguration.

President	Washington	J. Adams	Jefferson	Madison	Monroe	J. Q. Adams
Age	57	61	57	57	58	57

What was their average age at inauguration? (*Source: Significant American Presidents of the United States*) $57\frac{5}{6}$ yr

■ *Check your answers on page A-3.*

MINDSTRETCHERS

WRITING

1. Do you think that there is anything improper about an improper fraction? Explain.

Yes. The part is greater than the whole, thus the definition of a fraction as "part of a whole" is invalid.

GROUPWORK

2. Working with a partner, determine how many fractions there are between the numbers 1 and 2.   There are an infinite number of fractions $\left(\frac{3}{2}, \frac{4}{3}, \frac{5}{4}, \ldots\right)$.

CRITICAL THINKING

3. Consider the three equivalent fractions shown. Note that the numerators and denominators are made up of the digits 1, 2, 3, 4, 5, 6, 7, 8, and 9—each appearing once.

$$\frac{3}{6} = \frac{7}{14} = \frac{29}{58}$$

a. Verify that these fractions are equivalent by making sure that their cross products are equal. $42 = 42$; $174 = 174$; $406 = 406$

b. Complete the following fractions to form another trio of equivalent fractions that use the same nine digits only once.

$$\frac{2}{4} = \frac{3}{6} = \frac{79}{158}$$

2.3 Adding and Subtracting Fractions

OBJECTIVES

- *To add and subtract fractions and mixed numbers*
- *To estimate sums and differences involving mixed numbers*
- *To solve word problems involving the addition or subtraction of fractions or mixed numbers*

In Section 2.2 we examined what fractions mean, how they are written and how they are compared. In the rest of this chapter, we discuss computations involving fractions, beginning with sums and differences.

Adding and Subtracting Like Fractions

Let's first discuss how to add and subtract like fractions. Suppose that we want to add $\frac{1}{5}$ and $\frac{3}{5}$. A diagram can help us understand what is involved. First we shade one-fifth of the diagram, then another three-fifths. We see in the diagram that the total shaded area is four-fifths, so $\frac{1}{5} + \frac{3}{5} = \frac{4}{5}$. Note that we added the original numerators to get the numerator of the answer but that *the denominator stayed the same.*

The diagram at the left shows how to subtract like fractions by computing $\frac{4}{5} - \frac{1}{5}$. If we shade four-fifths of the diagram and then remove the shading in one-fifth, three-fifths remain shaded. Therefore $\frac{4}{5} - \frac{1}{5} = \frac{3}{5}$. Note that we could have gotten this answer simply by subtracting numerators without changing the denominator.

The following rule summarizes how to add or subtract fractions, *provided that they have the same denominator.*

To Add (or Subtract) Like Fractions

- first add (or subtract) the numerators,
- then use the given denominator, and
- finally write the answer in simplest form.

EXAMPLE 1	PRACTICE 1
Add: $\frac{7}{12} + \frac{2}{12}$	Find the sum of $\frac{7}{15}$ and $\frac{3}{15}$. $\frac{2}{3}$

Add the numerators.
↓

Solution Applying the rule, we get $\frac{7}{12} + \frac{2}{12} = \frac{7+2}{12} = \frac{9}{12}$, or $\frac{3}{4}$.

↑   ↑
Keep the same denominator.   Simplest form

| TIP | Be careful *not* to add the denominators when adding like fractions. |

EXAMPLE 2

Add: $\dfrac{12}{16}, \dfrac{3}{16},$ and $\dfrac{9}{16}$

Solution

$$\underset{\text{Simplify.}}{} \qquad \underset{\substack{\text{Answer as a} \\ \text{mixed number}}}{}$$

$$\dfrac{12}{16} + \dfrac{3}{16} + \dfrac{9}{16} = \dfrac{24}{16} = \dfrac{3}{2}, \quad \text{or} \quad 1\dfrac{1}{2}$$

So the sum of $\dfrac{12}{16}, \dfrac{3}{16},$ and $\dfrac{9}{16}$ is $1\dfrac{1}{2}$.

PRACTICE 2

Add: $\dfrac{13}{40} + \dfrac{11}{40} + \dfrac{23}{40}$ $1\dfrac{7}{40}$

EXAMPLE 3

Find the difference between $\dfrac{11}{7}$ and $\dfrac{3}{7}$.

Solution

$$\underset{\text{Subtract the numerators.}}{}$$

$$\dfrac{11}{7} - \dfrac{3}{7} = \dfrac{11 - 3}{7} = \dfrac{8}{7}, \quad \text{or} \quad 1\dfrac{1}{7}$$

$$\underset{\text{Keep the same denominator.}}{}$$

PRACTICE 3

$\dfrac{19}{20} - \dfrac{11}{20} = ?$ $\dfrac{2}{5}$

EXAMPLE 4

In the following diagram, how far is it from the college to the library via city hall?

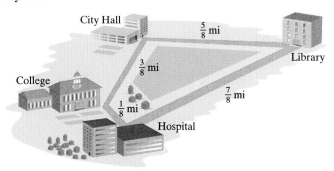

City Hall

$\dfrac{5}{8}$ mi

Library

$\dfrac{3}{8}$ mi

College

$\dfrac{7}{8}$ mi

$\dfrac{1}{8}$ mi

Hospital

Solution Examining the diagram, we see that

■ the distance from the college to city hall is $\dfrac{3}{8}$ mi and that

■ the distance from city hall to the library is $\dfrac{5}{8}$ mi.

To find the distance from the college to the library via city hall, we add.

$$\dfrac{3}{8} + \dfrac{5}{8} = \dfrac{3 + 5}{8} = \dfrac{8}{8} = 1$$

The distance is 1 mi.

PRACTICE 4

Using the diagram in Example 4, decide which route from the college to the library is shorter—via city hall or via the hospital. Explain.   Neither; they are both 1 mi.

EXAMPLE 5

According to one study, $\frac{3}{5}$ of the people who exercise regularly live at least to age 70, in contrast to only $\frac{1}{5}$ of the people who do not exercise regularly. What is the difference between these two fractions?

Solution Subtracting, we get $\frac{3}{5} - \frac{1}{5} = \frac{2}{5}$. Therefore the difference between these fractions is $\frac{2}{5}$. We can check our answer by adding $\frac{2}{5}$ to $\frac{1}{5}$ to get $\frac{3}{5}$.

PRACTICE 5

You purchase $\frac{3}{4}$ yd of material. If your pattern requires only $\frac{1}{4}$ yd, how much material will be left over?　$\frac{1}{2}$ yd

Adding and Subtracting Unlike Fractions

Adding (or subtracting) **unlike fractions** is more complicated than adding (or subtracting) like fractions. An extra step is required: changing the unlike fractions to equivalent like fractions. For instance, suppose that we want to add $\frac{1}{10}$ and $\frac{2}{15}$. Even though we can use any common denominator for these fractions, let's use their *least* common denominator to find equivalent fractions.

$$10 = 2 \cdot 5$$

$$15 = 3 \cdot 5$$

$$\text{LCD} = 2 \cdot 3 \cdot 5 = 30$$

Let's rewrite the fractions vertically as equivalent fractions with the denominator 30.

$$\frac{1}{10} = \frac{1 \cdot 3}{10 \cdot 3} = \frac{3}{30}$$

$$+\frac{2}{15} = \frac{2 \cdot 2}{15 \cdot 2} = \frac{4}{30}$$

Now we add the equivalent like fractions.

$$\frac{3}{30}$$

$$+\frac{4}{30}$$

$$\frac{7}{30}$$

So $\frac{1}{10} + \frac{2}{15} = \frac{7}{30}$.

We can also add and subtract unlike fractions horizontally.

$$\frac{1}{10} + \frac{2}{15} = \frac{3}{30} + \frac{4}{30} = \frac{3+4}{30} = \frac{7}{30}$$

To Add (or Subtract) Unlike Fractions

■ rewrite the fractions as equivalent fractions with a common denominator, usually the LCD;

■ add (or subtract) the numerators, keeping the same denominator; and

■ write the answer in simplest form.

EXAMPLE 6

Add: $\frac{5}{12} + \frac{5}{16}$

Solution First, we find the LCD, which is 48. After finding equivalent fractions, we add the numerators, keeping the same denominator.

$$\frac{5}{12} = \frac{20}{48}$$
$$+\frac{5}{16} = +\frac{15}{48}$$
$$\frac{35}{48} \leftarrow \text{Already in lowest terms}$$

PRACTICE 6

Add: $\frac{11}{12} + \frac{3}{4}$ $\quad 1\frac{2}{3}$

EXAMPLE 7

Subtract $\frac{1}{12}$ from $\frac{1}{3}$.

Solution Because 3 is a factor of 12, the LCD is 12. Again, let's set up the problem vertically.

$$\frac{1}{3} = \frac{4}{12}$$
$$-\frac{1}{12} = -\frac{1}{12} \quad \text{Subtract the numerators, keeping the same denominator.}$$
$$\frac{3}{12} = \frac{1}{4} \quad \text{Reduce } \frac{3}{12} \text{ to lowest terms.}$$

PRACTICE 7

Calculate: $\frac{4}{5} - \frac{1}{2}$ $\quad \frac{3}{10}$

EXAMPLE 8

Combine: $\frac{1}{3} + \frac{1}{6} - \frac{3}{10}$

Solution We add $\frac{1}{3}$ and $\frac{1}{6}$ and then subtract $\frac{3}{10}$ from this sum. The LCD of $\frac{1}{3}$ and $\frac{1}{6}$ is 6.

PRACTICE 8

Combine: $\frac{1}{3} - \frac{2}{9} + \frac{7}{8}$ $\quad \frac{71}{72}$

$$\frac{1}{3} = \frac{2}{6}$$

$$+\frac{1}{6} = +\frac{1}{6}$$

$$\frac{3}{6}, \quad \text{or} \quad \frac{1}{2}$$

Next, we subtract $\frac{3}{10}$ from $\frac{1}{2}$. Their LCD is 10.

$$\frac{1}{2} = \frac{5}{10}$$

$$-\frac{3}{10} = -\frac{3}{10}$$

$$\frac{2}{10}, \quad \text{or} \quad \frac{1}{5}$$

So $\frac{1}{3} + \frac{1}{6} - \frac{3}{10} = \frac{1}{5}$. Do you see another way of solving this problem, using the LCD of all three fractions?

Have students write their answers in a journal.

EXAMPLE 9

Find the perimeter of the figure.

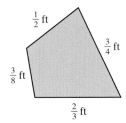

Solution Recall that the perimeter of a figure is the sum of the lengths of its sides.

$$\text{Perimeter} = \frac{3}{8} + \frac{1}{2} + \frac{3}{4} + \frac{2}{3}$$

$$\frac{3}{8} = \frac{9}{24}$$
$$\frac{1}{2} = \frac{12}{24}$$
$$\frac{3}{4} = \frac{18}{24}$$
$$+\frac{2}{3} = +\frac{16}{24}$$
$$\frac{55}{24}, \quad \text{or} \quad 2\frac{7}{24}$$

The perimeter of the figure is $2\frac{7}{24}$ feet.

PRACTICE 9

What is the perimeter of the figure?   2 mi

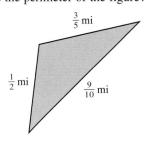

Adding and Subtracting Mixed Numbers with the Same Denominator

Now let's consider how to add and subtract **mixed numbers** with the same denominator. We start with addition.

Suppose, for instance, that we want to add $1\frac{1}{5}$ and $2\frac{1}{5}$. Let's draw a diagram to represent this sum.

We can rearrange the elements of the diagram by combining the whole numbers and the fractions separately.

This diagram shows that the sum is $3\frac{2}{5}$.

Note that we can also write and solve this problem vertically.

$$
\begin{array}{r}
1\frac{1}{5} \\
+\, 2\frac{1}{5} \\
\hline
3\frac{2}{5}
\end{array}
$$

Similarly, we can subtract mixed numbers that have the same denominator.

$$
\begin{array}{r}
3\frac{2}{5} \\
-\, 2\frac{1}{5} \\
\hline
1\frac{1}{5}
\end{array}
$$

Have students write their answers in a journal.

Can you explain how the preceding diagrams confirm this answer?

To Add (or Subtract) Mixed Numbers That Have the Same Denominator

■ add (or subtract) the fractions,
■ add (or subtract) the whole numbers, and
■ write the answer in simplest form.

EXAMPLE 10

Add: $8\dfrac{5}{9} + 10\dfrac{7}{9}$

Solution

$$8\dfrac{5}{9}$$

$$+10\dfrac{7}{9}$$

$$18\dfrac{12}{9} \;\leftarrow\; \text{Sum of the fractions}$$

⌐— Sum of the whole numbers

Because $\dfrac{12}{9} = 1\dfrac{3}{9}$, $18\dfrac{12}{9} = 18 + 1\dfrac{3}{9} = 19\dfrac{3}{9} = 19\dfrac{1}{3}$

Our answer is $19\dfrac{1}{3}$.

PRACTICE 10

Add: $25\dfrac{3}{10} + 9\dfrac{9}{10}$ $\quad 35\dfrac{1}{5}$

> Remind students that a mixed number is a sum. $18\dfrac{12}{9}$ means $18 + \dfrac{12}{9}$.

EXAMPLE 11

Find the sum of $3\dfrac{3}{4}$, $2\dfrac{1}{4}$, and $4\dfrac{1}{4}$.

Solution

$$3\dfrac{3}{4}$$

$$2\dfrac{1}{4} \qquad \text{Add the fractions and then add the whole numbers.}$$

$$+4\dfrac{1}{4}$$

$$9\dfrac{5}{4} = 10\dfrac{1}{4} \qquad \text{Since } \dfrac{5}{4} = 1\dfrac{1}{4}, \text{ we get}$$

$$9\dfrac{5}{4} = 9 + 1\dfrac{1}{4} = 10\dfrac{1}{4}.$$

PRACTICE 11

Find the sum of $2\dfrac{5}{16}$, $1\dfrac{3}{16}$, and 4. $\quad 7\dfrac{1}{2}$

EXAMPLE 12

Subtract: $4\dfrac{5}{6} - 2\dfrac{1}{6}$

Solution We set up the problem vertically.

$$4\dfrac{5}{6} \qquad \text{Subtract the fractions and then subtract the whole numbers.}$$

$$-2\dfrac{1}{6}$$

$$2\dfrac{4}{6} \qquad \text{Simplify the answer.}$$

Because $\dfrac{4}{6} = \dfrac{2}{3}$, $2\dfrac{4}{6} = 2\dfrac{2}{3}$. Therefore the difference is $2\dfrac{2}{3}$.

PRACTICE 12

Subtract $5\dfrac{3}{10}$ from $9\dfrac{7}{10}$. $\quad 4\dfrac{2}{5}$

EXAMPLE 13

The value of stock shares is normally expressed in dollars as a mixed number or fraction. Suppose that one morning the stock of a certain company opened at a price of $\$100\dfrac{5}{8}$ per share. By the end of the day, the price had risen by $\$5\dfrac{1}{8}$. What was the price of the stock per share by the end of the day?

Solution The price of the stock rose by $\$5\dfrac{1}{8}$, so we must add $\$100\dfrac{5}{8}$ and $\$5\dfrac{1}{8}$.

$$
\begin{array}{r}
100\dfrac{5}{8} \\[2mm]
+\,5\dfrac{1}{8} \\[1mm]
\hline
105\dfrac{6}{8}, \quad \text{or} \quad 105\dfrac{3}{4}
\end{array}
$$

At closing, then, the price of the stock was $\$105\dfrac{3}{4}$ per share.

PRACTICE 13

In a horse race, the winner beat the second-place horse by $1\dfrac{1}{2}$ lengths, and the second-place horse finished $2\dfrac{1}{2}$ lengths ahead of the third-place horse. By how many lengths did the third-place horse lose?   4 lengths

EXAMPLE 14

A movie lasts $2\dfrac{3}{4}$ hr. After an hour-and-a-quarter has passed, how much longer will the movie run?

Solution This question asks us to subtract $1\dfrac{1}{4}$ from $2\dfrac{3}{4}$.

$$
\begin{array}{r}
2\dfrac{3}{4} \\[2mm]
-\,1\dfrac{1}{4} \\[1mm]
\hline
1\dfrac{2}{4}, \quad \text{or} \quad 1\dfrac{1}{2}
\end{array}
$$

So there is still an hour-and-a-half of the movie to see.

PRACTICE 14

A photograph is displayed in a frame. The height of the frame is $7\dfrac{3}{16}$ in. and that of the photo is $5\dfrac{1}{16}$ in. What is the difference in their heights?

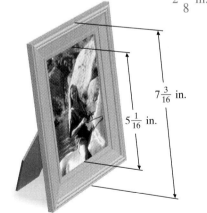

$2\dfrac{1}{8}$ in.

$7\frac{3}{16}$ in.

$5\frac{1}{16}$ in.

Adding and Subtracting Mixed Numbers with Different Denominators

We have previously shown that, when we add (or subtract) fractions with different denominators, we must first change the unlike fractions to equivalent like fractions. The same applies to adding (or subtracting) mixed numbers that have different denominators.

EXAMPLE 15

Find the sum of $3\frac{1}{5}$ and $7\frac{2}{3}$.

Solution The denominators have no common factor other than 1. Therefore the least common denominator is the product of 5 and 3, or 15.

$$3\frac{1}{5} = 3\frac{3}{15}$$
$$+7\frac{2}{3} = +7\frac{10}{15}$$

Now we use the rule for adding mixed numbers that have the same denominator.

$$3\frac{1}{5} = 3\frac{3}{15}$$
$$+7\frac{2}{3} = +7\frac{10}{15}$$
$$10\frac{13}{15}$$

Add the fractions and then add the whole numbers.

The sum of $3\frac{1}{5}$ and $7\frac{2}{3}$ is $10\frac{13}{15}$.

PRACTICE 15

Add $4\frac{1}{8}$ to $3\frac{1}{2}$. $7\frac{5}{8}$

Subtracting mixed numbers that have different denominators is similar to adding them.

EXAMPLE 16

Subtract $2\frac{7}{100}$ from $5\frac{9}{10}$.

Solution As usual, we use the LCD (which is 100) to find equivalent fractions. Then we subtract the equivalent mixed numbers with the same denominator. Again, let's set up the problem vertically.

$$5\frac{9}{10} = 5\frac{90}{100}$$
$$-2\frac{7}{100} = -2\frac{7}{100}$$
$$3\frac{83}{100}$$

Subtract the fractions and then subtract the whole numbers.

This fraction cannot be reduced.

Therefore the answer is $3\frac{83}{100}$.

PRACTICE 16

Calculate: $8\frac{2}{3} - 4\frac{1}{6}$ $4\frac{1}{2}$

EXAMPLE 17

Find the sum of $1\frac{2}{3}$, $8\frac{1}{4}$, and $3\frac{4}{5}$.

Solution Set up the problem vertically and use the LCD, which is 60.

$$
\begin{array}{rl}
1\dfrac{2}{3} = & 1\dfrac{40}{60} \\[2mm]
8\dfrac{1}{4} = & 8\dfrac{15}{60} \\[2mm]
+3\dfrac{4}{5} = & +3\dfrac{48}{60} \\[1mm]
\hline
& 12\dfrac{103}{60}
\end{array}
$$

Add the like fractions and then add the whole numbers.

Rewrite the answer.

Since $\dfrac{103}{60} = 1\dfrac{43}{60}$, $12\dfrac{103}{60} = 12 + 1\dfrac{43}{60} = 13\dfrac{43}{60}$.

PRACTICE 17

What is the sum of $5\frac{5}{8}$, $3\frac{1}{6}$, and $2\frac{5}{12}$?

$11\dfrac{5}{24}$

EXAMPLE 18

Find the missing dimension.

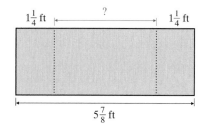

Solution The total length of the figure is $5\frac{7}{8}$ feet. To find the missing dimension, we need to add $1\frac{1}{4}$ feet and $1\frac{1}{4}$ feet and then subtract this sum from $5\frac{7}{8}$.

$$
\begin{array}{rl}
& 1\dfrac{1}{4} \\[2mm]
+ & 1\dfrac{1}{4} \\[1mm]
\hline
& 2\dfrac{2}{4} = 2\dfrac{1}{2}
\end{array}
\qquad
\begin{array}{rl}
5\dfrac{7}{8} = & 5\dfrac{7}{8} \\[2mm]
-2\dfrac{1}{2} = & -2\dfrac{4}{8} \\[1mm]
\hline
& 3\dfrac{3}{8}
\end{array}
$$

The missing dimension is $3\frac{3}{8}$ feet. We can check this answer by adding $1\frac{1}{4}$, $3\frac{3}{8}$, and $1\frac{1}{4}$, getting $5\frac{7}{8}$.

PRACTICE 18

The figure shown is called a **trapezoid**. Suppose that this trapezoid's perimeter is $20\frac{1}{2}$ mi. How long is the left side?

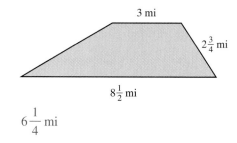

Subtracting Mixed Numbers with Renaming (Borrowing)

Recall from our discussion of subtracting whole numbers that, in problems in which a digit in the subtrahend is larger than the corresponding digit in the minuend, we need to borrow.

$$
\begin{array}{r}
\overset{2}{\cancel{3}}\ \overset{1}{2}\ 9 \\
-\ 8\ 7 \\
\hline
2\ 4\ 2
\end{array}
$$

A similar situation can arise when we are subtracting mixed numbers. If the fraction on the bottom is larger than the one on top, we *rename* (or *borrow from*) the whole number on top.

> ### To Add (or Subtract) Mixed Numbers with Different Denominators
> - rewrite the fractions as equivalent fractions with a common denominator, usually the LCD;
> - when subtracting, rename (or borrow from) the whole number on top if the fraction on the bottom is larger than the fraction on top;
> - add (or subtract) the numerators, keeping the same denominator;
> - add (or subtract) the whole numbers; and
> - write the answer in simplest form.

EXAMPLE 19	PRACTICE 19
Subtract: $6 - 1\dfrac{1}{3}$	Subtract: $9 - 7\dfrac{5}{7}$ $\quad 1\dfrac{2}{7}$

Solution Let's rewrite the problem vertically.

$$
\begin{array}{r}
6 \\
-1\dfrac{1}{3} \\
\hline
\end{array}
$$
There is no fraction on top from which to subtract $\dfrac{1}{3}$.

$$
\begin{array}{r}
5\dfrac{3}{3} \\
-1\dfrac{1}{3} \\
\hline
\end{array}
$$
Rename 6 as $5 + 1$, or $5 + \dfrac{3}{3}$, or $5\dfrac{3}{3}$.

$$
\begin{array}{r}
5\dfrac{3}{3} \\
-1\dfrac{1}{3} \\
\hline
4\dfrac{2}{3}
\end{array}
$$
Now subtract the mixed numbers.

So $6 - 1\dfrac{1}{3} = 4\dfrac{2}{3}$.

As in any subtraction problem, we can check our answer by addition.

$$4\dfrac{2}{3} + 1\dfrac{1}{3} = 5\dfrac{3}{3} = 6$$

> **TIP**
>
> Recall that $1 = \dfrac{2}{2} = \dfrac{3}{3} = \dfrac{4}{4} = \dfrac{5}{5}$ and so on. That is, any fraction having the same numerator and denominator (both nonzero) equals 1.

EXAMPLE 20	PRACTICE 20

EXAMPLE 20

Compute: $13\dfrac{2}{9} - 7\dfrac{8}{9}$

Solution First, we write the problem vertically.

$$13\dfrac{2}{9}$$
$$-7\dfrac{8}{9}$$

Because $\dfrac{8}{9}$ is larger than $\dfrac{2}{9}$, we need to rename $13\dfrac{2}{9}$.

$$13\dfrac{2}{9} = 12 + 1 + \dfrac{2}{9} = 12 + \dfrac{9}{9} + \dfrac{2}{9} = 12\dfrac{11}{9}$$

$$12\dfrac{11}{9}$$
$$-7\dfrac{8}{9}$$

Finally, we subtract and then write the answer in simplest form.

$$13\dfrac{2}{9} = 12\dfrac{11}{9}$$
$$-7\dfrac{8}{9} = -7\dfrac{8}{9}$$
$$\overline{\qquad\qquad 5\dfrac{3}{9}, \quad \text{or} \quad 5\dfrac{1}{3}}$$

PRACTICE 20

Find the difference between $15\dfrac{1}{12}$ and $9\dfrac{11}{12}$.

$5\dfrac{1}{6}$

EXAMPLE 21

Find the difference between $10\dfrac{1}{4}$ and $1\dfrac{5}{12}$.

Solution First, we write the equivalent fractions, using the LCD.

$$10\dfrac{1}{4} = 10\dfrac{3}{12}$$
$$-1\dfrac{5}{12} = -1\dfrac{5}{12}$$

PRACTICE 21

Find the difference between $16\dfrac{2}{5}$ and $3\dfrac{7}{8}$.

$12\dfrac{21}{40}$

Then, we subtract by renaming.

$$10\frac{3}{12} = 9\frac{15}{12} \quad \text{Since } 10\frac{3}{12} = 9 + \frac{12}{12} + \frac{3}{12} = 9\frac{15}{12}$$

$$-1\frac{5}{12} = -1\frac{5}{12}$$

$$\overline{\phantom{-1\frac{5}{12}} \ 8\frac{10}{12}}$$

Simplifying the answer, we get $8\frac{5}{6}$.

EXAMPLE 22

You hike from point A to point B along a trail. If the entire trail is $1\frac{1}{5}$ mi long, will you have more or less than $\frac{1}{2}$ mi left to hike when you get to point B?

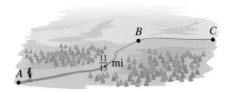

Solution First we must find the difference between the length of the entire trail, $1\frac{1}{5}$ mi, and the distance hiked, $\frac{11}{15}$ mi.

$$1\frac{1}{5} = 1\frac{3}{15} = \frac{18}{15}$$

$$-\frac{11}{15} = -\frac{11}{15} = -\frac{11}{15}$$

$$\overline{\phantom{-\frac{11}{15}} \ \frac{7}{15}}$$

Find the equivalent fractions using the LCD, 15. Then subtract by renaming $1\frac{3}{15}$ as $\frac{15}{15} + \frac{3}{15}$, or $\frac{18}{15}$.

The distance left to hike is $\frac{7}{15}$ mi. Finally, we compare $\frac{7}{15}$ mi and $\frac{1}{2}$ mi.

$$\frac{7}{15} = \frac{14}{30} \qquad \frac{1}{2} = \frac{15}{30}$$

Because $14 < 15$, $\frac{14}{30} < \frac{15}{30}$. Therefore $\frac{7}{15} < \frac{1}{2}$, and you have less than $\frac{1}{2}$ mi left to hike when you get to point B.

PRACTICE 22

You purchased a roll of wallpaper that unrolls to $30\frac{1}{2}$ yd long. Your contractor uses $26\frac{7}{8}$ yd from the roll to paper a room. Is there enough paper left on the roll for a job that requires 4 yd of paper?

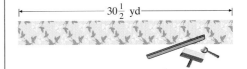

No, there will be only $3\frac{5}{8}$ yd left.

Estimating Sums and Differences of Mixed Numbers

When adding or subtracting mixed numbers, we can check by *estimating*, determining whether our estimate and our answer are close. Note that when we round mixed numbers, we round to the nearest whole number.

Checking a Sum by Estimating

$$1\frac{1}{5} \rightarrow \quad 1 \qquad \text{Because } \frac{1}{5} < \frac{1}{2}, \text{ round } down \text{ to the whole number 1.}$$

$$+2\frac{3}{5} \rightarrow +3 \qquad \text{Because } \frac{3}{5} > \frac{1}{2}, \text{ round } up \text{ to the whole number 3.}$$

$$3\frac{4}{5} \qquad 4 \qquad \text{Our answer, } 3\frac{4}{5}, \text{ is close to 4, the sum of the rounded addends (1 and 3).}$$

Checking a Difference by Estimating

$$3\frac{2}{5} \rightarrow \quad 3 \qquad \text{Because } \frac{2}{5} < \frac{1}{2}, \text{ round } down \text{ to 3.}$$

$$-1\frac{1}{5} \rightarrow -1 \qquad \text{Round } down \text{ to 1.}$$

$$2\frac{1}{5} \qquad 2 \qquad \text{Our answer, } 2\frac{1}{5}, \text{ is close to 2, the difference of the rounded numbers (3 and 1).}$$

EXAMPLE 23

Combine and check: $5\frac{1}{3} + 2\frac{4}{5} - 1\frac{1}{10}$

Solution We begin by adding the first two mixed numbers:

$$5\frac{1}{3} = \quad 5\frac{5}{15}$$
$$+2\frac{4}{5} = +2\frac{12}{15}$$
$$7\frac{17}{15}, \quad \text{or} \quad 8\frac{2}{15}$$

Next, we subtract $1\frac{1}{10}$ from this sum.

$$8\frac{2}{15} = \quad 8\frac{4}{30}$$
$$-1\frac{1}{10} = -1\frac{3}{30}$$
$$7\frac{1}{30}$$

So $5\frac{1}{3} + 2\frac{4}{5} - 1\frac{1}{10} = 7\frac{1}{30}$.

Now let's check this answer by estimation.

$$5\frac{1}{3} + 2\frac{4}{5} - 1\frac{1}{10}$$
$$\downarrow \qquad \downarrow \qquad \downarrow$$
$$5 + 3 - 1 = 7$$

The estimate, 7, is sufficiently close to $7\frac{1}{30}$ to confirm our answer.

PRACTICE 23

Compute $8\frac{1}{4} - 3\frac{4}{5} + 1\frac{9}{10}$. Then check.

$6\frac{7}{20}; 8 - 4 + 2 = 6$

Exercises 2.3

Add and simplify.

1. $\dfrac{2}{3} + \dfrac{2}{3}$

$1\dfrac{1}{3}$

2. $\dfrac{5}{7} + \dfrac{3}{7}$

$1\dfrac{1}{7}$

3. $\dfrac{3}{5} + \dfrac{3}{5}$

$1\dfrac{1}{5}$

4. $\dfrac{5}{9} + \dfrac{8}{9}$

$1\dfrac{4}{9}$

5. $\dfrac{5}{8} + \dfrac{5}{8}$

$1\dfrac{1}{4}$

6. $\dfrac{7}{10} + \dfrac{9}{10}$

$1\dfrac{3}{5}$

7. $\dfrac{11}{12} + \dfrac{7}{12}$

$1\dfrac{1}{2}$

8. $\dfrac{71}{100} + \dfrac{79}{100}$

$1\dfrac{1}{2}$

9. $\dfrac{1}{5} + \dfrac{1}{5} + \dfrac{2}{5}$

$\dfrac{4}{5}$

10. $\dfrac{1}{7} + \dfrac{3}{7} + \dfrac{2}{7}$

$\dfrac{6}{7}$

11. $\dfrac{3}{20} + \dfrac{1}{20} + \dfrac{8}{20}$

$\dfrac{3}{5}$

12. $\dfrac{1}{10} + \dfrac{3}{10} + \dfrac{1}{10}$

$\dfrac{1}{2}$

13. $\dfrac{2}{3} + \dfrac{1}{2}$

$1\dfrac{1}{6}$

14. $\dfrac{1}{4} + \dfrac{2}{5}$

$\dfrac{13}{20}$

15. $\dfrac{1}{2} + \dfrac{3}{8}$

$\dfrac{7}{8}$

16. $\dfrac{1}{6} + \dfrac{2}{3}$

$\dfrac{5}{6}$

17. $\dfrac{7}{10} + \dfrac{7}{100}$

$\dfrac{77}{100}$

18. $\dfrac{5}{6} + \dfrac{1}{12}$

$\dfrac{11}{12}$

19. $\dfrac{4}{5} + \dfrac{1}{8}$

$\dfrac{37}{40}$

20. $\dfrac{3}{4} + \dfrac{3}{7}$

$1\dfrac{5}{28}$

21. $\dfrac{2}{9} + \dfrac{5}{8}$

$\dfrac{61}{72}$

22. $\dfrac{1}{5} + \dfrac{1}{6}$

$\dfrac{11}{30}$

23. $\dfrac{3}{8} + \dfrac{1}{6}$

$\dfrac{13}{24}$

24. $\dfrac{7}{10} + \dfrac{2}{15}$

$\dfrac{5}{6}$

25. $\dfrac{4}{9} + \dfrac{5}{6}$

$1\dfrac{5}{18}$

26. $\dfrac{9}{10} + \dfrac{4}{5}$

$1\dfrac{7}{10}$

27. $\dfrac{87}{100} + \dfrac{3}{10}$

$1\dfrac{17}{100}$

28. $\dfrac{7}{20} + \dfrac{3}{4}$

$1\dfrac{1}{10}$

29. $\dfrac{1}{4}$ hr $+ \dfrac{1}{2}$ hr

$\dfrac{3}{4}$ hr

30. $\dfrac{1}{16}$ in. $+ \dfrac{1}{8}$ in.

$\dfrac{3}{16}$ in.

31. $\dfrac{1}{4}$ mi $+ \dfrac{1}{6}$ mi

$\dfrac{5}{12}$ mi

32. $\dfrac{1}{3}$ yd $+ \dfrac{1}{6}$ yd

$\dfrac{1}{2}$ yd

33. $\dfrac{1}{2} + \dfrac{1}{3} + \dfrac{1}{4}$

$1\dfrac{1}{12}$

34. $\dfrac{1}{3} + \dfrac{1}{4} + \dfrac{1}{6}$

$\dfrac{3}{4}$

35. $\dfrac{7}{8} + \dfrac{1}{5} + \dfrac{1}{4}$

$1\dfrac{13}{40}$

36. $\dfrac{1}{10} + \dfrac{2}{5} + \dfrac{5}{6}$

$1\dfrac{1}{3}$

Add and simplify. Then check by estimating.

37. $1\frac{1}{3} + 2\frac{1}{3}$
$3\frac{2}{3}$; $1 + 2 = 3$

38. $4\frac{1}{5} + 2\frac{3}{5}$
$6\frac{4}{5}$; $4 + 3 = 7$

39. $8\frac{1}{10} + 7\frac{3}{10}$
$15\frac{2}{5}$; $8 + 7 = 15$

40. $6\frac{1}{12} + 4\frac{1}{12}$
$10\frac{1}{6}$; $6 + 4 = 10$

41. $7\frac{3}{10} + 6\frac{9}{10}$
$14\frac{1}{5}$; $7 + 7 = 14$

42. $8\frac{2}{3} + 6\frac{2}{3}$
$15\frac{1}{3}$; $9 + 7 = 16$

43. $5\frac{1}{6} + 9\frac{5}{6}$
15; $5 + 10 = 15$

44. $2\frac{3}{10} + 7\frac{9}{10}$
$10\frac{1}{5}$; $2 + 8 = 10$

45. $4\frac{1}{4} + 3\frac{3}{4}$
8; $4 + 4 = 8$

46. $1\frac{1}{2} + 11\frac{1}{2}$
13; $2 + 12 = 14$

47. $2\frac{4}{5} + 2\frac{3}{5}$
$5\frac{2}{5}$; $3 + 3 = 6$

48. $3\frac{7}{12} + 12\frac{7}{12}$
$16\frac{1}{6}$; $4 + 13 = 17$

49. $37\frac{1}{2} + 5\frac{1}{3}$
$42\frac{5}{6}$; $38 + 5 = 43$

50. $19\frac{2}{9} + 20\frac{3}{10}$
$39\frac{47}{90}$; $19 + 20 = 39$

51. $8\frac{1}{12} + 6\frac{5}{6}$
$14\frac{11}{12}$; $8 + 7 = 15$

52. $2\frac{2}{15} + 3\frac{3}{100}$
$5\frac{49}{300}$; $2 + 3 = 5$

53. $5\frac{1}{4} + 5\frac{1}{6}$
$10\frac{5}{12}$; $5 + 5 = 10$

54. $17\frac{3}{8} + 20\frac{1}{5}$
$37\frac{23}{40}$; $17 + 20 = 37$

55. $3\frac{1}{3} + \frac{2}{5}$
$3\frac{11}{15}$; $3 + 0 = 3$

56. $4\frac{7}{10} + \frac{7}{20}$
$5\frac{1}{20}$; $5 + 0 = 5$

57. $8\frac{1}{5} + 5\frac{2}{3}$
$13\frac{13}{15}$; $8 + 6 = 14$

58. $4\frac{1}{9} + 20\frac{7}{10}$
$24\frac{73}{90}$; $4 + 21 = 25$

59. $\frac{2}{3} + 6\frac{1}{8}$
$6\frac{19}{24}$; $1 + 6 = 7$

60. $\frac{1}{6} + 3\frac{2}{5}$
$3\frac{17}{30}$; $0 + 3 = 3$

61. $9\frac{2}{3} + 10\frac{7}{12}$
$20\frac{1}{4}$; $10 + 11 = 21$

62. $20\frac{3}{5} + 4\frac{1}{2}$
$25\frac{1}{10}$; $21 + 5 = 26$

63. $65\frac{7}{10} + 30\frac{57}{100}$
$96\frac{27}{100}$; $66 + 31 = 97$

64. $1\frac{951}{1,000} + 1\frac{3}{10}$
$3\frac{251}{1,000}$; $2 + 1 = 3$

65. $6\frac{1}{10} + 3\frac{93}{100}$
$10\frac{3}{100}$; $6 + 4 = 10$

66. $4\frac{8}{9} + 5\frac{1}{3}$
$10\frac{2}{9}$; $5 + 5 = 10$

67. $22\frac{3}{5} + 22\frac{9}{10}$
$45\frac{1}{2}$; $23 + 23 = 46$

68. $4\frac{1}{2} + 6\frac{7}{8}$
$11\frac{3}{8}$; $5 + 7 = 12$

69. $10\frac{5}{6} + 8\frac{1}{4}$
$19\frac{1}{12}$; $11 + 8 = 19$

70. $9\frac{1}{2} + 1\frac{3}{4}$
$11\frac{1}{4}$; $10 + 2 = 12$

71. $30\frac{21}{100} + 5\frac{17}{20}$
$36\frac{3}{50}$; $30 + 6 = 36$

72. $8\frac{3}{10} + 2\frac{321}{1,000}$
$10\frac{621}{1,000}$; $8 + 2 = 10$

73. $80\frac{1}{3} + \frac{3}{4} + 10\frac{1}{2}$
$91\frac{7}{12}$; $80 + 1 + 11 = 92$

74. $\frac{1}{3} + 25\frac{7}{24} + 100\frac{1}{2}$
$126\frac{1}{8}$; $0 + 25 + 101 = 126$

75. $5\frac{1}{2}$ hr $+ 6\frac{1}{4}$ hr $+ 3\frac{1}{2}$ hr

$15\frac{1}{4}$ hr; $6 + 6 + 4 = 16$ hr

76. $2\frac{1}{2}$ days $+ 4\frac{3}{4}$ days $+ 6\frac{1}{2}$ days

$13\frac{3}{4}$ days; $3 + 5 + 7 = 15$ days

77. $2\frac{1}{3}$ ft $+ 2\frac{1}{4}$ ft $+ 2\frac{1}{6}$ ft

$6\frac{3}{4}$ ft; $2 + 2 + 2 = 6$ ft

78. $4\frac{1}{8}$ lb $+ 4\frac{3}{16}$ lb $+ 4\frac{3}{4}$ lb

$13\frac{1}{16}$ lb; $4 + 4 + 5 = 13$ lb

79. $6\frac{7}{8}$ in. $+ 2\frac{3}{4}$ in. $+ 1\frac{1}{5}$ in.

$10\frac{33}{40}$ in.; $7 + 3 + 1 = 11$ in.

80. $1\frac{2}{3}$ mi $+ 5\frac{5}{6}$ mi $+ 3\frac{1}{4}$ mi

$10\frac{3}{4}$ mi; $2 + 6 + 3 = 11$ mi

81. $\frac{2}{3}$ min $+ 5\frac{1}{2}$ min $+ 4\frac{7}{8}$ min

$11\frac{1}{24}$ min; $1 + 6 + 5 = 12$ min

82. $2\frac{1}{2}$ qt $+ 5\frac{1}{4}$ qt $+ 3\frac{5}{8}$ qt

$11\frac{3}{8}$ qt; $3 + 5 + 4 = 12$ qt

83. $\$20\frac{1}{16} + \$1\frac{1}{8} + \$1\frac{1}{2}$

$\$22\frac{11}{16}$; $20 + 1 + 2 = \$23$

84. $4\frac{2}{3}$ yd $+ 2\frac{11}{36}$ yd $+ 1\frac{1}{2}$ yd

$8\frac{17}{36}$ yd; $5 + 2 + 2 = 9$ yd

Subtract and simplify. Then check either by estimating or by adding.

85. $\frac{4}{5} - \frac{3}{5}$ $\frac{1}{5}$

86. $\frac{7}{9} - \frac{5}{9}$ $\frac{2}{9}$

87. $\frac{7}{10} - \frac{3}{10}$ $\frac{2}{5}$

88. $\frac{11}{12} - \frac{5}{12}$ $\frac{1}{2}$

89. $\frac{23}{100} - \frac{7}{100}$ $\frac{4}{25}$

90. $\frac{39}{100} - \frac{17}{100}$ $\frac{11}{50}$

91. $\frac{3}{2} - \frac{1}{2}$ 1

92. $\frac{6}{5} - \frac{4}{5}$ $\frac{2}{5}$

93. $\frac{3}{4} - \frac{1}{4}$ $\frac{1}{2}$

94. $\frac{7}{9} - \frac{4}{9}$ $\frac{1}{3}$

95. $\frac{12}{5} - \frac{2}{5}$ 2

96. $\frac{1}{8} - \frac{1}{8}$ 0

97. $\frac{2}{3}$ day $- \frac{1}{3}$ day $\frac{1}{3}$ day

98. $\$\frac{9}{10} - \$\frac{1}{10}$ $\$\frac{4}{5}$

99. $\frac{3}{4} - \frac{2}{3}$ $\frac{1}{12}$

100. $\frac{2}{5} - \frac{1}{6}$ $\frac{7}{30}$

101. $\dfrac{4}{9} - \dfrac{1}{6}$

$\dfrac{5}{18}$

102. $\dfrac{5}{8} - \dfrac{1}{6}$

$\dfrac{11}{24}$

103. $\dfrac{9}{10} - \dfrac{3}{100}$

$\dfrac{87}{100}$

104. $\dfrac{4}{5} - \dfrac{3}{4}$

$\dfrac{1}{20}$

105. $\dfrac{21}{100} - \dfrac{1}{10}$

$\dfrac{11}{100}$

106. $\dfrac{5}{6} - \dfrac{1}{8}$

$\dfrac{17}{24}$

107. $\dfrac{4}{7} - \dfrac{1}{2}$

$\dfrac{1}{14}$

108. $\dfrac{7}{10} - \dfrac{3}{100}$

$\dfrac{67}{100}$

109. $\dfrac{2}{5} - \dfrac{2}{9}$

$\dfrac{8}{45}$

110. $\dfrac{1}{2} - \dfrac{1}{4}$

$\dfrac{1}{4}$

111. $\dfrac{2}{3} - \dfrac{1}{6}$

$\dfrac{1}{2}$

112. $\dfrac{5}{8} - \dfrac{3}{10}$

$\dfrac{13}{40}$

113. $\dfrac{4}{9} - \dfrac{3}{8}$

$\dfrac{5}{72}$

114. $\dfrac{1}{2} - \dfrac{1}{5}$

$\dfrac{3}{10}$

115. $\dfrac{3}{4} - \dfrac{3}{5}$

$\dfrac{3}{20}$

116. $\dfrac{11}{12} - \dfrac{1}{3}$

$\dfrac{7}{12}$

117. $\dfrac{2}{11} - \dfrac{1}{6}$

$\dfrac{1}{66}$

118. $\dfrac{6}{8} - \dfrac{1}{2}$

$\dfrac{1}{4}$

119. $\dfrac{5}{6} - \dfrac{2}{3}$

$\dfrac{1}{6}$

120. $\dfrac{5}{10} - \dfrac{2}{5}$

$\dfrac{1}{10}$

121. $5\dfrac{3}{7} - 1\dfrac{1}{7}$

$4\dfrac{2}{7}$

122. $6\dfrac{2}{3} - 1\dfrac{1}{3}$

$5\dfrac{1}{3}$

123. $8\dfrac{4}{5} - 1\dfrac{2}{5}$

$7\dfrac{2}{5}$

124. $6\dfrac{9}{10} - 2\dfrac{2}{10}$

$4\dfrac{7}{10}$

125. $3\dfrac{7}{8} - 2\dfrac{1}{8}$

$1\dfrac{3}{4}$

126. $10\dfrac{5}{6} - 2\dfrac{5}{6}$

8

127. $20\dfrac{1}{2} - \dfrac{1}{2}$

20

128. $7\dfrac{3}{4} - \dfrac{1}{4}$

$7\dfrac{1}{2}$

129. $8\dfrac{1}{10} - 4$

$4\dfrac{1}{10}$

130. $5\dfrac{3}{8} - 1$

$4\dfrac{3}{8}$

131. $14\dfrac{1}{2} - 5$

$9\dfrac{1}{2}$

132. $7\dfrac{11}{12} - 3$

$4\dfrac{11}{12}$

133. $2\dfrac{1}{3} - 2$

$\dfrac{1}{3}$

134. $6\dfrac{9}{10} - 6$

$\dfrac{9}{10}$

135. $7\dfrac{5}{8} - 3$

$4\dfrac{5}{8}$

136. $11\dfrac{7}{8} - 10$

$1\dfrac{7}{8}$

137. $6 - 2\dfrac{2}{3}$

$3\dfrac{1}{3}$

138. $4 - 1\dfrac{1}{5}$

$2\dfrac{4}{5}$

139. $8 - 4\dfrac{7}{10}$

$3\dfrac{3}{10}$

140. $2 - 1\dfrac{1}{2}$

$\dfrac{1}{2}$

141. $10 - 3\frac{2}{3}$

$6\frac{1}{3}$

142. $5 - 4\frac{9}{10}$

$\frac{1}{10}$

143. $6 - \frac{1}{2}$

$5\frac{1}{2}$

144. $9 - \frac{3}{4}$

$8\frac{1}{4}$

145. $7\frac{1}{4} - 2\frac{3}{4}$

$4\frac{1}{2}$

146. $5\frac{1}{10} - 2\frac{3}{10}$

$2\frac{4}{5}$

147. $6\frac{1}{8} - 2\frac{7}{8}$

$3\frac{1}{4}$

148. $3\frac{1}{5} - 1\frac{4}{5}$

$1\frac{2}{5}$

149. $12\frac{2}{5} - 3\frac{3}{5}$

$8\frac{4}{5}$

150. $3\frac{7}{10} - \frac{9}{10}$

$2\frac{4}{5}$

151. $8\frac{1}{3} - 1\frac{2}{3}$

$6\frac{2}{3}$

152. $2\frac{1}{5} - \frac{4}{5}$

$1\frac{2}{5}$

153. $13\frac{1}{2} - 5\frac{2}{3}$

$7\frac{5}{6}$

154. $7\frac{1}{10} - 2\frac{1}{7}$

$4\frac{67}{70}$

155. $9\frac{3}{8} - 5\frac{5}{6}$

$3\frac{13}{24}$

156. $2\frac{1}{10} - 1\frac{27}{100}$

$\frac{83}{100}$

157. $20\frac{2}{9} - 4\frac{5}{6}$

$15\frac{7}{18}$

158. $9\frac{13}{100} - 6\frac{7}{10}$

$2\frac{43}{100}$

159. $3\frac{4}{5} - \frac{5}{6}$

$2\frac{29}{30}$

160. $1\frac{2}{8} - \frac{2}{6}$

$\frac{11}{12}$

161. $1\frac{3}{4}$ mi $- 1\frac{1}{2}$ mi

$\frac{1}{4}$ mi

162. $2\frac{1}{2}$ hr $- 1\frac{3}{4}$ hr

$\frac{3}{4}$ hr

163. $10\frac{1}{12}$ ft $- 4\frac{2}{3}$ ft

$5\frac{5}{12}$ ft

164. $\$7\frac{1}{4} - \$1\frac{5}{16}$

$\$5\frac{15}{16}$

165. $22\frac{7}{8}$ acres $- 8\frac{9}{10}$ acres

$13\frac{39}{40}$ acres

166. $9\frac{1}{10}$ ¢ $- 3\frac{1}{2}$ ¢

$5\frac{3}{5}$ ¢

167. $2\frac{1}{2}$ pints $- 1\frac{3}{4}$ pints

$\frac{3}{4}$ pint

168. $3\frac{1}{4}$ oz $- 2\frac{5}{16}$ oz

$\frac{15}{16}$ oz

Combine and simplify.

169. $\frac{5}{8} + \frac{9}{10} - \frac{1}{4}$ $1\frac{11}{40}$

170. $\frac{2}{3} - \frac{1}{5} + \frac{1}{2}$ $\frac{29}{30}$

171. $3\frac{1}{2} + 1\frac{1}{2} - \frac{1}{2}$ $4\frac{1}{2}$

172. $6\frac{9}{10} - 2\frac{1}{10} + \frac{1}{10}$ $4\frac{9}{10}$

173. $12\frac{1}{6} + 5\frac{9}{10} - 1\frac{3}{10}$ $16\frac{23}{30}$

174. $7\frac{1}{3} - 2\frac{4}{5} - 1\frac{1}{3}$ $3\frac{1}{5}$

175. $4\dfrac{1}{10} + 2\dfrac{9}{10} - 3\dfrac{3}{4}$ $\quad 3\dfrac{1}{4}$

176. $15\dfrac{1}{2} - 3\dfrac{4}{5} - 6\dfrac{1}{2}$ $\quad 5\dfrac{1}{5}$

177. $19\dfrac{1}{6} - 8\dfrac{9}{10} - \dfrac{1}{5}$ $\quad 10\dfrac{1}{15}$

178. $20\dfrac{1}{10} - \dfrac{1}{20} - 1\dfrac{1}{2}$ $\quad 18\dfrac{11}{20}$

Applications

Solve. Write your answer in simplest form.

179. You need a nail that will go through a $\dfrac{3}{4}$-in. door and stick out an extra $\dfrac{1}{8}$-in. How long should the nail be? $\frac{7}{8}$ in.

180. During an experiment, a chemist mixed $\dfrac{1}{2}$ oz of one solution with $\dfrac{2}{3}$ oz of another solution. What is the weight of the combined solution? $1\frac{1}{6}$ oz

181. The first game of a baseball doubleheader lasted $2\dfrac{1}{4}$ hr. The second game began after a $\dfrac{1}{4}$ hr break and lasted $2\dfrac{1}{2}$ hr. How long did the doubleheader take to play? 5 hr

182. Three candidates competed in an election. The winner got $\dfrac{5}{8}$ of the votes, and the second-place candidate got $\dfrac{1}{4}$ of the votes. If the rest of the votes went to the third candidate, what fraction of the votes did she get? (*Hint*: What would be the sum of the fractions of the votes for all three candidates?) $\frac{1}{8}$

183. Find the perimeter of the figure shown. $35\frac{3}{8}$ mi

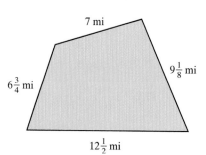

184. In the hallway pictured, how much greater is the length than the width? $5\frac{5}{6}$ ft

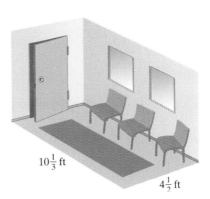

$10\frac{1}{3}$ ft

$4\frac{1}{2}$ ft

185. In testing a new drug, doctors found that $\frac{1}{2}$ of the patients given the drug improved, $\frac{2}{5}$ showed no change in their condition, and the remainder got worse. What fraction got worse? $\frac{1}{10}$

186. In a three-candidate election, one candidate received $\frac{1}{3}$ of the votes and another got $\frac{1}{2}$ of the votes. What fraction of the votes did the third candidate receive? $\frac{1}{6}$

187. Suppose that 4 packages are placed on a scale, as shown. If the scale balances, how heavy is the small package on the right? 1 lb

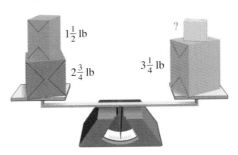

$1\frac{1}{2}$ lb

$2\frac{3}{4}$ lb

?

$3\frac{1}{4}$ lb

188. If the scale pictured balances, how heavy is the small package on the left? 2 lb

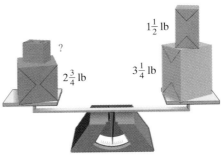

?

$2\frac{3}{4}$ lb

$1\frac{1}{2}$ lb

$3\frac{1}{4}$ lb

■ *Check your answers on page A-3.*

MINDSTRETCHERS

GROUPWORK

1. Working with a partner, complete the following magic square in which each row, column, and diagonal adds up to the same number.

$1\frac{1}{4}$	$\frac{2}{3}$	$1\frac{1}{12}$
$\frac{5}{6}$	1	$1\frac{1}{6}$
$\frac{11}{12}$	$1\frac{1}{3}$	$\frac{3}{4}$

MATHEMATICAL REASONING

2. A fraction with 1 as the numerator is called a **unit fraction**. For example, $\frac{1}{7}$ is a unit fraction. Write $\frac{3}{7}$ as the sum of three unit fractions, using no unit fraction more than once.

$$\frac{3}{7} = \frac{1}{28} + \frac{1}{7} + \frac{1}{4}$$

WRITING

3. Consider the following two ways of subtracting $2\frac{4}{5}$ from $4\frac{1}{5}$.

Method 1	**Method 2**

$$4\frac{1}{5} = 3 + \frac{5}{5} + \frac{1}{5} = \quad 3\frac{6}{5}$$
$$-2\frac{4}{5} \qquad\qquad\qquad = -2\frac{4}{5}$$
$$\overline{\qquad\qquad\qquad\qquad 1\frac{2}{5}}$$

$$4\frac{1}{5} = 4 + \frac{1}{5} + \frac{1}{5} = \quad 4\frac{2}{5}$$
$$-2\frac{4}{5} = 2 + \frac{4}{5} + \frac{1}{5} = -3$$
$$\overline{\qquad\qquad\qquad\qquad 1\frac{2}{5}}$$

a. Explain the difference between the two methods.   In Method 1 we "borrow" from the whole number in the minuend so that the fraction in the minuend is big enough to subtract the fraction in the subtrahend. In Method 2 we add a fraction to the subtrahend, making it a whole number. We add that same fraction to the minuend and then subtract.

b. Explain which method you prefer.   Answers may vary.

c. Explain why you prefer that method.   Answers may vary.

2.4 Multiplying and Dividing Fractions

OBJECTIVES

■ *To multiply and divide fractions and mixed numbers*

■ *To estimate products and quotients involving mixed numbers*

■ *To solve word problems involving the multiplication or division of fractions or mixed numbers*

This section begins with a discussion of multiplying fractions. We then move on to multiplying mixed numbers and conclude with dividing fractions and mixed numbers.

Multiplying Fractions

Many situations require us to multiply fractions. For instance, suppose that a mixture in a chemistry class calls for $\frac{4}{5}$ g of sodium chloride. If we make only $\frac{2}{3}$ of that mixture, we need $\frac{2}{3}$ of $\frac{4}{5}$, that is, $\frac{2}{3} \times \frac{4}{5}$ g of sodium chloride.

> Remind students that *of* generally means *times*. If they have some $5 bills and they give you 3 of them, their gift is worth 3 × $5, or $15.

To illustrate how to find this product, we diagram these two fractions.

$\frac{4}{5}$

$\frac{2}{3}$

In the following diagram, we are taking $\frac{2}{3}$ of the $\frac{4}{5}$.

> Point out that a good way to interpret the multiplication of fractions is taking *part of a part*.

Note that we divided the whole into 15 parts and that our product, containing 8 of the 15 small squares, represents the double-shaded region. The answer is therefore $\frac{8}{15}$ of the original whole, which we can compute as follows.

$$\frac{2}{3} \times \frac{4}{5} = \frac{8}{15}$$

The numerator and denominator of the answer are the products of the original numerators and denominators.

To Multiply Fractions

■ first multiply the numerators,

■ then multiply the denominators, and

■ finally write the answer in simplest form.

EXAMPLE 1

Multiply: $\dfrac{7}{8} \cdot \dfrac{9}{10}$

Solution

Multiply the numerators.

$$\dfrac{7}{8} \cdot \dfrac{9}{10} = \dfrac{7 \cdot 9}{8 \cdot 10} = \dfrac{63}{80}$$

Multiply the denominators.

PRACTICE 1

Find the product of $\dfrac{3}{4}$ and $\dfrac{5}{7}$. $\dfrac{15}{28}$

EXAMPLE 2

What is $\dfrac{3}{8}$ of 10?

Solution Finding $\dfrac{3}{8}$ of 10 means multiplying $\dfrac{3}{8}$ by 10.

$$\dfrac{3}{8} \times 10 = \dfrac{3}{8} \times \dfrac{10}{1} = \dfrac{3 \times 10}{8 \times 1} = \dfrac{30}{8} = \dfrac{15}{4}, \text{ or } 3\dfrac{3}{4}$$

PRACTICE 2

What is $\dfrac{2}{3}$ of 30?   20

> Emphasize that, when multiplying a whole number by a fraction, we first write the whole number as a fraction with a denominator of 1.

In Example 2, we multiplied the two fractions first and then simplified the answer. It is preferable, however, to reverse these steps: Simplify first and then multiply. By first simplifying, which is called *canceling*, we divide *any* numerator and *any* denominator by a common factor. Canceling before multiplying allows us to work with smaller numbers and still gives us the same answer.

EXAMPLE 3

Find the product of $\dfrac{4}{9}$ and $\dfrac{5}{8}$.

Solution Divide the numerator (4) and the denominator (8) by the same number (4).

$$\dfrac{4}{9} \times \dfrac{5}{8} = \dfrac{\overset{1}{4}}{9} \times \dfrac{5}{\underset{2}{8}} = \dfrac{1 \times 5}{9 \times 2} = \dfrac{5}{18}$$

Multiply the resulting fractions $\left(\dfrac{1}{9} \text{ and } \dfrac{5}{2}\right)$.

PRACTICE 3

Multiply: $\dfrac{7}{10} \cdot \dfrac{5}{11}$ $\dfrac{7}{22}$

> Show students that another way to solve this problem is to write the numerators and denominators as products of primes and then to simplify.

EXAMPLE 4

Multiply: $\dfrac{9}{8} \times \dfrac{6}{5} \times \dfrac{7}{9}$

Solution We cancel and then multiply.

$$\dfrac{9}{8} \times \dfrac{6}{5} \times \dfrac{7}{9} = \dfrac{\overset{1}{\cancel{9}}}{\underset{4}{\cancel{8}}} \times \dfrac{\overset{3}{\cancel{6}}}{5} \times \dfrac{7}{\underset{1}{\cancel{9}}} = \dfrac{21}{20}, \quad \text{or} \quad 1\dfrac{1}{20}$$

PRACTICE 4

Multiply: $\dfrac{7}{27} \cdot \dfrac{9}{4} \cdot \dfrac{8}{21}$ $\dfrac{2}{9}$

EXAMPLE 5

At a college, $\dfrac{3}{5}$ of the students take a math course. Of these students, $\dfrac{1}{6}$ take elementary algebra. What fraction of the students in the college take elementary algebra?

Solution We must find $\dfrac{1}{6}$ of $\dfrac{3}{5}$.

$$\dfrac{1}{6} \times \dfrac{3}{5} = \dfrac{1}{\underset{2}{\cancel{6}}} \times \dfrac{\overset{1}{\cancel{3}}}{5} = \dfrac{1 \times 1}{2 \times 5} = \dfrac{1}{10}$$

One-tenth of the students in the college take elementary algebra.

PRACTICE 5

A flight from New York to Los Angeles took 7 hr. With the help of the jet stream, the return trip took $\dfrac{3}{4}$ the time. How long did the trip from Los Angeles to New York take? $5\dfrac{1}{4}$ hr

EXAMPLE 6

Suppose that you spend $\dfrac{3}{8}$ of your monthly salary on rent. If your salary is $960, how much do you have left after paying the rent?

Solution Apply the strategy of breaking the question into two parts.

■ First, find $\dfrac{3}{8}$ of $960.

■ Then, subtract that result from $960.

Thus you can solve this problem by computing $960 - \left(\dfrac{3}{8} \times 960 \right)$.

$$960 - \left(\dfrac{3}{8} \times 960 \right) = 960 - \left(\dfrac{3}{\underset{1}{\cancel{8}}} \times \dfrac{\overset{120}{\cancel{960}}}{1} \right)$$

$$= 960 - 360 = 600$$

You have $600 left after paying the rent.

PRACTICE 6

Suppose that your aunt left one-fifth of her estate to each of her three children and the rest to a favorite charity. If her estate was valued at $95,000, how much went to charity? $38,000

> Show students that an alternative approach to solving this problem is to compute $\dfrac{5}{8}$ of 960.

Multiplying Mixed Numbers

Some situations require us to multiply mixed numbers. For instance, suppose that your regular hourly wage is $\$7\frac{1}{2}$ and that you make time-and-a-half for working overtime. To find your overtime hourly wage, you need to multiply $1\frac{1}{2}$ by $7\frac{1}{2}$. The key here is to first rewrite each mixed number as an improper fraction.

$$1\frac{1}{2} \times 7\frac{1}{2} = \frac{3}{2} \times \frac{15}{2} = \frac{45}{4}, \quad \text{or} \quad 11\frac{1}{4}$$

So you make $\$11\frac{1}{4}$ an hour overtime.

To Multiply Mixed Numbers
- change each mixed number to its equivalent improper fraction,
- follow the steps for multiplying fractions, and
- write the answer in simplest form.

EXAMPLE 7

Multiply $2\frac{1}{5}$ by $1\frac{1}{4}$.

Solution $2\frac{1}{5} \times 1\frac{1}{4} = \frac{11}{5} \times \frac{5}{4}$ First, rewrite each mixed number as an improper fraction.

$$= \frac{11 \times \overset{1}{\cancel{5}}}{\cancel{5} \times 4}$$ Then, cancel and multiply.

$$= \frac{11}{4}, \quad \text{or} \quad 2\frac{3}{4}$$ Finally, simplify the answer.

PRACTICE 7

Find the product of $3\frac{3}{4}$ and $2\frac{1}{10}$. $7\frac{7}{8}$

EXAMPLE 8

Multiply: $\left(4\frac{3}{8}\right)\left(4\right)\left(2\frac{2}{5}\right)$

Solution $\left(4\frac{3}{8}\right)\left(4\right)\left(2\frac{2}{5}\right) = \left(\frac{35}{8}\right)\left(\frac{4}{1}\right)\left(\frac{12}{5}\right)$

$$= \left(\frac{\overset{7}{\cancel{35}}}{\underset{2}{\cancel{8}}}\right)\left(\frac{\overset{1}{\cancel{4}}}{1}\right)\left(\frac{\overset{6}{\cancel{12}}}{\underset{1}{\cancel{5}}}\right) = 42$$

PRACTICE 8

Multiply: $\left(1\frac{3}{4}\right)\left(5\frac{1}{3}\right)\left(3\right)$ 28

Note in this problem that, although there are several ways to cancel, the answer always comes out the same.

EXAMPLE 9

A nurse gave a patient $2\frac{1}{2}$ tablets of the medication atropine sulfate. If each tablet contains $\frac{1}{30}$ grain (gr) of this medication, how much atropine sulfate did the patient receive?

Solution To find the total amount of atropine sulfate administered, we need to multiply $2\frac{1}{2}$ by $\frac{1}{30}$.

$$2\frac{1}{2} \times \frac{1}{30} = \frac{\overset{1}{\cancel{5}}}{2} \times \frac{1}{\underset{6}{\cancel{30}}} = \frac{1}{12}$$

Therefore the nurse gave the patient $\frac{1}{12}$ gr of atropine sulfate.

PRACTICE 9

A recipe for New England clam chowder calls for $2\frac{1}{4}$ lb butter. If you prepare $\frac{1}{12}$ of the quantity described in the recipe, how much butter will you need? $\frac{3}{16}$ lb

EXAMPLE 10

A lawn surrounding a garden is to be installed, as depicted in the following drawing.

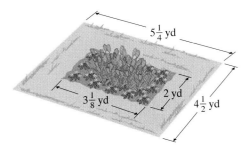

How many square yards of turf will we need to cover the lawn?

Solution Let's break this problem into three steps. First, we find the area of the rectangle with dimensions $5\frac{1}{4}$ yd and $4\frac{1}{2}$ yd. Then, we find the area of the small rectangle whose length and width are $3\frac{1}{8}$ yd and 2 yd, respectively. Finally, we subtract the area of the small rectangle from the area of the large rectangle.

Step 1. $5\frac{1}{4} \times 4\frac{1}{2} = \frac{21}{4} \times \frac{9}{2}$

$$= \frac{189}{8}, \quad \text{or} \quad 23\frac{5}{8}$$

The area of the large rectangle is $23\frac{5}{8}$ sq yd.

Step 2. $3\frac{1}{8} \times 2 = \frac{25}{8} \times \frac{2}{1}$

$$= \frac{25}{4}, \quad \text{or} \quad 6\frac{1}{4}$$

The area of the small rectangle is $6\frac{1}{4}$ sq yd.

continued

PRACTICE 10

How much greater is the area of a sheet of legal-size paper than a sheet of letter-size paper? $25\frac{1}{2}$ sq in.

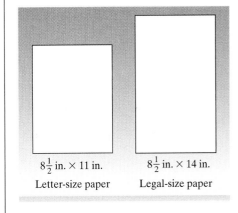

$8\frac{1}{2}$ in. × 11 in. $8\frac{1}{2}$ in. × 14 in.

Letter-size paper   Legal-size paper

Step 3.
$$23\frac{5}{8} = 23\frac{5}{8}$$
$$-6\frac{1}{4} = -6\frac{2}{8}$$
$$\overline{\qquad\qquad 17\frac{3}{8}}$$

The area of the lawn is therefore $17\frac{3}{8}$ sq yd.

We will need $17\frac{3}{8}$ sq yd of turf for the lawn.

Dividing Fractions

We now turn to quotients, beginning with dividing a fraction by a whole number. Suppose, for instance, that you want to share $\frac{1}{3}$ of a pizza with a friend, that is, to divide the $\frac{1}{3}$ into two equal parts. What part of the whole pizza will each of you receive?

This diagram shows $\frac{1}{3}$ of a pizza.

If we split the third into two equal parts, each part is $\frac{1}{6}$ of the pizza.

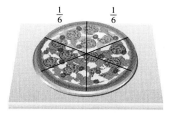

You and your friend will each get $\frac{1}{6}$ of the whole pizza, which you can compute as follows.

$$\frac{1}{3} \div 2 = \frac{1}{6}$$

Note that dividing a number by 2 is the same as taking $\frac{1}{2}$ of it. This equivalence suggests the procedure for dividing fractions shown on the next page.

Divisor

$$\frac{1}{3} \div 2 = \frac{1}{3} \div \frac{2}{1} = \frac{1}{3} \times \frac{1}{2} = \frac{1 \times 1}{3 \times 2} = \frac{1}{6}$$

$\frac{2}{1}$ and $\frac{1}{2}$ are reciprocals.

You may want to justify this procedure as follows.

$$\frac{\frac{1}{3}}{\frac{2}{1}} = \frac{\frac{1}{3} \cdot \frac{1}{2}}{\frac{2}{1} \cdot \frac{1}{2}} = \frac{1}{6}$$

This procedure involves *inverting*, or finding the *reciprocal* of the divisor. The reciprocal is found by switching the numerator and denominator.

To Divide Fractions

■ change the divisor to its reciprocal,
■ *multiply* the resulting fractions, and
■ simplify the answer.

EXAMPLE 11

Divide: $\frac{4}{5} \div \frac{3}{10}$

Solution $\frac{4}{5} \div \frac{3}{10} = \frac{4}{5} \times \frac{10}{3} = \frac{4 \times 2}{1 \times 3} = \frac{8}{3}$, or $2\frac{2}{3}$

$\frac{3}{10}$ and $\frac{10}{3}$ are reciprocals.

As in any division problem, we can check our answer by multiplying it by the divisor.

$$\frac{8}{3} \times \frac{3}{10} = \frac{4}{5}$$

Because $\frac{4}{5}$ is the dividend, we have confirmed our answer.

PRACTICE 11

Divide: $\frac{3}{4} \div \frac{1}{8}$ 6

TIP In a division problem, the fraction to the right of the division sign is the divisor. *Always invert the divisor*—the second fraction—not the dividend—the first fraction.

EXAMPLE 12

What is $\frac{4}{7}$ divided by 20?

Solution $\frac{4}{7} \div 20 = \frac{4}{7} \times \frac{1}{20} = \frac{1 \times 1}{7 \times 5} = \frac{1}{35}$

Invert $\frac{20}{1}$ and multiply.

PRACTICE 12

Compute the following quotient: $5 \div \frac{5}{8}$ 8

EXAMPLE 13

To stop the developing process, photographers use a chemical called stop bath. Suppose that a photographer needs $\frac{1}{4}$ bottle of stop bath for each roll of film. If the photographer has $\frac{2}{3}$ bottle of stop bath left, can he develop 3 rolls of film?

Solution We want to find out how many $\frac{1}{4}$'s there are in $\frac{2}{3}$, that is, to compute $\frac{2}{3} \div \frac{1}{4}$.

$$\frac{2}{3} \div \frac{1}{4} = \frac{2}{3} \times \frac{4}{1} = \frac{8}{3}, \quad \text{or} \quad 2\frac{2}{3}$$

Find the reciprocal of the divisor $\left(\frac{1}{4}\right)$ and then multiply.

So the photographer cannot develop 3 rolls of film.

PRACTICE 13

A house is built on ground that is sinking $\frac{3}{4}$ in./yr. How many years will it take the house to sink 2 in.? $\quad 2\frac{2}{3}$ yr

Dividing Mixed Numbers

Dividing mixed numbers is similar to dividing fractions, except that there is an additional step.

To Divide Mixed Numbers

■ change the mixed numbers to their equivalent improper fractions,
■ follow the steps for dividing fractions, and
■ simplify the answer.

EXAMPLE 14

Find $9 \div 2\frac{7}{10}$.

Solution $9 \div 2\frac{7}{10} = \frac{9}{1} \div \frac{27}{10}$ First, rewrite all mixed numbers as improper fractions.

$$= \frac{\overset{1}{9}}{1} \times \frac{10}{\underset{3}{27}}$$ Then, invert and multiply.

$$= \frac{10}{3}, \quad \text{or} \quad 3\frac{1}{3}$$

PRACTICE 14

Divide: $6 \div 3\frac{3}{4} \quad 1\frac{3}{5}$

EXAMPLE 15

What is $4\frac{1}{2} \div 2\frac{1}{2}$?

Solution $4\frac{1}{2} \div 2\frac{1}{2} = \frac{9}{2} \div \frac{5}{2} = \frac{9}{2} \times \frac{\overset{1}{2}}{5} = \frac{9}{5}$, or $1\frac{4}{5}$.

Invert and multiply.

PRACTICE 15

Divide $5\frac{3}{7}$ by $2\frac{3}{8}$. $2\frac{2}{7}$

EXAMPLE 16

There are $6\frac{3}{4}$ yd of silk in a roll. If it takes $\frac{3}{4}$ yd to make one designer tie, how many ties can be made from the roll?

Solution The question is: How many $\frac{3}{4}$'s fit into $6\frac{3}{4}$? It tells us that we must divide.

$6\frac{3}{4} \div \frac{3}{4} = \frac{\overset{9}{27}}{\underset{1}{4}} \times \frac{\overset{1}{4}}{\underset{1}{3}} = 9$

So 9 ties can be made from the roll of silk.

PRACTICE 16

According to a newspaper advertisement for a "diet shake," a man lost 33 lb in $5\frac{1}{2}$ months. How much weight did he lose per month?   6 lb

Estimating Products and Quotients of Mixed Numbers

As with adding or subtracting mixed numbers, it is important to check our answers when multiplying or dividing. We can check a product or a quotient of mixed numbers by estimating the answer and then confirming that our estimate and answer are reasonably close.

Checking a Product by Estimating

$2\frac{1}{5} \times 7\frac{2}{3} = \frac{11}{5} \times \frac{23}{3} = \frac{253}{15}$, or $16\frac{13}{15}$ Verify that the answer $\left(16\frac{13}{15}\right)$ and the estimate (16) are close.

$2 \times 8 = 16$

Because $16\frac{13}{15}$ is near 16, $16\frac{13}{15}$ is a reasonable answer.

Checking a Quotient by Estimating

$6\frac{1}{4} \div 2\frac{7}{10} = \frac{25}{4} \div \frac{27}{10} = \frac{25}{\underset{2}{4}} \times \frac{\overset{5}{10}}{27} = \frac{125}{54}$, or $2\frac{17}{54}$ Verify that the answer $\left(2\frac{17}{54}\right)$ and the estimate (2) are close.

$6 \div 3 = 2$

Because 2 is near $2\frac{17}{54}$, $2\frac{17}{54}$ is a reasonable answer.

EXAMPLE 17

Simplify and check: $3\frac{3}{4} \times 5\frac{1}{3} \div 2\frac{7}{9}$

Solution Following the order of operations rule, we work from left to right, multiplying the first two mixed numbers.

$$3\frac{3}{4} \times 5\frac{1}{3} = \frac{\overset{5}{\cancel{15}}}{\underset{1}{\cancel{4}}} \times \frac{\overset{4}{\cancel{16}}}{\underset{1}{\cancel{3}}} = 20$$

Then we divide 20 by $2\frac{7}{9}$ to get the answer.

$$20 \div 2\frac{7}{9} = \frac{20}{1} \div \frac{25}{9} = \frac{\overset{4}{\cancel{20}}}{1} \times \frac{9}{\underset{5}{\cancel{25}}} = \frac{36}{5}, \quad \text{or} \quad 7\frac{1}{5}$$

Now let's check by estimation.

$$3\frac{3}{4} \times 5\frac{1}{3} \div 2\frac{7}{9}$$
$$\downarrow \qquad \downarrow \qquad \downarrow$$
$$4 \times 5 \div 3 = 20 \div 3 \approx 7$$

Our answer, $7\frac{1}{5}$, and our estimate, 7, are reasonably close, confirming the answer.

PRACTICE 17

Compute: $5\frac{3}{5} \div 2\frac{1}{10} \times 2\frac{1}{4}$ 6

Exercises 2.4

Multiply.

1. $\dfrac{1}{3} \times \dfrac{2}{5}$

$\dfrac{2}{15}$

2. $8 \times \dfrac{1}{2}$

$\dfrac{7}{16}$

3. $\left(\dfrac{5}{8}\right)\left(\dfrac{2}{3}\right)$

$\dfrac{5}{12}$

4. $\left(\dfrac{3}{10}\right)\left(\dfrac{1}{4}\right)$

$\dfrac{3}{40}$

5. $\left(\dfrac{3}{4}\right)^2$

$\dfrac{9}{16}$

6. $\left(\dfrac{1}{8}\right)^2$

$\dfrac{1}{64}$

7. $\dfrac{4}{5} \times \dfrac{2}{5}$

$\dfrac{8}{25}$

8. $\dfrac{1}{2} \times \dfrac{3}{2}$

$\dfrac{3}{4}$

9. $\dfrac{7}{8} \times \dfrac{5}{4}$

$\dfrac{35}{32} = 1\dfrac{3}{32}$

10. $\dfrac{20}{3} \times \dfrac{2}{7}$

$\dfrac{40}{21} = 1\dfrac{19}{21}$

11. $\dfrac{5}{2} \cdot \dfrac{9}{8}$

$\dfrac{45}{16} = 2\dfrac{13}{16}$

12. $\dfrac{11}{10} \cdot \dfrac{9}{5}$

$\dfrac{99}{50} = 1\dfrac{49}{50}$

13. $\left(\dfrac{2}{5}\right)\left(\dfrac{5}{9}\right)$

$\dfrac{2}{9}$

14. $\left(\dfrac{4}{5}\right)\left(\dfrac{1}{4}\right)$

$\dfrac{1}{5}$

15. $\dfrac{7}{9} \times \dfrac{3}{4}$

$\dfrac{7}{12}$

16. $\dfrac{4}{5} \times \dfrac{1}{2}$

$\dfrac{2}{5}$

17. $\left(\dfrac{1}{8}\right)\left(\dfrac{6}{10}\right)$

$\dfrac{3}{40}$

18. $\left(\dfrac{4}{6}\right)\left(\dfrac{3}{8}\right)$

$\dfrac{1}{4}$

19. $\dfrac{10}{9} \times \dfrac{93}{100}$

$\dfrac{31}{30} = 1\dfrac{1}{30}$

20. $\dfrac{12}{5} \times \dfrac{15}{4}$

9

21. $5 \cdot \dfrac{1}{3}$

$\dfrac{5}{3} = 1\dfrac{2}{3}$

22. $\dfrac{1}{5} \cdot 6$

$\dfrac{6}{5} = 1\dfrac{1}{5}$

23. $\dfrac{2}{5} \times 7$

$\dfrac{14}{5} = 2\dfrac{4}{5}$

24. $\dfrac{3}{4} \times 9$

$\dfrac{27}{4} = 6\dfrac{3}{4}$

25. $\dfrac{2}{3} \times 20$

$\dfrac{40}{3} = 13\dfrac{1}{3}$

26. $\dfrac{5}{6} \times 5$

$\dfrac{25}{6} = 4\dfrac{1}{6}$

27. $\left(\dfrac{10}{3}\right)(4)$

$\dfrac{40}{3} = 13\dfrac{1}{3}$

28. $\dfrac{5}{3} \times 8$

$\dfrac{40}{3} = 13\dfrac{1}{3}$

29. $\dfrac{1}{2} \times 8$

4

30. $\dfrac{3}{4} \times 12$

9

31. $\dfrac{7}{8} \cdot 10$

$\dfrac{35}{4} = 8\dfrac{3}{4}$

32. $100 \cdot \dfrac{2}{5}$

40

33. $18 \cdot \dfrac{2}{9}$

4

34. $20 \cdot \dfrac{2}{5}$

8

35. $\dfrac{2}{3} \times 6$

4

36. $\dfrac{1}{6} \times 9$

$\dfrac{3}{2} = 1\dfrac{1}{2}$

37. $\dfrac{2}{5} \cdot 1\dfrac{1}{3}$

$\dfrac{8}{15}$

38. $5\dfrac{1}{2} \cdot \dfrac{1}{3}$

$1\dfrac{5}{6}$

39. $\dfrac{1}{3} \times 1\dfrac{1}{3}$

$\dfrac{4}{9}$

40. $5\dfrac{5}{6} \times \dfrac{1}{2}$

$2\dfrac{11}{12}$

41. $\left(\dfrac{7}{8}\right)\left(1\dfrac{1}{2}\right)$

$1\dfrac{5}{16}$

42. $\left(4\dfrac{1}{3}\right)\left(\dfrac{1}{5}\right)$

$\dfrac{13}{15}$

43. $\dfrac{1}{4} \times 8\dfrac{1}{2}$

$2\dfrac{1}{8}$

44. $\dfrac{1}{3} \times 2\dfrac{1}{5}$

$\dfrac{11}{15}$

45. $\left(\dfrac{5}{6}\right)\left(1\dfrac{1}{9}\right)$

$\dfrac{25}{27}$

46. $\left(\dfrac{9}{10}\right)\left(2\dfrac{1}{7}\right)$

$1\dfrac{13}{14}$

47. $\dfrac{1}{2} \times 5\dfrac{1}{3}$

$2\dfrac{2}{3}$

48. $4\dfrac{1}{2} \times \dfrac{2}{3}$

3

49. $\dfrac{4}{5} \cdot 1\dfrac{1}{4}$

1

50. $\dfrac{3}{8} \cdot 5\dfrac{1}{3}$

2

51. $\left(\dfrac{3}{16}\right)\left(4\dfrac{2}{3}\right)$

$\dfrac{7}{8}$

52. $\left(\dfrac{7}{9}\right)\left(2\dfrac{1}{4}\right)$

$1\dfrac{3}{4}$

53. $1\dfrac{1}{7} \times 1\dfrac{1}{5}$

$1\dfrac{13}{35}$

54. $2\dfrac{1}{3} \times 1\dfrac{1}{2}$

$3\dfrac{1}{2}$

55. $\left(2\dfrac{1}{10}\right)^2$

$4\dfrac{41}{100}$

56. $\left(1\dfrac{1}{2}\right)^2$

$2\dfrac{1}{4}$

57. $3\dfrac{9}{10} \cdot 2$

$7\dfrac{4}{5}$

58. $5 \cdot 1\dfrac{1}{2}$

$7\dfrac{1}{2}$

59. $100 \times 3\dfrac{3}{4}$

375

60. $1\dfrac{5}{6} \times 20$

$36\dfrac{2}{3}$

61. $1\dfrac{1}{2} \times 5\dfrac{1}{3}$

8

62. $5\dfrac{1}{4} \times 1\dfrac{1}{9}$

$5\dfrac{5}{6}$

63. $\left(2\dfrac{1}{2}\right)\left(1\dfrac{1}{5}\right)$

3

64. $\left(1\dfrac{3}{10}\right)\left(2\dfrac{4}{9}\right)$

$3\dfrac{8}{45}$

65. $12\dfrac{1}{2} \cdot 3\dfrac{1}{3}$

$41\dfrac{2}{3}$

66. $5\dfrac{1}{10} \cdot 1\dfrac{2}{3}$

$8\dfrac{1}{2}$

67. $66\dfrac{2}{3} \times 1\dfrac{7}{10}$

$113\dfrac{1}{3}$

68. $37\dfrac{1}{2} \times 1\dfrac{3}{5}$

60

69. $1\dfrac{5}{9} \times \dfrac{3}{8} \times 2$

$1\dfrac{1}{6}$

70. $\dfrac{1}{8} \times 2\dfrac{1}{4} \times 6$

$1\dfrac{11}{16}$

71. $\left(\dfrac{1}{2}\right)^2\left(2\dfrac{1}{3}\right)$

$\dfrac{7}{12}$

72. $\left(1\dfrac{1}{4}\right)^2\left(\dfrac{1}{5}\right)$

$\dfrac{5}{16}$

73. $\dfrac{4}{5} \times \dfrac{7}{8} \times 1\dfrac{1}{10}$

$\dfrac{77}{100}$

74. $8\dfrac{1}{3} \times \dfrac{3}{10} \times \dfrac{5}{6}$

$2\dfrac{1}{12}$

75. $\left(1\dfrac{1}{2}\right)^3$

$3\dfrac{3}{8}$

76. $\left(2\dfrac{1}{2}\right)^3$

$15\dfrac{5}{8}$

Divide.

Point out that changing the order of the fractions in a division problem changes the answer to its reciprocal.

77. $\dfrac{3}{5} \div \dfrac{2}{3}$

$\dfrac{9}{10}$

78. $\dfrac{2}{3} \div \dfrac{3}{5}$

$1\dfrac{1}{9}$

79. $\dfrac{4}{5} \div \dfrac{7}{8}$

$\dfrac{32}{35}$

80. $\dfrac{7}{8} \div \dfrac{4}{5}$

$1\dfrac{3}{32}$

81. $\dfrac{1}{2} \div \dfrac{1}{7}$

$3\dfrac{1}{2}$

82. $\dfrac{1}{7} \div \dfrac{1}{2}$

$\dfrac{2}{7}$

83. $\dfrac{5}{9} \div \dfrac{1}{8}$

$4\dfrac{4}{9}$

84. $\dfrac{1}{8} \div \dfrac{5}{9}$

$\dfrac{9}{40}$

85. $\dfrac{4}{5} \div \dfrac{8}{15}$

$1\dfrac{1}{2}$

86. $\dfrac{3}{10} \div \dfrac{6}{5}$

$\dfrac{1}{4}$

87. $\dfrac{7}{8} \div \dfrac{3}{8}$

$2\dfrac{1}{3}$

88. $\dfrac{10}{3} \div \dfrac{5}{6}$

4

89. $\dfrac{9}{10} \div \dfrac{3}{4}$

$1\dfrac{1}{5}$

90. $\dfrac{5}{6} \div \dfrac{1}{3}$

$2\dfrac{1}{2}$

91. $\dfrac{1}{10} \div \dfrac{2}{5}$

$\dfrac{1}{4}$

92. $\dfrac{3}{4} \div \dfrac{6}{5}$

$\dfrac{5}{8}$

93. $\dfrac{2}{3} \div 7$

$\dfrac{2}{21}$

94. $\dfrac{1}{2} \div 5$

$\dfrac{1}{10}$

95. $\dfrac{1}{5} \div 6$

$\dfrac{1}{30}$

96. $\dfrac{2}{11} \div 2$

$\dfrac{1}{11}$

97. $\dfrac{3}{5} \div 6$

$\dfrac{1}{10}$

98. $\dfrac{7}{10} \div 10$

$\dfrac{7}{100}$

99. $\dfrac{2}{3} \div 6$

$\dfrac{1}{9}$

100. $\dfrac{1}{20} \div 2$

$\dfrac{1}{40}$

101. $8 \div \dfrac{1}{5}$

40

102. $8 \div \dfrac{2}{9}$

36

103. $7 \div \dfrac{3}{7}$

$16\dfrac{1}{3}$

104. $10 \div \dfrac{2}{5}$

25

105. $4 \div \dfrac{3}{10}$

$13\dfrac{1}{3}$

106. $10 \div \dfrac{2}{3}$

15

107. $1 \div \dfrac{1}{7}$

7

108. $3 \div \dfrac{1}{8}$

24

109. $2\dfrac{5}{6} \div \dfrac{3}{7}$

$6\dfrac{11}{18}$

110. $5\dfrac{1}{9} \div \dfrac{2}{3}$

$7\dfrac{2}{3}$

111. $1\dfrac{1}{3} \div \dfrac{4}{5}$

$1\dfrac{2}{3}$

112. $7\dfrac{1}{10} \div \dfrac{1}{2}$

$14\dfrac{1}{5}$

Remind students that dividing a large number by a small number gives an answer greater than 1 and that dividing a small number by a large number gives an answer less than 1.

113. $8\dfrac{5}{6} \div \dfrac{9}{10}$

$9\dfrac{22}{27}$

114. $6\dfrac{1}{2} \div \dfrac{1}{2}$

13

115. $20\dfrac{1}{10} \div \dfrac{1}{5}$

$100\dfrac{1}{2}$

116. $15\dfrac{2}{3} \div \dfrac{5}{6}$

$18\dfrac{4}{5}$

117. $\dfrac{1}{6} \div 2\dfrac{1}{7}$

$\dfrac{7}{90}$

118. $\dfrac{2}{7} \div 1\dfrac{1}{3}$

$\dfrac{3}{14}$

119. $\dfrac{1}{2} \div 2\dfrac{3}{5}$

$\dfrac{5}{26}$

120. $\dfrac{3}{4} \div 3\dfrac{1}{9}$

$\dfrac{27}{112}$

121. $4 \div 1\dfrac{1}{4}$

$3\dfrac{1}{5}$

122. $7 \div 1\dfrac{9}{10}$

$3\dfrac{13}{19}$

123. $2\dfrac{1}{10} \div 20$

$\dfrac{21}{200}$

124. $5\dfrac{6}{7} \div 14$

$\dfrac{41}{98}$

125. $2\dfrac{1}{2} \div 3\dfrac{1}{7}$ **126.** $3\dfrac{1}{7} \div 2\dfrac{1}{2}$ **127.** $8\dfrac{1}{10} \div 5\dfrac{3}{4}$ **128.** $1\dfrac{7}{10} \div 5\dfrac{1}{8}$

$\dfrac{35}{44}$ $1\dfrac{9}{35}$ $1\dfrac{47}{115}$ $\dfrac{68}{205}$

129. $2\dfrac{1}{3} \div 4\dfrac{1}{2}$ **130.** $8\dfrac{1}{6} \div 2\dfrac{1}{2}$ **131.** $6\dfrac{3}{8} \div 2\dfrac{5}{6}$ **132.** $1\dfrac{2}{3} \div 1\dfrac{2}{5}$

$\dfrac{14}{27}$ $3\dfrac{4}{15}$ $2\dfrac{1}{4}$ $1\dfrac{4}{21}$

Simplify.

133. $\dfrac{2}{3} \times 1\dfrac{1}{3} + \dfrac{1}{2}$ **134.** $\dfrac{4}{5} \cdot 8 + \dfrac{9}{10}$ **135.** $5 - \dfrac{1}{3} \times \dfrac{2}{5}$ **136.** $3 \div \dfrac{2}{5} - 2\dfrac{1}{3}$

$1\dfrac{7}{18}$ $7\dfrac{3}{10}$ $4\dfrac{13}{15}$ $5\dfrac{1}{6}$

137. $\dfrac{1}{5} + 2\dfrac{3}{4} \times \dfrac{1}{8}$ **138.** $\dfrac{3}{8} \cdot \dfrac{1}{2} - \dfrac{1}{10}$ **139.** $4 \times \dfrac{2}{9} \div \dfrac{3}{4}$ **140.** $6 \times \dfrac{1}{4} \div 5$

$\dfrac{87}{160}$ $\dfrac{7}{80}$ $1\dfrac{5}{27}$ $\dfrac{3}{10}$

141. $3\dfrac{1}{2} \times 6 \div 5$ **142.** $4 \cdot \dfrac{2}{3} - 1\dfrac{1}{8}$ **143.** $10 \times \dfrac{1}{8} \times 2\dfrac{1}{2}$ **144.** $\dfrac{1}{3} \div \dfrac{1}{6} \times \dfrac{2}{3}$

$4\dfrac{1}{5}$ $1\dfrac{13}{24}$ $3\dfrac{1}{8}$ $1\dfrac{1}{3}$

Applications

Solve and check.

145. Most days, you swim $\dfrac{3}{4}$ mi. This morning, you only swam $\dfrac{1}{2}$ that distance. How far did you swim this morning? $\frac{3}{8}$ mi

146. On a test, you answered $\dfrac{4}{5}$ of the questions. Of the questions that you answered, you got $\dfrac{3}{4}$ correct. What fraction of the questions on the test did you get right? $\frac{3}{5}$

147. The house that you want to buy is selling for \$160,000. You need to put $\dfrac{1}{20}$ of the selling price down, and take out a mortgage for the rest. How much money do you need to put down?
\$8,000

148. There is a rule of thumb that you should spend no more than $\dfrac{1}{4}$ of your income on rent. If you make \$24,000 a year, what is the most you should spend per month on rent according to this rule?   \$500

149. You learn in an astronomy course that a first-magnitude star is $2\frac{1}{2}$ times as bright as a second-magnitude star, which in turn is $2\frac{1}{2}$ times as bright as a third-magnitude star. How many times as bright as a third-magnitude star is a first-magnitude star? $6\frac{1}{4}$

150. Which of these rooms has the larger area? The room that measures $15\frac{1}{2}$ ft by 12 ft.

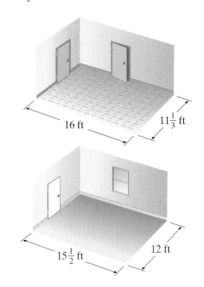

16 ft $11\frac{1}{3}$ ft

$15\frac{1}{2}$ ft   12 ft

151. Find the cost of buying carpeting at $\$7\frac{1}{2}$/sq ft for the hallway shown. $\$191\frac{1}{4}$

$8\frac{1}{2}$ ft

3 ft

152. Some people believe that gasohol is superior to plain gasoline as an automotive fuel. Gasohol is a mixture of gasoline $\left(\frac{9}{10}\right)$ and ethyl alcohol $\left(\frac{1}{10}\right)$. How much gasoline is there in $10\frac{1}{2}$ gal of gasohol? $9\frac{9}{20}$ gal

153. You own $1,000 worth of stocks and $1,000 worth of bonds. If the value of your stocks declines by $\frac{1}{4}$ and the value of your bonds declines by $\frac{1}{8}$, what is the total value of your stocks and bonds? $750 stocks, $875 bonds

154. A scientist is investigating the effect of coldness on human skin. In one of the scientist's experiments, the temperature starts at 70°F and drops by $\frac{1}{10}°$ every 2 min. What is the temperature after 6 min? $69\frac{7}{10}°$F

155. A trip to a nearby island takes $3\frac{1}{2}$ hr by boat and $\frac{1}{2}$ hr by airplane. How many times as fast as the boat is the plane?   7 times

156. Each dose of aspirin weighs $\frac{3}{4}$ gr. If you have 9 gr of aspirin on hand, how many doses can you administer?   12 doses

157. You develop your own film. Each roll takes $\frac{1}{8}$ bottle of developer solution. If you have a bottle of solution that is $\frac{3}{4}$ full, how many rolls of film can you develop?
6 rolls

158. A university fund-raising campaign collected $500 million in $3\frac{1}{2}$ yr. What was the average amount collected per year? $142\frac{6}{7}$ million

■ *Check your answers on page A-4.*

MINDSTRETCHERS

WRITING

1. Every number except 0 has a reciprocal. Explain why 0 does not have a reciprocal.
The reciprocal of 0 would have to be $\frac{1}{0}$, which does not exist. The product of any number and its reciprocal is 1, but 0 times any number is 0.

GROUPWORK

2. In the following magic square, the *product* of every row, column, and diagonal is 1. Working with a partner, complete the square.

3	$\frac{1}{6}$	2
$\frac{2}{3}$	1	$1\frac{1}{2}$
$\frac{1}{2}$	6	$\frac{1}{3}$

PATTERNS

3. Find the product: $1\frac{1}{2} \cdot 1\frac{1}{3} \cdot 1\frac{1}{4} \cdots 1\frac{1}{99} \cdot 1\frac{1}{100}$  $\frac{101}{2} = 50\frac{1}{2}$

KEY CONCEPTS and SKILLS

☐ = CONCEPT ☐ = SKILL

CONCEPT/SKILL	DESCRIPTION	EXAMPLES
[2.1] Prime number	A whole number that has exactly two different factors: itself and 1.	2, 3, 5
[2.1] Composite number	A whole number that has more than two factors. (The numbers 0 and 1 are neither prime nor composite.)	4, 8, 9
[2.1] Prime factorization	A whole number written as the product of its prime factors.	$30 = 2 \cdot 3 \cdot 5$
[2.1] Least common multiple (LCM)	The smallest nonzero number that is a multiple of two or more given numbers.	The LCM of 30 and 45 is 90.
[2.1] To compute the least common multiple (LCM)	■ Find the prime factorization of each number. ■ Identify the prime factors that appear in each factorization. ■ Multiply these prime factors, using each factor the greatest number of times that it occurs in any of the factorizations.	$20 = 2 \cdot 2 \cdot 5$ $= 2^2 \cdot 5$ $30 = 2 \cdot 3 \cdot 5$ The LCM of 20 and 30 is $2^2 \cdot 3 \cdot 5$, or 60.
[2.2] Fraction	Any number that can be written in the form $\frac{a}{b}$, where a and b are whole numbers and b is not zero.	$\frac{3}{11}, \frac{9}{5}$
[2.2] Proper fraction	A fraction less than 1, that is, a fraction whose numerator is smaller than its denominator.	$\frac{2}{7}, \frac{1}{2}$
[2.2] Mixed number	A number greater than 1 with a whole number part and a fractional part.	$5\frac{1}{3}, 4\frac{5}{6}$
[2.2] Improper fraction	A fraction greater than or equal to 1, that is, a fraction whose numerator is larger than or equal to its denominator.	$\frac{9}{4}, \frac{13}{5}$
[2.2] To change a mixed number to an improper fraction	■ Multiply the denominator of the fraction by the whole number portion of the mixed number. ■ Add the numerator of the fraction to that product. ■ Write the sum over the original denominator to form the improper fraction.	$4\frac{2}{3} = \frac{3 \times 4 + 2}{3}$ $= \frac{14}{3}$
[2.2] To change an improper fraction to a mixed number	■ Divide the numerator by the denominator. ■ If there is a remainder, write it over the original denominator.	$\frac{14}{3} = 4\frac{2}{3}$
[2.2] To find an equivalent fraction	Multiply the numerator and denominator by the same whole number; that is $\frac{a}{b} = \frac{a \cdot n}{b \cdot n}$, where neither b nor n is 0.	$\frac{3}{4} = \frac{3 \cdot 2}{4 \cdot 2} = \frac{6}{8}$
[2.2] To simplify a fraction	Divide its numerator and denominator by the same number; that is, $\frac{a}{b} = \frac{a \div n}{b \div n}$, where neither b nor n is 0.	$\frac{6}{8} = \frac{6 \div 2}{8 \div 2} = \frac{3}{4}$
[2.2] Like fractions	Fractions having the same denominator.	$\frac{2}{5}, \frac{3}{5}$

continued

CONCEPT/SKILL	DESCRIPTION	EXAMPLES
[2.2] Unlike fractions	Fractions having different denominators.	$\dfrac{3}{5}, \dfrac{3}{10}$
[2.2] To compare fractions	■ For like fractions, compare their numerators. ■ For unlike fractions, express them as equivalent fractions having the same denominator and then compare their numerators.	$\dfrac{6}{8}, \dfrac{7}{8} \quad 7 > 6$ $\dfrac{2}{3}, \dfrac{12}{15}$ or $\dfrac{10}{15}, \dfrac{12}{15} \quad 12 > 10$
[2.2] Least common denominator (LCD)	The least common multiple of the denominators of a given set of fractions.	The LDC of $\dfrac{11}{30}$ and $\dfrac{7}{45}$ is 90.
[2.3] To add (or subtract) like fractions	■ Add (or subtract) the numerators. ■ Use the given denominator. ■ Write the answer in simplest form.	$\dfrac{1}{8} + \dfrac{1}{8} = \dfrac{2}{8} = \dfrac{1}{4}$ $\dfrac{3}{8} - \dfrac{1}{8} = \dfrac{2}{8} = \dfrac{1}{4}$
[2.3] To add (or subtract) unlike fractions	■ Rewrite the fractions as equivalent fractions with a common denominator, usually the LCD. ■ Add (or subtract) the numerators, keeping the same denominator. ■ Write the answer in simplest form.	$\dfrac{2}{3} + \dfrac{1}{2} = \dfrac{4}{6} + \dfrac{3}{6}$ $= \dfrac{7}{6},$ or $1\dfrac{1}{6}$ $\dfrac{5}{12} - \dfrac{1}{6}$ $= \dfrac{5}{12} - \dfrac{2}{12} = \dfrac{3}{12},$ or $\dfrac{1}{4}$
[2.3] To add (or subtract) mixed numbers	■ Rewrite the fractions as equivalent fractions with a common denominator, usually the LCD. ■ When subtracting, rename (or borrow from) the whole number on top if the fraction on the bottom is larger than the fraction on top. ■ Add (or subtract) the numerators, keeping the same denominator. ■ Add (or subtract) the whole numbers. ■ Write the answer in simplest form.	$2\dfrac{1}{3}$ $+6\dfrac{2}{3}$ $\overline{8\dfrac{3}{3} = 9}$ $4\dfrac{1}{5} = \quad 3\dfrac{6}{5}$ $-1\dfrac{2}{5} = -1\dfrac{2}{5}$ $\overline{\phantom{-1\dfrac{2}{5} =} 2\dfrac{4}{5}}$
[2.4] To multiply fractions	■ Multiply the numerators. ■ Multiply the denominators. ■ Write the answer in simplest form.	$\dfrac{1}{2} \cdot \dfrac{3}{5} = \dfrac{3}{10}$

CONCEPT/SKILL	DESCRIPTION	EXAMPLES
[2.4] To multiply mixed numbers	■ Change each mixed number to its equivalent improper fraction. ■ Follow the steps for multiplying fractions.	$2\frac{1}{2} \cdot 1\frac{2}{3} = \frac{5}{2} \cdot \frac{5}{3}$ $= \frac{25}{6}$, or $4\frac{1}{6}$
[2.4] Reciprocal	The fraction $\frac{b}{a}$ formed by switching the numerator and denominator of the fraction $\frac{a}{b}$.	The reciprocal of $\frac{4}{3}$ is $\frac{3}{4}$.
[2.4] To divide fractions	■ Change the divisor to its reciprocal, and *multiply* the resulting fractions. ■ Simplify the answer.	$\frac{2}{5} \div \frac{3}{7} = \frac{2}{5} \cdot \frac{7}{3}$ $= \frac{14}{15}$
[2.4] To divide mixed numbers	■ Change the mixed numbers to their equivalent improper fractions. ■ Follow the steps for dividing fractions.	$2\frac{1}{2} \div 1\frac{1}{3} =$ $\frac{5}{2} \div \frac{4}{3} =$ $\frac{5}{2} \cdot \frac{3}{4} = \frac{15}{8}$, or $1\frac{7}{8}$

HISTORICAL NOTE

In societies throughout the world and across the centuries, people have written fractions in strikingly different ways. In ancient Greece, for example, the fraction $\frac{1}{4}$ was written Δ'' where Δ (read "delta") is the fourth letter of the Greek alphabet.

At one time, people wrote the numerator and denominator of fractions in Roman numerals, as shown at the left in a page from a sixteenth-century German book. In today's notation, the last fraction is $\frac{200}{460}$.

Source: David Eugene Smith and Jekuthiel Ginsburg, *Numbers and Numerals, A Story Book for Young and Old* (New York: Bureau of Publications, Teachers College, Columbia University, 1937).

Chapter 2 Review Exercises

To help you review this chapter, solve these problems.

[2.1]

Find all the factors of each number.

1. 150   1, 2, 3, 5, 6, 10,
15, 25, 30, 50, 75, 150

2. 180   1, 2, 3, 4, 5, 6, 9, 10,
12, 15, 18, 20, 30, 36, 45, 60,
90, 180

3. 57   1, 3, 19, 57

4. 72   1, 2,
3, 4, 6, 8, 9, 12
18, 24, 36, 72

Indicate whether each number is prime or composite.

5. 23
Prime

6. 33
Composite

7. 87
Composite

8. 67
Prime

Write the prime factorization of each number, using exponents.

9. 36
$2^2 \times 3^2$

10. 75
3×5^2

11. 200
$2^3 \times 5^2$

12. 54
2×3^3

Find the LCM.

13. 6 and 8
24

14. 5 and 10
10

15. 24 and 36
72

16. 15 and 20
60

[2.2]

In each diagram, identify the fraction that the shaded portion represents.

17. $\dfrac{1}{3}$

18. $\dfrac{2}{4} = \dfrac{1}{2}$

19. $\dfrac{3}{6} = \dfrac{1}{2}$

20.  $\dfrac{6}{12} = \dfrac{1}{2}$

Write each mixed number as an improper fraction.

21. $7\dfrac{2}{3}$
$\dfrac{23}{3}$

22. $1\dfrac{4}{5}$
$\dfrac{9}{5}$

23. $9\dfrac{1}{10}$
$\dfrac{91}{10}$

24. $8\dfrac{3}{7}$
$\dfrac{59}{7}$

Write each improper fraction as a mixed number.

25. $\dfrac{13}{2}$

$6\dfrac{1}{2}$

26. $\dfrac{14}{3}$

$4\dfrac{2}{3}$

27. $\dfrac{11}{4}$

$2\dfrac{3}{4}$

28. $\dfrac{23}{5}$

$4\dfrac{3}{5}$

Find n and check.

29. $\dfrac{4}{5} = \dfrac{n}{15}$

$n = 12$

30. $\dfrac{2}{7} = \dfrac{n}{14}$

$n = 4$

31. $\dfrac{1}{2} = \dfrac{n}{10}$

$n = 5$

32. $\dfrac{9}{10} = \dfrac{n}{30}$

$n = 27$

Simplify.

33. $\dfrac{14}{28}$

$\dfrac{1}{2}$

34. $\dfrac{15}{21}$

$\dfrac{5}{7}$

35. $\dfrac{30}{45}$

$\dfrac{2}{3}$

36. $\dfrac{54}{72}$

$\dfrac{3}{4}$

37. $5\dfrac{2}{4}$

$5\dfrac{1}{2}$

38. $8\dfrac{10}{15}$

$8\dfrac{2}{3}$

39. $6\dfrac{12}{42}$

$6\dfrac{2}{7}$

40. $8\dfrac{45}{63}$

$8\dfrac{5}{7}$

Insert the appropriate sign: $<$, $=$, or $>$.

41. $\dfrac{5}{8} \;>\; \dfrac{3}{8}$

42. $\dfrac{5}{6} \;>\; \dfrac{1}{6}$

43. $\dfrac{2}{3} \;<\; \dfrac{4}{5}$

44. $\dfrac{9}{10} \;>\; \dfrac{7}{8}$

45. $\dfrac{3}{4} \;>\; \dfrac{5}{8}$

46. $\dfrac{7}{10} \;>\; \dfrac{5}{9}$

47. $3\dfrac{1}{5} \;>\; 1\dfrac{9}{10}$

48. $5\dfrac{1}{8} \;>\; 5\dfrac{1}{9}$

Arrange in increasing order.

49. $\dfrac{2}{7}, \dfrac{3}{8}, \dfrac{1}{2}$

$\dfrac{2}{7}, \dfrac{3}{8}, \dfrac{1}{2}$

50. $\dfrac{1}{5}, \dfrac{1}{3}, \dfrac{2}{15}$

$\dfrac{2}{15}, \dfrac{1}{5}, \dfrac{1}{3}$

51. $\dfrac{4}{5}, \dfrac{9}{10}, \dfrac{3}{4}$

$\dfrac{3}{4}, \dfrac{4}{5}, \dfrac{9}{10}$

52. $\dfrac{7}{8}, \dfrac{7}{9}, \dfrac{13}{18}$

$\dfrac{13}{18}, \dfrac{7}{9}, \dfrac{7}{8}$

[2.3]

Add and simplify.

53. $\dfrac{1}{9} + \dfrac{4}{9}$

$\dfrac{5}{9}$

54. $\dfrac{7}{8} + \dfrac{3}{8}$

$\dfrac{10}{8} = 1\dfrac{1}{4}$

55. $\dfrac{7}{10} + \dfrac{4}{10}$

$\dfrac{11}{10} = 1\dfrac{1}{10}$

117. $1\frac{1}{3} \cdot 4\frac{1}{2}$

6

118. $3\frac{1}{4} \cdot 5\frac{2}{3}$

$18\frac{5}{12}$

119. $6\frac{3}{4} \times 1\frac{1}{4}$

$8\frac{7}{16}$

120. $8\frac{1}{2} \times 2\frac{1}{2}$

$21\frac{1}{4}$

121. $\frac{7}{8} \times 1\frac{1}{5} \times \frac{3}{7}$

$\frac{9}{20}$

122. $1\frac{3}{8} \times \frac{10}{11} \times 1\frac{1}{4}$

$1\frac{9}{16}$

123. $\left(3\frac{1}{3}\right)^3$

$37\frac{1}{27}$

124. $\left(1\frac{1}{2}\right)^3$

$3\frac{3}{8}$

Find the reciprocal.

125. $\frac{2}{3}$

$\frac{3}{2}$

126. $1\frac{1}{2}$

$\frac{2}{3}$

127. 8

$\frac{1}{8}$

128. $\frac{1}{4}$

4

Divide and simplify.

129. $\frac{7}{8} \div 5$

$\frac{7}{40}$

130. $\frac{5}{9} \div 9$

$\frac{5}{81}$

131. $\frac{2}{3} \div 5$

$\frac{2}{15}$

132. $\frac{1}{100} \div 2$

$\frac{1}{200}$

133. $\frac{1}{2} \div \frac{2}{3}$

$\frac{3}{4}$

> Have students note that the answers to Exercises 133 and 134 are different. Also point out that neither solution involves cancellation.

134. $\frac{2}{3} \div \frac{1}{2}$

$1\frac{1}{3}$

135. $6 \div \frac{1}{5}$

30

136. $7 \div \frac{4}{5}$

$8\frac{3}{4}$

137. $\frac{7}{8} \div \frac{3}{4}$

$1\frac{1}{6}$

138. $\frac{9}{10} \div \frac{1}{2}$

$1\frac{4}{5}$

139. $\frac{3}{5} \div \frac{3}{10}$

2

140. $\frac{2}{3} \div \frac{1}{6}$

4

141. $3\frac{1}{2} \div 2$

$1\frac{3}{4}$

142. $2 \div 3\frac{1}{2}$

$\frac{4}{7}$

143. $6\frac{1}{3} \div 4$

$1\frac{7}{12}$

144. $4 \div 6\frac{1}{3}$

$\frac{12}{19}$

145. $8\frac{1}{4} \div 1\frac{1}{2}$

$5\frac{1}{2}$

146. $3\frac{2}{5} \div 1\frac{1}{3}$

$2\frac{11}{20}$

147. $4\frac{1}{2} \div 2\frac{1}{4}$

2

148. $7\frac{1}{5} \div 2\frac{2}{5}$

3

Mixed Applications

Solve. Write your answer in simplest form.

149. The Summer Olympic Games are held during each year divisible by 4. Were the Olympic Games held in 1990?   No

150. What is the smallest amount of money that you can pay in both all quarters and all dimes?   50¢

151. Eight of the 32 human teeth are incisors. What fraction of human teeth are incisors? (*Source:* Ilsa Goldsmith, *Human Anatomy for Children*) $\frac{1}{4}$

152. The planets in the solar system consist of Earth, two planets closer to the sun than Earth, and six planets farther from the sun than Earth. What fraction of the planets in the solar system are closer than Earth to the sun? (*Source:* Patrick Moore, *Astronomy for the Beginner*) $\frac{2}{9}$

153. Of the 70,000 people who started a march, 10,000 finished. What fraction of the marchers finished? $\frac{1}{7}$

154. The President of the United States makes an annual salary of $200,000 in contrast to the Vice President who makes $181,400. Is the Vice President's salary more or less than 9/10 of the President's salary? (*Source:* World Almanac 2000)   More

155. A Filmworks camera has a shutter speed of $\dfrac{1}{8,000}$ sec, and a Lensmax camera has a shutter speed of $\dfrac{1}{6,000}$ sec. Which shutter is faster? (Hint: The faster shutter has the smaller shutter speed.)   The Filmworks camera

156. It is estimated that 1 of every 10 people in the world are left-handed. Is more or less than half the population left-handed?   Less

157. An insurance company reimbursed you $275 on a dental bill of $700. Did you get more or less than $\dfrac{1}{3}$ of your money back? Explain.   You got back more than $\frac{1}{3}$ because $\frac{275}{700} = \frac{11}{28} = \frac{33}{84}$ which is greater than $\frac{1}{3} = \frac{28}{84}$.

158. A union goes on strike if at least $\dfrac{2}{3}$ of the workers voting support the strike call. If 23 of the 32 voting workers support a strike, should a strike be declared? Explain.   Yes it should because $\frac{23}{32}$ is greater than $\frac{2}{3}$. $\frac{23}{32} = \frac{69}{96}$ whereas $\frac{2}{3} = \frac{64}{96}$

159. You are serving on a grand jury with 23 jurors. Sixteen jurors are needed for a quorum, and a vote of 12 jurors is needed to indict.

a. What fraction of the full jury is needed to indict? $\frac{12}{23}$

b. Suppose that 16 jurors are present. What fraction is needed to indict? $\frac{3}{4}$

160. In a tennis match, Alisa Gregory went to the net 12 times, winning the point 7 times. By contrast, Monica Yates won the point 4 of the 6 times that she went to the net.

a. Which player went to the net more often?
Alisa Gregory

b. Which player had a better rate of winning points at the net? Monica Yates

161. The following chart is a record of the amount of time (in hours) that two employees spent working the past weekend. Complete the chart.

Employee	Saturday	Sunday	Total
L. Chavis	$7\frac{1}{2}$	$4\frac{1}{4}$	$11\frac{3}{4}$
M. Young	$5\frac{3}{4}$	$6\frac{1}{2}$	$12\frac{1}{4}$
Total	$13\frac{1}{4}$	$10\frac{3}{4}$	24

162. Complete the following chart.

Worker	Hours per Day	Days Worked	Total Hours	Wage per Hour	Gross Pay
Scott	5	3	15	$7	$105.00
Janis	$7\frac{1}{4}$	4	29	$10	$290.00
Erwin	$4\frac{1}{2}$	$5\frac{1}{2}$	$24\frac{3}{4}$	$9	$222.75

163. The opening value todays of a share of stock in the Block Corporation was $\$5\frac{15}{16}$. The closing value was $\$4\frac{7}{8}$. What was the change in value? $1\frac{1}{16}$

164. You are driving to a town 15 mi away. If you have already driven $3\frac{1}{4}$ mi, how far away is the town?
$11\frac{3}{4}$ mi

165. A sea otter eats about $\frac{1}{5}$ of its body weight each day. How much will a 35-lb otter eat in a day?
(*Source:* Karl W. Kenyon, *The Sea Otter in the Eastern Pacific Ocean*) 7 lb

166. At a florist's, you buy a dozen roses for your mother on Mother's Day. The regular price of roses is $27 a dozen, but the florist has a "one-third-off" sale. How much do you pay for the roses? $18

167. You are on an airplane flying $1\frac{1}{2}$ times the speed of sound. If sound travels at about 1,000 fps, at what speed is the plane flying?   1,500 fps

168. On each page of a report, there are about 200 words. If the report is $10\frac{1}{2}$ pages long, how many words does it contain?   2,100

169. A rule of thumb for planting lily bulbs is to plant them 3 times as deep as they are wide. How deep should you plant a lily bulb that is $2\frac{1}{2}$ in. wide?
$7\frac{1}{2}$ in.

170. When you stand upright, the pressure per square inch on your hip joint is about $2\frac{1}{2}$ times your body weight. If you weigh 200 lb, what is that pressure?
500 lb/sq in.

171. In Roseville, 40 of every 1,000 people who want to work are unemployed, in contrast to 8 of every 100 people in Georgetown. How many times as great is the unemployment rate in Georgetown compared to that in Roseville?   2 times

172. You want to buy as many goldfish as possible for your new fish tank. A rule of thumb is that the total length of fish, in inches, should be less than the capacity of the tank in gallons. If you have a 10-gal tank and goldfish average $\frac{1}{2}$ in. in length, how many fish should you buy?   19

173. According to a newspaper advertisement, a man on a diet lost 60 lb in $5\frac{1}{2}$ mo. How much weight did he lose per month? $10\frac{10}{11}$ lb

174. A metal alloy is made by combining $\frac{1}{4}$ oz of copper with $\frac{2}{3}$ oz of tin. Find the alloy's total weight.
$\frac{11}{12}$ oz

■ *Check your answers on page A-4.*

Chapter 2 Posttest

To see if you have already mastered the topics in this chapter, take this test.

1. List all the factor of 75.   1, 3, 5, 15, 25, 75

2. Write 60 as the product of prime factors. $2^2 \times 3 \times 5$

3. The diagram illustrates what fraction? $\frac{4}{9}$

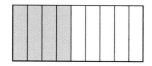

4. Write 12 as an improper fraction. $\frac{12}{1}$

5. Express $\dfrac{41}{4}$ as a mixed number. $10\frac{1}{4}$

6. Write $\dfrac{875}{1,000}$ in simplest form. $\frac{7}{8}$

7. Which is smaller, $\dfrac{2}{3}$ or $\dfrac{5}{10}$? $\frac{5}{10}$

8. What is the LCD for $\dfrac{3}{8}$ and $\dfrac{1}{12}$?   24

Add and simplify.

9. $\dfrac{2}{3} + \dfrac{1}{8} + \dfrac{3}{4}$ $1\frac{13}{24}$

10. $6\dfrac{7}{8} + 1\dfrac{3}{10}$ $8\frac{7}{40}$

Subtract and simplify.

11. $6 - 1\dfrac{5}{7}$ $4\frac{2}{7}$

12. $10\dfrac{1}{6} - 4\dfrac{2}{5}$ $5\frac{23}{30}$

Multiply and simplify.

13. $\left(\dfrac{1}{9}\right)^2$ $\frac{1}{81}$

14. $2\dfrac{2}{3} \times 4\dfrac{1}{2}$ 12

Divide and simplify.

15. $2\dfrac{1}{3} \div 3$ $\frac{7}{9}$

16. $3\dfrac{3}{8} \div 2\dfrac{1}{4}$ $1\frac{1}{2}$

Identify the place occupied by the underlined digit.

57. 25.7̲1   Tenths   **58.** 3.00̲2   Thousandths

59. 8.18̲3   Hundredths   **60.** 4̲9.771   Tens

61. 1,077.04̲2 Thousandths   **62.** 8,145.7̲5 Tenths

63. $253.7̲2   Ones   **64.** $7,571.3̲9 Hundredths

Between each pair of numbers, insert the appropriate sign, <, =, or >, to make a true statement.

65. 3.21 > 2.5   **66.** 8.66 > 4.952

67. 0.71 < 0.8   **68.** 1.2 < 1.38

69. 9.123 > 9.11   **70.** 0.5 < 0.52

71. 4 = 4.000   **72.** 7.60 = 7.6

73. 8.125 ft < 8.2 ft   **74.** 2.45 lb < 2.5 lb

Rearrange each group of numbers from smallest to largest.

75. 7.1, 7, 7.07   7, 7.07, 7.1   **76.** 0.002, 0.2, 0.02   0.002, 0.02, 0.2

77. 5.001, 4.9, 5.2   4.9, 5.001, 5.2   **78.** 3.85, 3.911, 2   2, 3.85, 3.911

79. 9.6 mi, 9.1 mi, 9.38 mi   **80.** 2.7 sec, 2.15 sec, 2 sec
9.1 mi, 9.38 mi, 9.6 mi   2 sec, 2.15 sec, 2.7 sec

Solve.

81. The jury awarded the plaintiff $1.85 million. Was this award more or less than $2.1 million?   Less

82. The more powerful an earthquake is, the higher its magnitude is on the Richter scale. Great earthquakes, such as the 1906 San Francisco earthquake, have magnitudes of 8.0 or higher. Is an earthquake with magnitude 7.8 considered to be a great earthquake? (**Source:** *New Encyclopedia Britannica*)   No

83. Last winter, your average daily heating bill was for 8.75 units of electricity. This winter, it was for 8.5 units. During which winter was the average higher?   Last winter

Solve. Write your answer in simplest form.

17. In your math class there are 12 men and 24 women. What fraction of the students in the class are men? $\frac{1}{3}$

18. Suppose that $\dfrac{2}{3}$ of a lemon pie is to be shared equally among 4 friends. How much will each friend get?
$\frac{1}{6}$ of the pie

19. Find the area of the rectangular floor pictured.
$90\frac{2}{3}$ sq ft

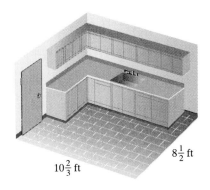

$10\frac{2}{3}$ ft $8\frac{1}{2}$ ft

20. Find the missing dimension in the diagram shown.
$3\frac{1}{3}$ in.

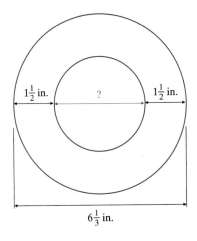

$1\frac{1}{2}$ in.   ? $1\frac{1}{2}$ in.

$6\frac{1}{3}$ in.

■ *Check your answers on page A-4.*

Cumulative Review Exercises

To help you review, solve the following:

1. Write in words: 5,000,315   Five million, three hundred, fifteen

2. Multiply: 5,814 × 100   581,400

3. Find the quotient: 89)80,812   908

4. Write $\frac{75}{100}$ in simplest form. $\frac{3}{4}$

5. Subtract: $8 - 1\frac{3}{5}$ $6\frac{2}{5}$

6. Find the sum: $1\frac{1}{2} + 4\frac{2}{3}$ $6\frac{1}{6}$

7. Which is larger, $\frac{1}{4}$, or $\frac{3}{8}$? $\frac{3}{8}$

8. A jury decided on punishments in an oil spill case. The jury ordered the captain of the oil barge to pay $5,000 in punitive damages and the oil company to pay $5 billion. The amount that the company had to pay is how many times the amount that the captain had to pay?   1 million times

9. The counter on a duplicating machine keeps track of the number of copies made on the machine. How many copies did you make if the counter showed 23,459 copies before you started duplicating and 24,008 after you finished?   549

10. You have two candles. A scented candle is 8 in. tall and burns $\frac{1}{2}$ in./hr. An unscented candle is 10 in. tall and burns $\frac{1}{3}$ in./hr. Which candle will last longer?
The unscented candle; $\frac{1}{2} > \frac{1}{3}$

■ *Check your answers on page A-4.*

Decimals

DECIMALS AND BLOOD TESTS

Blood tests reveal a great deal about a person's health—whether to reduce the cholesterol level to lower the risk of heart disease, or raise the red blood cell count to prevent anemia. And blood tests identify a variety of diseases, including AIDS or mononucleosis.

Blood analyses are typically carried out in clinical laboratories. Technicians in these labs run giant machines which perform thousands of blood tests per hour.

What these blood tests, known as "chemistries," actually do is to analyze blood for a variety of substances such as calcium or creatinine.

In any blood test, doctors look for abnormal levels of the substance being measured. For instance, the normal range on the creatinine test is typically from 0.7 to 1.5 milligrams (mg) per unit of blood. A high level may mean kidney disease; a low level, muscular dystrophy.

The normal range on the calcium test may be 9.0 to 10.5 mg per unit of blood. A result outside this range is a clue for any of several diseases.

Chapter 3 Pretest

To see if you have already mastered the topics in this chapter, take this test.

1. In the number 27.081, what place does the 8 occupy?
 Hundredths

2. Write in words: 4.012   Four and twelve thousandths

3. Round 3.079 to the nearest tenth.   3.1

4. Which is largest: 0.00212, 0.0029, or 0.000888?
 0.0029

Evaluate.

5. 7.02 + 3.5 + 11   21.52

6. 2.37 + 5.0038   7.3738

7. 13.79 − 2.1   11.69

8. 9 − 2.7 + 3.51   9.81

9. 8.3 × 1,000   8,300

10. 8.01 × 2.3   18.423

11. $(0.12)^2$ 0.0144

12. 5 + (3 × 0.7)   7.1

13. 6.05 ÷ 1,000   0.00605

14. $\dfrac{9.81}{0.3}$ 32.7

Express as a decimal.

15. $\dfrac{7}{8}$ 0.875

16. $2\dfrac{5}{6}$, rounded to the nearest hundredth.   2.83

Solve.

17. You learn in a science course that an acid is stronger if it has a lower "pH value." Which is stronger, an acid with a pH value of 3.7 or one with a pH value of 2.95?
 One with a pH value of 2.95

18. During the last four quarters, a corporation made profits of $3.7 million, $1.8 million, $2 million, and $0.5 million. What was the total profit for the year?
 $8 million

19. A serving of iceberg lettuce contains 3.6 mg of Vitamin C, whereas romaine lettuce contains 11.9 mg of Vitamin C. How many times as rich in Vitamin C is romaine as iceberg lettuce? Round your answer to the nearest whole number. (*Source: The Concise Encyclopedia of Foods and Nutrition*)   3 times

20. Suppose that a long distance telephone call costs $0.85 for the first 3 min and $0.17 for each additional minute. What is the cost of a 20-min call? $3.74

■ *Check your answers on page A-4.*

162

3.1 Introduction to Decimals

OBJECTIVES

- ■ *To read and write decimals*
- ■ *To find the fraction equivalent to a decimal*
- ■ *To compare decimals*
- ■ *To round decimals*
- ■ *To solve word problems using decimals*

What Decimals Are and Why They Are Important

Decimal notation is in common use. When we say that the price of a book is $32.75, that the length of a table is 1.8 m, or that the answer displayed on a calculator is 5.007, we are using decimals.

A number written as a **decimal** has

- ■ a whole number part, which precedes the decimal point, and

- ■ a fractional part, which follows the decimal point.

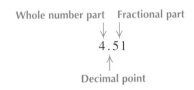

A decimal without a decimal point shown is understood to have one at the right end and is the same as a whole number. For instance, 3 and 3. are the same number.

The fractional part of any decimal is special in that its denominator is a power of 10, such as 10, 100, or 1,000. The use of the word *decimal* reminds us of the importance of the number 10 in this notation, just as decade means 10 years or December meant the 10th month of the year (which it was for the early Romans).

> Use this opportunity to review the meaning of powers of ten.

In many problems, we can choose to work with either decimals or fractions. Therefore we need to know how to work with both if we are to use the easier approach to solve a particular problem.

Decimal Places

Each digit in a decimal has a place value. The place value system for decimals is an extension of the place value system for whole numbers.

The places to the right of the decimal point are called **decimal places**. For instance, the number 64.149 is said to have three decimal places.

Recall that, for a whole number, place values are powers of ten: 1, 10, 100, ... By contrast, each place value for the fractional part of a decimal is the reciprocal of a power of ten:

$$\frac{1}{10}, \frac{1}{100}, \frac{1}{1,000}, \dots .$$

The first decimal place after the decimal point is the tenths place. Working to the right, the next decimal places are the hundredths place, the thousandths place, and so on.

> Tell students that going from left to right, the place values get smaller.
>
> $$1,000 \quad 100 \quad 10 \quad 1 \quad \frac{1}{10} \quad \frac{1}{100} \quad \frac{1}{1,000}$$
>
> Smaller

The following table shows the place values in the decimals 0.54 and 0.30716.

Ones	·	Tenths	Hundredths	Thousandths	Ten thousandths	Hundred thousandths
0	·	5	4			
0	·	3	0	7	1	6

The next table shows the place values for the decimals 7,204.5 and 513.285.

Thousands	Hundreds	Tens	Ones	·	Tenths	Hundredths	Thousandths
7	2	0	4	·	5		
	5	1	3	·	2	8	5

EXAMPLE 1

In each decimal, underline the digit that occupies the tenths place.

a. 0.7   **b.** 0.25   **c.** 3.09   **d.** 0.6947   **e.** 33.901

Solution For any decimal, the *first place* to the right of the decimal point is the tenths place.

a. 0.7   **b.** 0.25   **c.** 3.09   **d.** 0.6947   **e.** 33.901

PRACTICE 1

Underline the digit in the hundredths place:

a. 0.36   **b.** 0.472   **c.** 0.0251

d. 897.43   **e.** 1,912.643

Changing Decimals to Fractions

Knowing the place value system is the key to understanding what decimals mean, how to read them, and how to write them.

■ The decimal 0.9 is just another way of writing $(0 \times 1) + \left(9 \times \dfrac{1}{10}\right)$, or $\dfrac{9}{10}$. This decimal is read the same as the equivalent fraction: "nine tenths."

■ The decimal 0.21 represents 2 tenths + 1 hundredth. This expression simplifies to the following.

$$\left(2 \times \frac{1}{10}\right) + \left(1 \times \frac{1}{100}\right) = \frac{2}{10} + \frac{1}{100} = \frac{20}{100} + \frac{1}{100}, \quad \text{or} \quad \frac{21}{100}$$

So 0.21 is read "twenty-one hundredths."

■ The decimal 0.149 stands for $\dfrac{149}{1,000}$.

$$\left(1 \times \frac{1}{10}\right) + \left(4 \times \frac{1}{100}\right) + \left(9 \times \frac{1}{1,000}\right) = \frac{149}{1,000}$$

So 0.149 is read "one hundred forty-nine thousandths."

Let's summarize these examples.

Point out that in each example the number of decimal places equals the number of zeros in the equivalent fraction's denominator.

3 decimal places

$$0.\overset{\frown}{149} = \frac{149}{1,000}$$

3 zeros

Decimal	Equivalent Fraction	Read as
0.9	$\dfrac{9}{10}$	Nine tenths
0.21	$\dfrac{21}{100}$	Twenty-one hundredths
0.149	$\dfrac{149}{1,000}$	One hundred forty-nine thousandths

Note that in each of these decimals, the fractional part is the same as the numerator of the equivalent fraction: $0.149 = \dfrac{149}{1,000}$.

Now we know how to rewrite any decimal as a fraction or a mixed number.

> **To Change a Decimal to the Equivalent Fraction or Mixed Number**
> - copy the nonzero whole number part of the decimal and drop the decimal point,
> - place the fractional part of the decimal in the numerator of the equivalent fraction, and
> - make the denominator of the equivalent fraction the same as the place value of the rightmost digit.

EXAMPLE 2

Express 1.87 as a fraction.

Solution This decimal is equivalent to a mixed number. The whole number part is 1. The fractional part (87) of the decimal is the numerator of the equivalent fraction. The decimal has two decimal places, so the fraction's denominator has two zeros (that is, it is 100).

$$1.87 = 1\frac{87}{100}$$

Do you see that the answer can also be written as $\dfrac{187}{100}$?

PRACTICE 2

The decimal 2.03 is equivalent to what fraction? $2\dfrac{3}{100}$

> Have students explain their responses in a journal.

EXAMPLE 3

What is the fractional equivalent of 0.75?

Solution We can write 0.75 as $\dfrac{75}{100}$, which simplifies to $\dfrac{3}{4}$.

PRACTICE 3

Write 0.875 as a fraction reduced to lowest terms. $\dfrac{7}{8}$

EXAMPLE 4

Find the equivalent fraction of each decimal.

a. 3.2   **b.** 3.200

Solution

a. 3.2 represents $3\dfrac{2}{10}$, or $3\dfrac{1}{5}$.

b. 3.200 equals $3\dfrac{200}{1,000}$, or $3\dfrac{1}{5}$.

PRACTICE 4

Express each decimal in fractional form.

a. 5.6 $5\dfrac{3}{5}$ **b.** 5.6000 $5\dfrac{3}{5}$

> **TIP** As in Example 4, adding zeros in the rightmost decimal places does not change a decimal's value. However, we generally write decimals without these extra zeros.

EXAMPLE 5

Write each decimal as a fraction.

a. 1.309   **b.** 1.39

Solution

a. $1.309 = 1\dfrac{309}{1,000}$

b. $1.39 = 1\dfrac{39}{100}$

PRACTICE 5

What fraction is equivalent to each decimal?

a. 7.006 $7\dfrac{3}{500}$ **b.** 7.6 $7\dfrac{3}{5}$

> Point out to students that
> 1.309 ≠ 1.39, showing that we
> cannot drop any 0 that serves as a
> placeholder.

Knowing how to change a decimal to its equivalent fraction also helps us read the decimal.

EXAMPLE 6

Write each decimal in words.

a. 0.319   **b.** 2.71   **c.** 0.08

Solution

a. $0.319 = \dfrac{319}{1,000}$. We read the decimal as "three hundred nineteen thousandths."

b. $2.71 = 2\dfrac{71}{100}$. We read the decimal as "two and seventy-one hundredths."

c. $0.08 = \dfrac{8}{100}$. We read the decimal as "eight hundredths." Note that we *do not simplify* the equivalent fraction when reading the decimal.

PRACTICE 6

Write each decimal in words.

a. 0.61
Sixty-one
hundredths

b. 4.923
Four and nine
hundred
twenty-three
thousandths

c. 7.05
Seven and five
hundredths

EXAMPLE 7

Write each number in decimal notation.

a. Seven tenths   **b.** Five and thirty-two thousandths

Solution

a. Since 7 is in the tenths place, the decimal is written as 0.7.

b.
The whole number preceding *and* is in the ones place. ⌐

The last digit of 32 is in the thousandths place. ↓

5 . 0 3 2

We replace *and* with the decimal point.

We need a 0 to hold the tenths place.

The answer is 5.032.

PRACTICE 7

Write each number in decimal notation.

a. Forty-three thousandths   0.043

b. Ten and twenty-six hundredths   10.26

> Point out that another way to set
> up this problem is to write
> _ . _ _ _ and then to fill in the
> blanks.

EXAMPLE 8

For hay fever, you take a decongestant pill weighing three hundredths of a gram. Write the equivalent decimal.

Solution "Three hundredths" is written 0.03, with the digit 3 in the hundredths place.

PRACTICE 8

The number pi (usually written π) is approximately three and fourteen hundredths. Write the equivalent decimal.   3.14

Comparing Decimals

Suppose that we want to compare two decimals—say, 0.6 and 0.7. The key is to rethink the problem in terms of fractions.

$$0.6 = \frac{6}{10} \qquad 0.7 = \frac{7}{10}$$

> Students may need to review how to compare like fractions.

Because $\frac{7}{10} > \frac{6}{10}$, $0.7 > 0.6$.

There is another way to compare decimals that is faster than converting the decimals to fractions.

> **To Compare Decimals**
> ■ rewrite the numbers vertically, lining up the decimal points;
> ■ working from left to right, compare the digits that have the same place value; and
> ■ identify the decimal with the largest digit in a given place as the largest decimal.

EXAMPLE 9

Which is larger, 0.729 or 0.75?

Solution First let's line up the decimal point.

$$\begin{array}{c} \downarrow \\ 0.729 \\ 0.75 \\ \uparrow \end{array}$$

We see that both decimals have a 0 in the ones place. We next compare the digits in the tenths place and see that, again, they are the same. Looking to the right in the hundredths place, we see that $5 > 2$. Therefore $0.75 > 0.729$. Note that the decimal with more digits is not necessarily the larger decimal.

PRACTICE 9

Compare: 0.83 and 0.8297   0.83 is larger.

EXAMPLE 10

The dial on an X-ray machine controls emissions that vary in time from 0.066 sec to 3.2 sec. Which emission time is longer?

Solution To compare the decimals, line them up vertically.

$$\begin{array}{c} \downarrow \\ 0.066 \\ 3.2 \\ \uparrow \end{array}$$

Working from left to right, we compare the digits in the ones place. We see that $3 > 0$, so 3.2 sec is a longer emission than 0.066 sec.

PRACTICE 10

The planet Venus has gravity 0.9 times that of Earth, and Uranus has gravity 1.2 times that of Earth. Which planet has the stronger gravity—Venus or Uranus? (*Source:* The Diagram Group, *Comparisons*)   Uranus

EXAMPLE 11

Rank from smallest to largest: 1.17, 1.2, and 0.99

Solution First, we line up the decimals.

$$\downarrow$$
$$1.17$$
$$1.2$$
$$0.99$$
$$\uparrow$$

In the ones place, the first two decimals have a 1 and the third decimal has a 0, so the third decimal is the smallest. To compare the first two decimals, we look at the tenths place: $2 > 1$, so $1.2 > 1.17$. Therefore the three decimals in increasing order are 0.99, 1.17, and 1.2.

PRACTICE 11

Rewrite in decreasing order: 3.5, 2.9, and 3.8   3.8, 3.5, 2.9

EXAMPLE 12

Plastic garbage bags come in three gauges (that is, thicknesses): 0.003 in., 0.0025 in., and 0.002 in. The three gauges are called lightweight, regular weight, and heavyweight. Which is the lightweight gauge?

Solution To find the smallest of the decimals, we first line up the decimal points.

$$\downarrow$$
$$0.003$$
$$0.0025$$
$$0.002$$
$$\uparrow$$

Working from left to right, we see that the three decimals have the same digits until the thousandths place, where $3 > 2$. Therefore 0.003 must be the heavyweight gauge. To compare 0.0025 and 0.002, we look at the ten thousandths place. The 5 is greater than the 0 that is understood to be there. So 0.0025 in. must be the regular-weight gauge, and 0.002 the lightweight gauge.

PRACTICE 12

The higher the energy efficiency rating (EER) of an air conditioner, the more efficiently it uses electricity. Which is least efficient, an air conditioner with a rating of 8.2, one with a rating of 9, or one with a rating of 8.1? (*Source: Consumer Guide Magazine*)   The one with the rating of 8.1, because $9 > 8.2 > 8.1$.

Rounding Decimals

Review with students the rounding of whole numbers.

As with whole numbers, we can round decimals to a given place value. For instance, suppose that we want to round the decimal 1.58 to the nearest tenth. The decimal 1.58 lies between 1.5 and 1.6, so one of these two numbers will be our answer—but which? To decide, let's take a look at a number line.

Do you see from this diagram that 1.58 is closer to 1.6 than to 1.5?

$$1.\underline{5}8 \approx 1.6$$

└─ Tenths place

Rounding a decimal to the nearest tenth means that the last digit lies in the tenths place.

The following table shows the relationship between the place to which we are rounding and the number of decimal places in our answer.

Rounding to the nearest	means that the rounded decimal has
tenth $\left(\dfrac{1}{10}\right)$	one decimal place.
hundredth $\left(\dfrac{1}{100}\right)$	two decimal places.
thousandth $\left(\dfrac{1}{1,000}\right)$	three decimal places.
ten thousandth $\left(\dfrac{1}{10,000}\right)$	four decimal places.

Note that the number of decimal places is the same as the number of zeros in the corresponding denominator.

The following shortcut makes rounding decimals easy.

To Round a Decimal to a Given Decimal Place
- underline the place to which you are rounding;
- look at the digit to the right of the underlined digit—the critical digit; if this digit is 5 or more, add 1 to the underlined digit; if it is less than 5, leave the underlined digit unchanged; and
- drop all decimal places to the right of the underlined digit.

Let's apply this method to the problem that we just considered—namely, rounding 1.58 to the nearest tenth.

⌐ Tenths place
↓
1.5̲8   Underline the digit 5, which occupies the tenths place.

1.5̲8 ≈ 1.6   The critical digit, 8, is 5 or more, so add 1 to the 5 and
      ↑   then drop all digits to its right.
Critical digit ⌐

The following examples illustrate this method of rounding.

EXAMPLE 13

Round 94.735 to

a. the nearest tenth.   **b.** two decimal places.

c. the nearest thousandth.   **d.** the nearest ten.

e. the nearest whole number.

PRACTICE 13

Round 748.0772 to

a. the nearest tenth.   748.1

b. the nearest hundredth.   748.08

c. three decimal places.   748.077

d. the nearest whole number.   748

e. the nearest hundred.   700

continued

Solution

a. First we underline the digit 7 in the tenths place: 94.7̲35 The critical digit, 3, is less than 5, so we do not add 1 to the underlined digit. Dropping all digits to the right of the 7, we get 94.7. Note that our answer has only one decimal place because we are rounding to the nearest tenth.

b. We need to round 94.735 to two decimal places (to the nearest hundredth).

$$94.73̲5 \approx 94.74$$

The critical digit is 5 or more. Add 1 to the underlined digit and drop the decimal place to the right.

c. 94.735̲ ≈ 94.735 because the critical digit to the right of the 5 is understood to be 0.

d. We are rounding 94.735 to the nearest ten (not tenth!), which is a whole number place.

$$9̲4.735 \approx 90$$

Because 4 < 5, keep 9 in the tens place, insert 0 in the ones place and drop all decimal places.

e. Rounding to the nearest whole number means rounding to the nearest one.

$$94̲.735 \approx 95$$

Because 7 > 5, change the 4 to 5 and drop all decimal places.

EXAMPLE 14

Round 3.982 to the nearest tenth.

Solution First, we underline the digit 9 in the tenths place and identify the critical digit: 3.9̲82. The critical digit, 8, is more than 5, so we add 1 to the 9, get 10, and write down the 0. We add the carried 1 to 3 getting 4, and drop the 8 and 2.

$$3.9̲82 \approx 4.0$$
Drop

The answer is 4.0. Note that we do not drop the 0 in the tenths place of the answer to indicate that we have rounded to that place.

PRACTICE 14

Round 7.2962 to two decimal places.
7.30

EXAMPLE 15

On the commodity market, prices are often quoted in terms of thousandths or even ten-thousandths of a dollar. Suppose that the market price of a pound of coffee is $2.0883. What is this price to the nearest cent?

Solution A cent is one-hundredth of a dollar. Therefore we need to round 2.0883 to the nearest hundredth.

$$2.08̲83 \approx 2.09$$

The price to the nearest cent is $2.09.

PRACTICE 15

In a car race, the winner was clocked at a speed of 162.962 mph. Round this speed to the nearest tenth of a mile per hour.
163.0

Exercises 3.1

Write each decimal in words.

1. 0.61   Sixty-one hundredths

2. 0.72   Seventy-two hundredths

3. 0.305   Three hundred five thousandths

4. 0.849   Eight hundred forty-nine thousandths

5. 0.6   Six tenths

6. 0.3   Three tenths

7. 5.72   Five and seventy-two hundredths

8. 3.89   Three and eighty-nine hundredths

9. 24.002   Twenty-four and two thousandths

10. 370.081   Three hundred seventy and eighty-one thousandths

For each decimal find the equivalent fraction or mixed number, reduced to lowest terms.

11. 0.6 $\frac{3}{5}$

12. 0.8 $\frac{4}{5}$

13. 0.39 $\frac{39}{100}$

14. 0.27 $\frac{27}{100}$

15. 1.5 $1\frac{1}{2}$

16. 3.2 $3\frac{1}{5}$

17. 8.000   8

18. 6.700 $6\frac{7}{10}$

19. 5.012 $5\frac{3}{250}$

20. 20.304 $20\frac{38}{125}$

Write each number in decimal notation.

21. Eight tenths   0.8

22. Three hundredths   0.03

23. One and forty-one thousandths   1.041

24. Five and sixty-three hundredths   5.63

25. Sixty and one hundredth   60.01

26. Eighteen and four thousandths   18.004

27. Four and one hundred seven thousandths   4.107

28. Ninety-two and seven hundredths   92.07

29. Three and two tenths meters   3.2 m

30. Ninety-eight and six tenths degrees   98.6°

Each statement involves decimals. Write each decimal in words.

31. A speed of 1 mph is equivalent to 1.467 fps.   One and four hundred sixty-seven thousandths

32. A chemistry text gives 55.847 as the atomic weight of iron.   Fifty-five and eight hundred forty-seven thousandths

33. Over two years, the average score on a college admissions exam increased from 18.7 to 18.8.   Eighteen and seven tenths to eighteen and eight tenths

34. One specimen of the largest known butterfly, Queen Alexandra's Birdwing, weighs 0.176 oz and has a wing span of 11.02 in.   One hundred seventy-six thousandths; eleven and two hundredths

35. The following table shows the number of people per square mile living on various continents in a recent year.

Continent	Number of people per square mile
Asia	301.3
Africa	55.9
Europe	268.2
North America	46.6
South America	43.6

Three hundred one and three tenths
Fifty-five and nine tenths
Two hundred sixty-eight and two tenths
Forty-six and six tenths
Forty-three and six tenths

36. The coefficient of friction is a measure of the amount of friction produced when one surface rubs against another. The following table gives these coefficients for various surfaces.

Materials	Coefficient of friction
Wood on wood	0.3
Steel on steel	0.15
Steel on wood	0.5
A rubber tire on a dry concrete road	0.7
A rubber tire on a wet concrete road	0.5

Three tenths
Fifteen hundredths
Five tenths
Seven tenths
Five tenths

(*Source: CRC Handbook of Chemistry and Physics*)

37. Bacteria are single-celled organisms that typically measure from 0.00001 in. to 0.00008 in. across.   One hundred thousandth; eight hundred thousandths

38. In one month, the consumer confidence index rose from 71.9 points to 80.2 points.   Seventy-one and nine tenths; eighty and two tenths

Write each number in decimal notation.

39. The area of a plot of land is one and two tenths acres.   1.2 acres

40. The "lead" in many mechanical pencils is seven tenths mm thick.   0.7 mm

41. At the first Indianapolis 500 Auto Race in 1911, the winning speed was seventy-four and fifty-nine hundredths mph. (*Source: Jack Fox, The Indianapolis 500*)   74.59 mph

42. According to the owner's manual, the voltage produced by your camcorder battery is nine and six tenths volts (v).   9.6 v

43. At sea level, the air pressure on each square inch of surface area is fourteen and seven tenths lb.   14.7 lb

44. The doctor prescribed a dosage of one hundred twenty-five thousandths mg of Prolixin. 0.125 mg

45. In 1796, there was a U.S. coin in circulation worth five thousandths of a dollar. $0.005

46. In preparing an injection, the nurse measured out one and eight tenths milliliters (ml) of sterile water.   1.8 ml

47. The electrical usage in your apartment last month amounted to three hundred fifty-two and one tenth kilowatt hours (kWh).   352.1 kWh

48. In one week, the Dow Jones Industrial Average fell by three and sixty-three hundredths points.   3.63 points

Underline the digit that occupies the given place.

49. 2.7̲8   Tenths place

50. 9.01̲   Hundredths place

51. 0̲.03   Ones place

52. 6.8̲35   Tenths place

53. 358.0̲2   Tenths place

54. 823.001̲   Thousandths place

55. 0.77̲2   Hundredths place

56. 1̲35.83   Hundreds place

84. Suppose that, to qualify for the Dean's List at your college, your grade point average (GPA) must be above 3.5. If your GPA was 3.475, did you make the Dean's List?   No

85. A person with reasonably good vision can see objects as small as 0.0004 in. long. Can such a person see a mite—a tiny bug that is 0.003 in. long?   Yes

86. At the Winter Olympic Games, the two top skaters in the 500-m race finished in 39.44 sec and 39.5 sec. Which time was better?   39.44 sec

87. As part of an annual checkup, you take a blood test. The normal range for a particular substance is 1.1 to 2.3. If you score 0.95, is your blood in the normal range?   No

88. Last year, an electronics factory released 1.8 million pounds of toxic gas into the air. During the same time, a food factory and a chemical factory released 1.4 and 1.48 million pounds of toxic emissions, respectively. Which of the three factories was the worst polluter?   The electronics factory was the worst.

Round as indicated.

89. 17.36 to the nearest tenth   17.4

90. 8.009 to two decimal places   8.01

91. 3.5905 to the nearest thousandth
3.591

92. 3.5902 to the nearest thousandth
3.590

93. 37.08 to one decimal place   37.1

94. 3.08 to the nearest whole number   3

95. 0.396 to the nearest hundredth   0.40

96. 0.978 to the nearest tenth   1.0

97. 7.0571 to two decimal places   7.06

98. 3.038 to one decimal place   3.0

99. 8.7 mi to the nearest mile   9 mi

100. $35.75 to the nearest dollar $36

Round.

101.

To the nearest	8.0714	0.9916
Tenth	8.1	1.0
Hundredth	8.07	0.99
Ten	10	0

102.

To the nearest	0.8166	72.3591
Tenth	0.8	72.4
Hundredth	0.82	72.36
Ten	0	70

Round to the indicated place.

103. A bank pays interest on all its accounts to the nearest cent. If the interest on an account is $57.0285, how much interest does the bank pay? $57.03

104. Your city's sales tax rate, expressed as a decimal, is 0.0825. What is this rate to the nearest hundredth?   0.08

105. According to the organizers of a lottery, the probability of winning the lottery is 0.0008. Round this probability to three decimal places.   0.001

106. One day last week, a particular foreign currency was worth 0.7574 U.S. dollars ($US). How much is this currency worth to the nearest tenth of a dollar?   0.8 $US

107. According to a recent survey, the cost of medical care is 1.77 times what it was a decade ago. Express this decimal to the nearest tenth.   1.8

108. The length of the Panama Canal is 50.7 mi. Round this length to the nearest mile. (***Source:** New Encyclopedia Britannica*)   51 mi

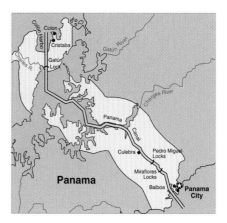

■ *Check your answers on page A-5.*

MINDSTRETCHERS

CRITICAL THINKING

1. For each question, either give the answer or explain why there is none.

a. Find the *smallest* decimal that when rounded to the nearest tenth is 7.5.

7.45

b. Find the *largest* decimal that when rounded to the nearest tenth is 7.5.

There is no specific "largest" decimal that answers this question. As long as the digit
after the 5 is a 4, we can simply keep placing a 9 to make the decimal larger (7.549,
7.5499, etc.).

WRITING

2. The next whole number after 7 is 8. What is the next decimal after 0.7? Explain.   There
is no *next* decimal after 0.7. The decimal 0.71 is close to 0.7, but 0.701 is closer. The
decimal 0.7001 is still closer, and so on.

GROUPWORK

3. Working with a partner, list fifteen numbers between 2.5 and 2.6.   Possible answer:
2.51, 2.52, 2.53, 2.54, 2.55, 2.56, 2.57, 2.58, 2.59, 2.511, 2.512, 2.513, 2.514, 2.515,
2.516

HISTORICAL NOTE

DISME:
The Art of Tenths,
OR;
Decimall Arithmetike,

Teaching how to performe all Computations
whatsoeuer, by whole Numbers without
Fractions , by the foure Principles of
Common Arithmeticke : namely, Addition, Substraction, Multiplication,
and Diuision.

Inuented by the excellent Mathematician,
Simon Steuin.

Published in English with some additions
by *Robert Norton,* Gent.

Imprinted at London by *S. S.* for *Hugh*
Astley, and are to be fold at his fhop at
Saint Magnus corner. 1 6 0 8,

In 1585 Simon Stevin, a Dutch engineer, published a book entitled *The Tenth* (*De Thiende* in the Flemish language, or *La Disme* in French) in which he presented a thorough account of decimals. Stevin sought to teach everyone "with an ease unheard of, all computations necessary between men by integers without fractions."

Stevin did not invent decimals; their history dates back thousands of years to ancient China, medieval Arabia, and Renaissance Europe. However, Stevin's writings popularized decimals, and also supported the notion of decimal coinage—as in American currency, where there are ten dimes to the dollar.

Source: Morris Kline, *Mathematics, A Cultural Approach* (Reading, Massachusetts: Addison-Wesley Publishing Company, 1962), p. 614.

3.2 Adding and Subtracting Decimals

OBJECTIVES

- *To add and subtract decimals*
- *To estimate the sum and difference of decimals*
- *To solve word problems involving the addition or subtraction of decimals*

In Section 3.1 we discussed the meaning of decimals and how to compare and round them. Now we turn our attention to computing with decimals, starting with sums and differences.

Adding Decimals

Adding decimals is similar to adding whole numbers: We add the digits in each place value position, carrying when necessary. Suppose that we want to find the sum of two decimals: $1.2 + 3.5$. First we rewrite the problem vertically, lining up the decimal points in the addends. Then we add as usual, inserting the decimal point below the other decimal points.

$$
\begin{array}{r}
\downarrow \\
1.2 \\
+3.5 \\
\hline
4.7 \\
\uparrow
\end{array}
\qquad \text{This addition is equivalent to} \qquad
\begin{array}{r}
1\frac{2}{10} \\
+3\frac{5}{10} \\
\hline
4\frac{7}{10}
\end{array}
$$

Note that, if we add the mixed numbers corresponding to the decimals, we get $4\frac{7}{10}$, which is equivalent to the answer 4.7.

To Add Decimals

- rewrite the numbers vertically, lining up the decimal points;
- add; and
- insert a decimal point in the answer below the other decimal points.

EXAMPLE 1

Add: $2.7 + 80.13 + 5.036$

Solution
$$
\begin{array}{r}
2.7 \\
80.13 \\
+5.036 \\
\hline
87.866
\end{array}
$$
Rewrite the addends with decimal points lined up vertically.

Add.

Insert the decimal point in the answer.

PRACTICE 1

$5.12 + 4.967 + 0.3$ 10.387

> Some students may prefer to insert 0's in these "ragged" decimals. For instance, Example 1 can be written as follows.
> $$
> \begin{array}{r}
> 2.700 \\
> 80.130 \\
> +5.036
> \end{array}
> $$

EXAMPLE 2

Evaluate: $2.367 + 5 + 0.143$

Solution Recall that 5 and 5. are the same.

$$
\begin{array}{r}
2.367 \\
5. \\
+0.143 \\
\hline
7.510 = 7.51
\end{array}
$$
Line up the decimal points and add.

Insert the decimal point in the answer.

We can drop the extra 0 at the right end.

PRACTICE 2

What is the sum of $7.31, $8, and $23.99?

$39.30

EXAMPLE 3

A runner's time was 0.06 sec longer than the world record of 21.71 sec. What was the runner's time?

Solution We need to compute the sum of the two numbers. The runner's time was 21.77 sec.

$$
\begin{array}{r}
0.06 \\
+21.71 \\
\hline
21.77
\end{array}
$$

Your child has the flu. This morning, his body temperature was 99.4°F. What was his temperature after it went up by 2.7°? 102.1°F

Subtracting Decimals

Now let's discuss subtracting decimals. As with addition, subtracting decimals is similar to subtracting whole numbers. To compute the difference between 12.83 and 4.2, for instance, we rewrite the problem vertically, lining up the decimal points. Then we subtract as usual. In the answer we insert a decimal point below the other decimal points.

$$
\begin{array}{r}
\downarrow \\
12.83 \\
-4.2 \\
\hline
8.63 \\
\uparrow
\end{array}
\quad \text{is equivalent to} \quad
\begin{array}{r}
12\dfrac{83}{100} = 12\dfrac{83}{100} \\[6pt]
-4\dfrac{2}{10} = -4\dfrac{20}{100} \\[6pt]
\hline
8\dfrac{63}{100}, \ \text{or} \ 8.63
\end{array}
$$

Again, note that if we subtract the equivalent mixed numbers, we get the same answer.

As in any subtraction problem, we can check this answer by adding the subtrahend (4.2) to the difference (8.63), confirming that we get the original minuend (12.83).

$$
\begin{array}{r}
8.63 \\
+4.2 \\
\hline
12.83
\end{array}
$$

To Subtract Decimals

■ first rewrite the numbers vertically, lining up the decimal points;
■ then subtract, adding extra zeros in the minuend if necessary for borrowing; and
■ finally, insert a decimal point in the answer, below the other decimal points.

EXAMPLE 4

Subtract: 5.03 − 2.11

Solution
$$
\begin{array}{r}
\downarrow \\
5.03 \\
-2.11 \\
\hline
2.92 \\
\uparrow
\end{array}
$$
Rewrite the problem with decimal points lined up vertically.
Subtract. Borrow when necessary.

Insert the decimal point in the answer.

To verify that our difference is correct, we check by addition.

Check:
$$
\begin{array}{r}
2.92 \\
+2.11 \\
\hline
5.03
\end{array}
$$

Find the difference: 71.38 − 25.17
46.21

EXAMPLE 5

65 is how much larger than 2.04?

Solution Recall that 65 and 65. are the same.

Insert zeros needed for borrowing.

$$
\begin{array}{r}
65.00 \\
-\,2.04 \\
\hline
62.96
\end{array}
$$

Line up the decimal points.

Subtract.

Insert the decimal point in the answer.

To check our answer (62.96), we add 2.04, verifying that the sum is 65. This check shows that 65 is 62.96 larger than 2.04.

PRACTICE 5

How much greater is $735 than $249.57?
$485.43

EXAMPLE 6

According to medical records, the average birth weight of a twin baby tends to be less than that of a single baby: 5.5 lb compared to 7.5 lb. How much heavier is the average single baby?

Solution We need to find the difference between 7.5 and 5.5.

$$
\begin{array}{r}
7.5 \\
-5.5 \\
\hline
2.0
\end{array}
$$

Checking by addition (5.5 + 2.0 = 7.5) confirms this answer. The average single baby is 2 lb heavier than the average twin baby.

PRACTICE 6

You are competing in the 28.5-mi swim around the island of Manhattan. After swimming 15 mi, how much farther do you have to go?   13.5 mi

EXAMPLE 7

Suppose that your salary is $350 a week, less deductions. The following table shows these deductions.

Deduction	Amount
Federal, state and city taxes	$100.80
Social security	13.50
Union fees	8.88

What is your take-home pay?

Solution You can break the question into two basic questions.
- How much money is deducted per week? The deductions ($100.80, $13.50, and $8.88) add up to $123.18 each week.
- How much of your salary is left after subtracting the deductions? The difference between $350 and $123.18 is $226.82, which is your take-home pay.

PRACTICE 7

You are at Town A and want to travel to Town B. Examine the map below and determine how much shorter it is to travel there directly instead of going by way of Town C.   2.1 mi

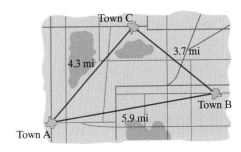

Estimating Sums and Differences

Being able to estimate in your head the sum or difference between two decimals is a useful skill, if only for checking an exact answer. To estimate, simply round the numbers to be added or subtracted and then carry out the operation on the rounded numbers.

EXAMPLE 8

Compute the sum: $0.17 + 0.4 + 0.083$. Check by estimation.

Solution First, we add. Then, to check, we round the addends—say, to the nearest tenth—and add the rounded numbers.

$$
\begin{array}{rcl}
0.17 & \approx & 0.2 \\
0.4 & \approx & 0.4 \\
+0.083 & \approx & +0.1 \\
\end{array}
$$

Exact sum   0.653   0.7   Estimated sum

Our exact sum is close to our estimated sum, and in fact, rounds to it. So we can have confidence that our answer, 0.653, is correct.

PRACTICE 8

Add 0.093, 0.008, and 0.762. Then check by estimating.   0.863

EXAMPLE 9

A movie budgeted at $7.25 million ended up costing $1.655 million more. Estimate the final cost of the movie.

Solution Let's round each number to the nearest million dollars.

$$
\begin{array}{rcl}
1.655 & \approx & 2 \text{ million} \\
7.25 & \approx & +7 \text{ million} \\
& & 9 \text{ million}
\end{array}
$$

Adding the rounded numbers, we see that the movie cost approximately $9 million.

PRACTICE 9

As an investment, you buy a painting for $2.3 million. A year later, you sell the painting at a profit of $1.75 million. Estimate the selling price. $4 million

---

EXAMPLE 10

Subtract: $0.713 - 0.082$. Then check by estimating.

Solution First we find the exact answer and then round to get an estimate.

$$
\begin{array}{rcl}
0.713 & \approx & 0.7 \\
-0.082 & \approx & -0.1 \\
\end{array}
$$

Exact difference   0.631   0.6   Estimated difference

Our exact answer, 0.631, is close to 0.6, so we can feel confident that it is correct.

PRACTICE 10

Compute: $0.17 - 0.091$. Use estimation to check.   0.079

EXAMPLE 11

When the chunnel connecting the United Kingdom and France was built, French and British construction workers dug from their respective countries. They met at the point shown on the map on the next page.

PRACTICE 11

From the deposit ticket shown on the next page, estimate the total amount deposited.

continued

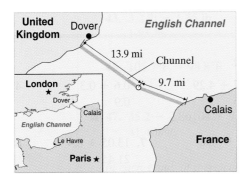

Estimate: <u>Possible answer: $480.00</u>

Estimate how much farther the British workers had dug than the French workers. (*Source: The New York Times*)

Solution We can round 13.9 to 14 and 9.7 to 10. The difference between 14 and 10 is 4, so the British workers dug about 4 miles farther than the French workers.

 ## Adding and Subtracting Decimals on a Calculator

When adding or subtracting decimals, press the ⬚ · key to enter the decimal point. If a sum or difference ends with a 0 in the rightmost decimal place, does your calculator drop the 0? If a sum or difference has no whole number part, does your calculator insert a 0?

EXAMPLE	**PRACTICE**

EXAMPLE

Compute: $2.7 + 4.1 + 9.2$

Solution

Input	2.7	+	4.1	+	9.2	=
Display	2.7	2.7	4.1	6.8	9.2	16.

   We can check this sum by estimating: $3 + 4 + 9 = 16$, which is the exact answer calculated.

PRACTICE

Find the sum:
$3.82 + 9.17 + 66.24$
79.23

EXAMPLE

Find the difference: $83.71 - 83.70002$

Solution

Input	83.71	−	83.70002	=
Display	83.71	83.71	83.70002	0.00998

   We can check this difference by adding: $0.00998 + 83.70002 = 83.71$.

PRACTICE

Compute: $5.00003 - 5.00001$
0.00002

Estimating Products

A good way to estimate mentally the product of decimals is to round each factor so that it has only one nonzero digit. Then multiply the rounded factors.

Suppose that we want to estimate the product of the decimals 19.0382 and 0.061.

$$
\begin{array}{r}
19.0382 \approx \quad 20 \\
\times 0.061 \approx \times 0.06 \\
\hline
01.20 = 1.2
\end{array}
$$

The estimated product, 1.20, has two decimal places—the total number of decimal places before we dropped the extra 0.

EXAMPLE 9

Multiply 0.703 by 0.087 and check your answer by estimating.

Solution First, we multiply the factors to find the exact product. Then, we round each factor and mentally multiply them.

$$
\begin{array}{r}
0.703 \approx \quad 0.7 \leftarrow \text{Rounded to have one nonzero digit} \\
\times \ 0.087 \approx \times 0.09 \leftarrow \text{Rounded to have one nonzero digit}
\end{array}
$$

Exact product   0.061161   0.063   Estimated product

Checking our exact answer, we see that the estimated product is fairly close.

PRACTICE 9

Find the product of 0.0037 × 0.092, using estimation to check.   0.0003404;
Estimate: 0.004 × 0.09 = 0.00036

EXAMPLE 10

Suppose that we budget $9,000 for the rent that our company must pay each year. Last year, the company's actual expenses were $28,712.55, with 0.39 of that amount going for rent. We want an estimate to decide whether we were over budget for the rent.

Solution We know that our company's rent amounts to the product of 0.39 and $28,712.55. We estimate this product.

$$
\begin{array}{r}
28{,}712.55 \approx 30{,}000 \\
\times \quad 0.39 \approx \times \ 0.4 \\
\hline
12000.0 = 12{,}000
\end{array}
$$

The amount going for rent was about $12,000. We had budgeted only $9,000 for rent, so we were over budget.

PRACTICE 10

The planet Earth travels through space at a speed of 18.6 mi/sec. Estimate how far Earth travels in 60 sec. (*Source:* The Diagram Group, *Comparisons*)   Possible answer: 1,140 mi

Multiplying Decimals on a Calculator

Multiply decimals on a calculator by entering each decimal as you would enter a whole number, but insert a decimal point as needed. If there are too many decimal places in your answer to fit in the display, investigate whether your calculator rounds the answer or simply cuts off the final digits.

EXAMPLE	PRACTICE
Compute $8,278.55 \times 0.875$, rounding your answer to the nearest hundredth. Then check the answer by estimating.	Find the product of 2,471.66 and 0.33, rounding to the nearest tenth. Check your answer. 815.6

Solution

Input	8278.55	$\times$	0.875	$=$
Display	8278.55	8278.55	0.875	7243.73125

Now 7,243.73125 rounded to the nearest hundredth is 7,243.73. Checking by estimating we get $8,000 \times 0.9$, or 7,200, which is close to our exact answer.

EXAMPLE	PRACTICE
Find $(1.9)^2$	Calculate: $(2.1)^3$ 9.261

Solution

Input	1.9	$\times$	1.9	$=$
Display	1.9	1.9	1.9	3.61

Now let's check by estimating. Since 1.9 rounded to the nearest whole number is 2, $(1.9)^2$ should be close to 2^2, or 4, which is close to our answer, 3.61.

Exercises 3.3

Insert a decimal point in each product. Check by estimating.

1. $2.356 \times 1.27 = 299212$ 2.99212

2. $97.26 \times 5.3 = 515478$ 515.478

3. $0.0019 \times 0.051 = 969$ 0.0000969

4. $0.0089 \times 0.0021 = 1869$ 0.00001869

5. $3{,}144 \times 0.065 = 204360$ 204.36

6. $837 \times 0.15 = 12555$ 125.55

7. $71.2 \times 35 = 24920$ 2,492.0

8. $0.002 \times 37 = 0074$ 0.074

9. $2.87 \times 1{,}000 = 287000$ 2,870.00

10. $492.31 \times 10 = 492310$ 4,923.10

11. $\$4.25 \times 0.173 = \$73525$ $0.73525

12. $11.2 \text{ ft} \times 0.75 = 8400 \text{ ft}$ 8.400 ft

Find the product. Then check by estimating.

13. 0.6
$\times 0.3$
0.18

14. 0.8
$\times 0.7$
0.56

15. 0.5
$\times 0.8$
0.40

16. 0.6
$\times 0.6$
0.36

17. 0.1
$\times 0.2$
0.02

18. 0.09
$\times\ 0.4$
0.036

19. 0.04
$\times 0.02$
0.0008

20. 0.03
$\times 0.01$
0.0003

21. 2.55
$\times\ 0.3$
0.765

22. 80.7
$\times\ 0.6$
48.42

23. 0.96
$\times\ 2.1$
2.016

24. 0.043
$\times\ 0.02$
0.00086

25. 38.01
$\times\ \ 0.2$
7.602

26. 1.22
$\times\ \ 3$
3.66

27. 125
$\times 0.004$
0.5

28. 0.003
$\times\ \ 1.7$
0.0051

29. 3.8×1.5
5.7

30. 9.51×0.7
6.657

31. 12.45×0.3
3.735

32. 72.558×0.2
14.5116

33. 13.74×11
151.14

34. $1{,}245 \times 2.5$
3,112.5

35. $(0.21)(0.4)$
0.084

36. $(0.3)(1.5)$
0.45

37. 83.127×100
8,312.7

38. 4.9×10
49

39. $0.0023 \times 10{,}000$
23

40. $0.0135 \times 1{,}000$
13.5

41. 0.7×10^2
70

42. 0.6×10^4
6,000

43. $(0.3)^2$
0.09

44. $(1.2)^2$
1.44

45. $(1.5)(0.6)(0.1)$
0.09

46. $(12)(3.5)(0.2)$
8.4

47. $(0.001)^3$
0.000000001

48. $(0.1)^4$
0.0001

49. $30 - 2.5 \times 1.7$
25.75

50. $8 + 4.1 \times 2$
16.2

51. $17 \text{ ft} \times 2.5$
42.5 ft

52. $5 \text{ hr} \times 0.75$
3.75 hr

53. $3.5 \text{ mi} \times 0.4$
1.4 mi

54. $9.1 \text{ m} \times 1,000$
9,100 m

55.
$$\begin{array}{r} 43.87 \\ \times 0.075 \\ \hline 3.29025 \end{array}$$

56.
$$\begin{array}{r} 18,275.33 \\ \times \quad 0.39 \\ \hline 7,127.3787 \end{array}$$

57.
$$\begin{array}{r} 99,125 \\ \times \quad 2.75 \\ \hline 272,593.75 \end{array}$$

58.
$$\begin{array}{r} 3.512 \\ \times \quad 1.47 \\ \hline 5.16264 \end{array}$$

Complete each table.

59.

Input	Output
1	$3.8 \times \mathbf{1} - 0.2 = 3.6$
2	$3.8 \times \mathbf{2} - 0.2 = 7.4$
3	$3.8 \times \mathbf{3} - 0.2 = 11.2$
4	$3.8 \times \mathbf{4} - 0.2 = 15$

60.

Input	Output
1	$7.5 \times \mathbf{1} + 0.4 = 7.9$
2	$7.5 \times \mathbf{2} + 0.4 = 15.4$
3	$7.5 \times \mathbf{3} + 0.4 = 22.9$
4	$7.5 \times \mathbf{4} + 0.4 = 30.4$

Each product is rounded to the nearest hundredth. In each group of three products, one is wrong. Use estimation to explain which product is incorrect.

61. a. $51.6 \times 0.813 = 419.51$ **b.** $2.93 \times 7.283 = 21.34$ **c.** $(5.004)^2 = 25.04$
a: $50 \times 1 = 50$

62. a. $0.004 \times 3.18 = 0.01$ **b.** $2.99 \times 0.287 = 0.86$ **c.** $(1.985)^3 = 10.82$
c: $2^3 = 8$

63. a. $4.913 \times 2.18 = 10.71$ **b.** $0.023 \times 0.71 = 0.16$ **c.** $(8.92)(1.0027) = 8.94$
b: $0.02 \times 0.7 = 0.014$

64. a. $\$138.28 \times 0.075 = \$10.37$ **b.** $(0.2715)^2 = 0.07$

  c. $0.77 \times \$6,005.79 = \$462.45$ c: $0.8 \times 6,000 = 4,800$

Solve. Then check your answer by estimating.

65. Sound travels at approximately 1,000 fps. If a jet plane is flying at Mach 2.9 (that is, 2.9 times the speed of sound), what is its speed?   2,900 fps

66. You take out a life insurance policy, paying $323.50 in premiums per year for 10 yr. How much did you pay altogether in premiums? $3,235.00

67. One steel company bought another for about $2.6 billion. Express this amount in standard form. $2,600,000,000.00

68. According to the first American census in 1790, the population of the United States was approximately 3.9 million. Write this number in standard form. (*Source: The Statistical History of the United States*)   3,900,000

69. The area of this circle is approximately 3.14×6.25 ft^2. Find this area to the nearest tenth.   19.6 ft^2

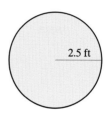

2.5 ft

70. Find the area (in m^2) of the room pictured.   16.43 m^2

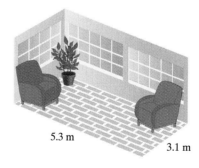

5.3 m

3.1 m

71. Over a 5-day period, you administered 10 tablets to a patient. If each tablet contained 0.125 mg of the drug Digoxin, how much Digoxin did you administer?   1.25 mg

72. Water weighs approximately 62.5 pounds per cubic foot (lb/ft^3). If a bathtub contains about 30 ft^3 of water, how much does the water in the bathtub weigh?   1,875 lb

73. The sales receipt for your purchases during a shopping trip reads as follows.

Purchase	Quantity	Unit Price	Price
Belt	1	$11.99	$ 11.99
Shirt	3	$16.95	$ 50.85
Total Price			$ 62.84

a. Complete the table.

b. If you pay for these purchases with four $20 bills, how much change should you get?
$17.16

74. On an electric bill, *usage* is the difference between the meter's *current reading* and the *previous reading* in kilowatt hours (kWh). The *amount due* is the product of the usage and the *rate per kWh*. Find the two missing quantities in the table, rounding to the nearest hundredth.

Previous Reading	750.07 kWh
Current Reading	1,115.14 kWh
Usage	365.07 kWh
Rate per kWh	$0.10
Amount Due	$36.51

▦ *Use a calculator to solve the following problems, giving (a) the operation(s) carried out in your solution, (b) the exact answer, and (c) an estimate.*

75. Scientists have discovered a relationship between the length of a person's bones and the person's overall height. For instance, a male's height can be predicted from the length of his femur bone by using the formula $(1.9 \times femur) + 32.0$. With this formula, estimate the height of the German giant Constantine whose femur measured 29.9 in.
(a) Multiplication and addition   (b) 88.81 in.   (c) 90 in.

76. In order to buy a $125,000 house, you put down $25,000 and take out a mortgage on the balance. To pay off the mortgage, you pay $877.57/mo for the following 360 mo. How much more will you end up paying for the house than the original price of $125,000?   (a) Multiplication, addition, subtraction   (b) $215,925.20   (c) $260,000

MINDSTRETCHERS

PATTERNS

1. When $(0.001)^{100}$ is multiplied out, how many decimal places will it have?   300

$(0.001)^2$ has 2×3 digits.

$(0.001)^3$ has 3×3 digits.

$(0.001)^4$ has 4×3 digits.

⋮

$(0.001)^{100}$ has 100×3 digits.

MATHEMATICAL REASONING

2. Give an example of two decimals

a. whose sum is greater than their product, and   Possible answer: 0.8 and 0.6; sum = 1.4, product = 0.48

b. whose product is greater than their sum.   Possible answer: 2.3 and 3.1; sum = 5.4, product = 7.13

GROUPWORK

3. In the product to the right, each letter stands for a different digit. Working with a partner, identify all the digits.   A = 2 or 5; B = 2 or 5; C = 1; D = 3

$$\begin{array}{r} A.B \\ \times B.A \\ \hline C\ D \end{array}$$

$$\begin{array}{r} 2.5 \\ \times 5.2 \\ \hline 13 \end{array} \quad \text{or} \quad \begin{array}{r} 5.2 \\ \times 2.5 \\ \hline 13 \end{array}$$

3.4 Dividing Decimals

In this section we first consider changing a fraction to its decimal equivalent, which involves both division and decimals. Then we move on to our main concern—the division of decimals.

Changing a Fraction to the Equivalent Decimal

Earlier in this chapter, we discussed how to change a decimal to its equivalent fraction. Now let's consider the opposite problem—how to change a fraction to its equivalent decimal.

When the denominator of a fraction is already a power of 10, the problem is simple. For example, the decimal equivalent of $\frac{43}{100}$ is 0.43.

But what about the more difficult problem of when the denominator is *not* a power of 10? A good strategy is to find an equivalent fraction that does have a power of 10 as its denominator. Consider, for instance, the fraction $\frac{3}{4}$. Since 4 is a factor of 100, we can easily find an equivalent fraction having a denominator of 100.

> Have students review how to find a missing part of two equivalent fractions.

$$\frac{3}{4} = \frac{3 \cdot 25}{4 \cdot 25} = \frac{75}{100} = 0.75$$

So 0.75 is the decimal equivalent of $\frac{3}{4}$.

There is a faster way to show that $\frac{3}{4}$ is the same as 0.75, without having to find an equivalent fraction.

Because $\frac{3}{4}$ can mean $3 \div 4$, we divide the numerator (3) by the denominator (4). Note that if we continue to divide to the hundredths place, there is no remainder. So this division also tells us that $\frac{3}{4}$ equals 0.75.

$$
\begin{array}{r}
0.75 \\
4\overline{)3.00} \\
\underline{0} \\
3\,0 \\
\underline{2\,8} \\
20 \\
\underline{20}
\end{array}
$$

⌐ Insert the decimal point directly above the point in the dividend.

← The decimal point is after the 3. Insert enough 0's to continue dividing as far as necessary.

To Change a Fraction to Its Decimal Equivalent

- divide the denominator of the fraction into the numerator, inserting to its right both a decimal point and enough zeros to get an answer either without a remainder or rounded to a given decimal place; and
- place a decimal point in the quotient directly above the decimal point in the dividend.

EXAMPLE 1

Express $\dfrac{1}{2}$ as a decimal.

Solution To find the decimal equivalent, we divide the fraction's numerator by its denominator.

$$\begin{array}{r} 0.5 \\ 2\overline{)1.0} \\ \underline{0} \\ 1\,0 \\ \underline{1\,0} \end{array}$$

Add a decimal point and a 0 to the right of the 1. Because there is no remainder in the tenths place, we can stop dividing.

So 0.5 is the decimal equivalent of $\dfrac{1}{2}$.

To check this answer, we verify that the fractional equivalent of 0.5 is $\dfrac{1}{2}$.

$$0.5 = \frac{5}{10} = \frac{1 \times \overset{1}{\cancel{5}}}{2 \times \underset{1}{\cancel{5}}} = \frac{1}{2}$$

Our answer checks.

PRACTICE 1

Write the fraction $\dfrac{3}{8}$ as a decimal.   0.375

EXAMPLE 2

Convert $2\dfrac{3}{5}$ to a decimal.

Solution Let's first change this mixed number to an improper fraction: $2\dfrac{3}{5} = \dfrac{13}{5}$. We can then change this improper fraction to a decimal by dividing its numerator by its denominator.

$$\begin{array}{r} 2.6 \\ 5\overline{)13.0} \\ \underline{10} \\ 3\,0 \\ \underline{3\,0} \end{array}$$

So 2.6 is the decimal form of $2\dfrac{3}{5}$.

We can check this answer by converting it back from a decimal to its mixed number form.

$$2.6 = 2\frac{6}{10} = 2\frac{3 \times \overset{1}{\cancel{2}}}{5 \times \underset{1}{\cancel{2}}} = 2\frac{3}{5} \quad \text{Our answer checks.}$$

PRACTICE 2

Write $7\dfrac{5}{8}$ as a decimal.   7.625

Point out that an alternative approach to this problem is to leave the whole number portion of the mixed number alone, changing the proper fraction to its decimal form, and then combining.

When converting some fractions to decimal notation, we keep getting a remainder as we divide. In this case, we round the answer to a given decimal place.

EXAMPLE 3

Convert $4\dfrac{8}{9}$ to a decimal, rounded to the nearest hundredth.

Solution First, we change this mixed number to an improper fraction. Then, we convert it to a decimal.

$$4\frac{8}{9} = \frac{44}{9} = 9\overline{)44.000}$$

4.88**8**, rounding to 4.89

← In order to round to the nearest hundredth, we must continue to divide to the thousandths place. So we insert three zeros.

$$\begin{array}{r} 36 \\ \overline{8\,0} \\ 7\,2 \\ \overline{80} \\ \underline{72} \\ \overline{80} \\ \underline{72} \end{array}$$

PRACTICE 3

Express $83\dfrac{1}{3}$ as a decimal, rounded to the nearest tenth.   83.3

Let's return for a moment to Example 3. Note that if instead of rounding we had continued to divide we would have gotten as our answer the *repeating decimal* 4.88888... (also written 4.$\overline{8}$). Can you think of any other fraction that is equivalent to a repeating decimal?

Let's now consider some word problems in which we need to convert fractions to decimals.

Have students explain their answers in a journal.

EXAMPLE 4

A share of stock sells for $5\frac{7}{8}$ dollars. Express this amount in dollars and cents, to the nearest cent.

Solution To solve, we must convert the mixed number $5\frac{7}{8}$ to a decimal.

$$5\frac{7}{8} = \frac{47}{8} = 8)\overline{47.000}$$

$$\begin{array}{r} 5.875 \approx 5.88 \\ \underline{40} \\ 7\,0 \\ \underline{6\,4} \\ 60 \\ \underline{56} \\ 40 \\ \underline{40} \end{array}$$

A share of this stock sells for $5.88, to the nearest cent.

PRACTICE 4

The gas nitrogen makes up about $\frac{39}{50}$ of the air in the atmosphere. Express this fraction as a decimal, rounded to the nearest tenth. (*Source: World of Scientific Discovery*)   0.8

Remind students that a cent is a hundredth of a dollar.

Dividing Decimals

Now let's turn our attention to dividing decimals, as in $0.006 \div 0.02$. To solve, we find an equivalent fraction whose denominator is a whole number.

$$0.006 \div 0.02 = \frac{0.006}{0.02} = \frac{0.006 \times 100}{0.02 \times 100} = \frac{000.6}{002.} = \frac{0.6}{2}$$

Some students may need to see this problem done fractionally to verify that the answer is correct.

We now can carry out the division, positioning the decimal point in the quotient directly above the decimal point in the dividend.

$$\begin{array}{r} 0.3 \\ 2)\overline{0.6} \\ \underline{0} \\ 6 \\ \underline{6} \end{array}$$ ← The answer has one decimal place.

In short,

$$002.)\overline{000.6}$$ with quotient 0.3

As in any division problem, we can check our answer by confirming that the product of the quotient and the *original divisor* equals the *original dividend*.

$$\begin{array}{r} 0.3 \\ \times 0.02 \\ \hline 0.006 \end{array}$$

To Divide Decimals

- move the decimal point in the divisor to the right end of the number,
- move the decimal point in the dividend the same number of places to the right as in the divisor,
- insert a decimal point in the quotient directly above the decimal point in the dividend, and
- divide the new dividend by the new divisor, adding zeros at the right end of the dividend as necessary.

EXAMPLE 5

What is 0.035 divided by 0.25?

Solution   Move the decimal point to the right end, making the divisor a whole number.

$$0.25\overline{)0.035}$$

Move the decimal point in the dividend the same number of places.

Finally, we divide 3.5 by 25, which gives us 0.14.

$$
\begin{array}{r}
0.14 \\
25\overline{)3.50} \\
2\,5 \\
\hline
1\,00 \\
1\,00 \\
\hline
\end{array}
$$

When checking our answer, we see that the product of the quotient and the original divisor is equal to the original dividend.

$$
\begin{array}{r}
0.14 \\
\times\,0.25 \\
\hline
0\,70 \\
02\,8 \\
\hline
0.0350 = 0.035
\end{array}
$$

PRACTICE 5

Divide and check: 2.706 ÷ 0.15   18.04;

$$
\begin{array}{r}
18.04 \\
\times 0.15 \\
\hline
2.706
\end{array}
$$

EXAMPLE 6

Find the quotient: 6 ÷ 0.0012. Check your answer.

Solution   The decimal point is moved four places to the right.

$$0.0012\overline{)6.0000}$$

To move the decimal point four places to the right, we must add four zeros as placeholders.

$$
\begin{array}{r}
5,000 \\
12\overline{)60,000}
\end{array}
$$

To check, we multiply.

$$
\begin{array}{r}
5,000 \\
\times\ \ 0.0012 \\
\hline
0006.0000 = 6
\end{array}
$$

Our answer checks.

PRACTICE 6

Divide $\dfrac{8.2}{0.004}$ and then check.   2,050;

$$
\begin{array}{r}
2,050 \\
\times 0.004 \\
\hline
8.200 = 8.2
\end{array}
$$

EXAMPLE 7

Divide and round to the nearest hundredth: $0.7)\overline{40.2}$. Then check.

Solution $0.7)\overline{40.2}$

$$
\begin{array}{r}
57.428 \approx 57.43 \text{ to the nearest hundredth} \\
7)\overline{402.000} \\
35 \\
\hline
52 \\
49 \\
\hline
3\,0 \\
2\,8 \\
\hline
20 \\
14 \\
\hline
60 \\
56 \\
\hline
4
\end{array}
$$

Check:

$$
\begin{array}{r}
57.43 \\
\times \ 0.7 \\
\hline
40.201 \approx 40.2
\end{array}
$$

Because we rounded our answer, the check gives us a product only approximately equal to the original dividend.

PRACTICE 7

Find the quotient of 8.07 and 0.11, rounded to the nearest tenth.   73.4

> Remind students how to *multiply* a decimal by a power of 10.

EXAMPLE 8

Compute and check: $8.319 \div 1,000$

Solution

$$
\begin{array}{r}
0.008319 \\
1,000)\overline{8.319000} \\
8\,000 \\
\hline
3190 \\
3000 \\
\hline
1900 \\
1000 \\
\hline
9000 \\
9000
\end{array}
$$

Check:

$$
\begin{array}{r}
0.008319 \\
\times \quad 1,000 \\
\hline
0008.319000 = 8.319
\end{array}
$$

Here the divisor is a power of 10 that ends in three zeros. Thus we could have gotten the answer more easily by simply moving the decimal point in the dividend *three places to the left.*

$$\frac{8.319}{1,000} = 0.008319$$

PRACTICE 8

Divide: $100)\overline{3.41}$ 0.0341

> Have students write in their journals the difference between the shortcuts for multiplying and for dividing by a power of 10.

As Example 8 illustrates, a shortcut for dividing a decimal by a power of 10 is to *move the decimal point to the left the same number of places as the power of 10 has zeros.*

EXAMPLE 9

Compute: $\dfrac{7.2}{100}$

PRACTICE 9

Calculate: $0.86 \div 1,000$ 0.00086

continued

Solution We are dividing by a power of 10 with two zeros, so we can find this quotient simply by moving the decimal point in 7.2 to the left two places.

$$\frac{7.2}{100} = .072, \quad \text{or} \quad 0.072$$

The quotient is 0.072.

Now let's try using these skills in some applications.

EXAMPLE 10

On a part-time job, you make $6.50 an hour. How long will it take you to make $312?

Solution The question asks for the quotient of 312 and 6.5.

$$6.5\overline{)312}$$

Now divide.
$$\begin{array}{r} 48 \\ 65\overline{)3,120} \\ \underline{2\ 60} \\ 520 \\ \underline{520} \end{array}$$

After checking (48 × 6.5 = 312), we can conclude that it will take 48 hours to make $312.

PRACTICE 10

The doctor ordered you to give 0.6 mg of a drug to a patient. If each tablet contains 0.15 mg of the drug, how many tablets should you administer?   4 tablets

---

EXAMPLE 11

You raise chickens using organic feed. Four of them weigh 3.2 lb, 3.5 lb, 2.9 lb, and 3.6 lb. How much less than the average weight of the four chickens was the weight of the lightest chicken?

Solution This question has two parts.

What is the average weight of the chickens?

$$\frac{3.2 + 3.5 + 2.9 + 3.6}{4} = \frac{13.2}{4} = 3.3$$

The average weight is 3.3 lb.

How much greater is this average (3.3 lb) than the weight of the lightest chicken (2.9 lb)?

$$3.3 - 2.9 = 0.4$$

The answer is 0.4 lb.

PRACTICE 11

In a diving competition, five judges scored each diver. To compute a diver's overall score, they averaged the diver's scores after first dropping the highest and lowest scores. What is the overall score of a diver who earns scores of 6.2, 6.7, 6.3, 6.4, and 6.5?   6.4

> Remind students how to compute an average.

Estimating Quotients

As we have shown, one way to check the quotient of two decimals is by multiplying. Another way is by estimating.

To check a decimal quotient by estimating, we can round each decimal to one nonzero digit and then mentally divide the rounded numbers. But we must be careful to position the decimal point correctly in our estimate.

EXAMPLE 12

Divide and check by estimating: $3.36 \div 0.021$

Solution $0.021\overline{\smash{)}3.360}$

We compute the exact answer.

$$
\begin{array}{r}
160 \\
21\overline{\smash{)}3{,}360} \\
\underline{2\ 1} \\
1\ 26 \\
\underline{1\ 26} \\
00 \\
\underline{00}
\end{array}
$$

So 160 is our quotient.

   Now let's check by estimating. Because $3.36 \approx 3$ and $0.021 \approx 0.02$, $3.36 \div 0.021 \approx 3 \div 0.02$. We mentally divide to get the estimate.

$$
\begin{array}{r}
1\ 50 \\
0.02\overline{\smash{)}3.00}
\end{array}
$$

Our estimate, 150, is reasonably close to our exact answer, 160, and so confirms it.

PRACTICE 12

Compute and check by estimation:
$8.229 \div 0.39$ 21.1; possible estimate: 20

Dividing on a Calculator

When dividing decimals on a calculator, be careful to enter the dividend first and then the divisor. Note that when the dividend is larger than the divisor, the quotient is greater than 1 and that when the dividend is smaller than the divisor, the quotient is less than 1.

EXAMPLE

Calculate $0.07 \div 0.3$ and then round to the nearest hundredth.

Solution

Input	0.07	÷	0.3	=
Display	0.07	0.07	0.3	0.23333333

The answer, when rounded to the nearest hundredth, is 0.23. As expected, the answer is less than 1 because $0.07 < 0.3$.

PRACTICE

Compute the following quotient, rounding to the nearest hundredth: $0.3 \div 0.07$
4.29

EXAMPLE

Divide $1.6\overline{\smash{)}8.6}$ and then round to the nearest tenth.

Solution

Input	8.6	÷	1.6	=
Display	8.6	8.6	1.6	5.375

The answer, when rounded to the nearest tenth, is 5.4. As expected, the answer is greater than 1 because $8.6 > 1.6$.

PRACTICE

Find the quotient $8.6\overline{\smash{)}1.6}$ and then round to the nearest tenth.   0.2

Exercises 3.4

Change to the equivalent decimal. Then check.

1. $\frac{11}{2}$ 5.5   **2.** $\frac{13}{5}$ 2.6   **3.** $\frac{21}{4}$ 5.25   **4.** $\frac{17}{8}$ 2.125

5. $\frac{37}{10}$ 3.7   **6.** $\frac{517}{100}$ 5.17   **7.** $1\frac{5}{8}$ 1.625   **8.** $10\frac{3}{4}$ 10.75

9. $2\frac{7}{8}$ 2.875   **10.** $8\frac{2}{5}$ 8.4   **11.** $21\frac{3}{100}$ 21.03   **12.** $60\frac{17}{100}$ 60.17

Divide and check.

13. $4\overline{)17}$ 4.25   **14.** $2\overline{)35}$ 17.5   **15.** $5\overline{)21}$ 4.2   **16.** $6\overline{)33}$ 5.5

17. $8\overline{)11}$ 1.375   **18.** $6\overline{)9}$ 1.5   **19.** $18\overline{)153}$ 8.5   **20.** $14\overline{)217}$ 15.5

Change to the equivalent decimal. Round to the nearest hundredth.

21. $\frac{2}{3}$ 0.67   **22.** $\frac{5}{6}$ 0.83   **23.** $\frac{7}{9}$ 0.78   **24.** $\frac{1}{3}$ 0.33

25. $3\frac{1}{9}$ 3.11   **26.** $2\frac{4}{7}$ 2.57   **27.** $5\frac{1}{16}$ 5.06   **28.** $10\frac{11}{32}$ 10.34

Divide. Express any remainder as a decimal rounded to the nearest tenth.

29. $7\overline{)23}$ 3.3   **30.** $9\overline{)41}$ 4.6   **31.** $11\overline{)3}$ 0.3   **32.** $13\overline{)2}$ 0.2

33. $3\overline{)911}$ 303.7   **34.** $7\overline{)208}$ 29.7   **35.** $7\overline{)46}$ 6.6   **36.** $3\overline{)82}$ 27.3

Insert the decimal point in the appropriate place.

37. $0.7\overline{)4\,1.1\,7\,4}$ ⟨5 8.8 2⟩   **38.** $3\overline{)0.0\,1\,7\,1}$ ⟨0.0 0 5 7⟩   **39.** $3.9\overline{)2\,6.9\,1}$ ⟨6.9⟩   **40.** $0.58\overline{)0.3\,8\,4\,5\,4}$ ⟨0.6 6 3⟩

Divide and check, either by multiplying or by estimating.

41. $4\overline{)3.72}$
0.93

42. $5\overline{)97.5}$
19.5

43. $8\overline{)23.1}$
2.8875

44. $2\overline{)0.0035}$
0.00175

45. $3\overline{)81.51}$
27.17

46. $9\overline{)0.252}$
0.028

47. $7\overline{)2.002}$
0.286

48. $6\overline{)24.042}$
4.007

49. $\dfrac{17.2}{4}$
4.3

50. $\dfrac{0.75}{5}$
0.15

51. $\dfrac{0.003}{2}$
0.0015

52. $\dfrac{1.04}{8}$
0.13

53. $8.65 \div 5$
1.73

54. $0.42 \div 3$
0.14

55. $11.5 \div 4$
2.875

56. $7.3 \div 2$
3.65

57. $0.8 \div 2$
0.4

58. $0.6 \div 3$
0.2

59. $3.52 \div 5$
0.704

60. $1.92 \div 4$
0.48

61. $4.7 \div 0.5$
9.4

62. $8.6 \div 0.2$
43

63. $5 \div 0.4$
12.5

64. $9 \div 0.6$
15

65. $0.03 \div 0.1$
0.3

66. $1.2 \div 0.01$
120

67. $0.38 \div 1.9$
0.2

68. $0.075 \div 0.25$
0.3

69. $95.2 \div 100$
0.952

70. $8.4 \div 1{,}000$
0.0084

71. $81.6 \div 10$
8.16

72. $513.22 \div 100$
5.1322

73. $2.7 \div 1{,}000$
0.0027

74. $0.082 \div 100$
0.00082

75. $4.95 \div 10$
0.495

76. $9.03 \div 1{,}000$
0.00903

Compute, rounding to the nearest tenth. Check, either by multiplying or by estimating.

77. $0.8\overline{)307.2}$
384.0

78. $0.4\overline{)81.9}$
204.8

79. $0.05\overline{)9}$
180.0

80. $0.7\overline{)42}$
60.0

81. $0.9\overline{)0.0057}$
0.0

82. $0.2\overline{)0.003}$
0.0

83. $0.01\overline{)98.02}$
9,802.0

84. $0.03\overline{)59.13}$
1,971.0

85. $\dfrac{0.057}{0.2}$
0.3

86. $\dfrac{3.69}{0.4}$
9.2

87. $\dfrac{4}{0.07}$
57.1

88. $\dfrac{8.1}{0.09}$
90.0

89. $\dfrac{87}{0.009}$
9,666.7

90. $\dfrac{23}{0.06}$
383.3

91. $\dfrac{8.312}{0.7}$
11.9

92. $\dfrac{4.227}{0.03}$
140.9

93. $6.45 \div 1.2$
5.4

94. $69.1 \div 2.4$
28.8

95. $0.8 \div 3.5$
0.2

96. $0.4 \div 2.7$
0.1

97. $35.77 \div 0.11$
325.2

98. $0.291 \div 1.7$
0.2

99. $961.2 \div 2.1$
457.7

100. $9.05 \div 3.2$
2.8

101. $8 + \dfrac{3.05}{5}$
8.6

102. $\dfrac{2.04}{3} + 1$
1.7

103. $\dfrac{8.1 \times 0.2}{0.4}$
4.1

104. $\dfrac{7.5}{2.5 \times 0.6}$
5.0

105. $49.071 \div 0.728$
67.4

106. $18.3 \div 7.96$
2.3

107. $3 \div 0.0721$
41.6

108. $100 \div 3.89$
25.7

Complete each table.

109.

Input	Output
1	$15 \div \mathbf{1} - 0.2 = 14.8$
2	$15 \div \mathbf{2} - 0.2 = 7.3$
3	$15 \div \mathbf{3} - 0.2 = 4.8$
4	$15 \div \mathbf{4} - 0.2 = 3.55$

110.

Input	Output
1	$7.5 \div \mathbf{1} + 0.4 = 7.9$
2	$7.5 \div \mathbf{2} + 0.4 = 4.15$
3	$7.5 \div \mathbf{3} + 0.4 = 2.9$
4	$7.5 \div \mathbf{4} + 0.4 = 2.275$

Each of the following quotients is rounded to the nearest hundredth. In each group of three quotients, one is wrong. Use estimation to explain which quotient is incorrect.

111. a. $5.7 \div 89 \approx 0.06$ **b.** $0.77 \div 0.0019 \approx 405.26$ **c.** $31.5 \div 0.61 \approx 516.39$

c: $32 \div 0.6 \approx 53.3$

112. a. $\dfrac{12.83}{5.88} \approx 0.22$ **b.** $\dfrac{2.771}{0.452} \approx 6.13$ **c.** $\dfrac{389.224}{1.79} \approx 217.44$

a: $\dfrac{13}{6} \approx 2.2$

113. a. $61.27 \div 0.057 \approx 1{,}074.91$ **b.** $0.614 \div 2.883 \approx 2.13$

c. $0.0035 \div 0.00481 \approx 0.73$ b: $0.6 \div 3 = 0.2$

114. a. $\$365 \div \$4.89 \approx 7.46$ **b.** $\$17{,}358.27 \div \$365 \approx 47.56$

c. $\$3{,}000 \div \$2.54 \approx 1{,}181.10$ a: $370 \div 5 = 74$

Solve and check.

115. A stalactite is an icicle-shaped mineral deposit that hangs from the roof of a cave. If it took a thousand years for a stalactite to grow to a length of 3.7 in., how much did it grow per year?   0.0037 in./yr

116. In a great earthquake, the damage to 100 houses was estimated at $12.7 million. What was the average damage per house? $0.127 million = $127,000

117. So far this season, the women's team has won 21 games and lost 14 games. The men's team has won 22 games and lost 18.

a. The women's team won what fraction of the games that it played? Express this fraction as a decimal. $\frac{21}{35}$; 0.6

b. The men's team won what fraction of its games? Express this fraction as a decimal. $\frac{22}{40}$; 0.55

c. Which team has a better record? Explain.   The women's team has a better record. Its winning percentage is 60% and the men's team is 55%.

118. Yesterday, 0.08 in. of rain fell. Today, $\frac{1}{4}$ in. of rain fell.

a. How much rain fell today, expressed as a decimal?   0.25 in.

b. Which day did more rain fall? Explain.   Today, because 0.25 > 0.08.

119. You road test three cars to see which gives the best gasoline mileage.

Car	Distance Driven (mi)	Gasoline Used (gal)	Miles/Gallon
A	18.6	1.6	11.625
B	7.8	0.6	13
C	23.4	1.2	19.5

a. For each car, compute how many miles it gets per gallon.

b. Which car gives the best mileage?   Car C

120. The following table shows the number of assists that three basketball players have handed out so far this season.

Player	No. of Games	No. of Assists	Average
A	82	926	11.3
B	78	836	10.7
C	81	807	10.0

a. Compute the average number of assists per game for each player, expressed as a decimal rounded to the nearest tenth.

b. Decide which player has the highest average.   Player A

121. You decide to buy stock that sells for $1.50 a share. How many shares can you buy for $3,000?   2,000 shares

122. Typically, the heaviest organ in the body is the skin, weighing about 9 lb. By contrast, the heart weighs approximately 0.7 lb. About how many times the weight of the heart is that of the skin? (*Source: World of Scientific Discovery*)   13 times

123. A light microscope can distinguish two points separated by 0.0005 mm, whereas an electron microscope can distinguish two points separated by 0.0000005 mm. How many times as powerful as the light microscope is the electron microscope? 1,000 times

124. Gold was recently discovered in both Northern Ireland and South Africa. In the Irish mine, each pound of ore has 0.003 oz of gold. The African ore has 0.00008 oz of gold per pound. How many times as rich as the African ore is the Irish ore?   37.5 times

125. The fastest lap time in a swimming meet was 7.9 sec. If one of the losing swimmers swam 3 laps with times of 8.2, 8, and 8.9 sec, how much longer was her average lap time than the fastest time, to the nearest tenth of a second?   0.5 sec

126. The ages of the five starters on an all-star basketball team are 21, 28, 30, 25, and 24. If the team's coach is 45, how much older is he than the average player?   19.4 yr

■ *Use a calculator to solve the the following problems, giving (a) the operation(s) carried out in your solution, (b) the exact answer, and (c) an estimate of the answer.*

127. Babe Ruth got 2,873 hits in 8,399 times at bat, resulting in a batting average of $\frac{2,873}{8,399}$, or approximately .342. Another great player, Ty Cobb, got 4,191 hits out of 11,429 times at bat. What was his batting average, expressed as a decimal rounded to the nearest thousandth? (Note that batting averages don't have a zero to the left of the decimal point because they can never be greater than 1.) (*Source: The Baseball Encyclopedia*)   (a) Division   (b) 0.367   (c) 0.4

128. Light travels at a speed of 186,000 mi/sec. If Earth is about 93,000,000 mi from the sun, how many seconds, to the nearest tenth of a second, does it take for light to reach Earth from the sun?   (a) Division   (b) 500 sec   (c) 500 sec

■ *Check your answers on page A-5.*

MINDSTRETCHERS

PATTERNS

1. In the *repeating decimal* 0.142847142847142847... , identify the 994th digit to the right of the decimal point.   8

GROUPWORK

2. In the following magic square, 3.375 is the *product* of the numbers in every row, column, and diagonal. Working with a partner, fill in the missing numbers.

4.5	0.25	3
1	1.5	2.25
0.75	9	0.5

WRITING

3. a. $0.5 \div 0.8 = ?$ 0.625

b. $0.8 \div 0.5 = ?$ 1.6

c. Find the product of your answers in parts (a) and (b). Explain how you could have predicted this product.   Each fraction is the reciprocal of the other, so their product must equal 1.

variable and 50 is a constant. In writing an algebraic expression, we usually omit any multiplication symbol: $50d$ means $50 \cdot d$.

KEY

CONCEP

[3.1] Dec

[3.1] Dec

[3.1] To
   to t
   fra
   nur

[3.1] To

[3.1] To
   a gi

[3.2] To

[3.2] To

[3.3] To

Definition

A **variable** is a letter that represents an unknown number.

A **constant** is a known number.

An **algebraic expression** contains one or more variables and may contain any number of constants.

There are many ways to translate an algebraic expression to words, as the following table indicates.

$x + 4$ translates to	$n - 3$ translates to	$\frac{3}{4} \cdot y$, or $\frac{3}{4}y$, translates to	$z \div 5$ translates to
■ x plus 4. ■ x increased by 4. ■ the sum of x and 4. ■ 4 more than x.	■ n minus 3. ■ n decreased by 3. ■ the difference between n and 3. ■ 3 less than n.	■ $\frac{3}{4}$ times y. ■ the product of $\frac{3}{4}$ and y. ■ $\frac{3}{4}$ of y.	■ z divided by 5. ■ the quotient of z and 5. ■ z over 5.

EXAMPLE 1

Translate each algebraic expression in the table to words.

Solution

Algebraic Expression	Translation
a. $\dfrac{p}{3}$	p divided by 3
b. $x - 4$	4 less than x
c. $5f$	5 times f
d. $2 + y$	The sum of 2 and y
e. $\dfrac{2}{3}a$	$\dfrac{2}{3}$ of a

PRACTICE 1

Translate each algebraic expression to words.

Algebraic Expression	Translation
a. $\dfrac{1}{2}p$	One-half of p
b. $5 - x$	x less than 5
c. $y \div 4$	y divided by 4
d. $n + 3$	3 more than n
e. $\dfrac{3}{5}b$	$\dfrac{3}{5}$ of b

EXAMPLE 2

Translate each word phrase in the table to an algebraic expression.

Solution

Word Phrase	Translation
a. Sixteen more than m	$m + 16$
b. The product of 5 and b	$5b$
c. The quotient of 6 and z	$6 \div z$
d. A number decreased by 4	$n - 4$, where n represents the number
e. Three-eighths of a number	$\dfrac{3}{8}n$, where n represents the number

PRACTICE 2

Express each word phrase as an algebraic expression.

Word Phrase	Translation
a. x plus 9	$x + 9$
b. Ten times y	$10y$
c. The difference between n and 7	$n - 7$
d. A number divided by 5	Possible answer: $n \div 5$
e. Two-fifths of a number	Possible answer: $\dfrac{2}{5}n$

Now let's look at word problems that involve translations.

EXAMPLE 3

Suppose that p partners share equally in the profits of a business. What is each partner's share if the profit was $2,000?

Solution Each partner should get the quotient of 2,000 and p, which can be written algebraically as $\dfrac{2,000}{p}$ dollars.

PRACTICE 3

Next weekend you want to study for your four classes. If you have h hr to study in all and you want to devote the same amount of time to each class, how much time will you study per class? $\dfrac{h}{4}$ hr

Evaluating Expressions

In this section we look at how to evaluate algebraic expressions. Let's begin with a simple example.

Suppose that your balance in a savings account is $200. If you then deposit d dollars, you will have $200 + d$ dollars in the account.

To evaluate the expression $200 + d$ for a particular value of d, we replace d with that number. You deposited $50, so we would replace d by 50:

$$200 + d = 200 + 50 = 250$$

Your new balance would be $250.

The following rule is helpful for evaluating expressions.

> **To Evaluate an Algebraic Expression**
> - replace each variable with the given number, and
> - carry out the computation.

Now let's consider some more examples.

| EXAMPLE 4 | PRACTICE 4 |

Evaluate each algebraic expression.

Solution

Algebraic Expression	Solution
a. $n + 8$, if $n = 15$	$15 + 8 = 23$
b. $9 - z$, if $z = 7.89$	$9 - 7.89 = 1.11$
c. $\frac{2}{3}r$, if $r = 18$	$\frac{2}{3} \times 18 = 12$
d. $y \div 4$, if $y = 3.6$	$3.6 \div 4 = 0.9$

Find the value of each algebraic expression.

Algebraic Expression	Solution
a. $\frac{s}{4}$, if $s = 100$	25
b. $0.2y$, if $y = 1.9$	0.38
c. $x - 4.2$, if $x = 9$	4.8
d. $25 + z$, if $z = 1.6$	26.6

The following examples illustrate how to write and evaluate expressions to solve word problems.

| EXAMPLE 5 | PRACTICE 5 |

Power consumption for a period of time is measured in watt-hours (Wh), where a Wh means 1 W of power for 1 hr. How many Wh of energy will a 60W bulb consume in h hr? In 3 hr?

Solution The expression that represents the number of Wh used in h hr is $60h$. So for $h = 3$, the number of Wh is $60 \cdot 3$, or 180. Therefore 180 Wh of energy would be used in 3 hr.

In a keyboarding class, you type 55 words per minute (wpm). How many words do you type in m min? In 30 min? $55m$ words in m min and 1650 words in 30 min

EXAMPLE 6

Suppose that there are 180 days in the local school year. How many days were you present at school if you were absent d days? 9 days?

Solution If d represents the number of days that you were absent, the expression $180 - d$ represents the number of days you were present. If you were absent 9 days, we substitute 9 for d in the expression:

$$180 - d = 180 - 9 = 171$$

So you were present 171 days.

At a coffee shop, your bill is $15.45 plus the tip. How much will you pay if you give a tip of t dollars? $3?   You pay 15.45 + t dollars and $18.45 for $t = $3.

HISTORICAL NOTE

Just as we solve an equation to identify an unknown number, we use a balance scale to determine an unknown weight. In this picture which dates from 3,400 years ago, an Egyptian weighs gold rings against a counterbalance in the form of a bull's head.

The balance scale is an ancient measuring device. These scales were used by Sumarians for weighing precious metals and gems at least 9,000 years ago.

Source: O.A.W. Dilke, *Mathematics and Measurement* (Berkeley: University of California Press/British Museum, 1987).

Exercises 4.1

Translate each expression to two different word phrases. Answers may vary.

1. $t + 3$
3 more than t; t plus 3

2. $2 + r$
r more than 2; The sum of r and 2

3. $c - 4$
c minus 4; 4 subtracted from c

4. $x - 5$
x minus 5; 5 less than x

5. $c \div 3$
c divided by 3; The quotient of c and 3

6. $\dfrac{z}{2}$
The quotient of z and 2; z divided by 2

7. $10s$
10 times s; The product of 10 and s

8. $11t$
The product of 11 and t; 11 times t

9. $y - 10$
y minus 10; 10 less than y

10. $w - 1$
1 less than w; w minus 1

11. $7a$
7 times a; The product of 7 and a

12. $4x$
4 multiplied by x; 4 times x

13. $x \div 4$
x divided by 4; The quotient of x and 4

14. $\dfrac{y}{5}$
y divided by 5; y over 5

15. $x - \dfrac{1}{2}$
x minus $\frac{1}{2}$; $\frac{1}{2}$ less than x

16. $x - \dfrac{1}{3}$
$\frac{1}{3}$ subtracted from x; x minus $\frac{1}{3}$

17. $\dfrac{1}{2}w$
$\frac{1}{2}$ times w; $\frac{1}{2}$ of w

18. $\dfrac{4}{5}y$
The product of $\frac{4}{5}$ and y; $\frac{4}{5}$ of y

19. $x - 2$
x minus 2; The difference between x and 2

20. $y - 6$
y decreased by 6; 6 less than y

21. $1 + x$
1 increased by x; x added to 1

22. $n + 7$
n plus 7; 7 more than n

23. $3p$
3 times p; The product of 3 and p

24. $2x$
Twice x; 2 multiplied by x

25. $n - 1.1$
n decreased by 1.1; n minus 1.1

26. $x - 6.5$
x minus 6.5; 6.5 less than x

27. $y \div 0.9$
y divided by 0.9; The quotient of y and 0.9

28. $\dfrac{n}{2.4}$
n over 2.4; n divided by 2.4

Translate each word phrase to an algebraic expression.

29. x plus 10 $\quad x + 10$

30. d increased by 12 $\quad d + 12$

31. 5 less than n $\quad n - 5$

32. b decreased by 3 $\quad b - 3$

33. The sum of y and 5 $\quad y + 5$

34. 4 more than x $\quad x + 4$

35. t divided by 6 $\quad$ Possible answer: $t \div 6$

36. r over 2 $\quad \dfrac{r}{2}$

37. The product of 10 and y $10y$

38. 5 times p $5p$

39. w minus 5 $w - 5$

40. The difference between n and 5
$n - 5$

41. n increased by 100 $n + 100$

42. The sum of x and 20 $x + 20$

43. The quotient of z and 3   Possible answer: $z \div 3$

44. n divided by 10   Possible answer: $\dfrac{n}{10}$

45. $\dfrac{2}{5}$ of x $\dfrac{2}{5}x$

46. The product of $\dfrac{1}{2}$ and y $\dfrac{1}{2}y$

47. Six subtracted from k $k - 6$

48. Eight less than z $z - 8$

49. Five more than a number $n + 5$

50. The sum of a number and 18 $n + 18$

51. The difference between a number and five and one-half
$n - 5\dfrac{1}{2}$

52. Three-fourths less than a number
$n - \dfrac{3}{4}$

Evaluate each expression.

53. $y + 7$, if $y = 19$
26

54. $3 + n$, if $n = 2.9$
5.9

55. $7 - x$, if $x = 4.5$
2.5

56. $y - 19$, if $y = 25$
6

57. $\dfrac{3}{4}p$, if $p = 20$
15

58. $\dfrac{4}{5}n$, if $n = 1\dfrac{1}{4}$
1

59. $x \div 2$, if $x = 2\dfrac{1}{3}$
$1\dfrac{1}{6}$

60. $\dfrac{n}{3}$, if $n = 7.5$
2.5

61. $p - 7.9$, if $p = 9$
1.1

62. $20.1 - y$, if $y = 7$
13.1

63. $x + \dfrac{5}{6}$, if $x = \dfrac{1}{6}$
1

64. $\dfrac{1}{3}y$, if $y = \dfrac{1}{2}$
$\dfrac{1}{6}$

Complete each table.

65.

x	$x + 8$
1	9
2	10
3	11
4	12

66.

x	$10 - x$
1	9
2	8
3	7
4	6

67.

n	$n - 0.2$
1	0.8
2	1.8
3	2.8
4	3.8

68.

b	$b + 2.5$
1	3.5
2	4.5
3	5.5
4	6.5

69.

x	$\frac{3}{4}x$
4	3
8	6
12	9
16	12

70.

n	$\frac{2}{3}n$
3	2
6	4
9	6
12	8

71.

z	$\frac{z}{2}$
2	1
4	2
6	3
8	4

72.

y	$\frac{y}{5}$
5	1
10	2
15	3
20	4

Applications

Solve.

73. Suppose that you have c classes. Your friend has two more classes than you. How many classes does your friend have? $c + 2$

74. When you take out a mortgage, each monthly payment has two parts. One part goes toward the principal and the other toward the interest. If the principal payment is $344.86 and the interest payment is i, write an algebraic expression for the total payment. $344.86 + i$ dollars

75. For the triangle shown, write an expression for the sum of the measures of the three angles. $30° + 90° + d°$, or $120° + d°$

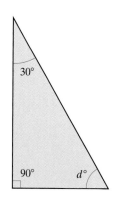

76. Write an expression for the sum of the lengths of the sides in the figure shown.

$8 + 6 + 5 + b$, or $19 + b$

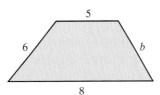

77. If you drive at a speed of r mph for t hr, you will travel a distance of $r \cdot t$ mi. How far will you travel at a speed of 50 mph in 3 hr?   150 mi

78. If a basketball player makes b baskets in a attempts, her field goal average is defined to be $\dfrac{b}{a}$. Find the field goal average of a player who made 12 baskets in 25 attempts.

$\dfrac{12}{25}$

79. Suppose that you sell magazine subscriptions and earn $2 for each subscription sold.

a. Write an expression for your earnings if you sell s subscriptions. $2s$ dollars

b. Find your earnings if you sell 14 subscriptions. $28

80. A computer repair technician charges $40/hr for labor.

a. Write an expression for the cost of h hr of work. $40h$ dollars

b. Find the cost of a computer repair that takes $2\dfrac{1}{2}$ hr. $100

■ *Check your answers on page A-6.*

MINDSTRETCHERS

CRITICAL THINKING

1. Consider the expression $x + x$. Why does this expression mean the same as the expression $2x$?   Adding any number to itself is equivalent to multiplying that number by 2.

_____

_____

GROUPWORK

2. Consider the areas of the following rectangles. For some values of x, the rectangle on the left has a larger area; for other values of x, the rectangle on the right is larger.

a. Find a value of x for which the rectangle on the left has the larger area.   Possible answers: 5, 6, 7, 8, 9, 10, ...

b. Find a value of x for which the area of the rectangle on the right is larger.   Possible answers: 1, 2, 3

c. Compare your findings with the members of your group.

WRITING

3. Algebra is universal; that is, it is used in all countries of the world regardless of the language spoken. If you know how to speak a language other than English, translate each of the following algebraic expressions to that language.   Answers may vary.

a. $7x$ _____

b. $x - 2$ _____

c. $3 + x$ _____

d. $\dfrac{x}{3}$ _____

4.2 Solving Addition and Subtraction Equations

OBJECTIVES

■ *To translate sentences to equations involving addition or subtraction*
■ *To solve addition and subtraction equations*
■ *To solve word problems involving equations with addition or subtraction*

What an Equation Is

An equation has a left-side expression, an equals sign, and a right-side expression.

$$\underbrace{x + 3}_{\substack{\text{Left}\\\text{side}}} \overset{\substack{\text{Equals}\\\text{sign}}}{=} \underbrace{5}_{\substack{\text{Right}\\\text{side}}}$$

Definition

An **equation** is a mathematical statement that two expressions are equal.

For example,

$$1 + 2 = 3$$
$$x - 5 = 6$$
$$2 + 7 + 3 = 12$$
$$3x = 9$$

are all equations.

Translating Sentences to Equations

In translating sentences to equations, certain words and phrases mean the same as the equals sign:

■ equals

■ is

■ is equal to

■ is the same as

■ yields

■ results in

Let's look at some examples of translating sentences to equations that involve addition or subtraction, and vice versa.

EXAMPLE 1

Translate each sentence in the table to an equation.

Solution

Sentence	Equation
a. The sum of y and 3 is equal to $7\frac{1}{2}$.	$y + 3 = 7\frac{1}{2}$
b. The difference between x and 9 is the same as 14.	$x - 9 = 14$
c. Increasing a number by 1.5 yields 3.	$n + 1.5 = 3$
d. Three less than a number is 10.	$n - 3 = 10$

PRACTICE 1

Write an equation for each phrase or sentence.

Sentence	Equation
a. n decreased by 5.1 is 9.	$n - 5.1 = 9$
b. y plus 5 is equal to 12.	$y + 5 = 12$
c. The difference between a number and 4 is the same as 12.	$n - 4 = 12$
d. Five more than a number is $7\frac{3}{4}$.	$n + 5 = 7\frac{3}{4}$

EXAMPLE 2

In your savings account the previous balance, P, plus a deposit of $7.50 equals the new balance of $43.25. Write an equation that represents this situation.

Solution Previous balance   plus   deposit   equals   new balance.

$\qquad\qquad\qquad P \qquad\quad + \qquad 7.50 \qquad = \qquad 43.25$

The equation is $P + 7.50 = 43.25$

PRACTICE 2

The sale price of a jacket is $49.95. This amount is $6 less than the regular price. Write an equation that represents this situation. $p - 6 = 49.95$, where p is the regular price.

Equations Involving Addition and Subtraction

Let's look at a simple example of using algebra to solve a word sentence. Suppose that you are told that five *more than some number* is equal to seven. You can find that number by solving the addition equation $x + 5 = 7$. To solve an equation means to find the number that, when substituted for the variable x, makes the equation a true statement.

To solve the equation $x + 5 = 7$, we can think of a balance scale like the one shown.

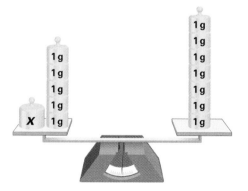

For the balance to remain level, whatever we do to one side we must also do to the other side. In this case, if we subtract 5 g from each side of the balance, we can conclude that the unknown weight, x, must be 2 g, as shown on the next page.

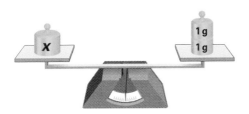

Similarly in the *subtraction equation* $y - 3 = 7$, if we add 3 to each side of the equation, we find that y equals 10.

In solving these and other equations, the key is to **isolate the variable**, that is, to get the variable alone on one side of the equation.

These examples suggest the following rule.

To Solve Addition or Subtraction Equations

- in an addition equation, subtract the same number from each side of the equation in order to isolate the variable on one side;

$$x + a = b$$
$$x + a - a = b - a$$
$$x = b - a$$

- in a subtraction equation, add the same number to each side of the equation in order to isolate the variable on one side; and

$$x - a = b$$
$$x - a + a = b + a$$
$$x = b + a$$

- in either case, check the solution by substituting the value of the unknown in the original equation to verify that the resulting equation is true.

Because addition and subtraction are **opposite operations**, one operation "undoes" the other. The following examples illustrate how to perform an opposite operation to each side of an equation when you are solving for the unknown.

EXAMPLE 3

Solve and check: $y + 9 = 17$

Solution

$$y + 9 = 17$$
$$y + 9 - 9 = 17 - 9 \qquad \text{Subtract 9 from each side of the equation.}$$
$$y + 0 = 8 \qquad 9 - 9 = 0 \text{ and } 17 - 9 = 8.$$
$$y = 8 \qquad y + 0 = y \text{ (any number added to 0 is the number).}$$

Check: $y + 9 = 17$
$$8 + 9 \stackrel{?}{=} 17 \qquad \text{Substitute 8 for } y \text{ in the original equation.}$$
$$17 \stackrel{\checkmark}{=} 17 \qquad \text{The equation is true, so 8 is the solution to the equation.}$$

Note that, because 9 was added to y in the original equation, we solved by subtracting 9 from both sides of the equation in order to isolate the variable.

PRACTICE 3

Solve and check: $x + 5 = 14$ $x = 9; 9 + 5 \stackrel{?}{=} 14, 14 \stackrel{\checkmark}{=} 14$

EXAMPLE 4

Solve and check: $n - 2.5 = 0.7$

Solution

$$n - 2.5 = 0.7$$

$$n - \underbrace{2.5 + 2.5}_{} = \underbrace{0.7 + 2.5}_{} \quad \text{Add 2.5 to each side of the equation.}$$

$$\underbrace{n - \quad 0}_{n} = \quad 3.2$$

$$n = 3.2$$

Check: $n - 2.5 = 0.7$

$$3.2 - 2.5 \overset{?}{=} 0.7 \quad \text{Substitute 3.2 for } n \text{ in the original equation.}$$

$$0.7 \overset{\checkmark}{=} 0.7$$

Can you explain why checking an answer is important?

Have students write their answers in a journal.

PRACTICE 4

Solve and check: $t - 0.9 = 1.8$ $\quad t = 2.7; 2.7 - 0.9 \overset{?}{=} 1.8, 1.8 \overset{\checkmark}{=} 1.8$

EXAMPLE 5

Solve and check: $x + \dfrac{1}{2} = 3\dfrac{1}{2}$

Solution

$$x + \frac{1}{2} = 3\frac{1}{2}$$

$$x + \frac{1}{2} - \frac{1}{2} = 3\frac{1}{2} - \frac{1}{2} \quad \text{Subtract } \frac{1}{2} \text{ from each side of the equation.}$$

$$x = 3$$

Check: $x + \dfrac{1}{2} = 3\dfrac{1}{2}$

$$3 + \frac{1}{2} \overset{?}{=} 3\frac{1}{2} \quad \text{Substitute 3 for } x \text{ in the original equation.}$$

$$3\frac{1}{2} \overset{\checkmark}{=} 3\frac{1}{2}$$

PRACTICE 5

Solve and check: $m + \dfrac{1}{4} = 5\dfrac{1}{2}$ $\quad m = 5\frac{1}{4}; 5\frac{1}{4} + \frac{1}{4} \overset{?}{=} 5\frac{1}{2}$

$$5\frac{2}{4} \overset{?}{=} 5\frac{1}{2}$$

$$5\frac{1}{2} \overset{\checkmark}{=} 5\frac{1}{2}$$

EXAMPLE 6

Write the sentences on the left in the table as equations. Then solve and check.

Solution

Sentence	Equation	Check
a. Fifteen is equal to y increased by 9.	$15 = 9 + y$ $15 - 9 = 9 - 9 + y$ $6 = y$	$15 \overset{?}{=} 9 + y$ $15 \overset{?}{=} 9 + 6$ $15 \overset{\checkmark}{=} 15$
b. Ten is equal to m decreased by 8.	$10 = m - 8$ $10 + 8 = m - 8 + 8$ $18 = m$	$10 \overset{?}{=} m - 8$ $10 \overset{?}{=} 18 - 8$ $10 \overset{\checkmark}{=} 10$

> Point out that the equations $15 = y + 9$ and $y + 9 = 15$ are equivalent.

Note that we isolated the variable on the right side of the equation instead of the left side. The result is the same.

PRACTICE 6

Translate each word sentence to an algebraic equation. Then solve and check.

a. 11 is 4 less than m. $11 = m - 4$; $m = 15$

b. The sum of 12 and n equals 21. $12 + n = 21$; $n = 9$

Word Problems Involving Addition and Subtraction Equations

Equations are often useful **mathematical models** of real-world situations, as the following examples show.

EXAMPLE 7

Suppose that a chemistry experiment requires you to find the mass of the water in a plastic bottle. If the mass of the bottle with water is 21.49 g and the mass of the empty bottle is 9.56 g, write an equation to find the mass of the water. Then solve and check.

Solution Recall that some problems can be solved by drawing a diagram. Let's use that strategy here.

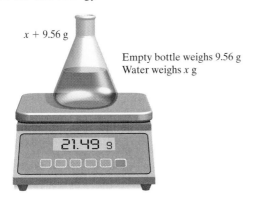

$x + 9.56$ g

Empty bottle weighs 9.56 g
Water weighs x g

PRACTICE 7

You have $809.46 in your savings account. This amount is $144.95 more than you had in your account yesterday. Write an equation to determine the previous amount. Then solve and check.
$809.46 = 144.95 + x$; $x = \$664.51$

continued

The diagram suggests the equation $21.49 = x + 9.56$, where x represents the mass of the water.

Solving this equation, we get:

$$21.49 = x + 9.56$$

$$21.49 - 9.56 = x + 9.56 - 9.56$$

$$11.93 = x$$

The weight of the water is 11.93 g.

Check: $21.49 = x + 9.56$

$$21.49 \overset{?}{=} 11.93 + 9.56$$

$$21.49 \overset{\checkmark}{=} 21.49$$

EXAMPLE 8

Because of a recession, the monthly output of a factory fell by $15\frac{1}{2}$ metric tons to $37\frac{7}{8}$ metric tons (1 metric ton equals 1,000 kg, or about 2,200 lb). To determine how many metric tons the monthly factory output was before the recession, write an equation. Then solve and check.

Solution Let x equal the original factory output, in metric tons. This gives us the equation:

$$x - 15\frac{1}{2} = 37\frac{7}{8}$$

Now we solve for the unknown:

$$x - 15\frac{1}{2} + 15\frac{1}{2} = 37\frac{7}{8} + 15\frac{1}{2}$$

$$x = 53\frac{3}{8}$$

The factory produced $53\frac{3}{8}$ metric tons monthly before the recession.

Check: $x - 15\frac{1}{2} = 37\frac{7}{8}$

$$53\frac{3}{8} - 15\frac{1}{2} \overset{?}{=} 37\frac{7}{8}$$

$$37\frac{7}{8} \overset{\checkmark}{=} 37\frac{7}{8}$$

PRACTICE 8

Yesterday the stock of a Fortune 500 company closed at $\$26\frac{1}{4}$, after losing $\$1\frac{5}{8}$.

Write an equation to determine how much the stock opened for at the beginning of the day. Then solve and check.

$$26\frac{1}{4} = x - 1\frac{5}{8}; x = \$27\frac{7}{8}$$

Remind students to follow the rules for adding and subtracting fractions in this example.

Exercises 4.2

For Extra Help

 Tape 5

 InterAct Math
Tutorial Software

 AWL Tutor Center

Student's Solutions
Manual

 www.awlonline.com/
akstbragg

Translate each sentence to an equation.

1. z minus 9 is 25. $z - 9 = 25$

2. x decreased by 7 yields 29.
$x - 7 = 29$

3. The sum of 7 and x is 25. $7 + x = 25$

4. m plus 19 equals 34. $m + 19 = 34$

5. t decreased by 3.1 equals 4.
$t - 3.1 = 4$

6. r minus 5 is equal to 6.4.
$r - 5 = 6.4$

7. $\frac{3}{2}$ increased by y yields $\frac{9}{2}$.
$\frac{3}{2} + y = \frac{9}{2}$

8. The sum of n and $2\frac{1}{3}$ is 8.

$n + 2\frac{1}{3} = 8$

9. $3\frac{1}{2}$ less than a number is equal to 7.

$n - 3\frac{1}{2} = 7$

10. The difference between a number and
$1\frac{1}{2}$ is the same as $7\frac{1}{4}$.

$n - 1\frac{1}{2} = 7\frac{1}{4}$

By answering yes or no, indicate whether the value of x shown is a solution to the given equation.

11.

Equation	Value of x	Solution?
a. $x + 1 = 9$	8	Yes
b. $x - 3 = 4$	5	No
c. $x + 0.2 = 5$	4.8	Yes
d. $x - \frac{1}{2} = 1$	$\frac{1}{2}$	No

12.

Equation	Value of x	Solution?
a. $x - 39 = 5$	44	Yes
b. $x - 2 = 6$	4	No
c. $x + 2.8 = 4$	1.2	Yes
d. $x - \frac{2}{3} = 1$	$1\frac{2}{3}$	Yes

Identify the operation to perform on each side of the equation to isolate the variable.

13. $x + 4 = 6$
Subtract 4

14. $x - 6 = 9$
Add 6

15. $x - 11 = 4$
Add 11

16. $x + 10 = 17$
Subtract 10

17. $x - 7 = 24$
Add 7

18. $x + 21 = 25$
Subtract 21

19. $3 = x + 2$
Subtract 2

20. $10 = x - 3$
Add 3

Solve and check.

21. $a - 7 = 24$
$a = 31$

22. $x - 9 = 13$
$x = 22$

23. $y + 19 = 21$
$y = 2$

24. $z + 23 = 31$
$z = 8$

25. $x - 2 = 10$
$x = 12$

26. $t - 4 = 19$
$t = 23$

27. $n + 9 = 13$
$n = 4$

28. $d + 12 = 12$
$d = 0$

29. $c - 14 = 33$
   $c = 47$

30. $a - 9 = 27$
   $a = 36$

31. $x - 3.4 = 9.6$
   $x = 13$

32. $m - 12.5 = 13.7$
   $m = 26.2$

33. $z + 2.4 = 5.3$
   $z = 2.9$

34. $t + 2.3 = 6.7$
   $t = 4.4$

35. $n - 8 = 0.9$
   $n = 8.9$

36. $c - 0.7 = 6$
   $c = 6.7$

37. $y + 8.1 = 9$

   $y = 0.9$

38. $a + 0.7 = 2$

   $a = 1.3$

39. $x + 2\frac{1}{3} = 9$

   $x = 6\frac{2}{3}$

40. $z + 3\frac{2}{5} = 11$

   $z = 7\frac{3}{5}$

41. $m - 1\frac{1}{3} = 4$

   $m = 5\frac{1}{3}$

42. $s - 4\frac{1}{2} = 8$

   $s = 12\frac{1}{2}$

43. $x + 3\frac{1}{4} = 7$

   $x = 3\frac{3}{4}$

44. $t + 1\frac{1}{2} = 5$

   $t = 3\frac{1}{2}$

45. $5 + m = 7$
   $m = 2$

46. $17 + d = 20$
   $d = 3$

47. $39 = y - 51$
   $y = 90$

48. $44 = c - 3$
   $c = 47$

49. $5 = y - 1\frac{1}{4}$

   $y = 6\frac{1}{4}$

50. $3 = t - 1\frac{2}{3}$

   $t = 4\frac{2}{3}$

51. $4 = n + 3\frac{1}{2}$

   $n = \frac{1}{2}$

52. $5 = a + 2\frac{1}{3}$

   $a = 2\frac{2}{3}$

53. $2.3 = x - 5.9$
   $x = 8.2$

54. $4.1 = d - 6.9$
   $d = 11$

55. $y - 7.01 = 12.9$
   $y = 19.91$

56. $x - 3.2 = 5.23$
   $x = 8.43$

▦ **57.** $x + 3.443 = 8$
   $x = 4.557$

▦ **58.** $x + 0.035 = 2.004$
   $x = 1.969$

▦ **59.** $2.986 = y - 7.265$
   $y = 10.251$

▦ **60.** $3.184 = y - 1.273$
   $y = 4.457$

Translate each sentence to an equation. Then solve and check.

61. Three more than *n* is 11. $n + 3 = 11$;
   $n = 8$

62. The sum of *x* and 15 is the same as 33.
   $x + 15 = 33$; $x = 18$

63. Six less than *y* equals 7. $y - 6 = 7$;
   $y = 13$

64. The difference between *t* and 4 yields 1.
   $t - 4 = 1$; $t = 5$

65. If 10 is added to *n*, the sum is 19.
   $n + 10 = 19$; $n = 9$

66. Twenty-five added to a number *m* gives
   a result of 53. $m + 25 = 53$; $m = 28$

67. *x* increased by 3.6 is equal to 9.
   $x + 3.6 = 9$; $x = 5.4$

68. *n* plus $3\frac{1}{2}$ equals 7. $n + 3\frac{1}{2} = 7$;

   $n = 3\frac{1}{2}$

69. A number minus $4\frac{1}{3}$ is the same as $2\frac{2}{3}$.

   $n - 4\frac{1}{3} = 2\frac{2}{3}$; $n = 7$

70. A number decreased by 1.6 is 5.9.
   $n - 1.6 = 5.9$; $n = 7.5$

Choose the equation that describes the situation.

71. After 6 mo of dieting and exercising, you lost $8\frac{1}{2}$ lb. If you now weigh 135 lb, what was your original weight?   Equation c

a. $w + 8\frac{1}{2} = 135$ **b.** $w - 126\frac{1}{2} = 8\frac{1}{2}$ **c.** $w - 8\frac{1}{2} = 135$ **d.** $w + 135 = 143\frac{1}{2}$

72. You have *d* dollars. After paying a bill of $15.25, you have $6.75 left. How many dollars did you have at first?   Equation b

a. $d + 15.25 = 22$ **b.** $d - 15.25 = 6.75$ **c.** $d - 6.75 = 15.25$ **d.** $d + 6.75 = 15.25$

73. Suppose that there are 200 megabytes (MB) of memory on your computer's hard drive. You store a document that takes up 8.7 MB. How many megabytes of memory are left?

a. $x - 8.7 = 200$ **b.** $x - 200 = 8.7$ **c.** $x + 200 = 28.7$ **d.** $x + 8.7 = 200$
Equation d

74. At a certain college, tuition costs a student $2,000 a semester. If the student received $1,250 in financial aid, how much more money does the student need for the semester's tuition?   Equation a

a. $x + 1,250 = 2,000$ **b.** $x + 2,000 = 3,250$

c. $x - 2,000 = 1,250$ **d.** $x - 1,250 = 2,000$

Applications

Write an equation. Solve and then check.

75. An article on Broadway shows reported that this week the box office receipts for a particular show were $621,000. If that amount was $13,000 less than last week's, how much money did the show take in last week? $621,000 = x - 13,000; x = \$634,000$

76. The first algebra textbook was written by the Arab mathematician Muhammad ibn Musa al-Khwarazmi. The title of that book, which gave rise to the word *algebra*, was *Aljabr wa'lmuqabalah*, meaning "the art of bringing together unknowns to match a known quantity." If the book appeared in the year 825, how many years ago was this? (***Source:*** R.V. Young, *Notable Mathematicians*) $x + 825 = $ year used; answer will vary depending on year used

77. In the triangle shown, angles *A* and *B* are complementary; that is, the sum of their measures is 90°. Find the number of degrees in $\angle B$.

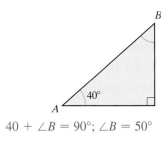

$40 + \angle B = 90°; \angle B = 50°$

78. In the following diagram, angles *x* and *y* are supplementary, that is, the sum of their measures is 180°. If $\angle x = 109°$, find the number of degrees in $\angle y$.

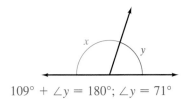

$109° + \angle y = 180°; \angle y = 71°$

■ *Check your answers on page A-7.*

MINDSTRETCHERS

GROUPWORK

1. Working with a partner, compare the equations $x - 4 = 6$ and $x - a = b$.

a. Use what you know about the first equation to solve the second equation for x.

$$
\begin{array}{lcl}
x - 4 = 6 & \longleftrightarrow & x - a = b \\
x - 4 + 4 = 6 + 4 & \longleftrightarrow & x - a + a = b + a \\
x = 10 & \longleftrightarrow & x = b + a
\end{array}
$$

b. What are the similarities and the differences between the two equations?

Possible answer: Similarities—Both equations involve subtracting a quantity from x.

Both equations are solved by addition.

Differences: In $x - 4 = 6$, the solution is an arithmetic expression. In $x - a = b$, the

solution is an algebraic expression. The solution of $x - a = b$ must be expressed as the

sum of two quantities.

c. Discuss your findings.

Answers may vary.

WRITING

2. Equations often serve as models for solving word problems. Write two different word problems corresponding to each following equation.

a. $x + 4 = 9$

■ Answers may vary.

■

b. $x - 1 = 5$

■ Answers may vary.

■

CRITICAL THINKING

3. In the magic square at the right, the sum of each row, column, and diagonal is the same. Find that sum and write and solve equations to get the values of f, g, h, r, and t.　Sum $= 30$; possible
equations:　$r + 21 = 30, r = 9$　$f + 17 = 30, f = 13$
　　　　　　$g + 22 = 30, g = 8$　$t + 23 = 30, t = 7$
　　　　　　$h + 18 = 30, h = 12$

f	6	11
g	10	h
r	14	t

4.3 Solving Multiplication and Division Equations

Objectives

- *To translate sentences to equations involving multiplication or division*
- *To solve multiplication and division equations*
- *To solve word problems involving equations with multiplication or division*

Translating Sentences to Equations

In order to translate sentences involving multiplication or division to equations, we must recall the key words that indicate when to multiply and when to divide.

EXAMPLE 1

Translate each sentence in the table to an equation.

Solution

	Sentence	Equation
a.	The product of 3 and x is equal to 0.6.	$3x = 0.6$
b.	The quotient of y and 4 is 15.	$\dfrac{y}{4} = 15$
c.	Two-thirds of a number is 9.	$\dfrac{2}{3}n = 9$
d.	One-half is equal to some number over 6.	$\dfrac{1}{2} = \dfrac{n}{6}$

PRACTICE 1

Write an equation for each sentence.

	Sentence	Equation
a.	Twice x is the same as 14.	$2x = 14$
b.	The quotient of a and 6 is 1.5.	$\dfrac{a}{6} = 1.5$
c.	Some number divided by 0.3 is equal to 1.	$\dfrac{n}{0.3} = 1$
d.	Ten is equal to one-half of some number.	$10 = \dfrac{1}{2}n$

EXAMPLE 2

A house that sold for $125,000 is twice its assessed value, x. Write an equation to represent this situation.

Solution $\underbrace{\text{Selling price of the house}}$ is $\underbrace{\text{twice its assessed value, } x.}$
$\qquad\qquad\quad 125{,}000 \qquad\quad = \qquad\qquad 2x$

The equation is $125{,}000 = 2x$.

PRACTICE 2

The area of a rectangle is equal to the product of its length (3 ft) and its width (w). The rectangle's area is 15 sq ft. Represent this relationship in an equation. $15 = 3w$

Equations Involving Multiplication and Division

As for addition equations, we can also solve *multiplication equations* by thinking of a balance scale like the one shown below at the left.

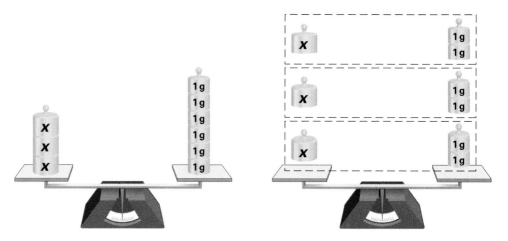

For example, consider the sentence "Three times some number x equals six," which translates to the multiplication equation $3x = 6$. We want to find the number for the variable x that, when substituted, makes this equation a true statement. To keep the balance level, whatever we do to one side we must do to the other side. In this case, dividing each side of the balance by 3 shows that in each group the unknown, x, must equal 2, as shown above at the right.

Similarly in the division equation $\dfrac{x}{4} = 3$, we can multiply each side of the equation by 4 and then conclude that x equals 12.

To Solve Multiplication or Division Equations

■ In a multiplication equation, divide by the same number on each side of the equation to isolate the variable on one side.

$$ax = b$$
$$\frac{\cancel{a}x}{\cancel{a}} = \frac{b}{a}$$
$$x = \frac{b}{a}$$

■ In a division equation, multiply by the same number on each side of the equation to isolate the variable on one side.

$$\frac{x}{a} = b$$
$$\cancel{a}\,\frac{x}{\cancel{a}} = ab$$
$$x = ab$$

■ In either case, check the solution by substituting the value of the unknown in the original equation to verify that the resulting equation is true.

Because multiplication and division are opposite operations, one "undoes" the other. The following examples show how to perform the opposite operation on each side of an equation to solve for the unknown.

EXAMPLE 3

Solve and check: $5x = 20$

Solution $5x = 20$

$\dfrac{5x}{5} = \dfrac{20}{5}$ Divide each side of the equation by 5: $\dfrac{5x}{5} = 1x$, or x.

$x = 4$

Check: $5x = 20$

$5(4) \overset{?}{=} 20$ Substitute 4 for x in the original equation.

$20 \overset{\checkmark}{=} 20$ The equation is true, so 4 is the solution to the original equation.

PRACTICE 3

Solve and check: $6x = 30$ $x = 5$

Have students explain in a journal why $1x = x$.

EXAMPLE 4

Solve and check: $5 = \dfrac{y}{2}$

Solution $5 = \dfrac{y}{2}$

$2 \cdot 5 = 2 \cdot \dfrac{y}{2}$ Multiply each side of the equation by 2: $2 \cdot \dfrac{y}{2} = \dfrac{2}{1} \cdot \dfrac{y}{2} = 1y$, or y.

$10 = y$

Check: $5 = \dfrac{y}{2}$

$5 \overset{?}{=} \dfrac{10}{2}$ Substitute 10 for y in the original equation.

$5 \overset{\checkmark}{=} 5$

PRACTICE 4

Solve and check: $1 = \dfrac{a}{6}$ $a = 6$

EXAMPLE 5

Solve and check: $0.2n = 4$

Solution $0.2n = 4$

$\dfrac{0.2n}{0.2} = \dfrac{4}{0.2}$ Divide each side by 0.2: $0.2\overline{)4.0}$, or 20.

$n = 20$

PRACTICE 5

Solve and check: $1.5x = 6$ $x = 4$

Remind students to follow the rule for dividing decimals in this example.

continued

Check: $0.2n = 4$

$$0.2(20) \overset{?}{=} 4 \quad \text{Substitute 20 for } n \text{ in the original equation.}$$

$$4.0 \overset{?}{=} 4$$

$$4 \overset{\checkmark}{=} 4$$

EXAMPLE 6

Solve and check: $\dfrac{m}{0.5} = 1.3$

Solution $\dfrac{m}{0.5} = 1.3$

$$(0.5)\dfrac{m}{0.5} = (1.3)\,(0.5) \quad \textbf{Multiply each side by 0.5.}$$

$$m = 0.65$$

Check: $\dfrac{m}{0.5} = 1.3$

$$\dfrac{0.65}{0.5} \overset{?}{=} 1.3 \quad \textbf{Substitute 0.65 for } m \text{ in the original equation.}$$

$$1.3 \overset{\checkmark}{=} 1.3$$

PRACTICE 6

Solve and check: $\dfrac{a}{2.4} = 1.2 \quad a = 2.88$

> Remind students to follow the rule for multiplying decimals in this example.

EXAMPLE 7

Solve and check: $\dfrac{2}{3}n = 6$

Solution $\dfrac{2}{3}n = 6$

$$\dfrac{2}{3}n \div \dfrac{2}{3} = 6 \div \dfrac{2}{3} \quad \textbf{Divide each side by } \dfrac{2}{3}.$$

$$\left(\dfrac{2}{3}n\right)\left(\dfrac{3}{2}\right) = 6\left(\dfrac{3}{2}\right) \quad \textbf{To divide by a number means to multiply by its reciprocal.}$$

$$n = 9$$

Check: $\dfrac{2}{3}n = 6$

$$\dfrac{2}{3}(9) \overset{?}{=} 6 \quad \textbf{Substitute 9 for } n \text{ in the original equation.}$$

$$\dfrac{2}{3}\left(\dfrac{9}{1}\right) \overset{?}{=} 6$$

$$6 \overset{\checkmark}{=} 6$$

PRACTICE 7

Solve and check: $\dfrac{3}{4}x = 12 \quad x = 16$

> An alternative method of solving Example 7 is to multiply both sides of the equation by the reciprocal of $\dfrac{2}{3}$.

> Remind students to follow the rule for dividing fractions in this example.

Word Problems Involving Multiplication and Division Equations

As in the case of addition and subtraction equations, multiplication and division equations can be useful mathematical models of real-world situations. To derive these models, we need to be able to translate word sentences to algebraic equations.

EXAMPLE 8

Write each sentence as an algebraic equation. Then solve and check.

Solution

Sentence	Equation	Check
a. Thirty-five is equal to the product of 5 and x.	$35 = 5x$ $\dfrac{35}{5} = \dfrac{5x}{5}$ $7 = x$	$35 = 5x$ $35 \overset{?}{=} 5(7)$ $35 \overset{\checkmark}{=} 35$
b. One equals p divided by 3.	$1 = \dfrac{p}{3}$ $3 \cdot 1 = 3 \cdot \dfrac{p}{3}$ $3 = p$	$1 = \dfrac{p}{3}$ $1 \overset{?}{=} \dfrac{3}{3}$ $1 \overset{\checkmark}{=} 1$

PRACTICE 8

Translate each sentence to an equation. Then solve and check.

Sentence	Equation	Check
a. Twelve is equal to the quotient of z and 6.	$12 = \dfrac{z}{6}$ $z = 72$	$12 \overset{?}{=} \dfrac{72}{6}$ $12 \overset{\checkmark}{=} 12$
b. Sixteen equals twice x.	$16 = 2x$ $8 = x$	$16 \overset{?}{=} 2(8)$ $16 \overset{\checkmark}{=} 16$

EXAMPLE 9

A baseball player runs 360 ft when hitting a home run.

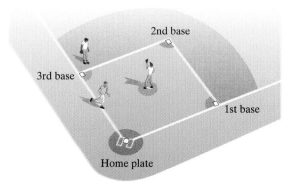

PRACTICE 9

The Pentagon is the headquarters of the U.S. Department of Defense.

continued

If the distances between successive bases on a baseball diamond are equal, how far is it from third base to home plate? Write an equation. Then solve and check.

Solution Let x equal the distance between successive bases. Since these distances are equal, $4x$ represents the distance around the bases.

But we know that the distance around the bases also equals 360 ft. So $4x = 360$, and we solve this equation for x.

$$4x = 360$$

$$\frac{4x}{4} = \frac{360}{4} \quad \text{Divide each side by 4.}$$

$$x = 90$$

The distance from third base to home plate is 90 ft.

Check: $4x = 360$

$$4(90) \overset{?}{=} 360 \quad \text{Substitute 90 for } x.$$

$$360 \overset{\checkmark}{=} 360$$

The distance around the Pentagon is about 1.6 km. If each side of the Pentagon is the same length, write an equation to find that length. Then solve and check. (*Source:* Gene Gurney, *The Pentagon*)

$1.6 = 5x;\ x = 0.32$ km

EXAMPLE 10

One-third of your monthly budget goes to food. If $320 goes to food, how much is your total monthly budget? Write an equation. Then solve and check.

Solution Let b equal your total monthly budget. One-third of this budget, or $320, goes for food, so we can write $\frac{1}{3}b = 320$. We solve this equation for b.

$$\frac{1}{3}b = 320$$

$$3 \cdot \frac{1}{3}b = 3 \cdot 320 \quad \text{Multiply each side by 3.}$$

$$b = 960$$

Your monthly budget is $960.

Check: $\frac{1}{3}b = 320$

$$\frac{1}{3} \cdot 960 \overset{?}{=} 320 \quad \text{Substitute 960 for } b.$$

$$320 \overset{\checkmark}{=} 320$$

PRACTICE 10

Six months after buying a used car, you sold it, taking a loss equal to $\frac{1}{5}$ of the car's original price. If your loss was $725, what was the original price? Write an equation. Then solve and check.

$\frac{1}{5}x = 725;\ x = \$3,625$

Exercises 4.3

For Extra Help

 Tape 5

 InterAct Math
Tutorial Software

 AWL Tutor Center

 Student's Solutions
Manual

 www.awlonline.com/
akstbragg

Translate each sentence to an equation.

1. Three-fourths of a number y is 12.

$\frac{3}{4}y = 12$

2. The product of $\frac{2}{3}$ and x is 20.

$\frac{2}{3}x = 20$

3. A number x divided by 7 is equal to $\frac{7}{2}$.

$\frac{x}{7} = \frac{7}{2}$

4. The quotient of z and 1.5 is 10.

$\frac{z}{1.5} = 10$

5. A third of x is 2.

$\frac{1}{3}x = 2$

6. Two times m is equal to 11. $2m = 11$

7. The quotient of x and 3 is equal to $\frac{1}{3}$.

$\frac{x}{3} = \frac{1}{3}$

8. A number n divided by 100 is 0.36.

$\frac{n}{100} = 0.36$

9. The product of 9 and z is the same as 27. $9z = 27$

10. Four-fifths of x is equal to 24.

$\frac{4}{5}x = 24$

Identify the operation to perform on each side of the equation to isolate the variable.

11. $3x = 15$

Divide by 3

12. $6y = 18$

Divide by 6

13. $\frac{x}{2} = 9$

Multiply by 2

14. $\frac{y}{6} = 1$

Multiply by 6

15. $\frac{3}{4}a = 21$

Divide by $\frac{3}{4}$
or multiply by $\frac{4}{3}$

16. $\frac{2}{3}m = 14$

Divide by $\frac{2}{3}$
or multiply by $\frac{3}{2}$

17. $1.5b = 15$

Divide by 1.5

18. $2.6x = 52$

Divide by 2.6

By answering yes or no, indicate whether the value of x shown is a solution to the given equation.

19.

Equation	Value of x	Solution?
a. $5x = 20$	4	Yes
b. $3x = 12$	36	No
c. $\frac{x}{4} = 8$	2	No
d. $\frac{x}{0.2} = 4$	8	No

20.

Equation	Value of x	Solution?
a. $\frac{x}{3} = 10$	30	Yes
b. $2.5x = 5$	2	Yes
c. $2x = \frac{1}{3}$	$\frac{1}{6}$	Yes
d. $\frac{x}{0.4} = 3$	12	No

Solve and check.

21. $5x = 30$

    $x = 6$

22. $8y = 8$

    $y = 1$

23. $\dfrac{x}{2} = 9$

    $x = 18$

24. $\dfrac{n}{9} = 3$

    $n = 27$

25. $36 = 9n$

    $n = 4$

26. $125 = 5x$

    $x = 25$

27. $\dfrac{x}{7} = 13$

    $x = 91$

28. $\dfrac{w}{10} = 21$

    $w = 210$

29. $1.7y = 6.8$

    $y = 4$

30. $0.5a = 7.5$

    $a = 15$

31. $2.1b = 42$

    $b = 20$

32. $1.5x = 45$

    $x = 30$

33. $\dfrac{m}{15} = 10.5$

    $m = 157.5$

34. $\dfrac{p}{10} = 12.1$

    $p = 121$

35. $\dfrac{t}{0.4} = 1$

    $t = 0.4$

36. $\dfrac{n}{0.5} = 6$

    $n = 3$

37. $\dfrac{2}{3}x = 1$

    $x = \dfrac{3}{2}$, or $1\dfrac{1}{2}$

38. $\dfrac{1}{8}n = 3$

    $n = 24$

39. $\dfrac{1}{4}x = 9$

    $x = 36$

40. $\dfrac{3}{7}t = 15$

    $t = 35$

41. $17t = 51$

    $t = 3$

42. $100x = 400$

    $x = 4$

43. $10y = 4$

    $y = \dfrac{2}{5}$

44. $100n = 50$

    $n = \dfrac{1}{2}$

45. $7 = \dfrac{n}{100}$

    $n = 700$

46. $40 = \dfrac{p}{10}$

    $p = 400$

47. $2.5 = \dfrac{x}{5}$

    $x = 12.5$

48. $4.6 = \dfrac{z}{2}$

    $z = 9.2$

49. $2 = 4x$

    $x = \dfrac{1}{2}$

50. $3 = 5x$

    $x = \dfrac{3}{5}$

51. $\dfrac{14}{3} = \dfrac{7}{9}m$

    $m = 6$

52. $\dfrac{4}{9} = \dfrac{2}{3}a$

    $a = \dfrac{2}{3}$

⊞ **53.** $3.14x = 21.3834$

    $x = 6.81$

⊞ **54.** $2.54x = 48.26$

    $x = 19$

⊞ **55.** $\dfrac{x}{1.414} = 3.5$

    $x = 4.949$

⊞ **56.** $\dfrac{x}{1.732} = 1.732$

    $x = 2.999824$

Translate each sentence to an equation. Then solve and check.

57. The product of 8 and n is 56.
    $8n = 56;\ n = 7$

58. The product of 12 and m is 3.
    $12m = 3;\ m = \dfrac{1}{4}$

59. Three-fourths of a number y is equal to 18. $\dfrac{3}{4}y = 18;\ y = 24$

60. One-third of a number x is 16.
    $\dfrac{1}{3}x = 16;\ x = 48$

61. A number x divided by 5 is 11.
    $\dfrac{x}{5} = 11;\ x = 55$

62. A number y divided by 100 is 10.
    $\dfrac{y}{100} = 10;\ y = 1{,}000$

63. Twice x is equal to 36. $2x = 36;$
    $x = 18$

64. Two times m is 100.
    $2m = 100;\ m = 50$

65. A third of x is 4. $\dfrac{1}{3}x = 4;\ x = 12$

66. Five-sevenths of y is 10.
    $\dfrac{5}{7}y = 10;\ y = 14$

67. A number n divided by 5 is equal to $1\frac{3}{5}$. $\frac{n}{5} = 1\frac{3}{5}$; $n = 8$

68. A number n divided by 14 is equal to $1\frac{1}{2}$. $\frac{n}{14} = 1\frac{1}{2}$; $n = 21$

69. The quotient of y and 2.5 is 10. $\frac{y}{2.5} = 10$; $y = 25$

70. x divided by 15 equals 3.6. $\frac{x}{15} = 3.6$; $x = 54$

Choose the equation that describes each situation.

71. Suppose that you spend \$20, which is $\frac{1}{4}$ of your total money, m. How much money did you have in the beginning?   Equation d

 a. $m - \frac{1}{4} = 20$ **b.** $4m = 20$ **c.** $m + \frac{1}{4} = 20$ **d.** $\frac{1}{4}m = 20$

72. Find the weight of a child if $\frac{1}{3}$ of her weight is 9 lb.   Equation b

 a. $3x = 9$ **b.** $\frac{1}{3}x = 9$ **c.** $x + 3 = 9$ **d.** $x + \frac{1}{3} = 9$

73. You plan to buy a CD player 8 weeks from now. If the CD player costs \$340, how much money must you save each week in order to buy it?   Equation a

 a. $8c = 340$ **b.** $c + 8 = 340$ **c.** $\frac{c}{8} = 340$ **d.** $c - 8 = 340$

74. The student government at your college sold tickets to a play. From the ticket sales, they collected \$300, which is twice the cost of the play. How much did the play cost?

 a. $\frac{n}{2} = 300$ **b.** $n - 2 = 300$ **c.** $2n = 300$ **d.** $n + 2 = 300$

  Equation c

Applications

Write an equation. Solve and then check.

75. In the square below, the perimeter is 60 units. Find the length of one side of the square. $4s = 60$; $s = 15$

76. Find the width of the rectangle shown below. $7w = 35$; $w = 5$ ft

Length = 7 ft

Area = 35 sq ft   Width = ?

77. An airplane drops 750 ft, or $\frac{1}{6}$ of its previous altitude. How high was the airplane before the drop?

$\frac{1}{6}x = 750$; $x = 4{,}500$ ft

78. On sale, a jacket costs \$66.50, which is $\frac{1}{4}$ of its original price. How much was the original price?   \$266

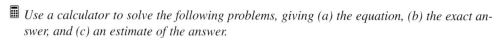 *Use a calculator to solve the following problems, giving (a) the equation, (b) the exact answer, and (c) an estimate of the answer.*

79. About 0.67 of a company's total budget goes to personnel. If the company's total budget is $34 million, what is its personnel cost to the nearest million dollars?   (a) Multiplication; (b) $23 million; (c) possible estimate: $21 million

80. An 18-carat gold bracelet is 0.75 pure gold. Suppose that an 18-carat gold bracelet contains 3.5 oz of gold. What is the total weight of the bracelet, rounded to the nearest tenth of an ounce?   (a) Division; (b) 4.7 oz; (c) possible estimate: 4 oz

■ *Check your answers on page A-7.*

MINDSTRETCHERS

WRITING

1. Write two different word problems that are applications of each equation.

a. $4x = 20$ Answers may vary.

■ _____

■ _____

b. $\dfrac{x}{2} = 5$ Answers may vary.

■ _____

■ _____

GROUPWORK

2. The equations $\dfrac{r}{7} = 2$ and $\dfrac{7}{r} = 2$ are similar in form. Working with a partner, answer the following questions.

a. How would you solve the first equation for r?   Multiply both sides by 7.

b. How can you use what you know about the first equation to solve the second equation for r?   Multiply both sides by r. Then divide both sides by 2.

c. What are the similarities and differences between the two equations?   Possible answers: Similarities—Both equations are division equations. To solve, begin by multiplying both sides by the denominator. Differences—In the first equation the variable is in the numerator, whereas in the second equation the variable is in the denominator. The first equation is solved by using multiplication, whereas the second equation is solved by using multiplication and division.

CRITICAL THINKING

3. In a magic square with four rows and four columns, the sum of the entries in each row, column, and diagonal is the same. If the entries are the consecutive whole numbers 1 through 16, what is the sum of the numbers in each diagonal?   34

KEY CONCEPTS and SKILLS

□ = CONCEPT   □ = SKILL

CONCEPT/SKILL	DESCRIPTION	EXAMPLE
[4.1] **Variable**	A letter that represents an unknown number.	x, y, t
[4.1] **Constant**	A known number.	$2, \dfrac{1}{3}, 5.6$
[4.1] **Algebraic expression**	An expression that contains one or more variables and may contain any number of constants.	$x + 3$
[4.1] **To evaluate an expression**	■ Replace each variable with a number. ■ Carry out the computation.	Evaluate $8 - x$ for $x = 3.5$: $8 - x = 8 - 3.5$, or 4.5
[4.2] **Equation**	A mathematical statement that two expressions are equal.	$2 + 4 = 6$, $x + 5 = 7$
[4.2] **To solve addition or subtraction equations**	■ In an addition equation, subtract the same number from each side of the equation, isolating the variable on one side.	$y + 9 = 15$ $y + 9 - 9 = 15 - 9$ $y = 6$ Check: $y + 9 = 15$ $6 + 9 \stackrel{?}{=} 15$ $15 \stackrel{\checkmark}{=} 15$
	■ In a subtraction equation, add the same number to each side of the equation, isolating the variable on one side. ■ In either case check the solution by substituting the value of the unknown in the original equation to verify that the resulting equation is true.	$w - 6\dfrac{1}{2} = 8$ $w - 6\dfrac{1}{2} + 6\dfrac{1}{2} = 8 + 6\dfrac{1}{2}$ $w = 14\dfrac{1}{2}$ Check: $w - 6\dfrac{1}{2} = 8$ $14\dfrac{1}{2} - 6\dfrac{1}{2} \stackrel{?}{=} 8$ $8 \stackrel{\checkmark}{=} 8$
[4.3] **To solve multiplication or division equations**	■ In a multiplication equation, divide by the same number on each side of the equation, isolating the variable on one side.	$1.3r = 26$ $\dfrac{1.3r}{1.3} = \dfrac{26}{1.3}$ $r = 20$ Check: $1.3r = 26$ $1.3(20) \stackrel{?}{=} 26$ $26 \stackrel{\checkmark}{=} 26$
	■ In a division equation, multiply by the same number on each side of the equation, isolating the variable on one side. ■ In either case check the solution by substituting the value of the unknown in the original equation to verify that the resulting equation is true.	$\dfrac{x}{7} = 8$ $7 \cdot \dfrac{x}{7} = 7 \cdot 8$ $x = 56$ Check: $\dfrac{x}{7} = 8$ $\dfrac{56}{7} \stackrel{?}{=} 8$ $8 \stackrel{\checkmark}{=} 8$

Chapter 4 Review Exercises

To help you review this chapter, solve these problems.

[4.1]

Translate each expression to words.

1. $x + 1$
  x plus 1

2. $y + 4$
  4 more than y

3. $w - 1$
  w minus 1

4. $s - 3$
  3 less than s

5. $\dfrac{c}{7}$
  c over 7

6. $\dfrac{a}{10}$
  The quotient of a and 10

7. $2x$
  2 times x

8. $6y$
  The product of 6 and y

9. $y \div 0.1$
  y divided by 0.1

10. $n \div 1.5$
  The quotient of n and 1.5

11. $\dfrac{1}{3}x$
  $\dfrac{1}{3}$ *of x*

12. $\dfrac{1}{10}w$
  $\dfrac{1}{10}$ *of w*

Translate each phrase to an expression.

13. Sixteen more than m
  $m + 16$

14. The sum of b and $\dfrac{1}{2}$
  $b + \dfrac{1}{2}$

15. y decreased by 1.4
  $y - 1.4$

16. Eight less than z
  $z - 8$

17. The quotient of 6 and z
  $\dfrac{6}{z}$

18. n divided by 2.5
  $n \div 2.5$

19. The product of a number and 3
  $3n$

20. Twelve times some number
  $12n$

Evaluate each expression.

21. $b + 8$, for $b = 4$
  12

22. $d + 12$, for $d = 7$
  19

23. $a - 5$, for $a = 5$
  0

24. $c - 9$, for $c = 15$
  6

25. $1.5x$, for $x = 0.2$
  0.3

26. $1.3t$, for $t = 5$
  6.5

27. $\dfrac{1}{2}n$, for $n = 3$
  $1\dfrac{1}{2}$

28. $\dfrac{1}{6}a$, for $a = 2\dfrac{1}{2}$
  $\dfrac{5}{12}$

29. $w - 9.6$, for $w = 10$
  0.4

30. $v - 3\dfrac{1}{2}$, for $v = 8$
  $4\dfrac{1}{2}$

31. $\dfrac{m}{1.5}$, for $m = 2.5$
  $1\dfrac{2}{3}$

32. $\dfrac{x}{0.2}$, for $x = 1.8$
  9

[4.2]

Solve and check.

33. $x + 11 = 20$
$x = 9$

34. $y + 15 = 24$
$y = 9$

35. $n - 19 = 7$
$n = 26$

36. $b - 12 = 8$
$b = 20$

37. $a + 2.5 = 6$
$a = 3.5$

38. $c + 1.6 = 9.1$
$c = 7.5$

39. $x - 1.8 = 9.2$
$x = 11$

40. $y - 1.4 = 0.6$
$y = 2$

41. $w + 1\frac{1}{2} = 3$
$w = 1\frac{1}{2}$

42. $s + \frac{2}{3} = 1$
$s = \frac{1}{3}$

43. $c - 1\frac{1}{4} = 5$
$c = 6\frac{1}{4}$

44. $p - 6 = 5\frac{2}{3}$
$p = 11\frac{2}{3}$

45. $7 = m + 2$
$m = 5$

46. $10 = n + 10$
$n = 0$

47. $39 = c - 39$
$c = 78$

48. $72 = y - 18$
$y = 90$

49. $38 + n = 49$
$n = 11$

50. $37 + x = 62$
$x = 25$

⊞ 51. $4.0875 + x = 35.136$
$x = 31.0485$

⊞ 52. $24.625 = m - 1.9975$
$m = 26.6225$

[4.2–4.3]

Translate each sentence to an equation.

53. n decreased by 19 is 35. $n - 19 = 35$

54. Thirty-seven less than t equals 234. $t - 37 = 234$

55. Nine increased by x is equal to $5\frac{1}{2}$. $9 + x = 5\frac{1}{2}$

56. Twenty-six more than s is $30\frac{1}{3}$. $s + 26 = 30\frac{1}{3}$

57. Twice y is 16. $2y = 16$

58. The product of t and 25 is 175. $25t = 175$

59. Thirty-four is equal to n divided by 19. $\frac{n}{19} = 34$

60. Seventeen is the quotient of z and 13. $\frac{z}{13} = 17$

61. One-third of a number equals 27. $\frac{1}{3}n = 27$

62. Two-fifths of a number equals 4. $\frac{2}{5}n = 4$

By answering yes or no, indicate whether the value of x shown is a solution to the given equation.

63.

Equation	Value of x	Solution?
a. $0.3x = 6$	2	No
b. $x - \dfrac{1}{2} = 1\dfrac{2}{3}$	$1\dfrac{1}{6}$	No
c. $\dfrac{x}{0.5} = 7$	35	No
d. $x + 0.1 = 3$	3.1	No

64.

Equation	Value of x	Solution?
a. $0.2x = 6$	30	Yes
b. $x + \dfrac{1}{2} = 1\dfrac{2}{3}$	$\dfrac{5}{6}$	No
c. $\dfrac{x}{0.2} = 4.1$	8.2	No
d. $x + 0.5 = 7.4$	2.4	No

[4.3]

Solve and check.

65. $2x = 10$

$x = 5$

66. $8t = 16$

$t = 2$

67. $\dfrac{a}{7} = 15$

$a = 105$

68. $\dfrac{n}{6} = 9$

$n = 54$

69. $9y = 81$

$y = 9$

70. $10r = 100$

$r = 10$

71. $\dfrac{w}{10} = 9$

$w = 90$

72. $\dfrac{x}{100} = 1$

$x = 100$

73. $1.5y = 30$

$y = 20$

74. $1.2a = 144$

$a = 120$

75. $\dfrac{1}{8}n = 4$

$n = 32$

76. $\dfrac{1}{2}b = 16$

$b = 32$

77. $\dfrac{m}{1.5} = 2.1$

$m = 3.15$

78. $\dfrac{z}{0.3} = 1.9$

$z = 0.57$

79. $100x = 40$

$x = \dfrac{2}{5}$

80. $10y = 5$

$y = \dfrac{1}{2}$

81. $0.3 = \dfrac{m}{4}$

$m = 1.2$

82. $1.4 = \dfrac{b}{7}$

$b = 9.8$

▣ **83.** $0.866x = 10.825$

$x = 12.5$

▣ **84.** $\dfrac{x}{0.707} = 2.1$

$x = 1.4847$

Mixed Applications

Write an expression for each problem. Then evaluate the expression for the given amount.

85. DVDs at a store are on sale for $12.99 each. How much will x DVDs cost?   8 DVDs? $12.99x$; $103.92

86. Suppose that you work 40 hr/week. What is your hourly wage if you earn d dollars per week? $382 per week?

$\dfrac{d}{40}$ dollars/hr; $9.55/hr

87. The local supermarket sells a certain fruit for 89¢/lb. How much will p pounds cost?   3 lb? $89p$ cents; $2.67

88. Suppose that you borrow $3,000 from a bank. You must pay the amount borrowed plus a finance charge. How much will you pay the bank if the finance charge is d dollars? $225? $3,000 + d$ dollars; $3,225

Write an equation. Then solve and check.

89. In the Olympics, a gymnast scored a total of 19.14 in two gymnastic events. If his score in one event was 9.59, what was his score in the other event?

$9.59 + x = 19.14; x = 9.55$

90. Hurricane Gilbert was one of the mightiest storms to hit the Western Hemisphere in the twentieth century. A newspaper reported that the hurricane left 500,000 people, or about one-fourth of the population of Jamaica, homeless. Approximately how many people lived in Jamaica? (*Source:* J.B. Elsner and A.B. Kara, *Hurricanes of The North Atlantic*)

$\frac{1}{4}x = 500{,}000; x = 2{,}000{,}000$

91. If the height of a stack of 10,000 one-dollar bills is 1,000 mm high, how thick is a one-dollar bill?
$10{,}000t = 1{,}000; t = 0.1$ mm

92. A bowler's final score is the sum of her handicap and scratch score (actual score). If a bowler has a final score of 225 and a handicap of 50, what was her scratch score? $225 = x + 50; x = 175$

93. Suppose that you weigh 286 lb on Jupiter. If this weight is $2\frac{3}{5}$ times your weight on Earth, how much do you weigh on Earth?
$2\frac{3}{5}w = 286; w = 110$ lb

94. The Nile River is about 1.8 times as long as the Missouri River. If the Nile is about 6,696 kilometers long, approximately how long is the Missouri?
(*Source:* Eliot Elisofon, *The Nile*)
$1.8x = 6{,}696; x = 3{,}720$ km

95. The normal body temperature is 98.6 °F. When you were ill, your temperature rose to 101 °F. This temperature is how many degrees above normal?
$98.6 + x = 101; x = 2.4$ °F

96. Stacey's charges $8 less than Maxine's for the same jacket. If the price in Stacey's is $42, what is the price in Maxine's? $y - 8 = 42; y = \$50$

■ *Check your answers on page A-7.*

Chapter 4 Posttest

To see if you have already mastered the topics in this chapter, take this test.

Write each expression in words.

1. $x + \dfrac{1}{2}$ Possible answer: x plus $\dfrac{1}{2}$

2. $\dfrac{a}{3}$ Possible answer: the quotient of a and 3

Translate each phrase to an expression.

3. 8 more than a number $n + 8$

4. The product of $\dfrac{1}{8}$ and p $\dfrac{1}{8}p$

Evaluate each expression.

5. $a - 1.5$, for $a = 1.5$ 0

6. $\dfrac{b}{9}$, for $b = 2\dfrac{1}{4}$ $\dfrac{1}{4}$

Translate each sentence to an equation.

7. The difference between x and 6 is $4\dfrac{1}{2}$.

$x - 6 = 4\dfrac{1}{2}$

8. The quotient of y and 8 is 3.2. $\dfrac{y}{8} = 3.2$

Solve and check.

9. $x + 10 = 10$

$x = 0$

10. $y - 6 = 6$

$y = 12$

11. $3n = 81$

$n = 27$

12. $\dfrac{a}{9} = 82$

$a = 738$

13. $m - 1.8 = 6$

$m = 7.8$

14. $1.5n = 75$

$n = 50$

15. $10x = 54$

$x = 5.4$

16. $\dfrac{n}{100} = 7.6$

$n = 760$

Write an equation. Then solve and check.

17. A recipe for seafood stew requires $2\dfrac{1}{4}$ lb of fish. After buying $1\dfrac{3}{4}$ lb of bluefish, you decide to fill out the recipe with codfish. How many pounds of codfish should you buy?

$1\dfrac{3}{4} + x = 2\dfrac{1}{4}; x = \dfrac{1}{2}$ lb

18. A newspaper reported that this year, 30,000 elephants— $\dfrac{1}{3}$ of all the elephants in a certain country—had been hunted down and killed for their ivory tusks. How many elephants were there at the beginning of the year?

$\dfrac{1}{3}x = 30{,}000; x = 90{,}000$ elephants

19. A student aide earns $7.25/hr. If the amount in his paycheck last month was $145, how many hours did he work? $7.25x = 145; x = 20$ hr

20. Suppose that you lost $14\dfrac{1}{2}$ lb over a 7-mo period. If you now weigh $156\dfrac{1}{2}$ lb, how much did you weigh before?

$x - 14\dfrac{1}{2} = 156\dfrac{1}{2}; x = 171$ lb

■ *Check your answers on page A-7.*

Cumulative Review Exercises

To help you review, solve the following.

1. Subtract: $8\frac{1}{4} - 2\frac{7}{8}$ $5\frac{3}{8}$

2. Find the quotient: $7.5 \div 1{,}000$ 0.0075

3. Decide whether 2 is a solution to the equation
$w + 3 = 5$ Yes

4. Solve and check: $9y = 27$ $y = 3$

5. Solve and check: $w + 2\frac{1}{3} = 4$ $w = 1\frac{2}{3}$

6. Solve and check: $n - 3.8 = 4$ $n = 7.8$

7. Solve and check: $\frac{x}{2} = 16$ $x = 32$

8. In animating a cartoon, artists drew 24 cartoons to appear during one second of screen time. How many cartoons did they have to draw to produce a 5-minute cartoon?   7,200

9. Your dental insurance reimbursed you $200 on a dental bill of $700. Did you get less or more than $\frac{1}{3}$ of your money back? Explain.   You got back $\frac{2}{7}$ of your money, which is less than $\frac{1}{3}$.

10. The weight of an object on the moon is only 0.167 times its weight on Earth. If a man weighs 150 lb on Earth, calculate his weight on the moon. (*Source:* Seymour Simon, *The Moon*)   25.05 lb

■ *Check your answers on page A-7.*

C H A P T E R **5**

Ratio and Proportion

RATIO AND PROPORTION IN PHARMACOLOGY

Many of the medicines that pharmacists dispense come in solutions. An example is aminophylline, a medicine that people with asthma take to ease their breathing.

To prepare a solution of aminophylline, pharmacists dissolve 250 mg of aminophylline for every 10 cubic centimeters (cc or cm^3) of sterile water.

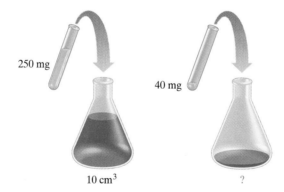

250 mg

40 mg

10 cm^3

?

In filling a prescription for, say, 40 mg of aminophylline, we must determine the amount of sterile water needed. Pharmacists use the concepts of ratio and proportion to establish how much of the sterile water to dispense.

Chapter 5 Pretest

To see if you have already mastered the topics in this chapter, take this test.

Write each ratio or rate in simplest form.

1. 6 to 8 $\dfrac{3}{4}$

2. 21 to 27 $\dfrac{7}{9}$

3. \$30 to \$18 $\dfrac{5}{3}$

4. 19 ft to 51 ft $\dfrac{19}{51}$

5. 12 dental assistants for every 6 dentists
$\dfrac{2 \text{ dental assistants}}{1 \text{ dentist}}$

6. 48 gallons (gal) of water in 15 min $\dfrac{16 \text{ gal}}{5 \text{ min}}$

7. 60 baskets in 180 attempts $\dfrac{1 \text{ basket}}{3 \text{ attempts}}$

8. 10 mg every 6 hr $\dfrac{5 \text{ mg}}{3 \text{ hr}}$

9. \$690 for 3 boxes of ceramic tiles $\dfrac{\$230}{1 \text{ box}}$

10. 15 pages in 75 min $\dfrac{1 \text{ page}}{5 \text{ min}}$

Determine whether each proportion is true or false.

11. $\dfrac{2}{3} = \dfrac{16}{24}$ True

12. $\dfrac{32}{20} = \dfrac{8}{3}$ False

Solve for x. Then check.

13. $\dfrac{6}{8} = \dfrac{x}{12}$ $x = 9$

14. $\dfrac{21}{x} = \dfrac{2}{3}$ $x = 31\dfrac{1}{2}$

15. $\dfrac{\frac{1}{2}}{4} = \dfrac{2}{x}$ $x = 16$

16. $\dfrac{x}{6} = \dfrac{8}{0.3}$ $x = 160$

Solve.

17. A contractor combines 80 lb of sand with 100 lb of gravel. In this mixture, what is the ratio of sand to gravel? $\dfrac{4}{5}$

18. At a local bulk laundry shop, you pay \$54 to clean 27 lb of laundry. At this rate, how much do you pay per pound? \$2/lb

19. On a college campus, the student to faculty ratio is 35 to 2. How many faculty members should the college have to maintain this ratio for a student body of 3,570? 204

20. The scale on a map is 3 in. to 31 mi. If two cities are 8.4 in. apart on the map, what is the actual distance, to the nearest mile, between the two cities? 87 mi

■ *Check your answers on page A-7.*

5.1 Introduction to Ratios

OBJECTIVES

■ *To write ratios of like quantities in simplest form*
■ *To write ratios of unlike quantities in simplest form*
■ *To solve word problems involving ratios*

What Ratios Are and Why They Are Important

We constantly need to compare quantities. Whether in sports, medicine, or business—to name just a few of numerous applications—we use **ratios** to make comparisons. Consider the ratios in the following examples.

■ The season record of the winning volleyball team was 16 and 4.

■ A nurse prepared a 1-to-25 boric acid solution.

■ The stock's price–earnings ratio was 13 to 1.

Can you think of other examples of ratios in your daily life?

> Have students explain their answers in a journal.

The preceding examples illustrate the following definition of a ratio.

Definition

A **ratio** is a comparison in terms of a quotient of two numbers a and b, usually written as $\frac{a}{b}$.

There are three basic ways to write a ratio. For instance, we can write the ratio 1 to 25 as:

$$1 \text{ to } 25 \qquad 1:25 \qquad \frac{1}{25}$$

> Explain to students that the fractional form of the ratio is the most useful form for solving ratio or proportion problems.

No matter which notation we use for this ratio, it is read "1 to 25."

Simplifying Ratios

Because the ratio of two numbers can be written as a fraction, we can say that, as with any fraction, a ratio is said to be in simplest form (or reduced to lowest terms) when 1 is the only common factor of the two numbers.

> Students may need to review simplifying fractions in Section 2.2.

Let's consider some examples of writing ratios in simplest form.

EXAMPLE 1	PRACTICE 1
Write the ratio 10 to 5 in simplest form.	Write the ratio 8 to 12 in simplest form.

EXAMPLE 1

Write the ratio 10 to 5 in simplest form.

Solution The ratio 10 to 5 expressed as a fraction is $\frac{10}{5}$.

$$\underset{\substack{\uparrow \\ \text{Reduce} \\ \text{by 5.}}}{\frac{10}{5}} = \underset{\substack{\uparrow \\ \text{Lowest} \\ \text{terms}}}{\frac{2}{1}}$$

So the ratio 10 to 5 is the same as the ratio 2 to 1. Note that the ratio 2 to 1 means that the first number is twice as large as the second number.

PRACTICE 1

Write the ratio 8 to 12 in simplest form.

$\frac{2}{3}$

> Point out to students that the ratio of 10 to 5 $\left(\frac{10}{5}\right)$ is not the same as the ratio of 5 to 10 $\left(\frac{5}{10}\right)$.

Frequently we deal with quantities that have units, such as months or feet. When both quantities have the same unit, they are called **like quantities**. In a ratio of like quantities, the units drop out.

EXAMPLE 2

Express the ratio 5 mo to 3 mo in simplest form.

Solution The ratio 5 mo to 3 mo expressed as a fraction is $\dfrac{5 \text{ mo}}{3 \text{ mo}}$.

Simplifying, we get $\dfrac{5}{3}$, which is already in lowest terms. Note that with ratios we do not rewrite improper fractions as mixed numbers because our answer must be a comparison of *two* numbers.

PRACTICE 2

Express in simplest form the ratio 9 ft to 5 ft. $\dfrac{9}{5}$

Examples 1 and 2 illustrate the following rule for simplifying a ratio.

> **To Write a Ratio in Simplest Form**
> ■ write the ratio as a fraction, and
> ■ reduce the fraction to simplest form, dropping any units given.

EXAMPLE 3

During a 7-month season, a championship baseball team won 100 games and lost 50 games. What was the ratio of the team's wins to games played?

Solution This is a two-step problem. First, we must find how many games the team played.

100 games won + 50 games lost = 150 games played

Then we write the ratio of the number of games won to the number of games played.

$$\frac{\text{Games won}}{\text{Games played}} = \frac{100}{150} = \frac{2}{3}$$

The ratio is 2 to 3, which means the team won 2 of every 3 games played.

PRACTICE 3

A coin had an original value of $9.50. If it increased in value to $12 in 3 yr, what is the ratio of the increase to the original value? $\dfrac{5}{19}$

Now let's compare **unlike quantities**, that is, quantities that have different units. Such a comparison is called a **rate**.

> **Definition**
>
> A **rate** is a ratio of unlike quantities.

For instance, suppose that your rate of pay is $52 for each 8 hr of work. Simplifying this rate, we get:

$$\frac{\$52}{8 \text{ hr}} = \frac{\$13}{2 \text{ hr}}$$

So you are paid $13 for every 2 hr that you worked. Note that the units are expressed as part of the answer.

This example suggests the following rule for simplifying a rate.

To Simplify a Rate
- ■ write the rate as a fraction, and
- ■ reduce the fraction to simplest form, *keeping the units*.

EXAMPLE 4

Simplify each rate.

a. 350 mi to 18 gal of gas

b. 18 trees to produce 2,000 lb of paper

Solution

a. 350 mi to 18 gal $= \dfrac{350 \text{ mi}}{18 \text{ gal}} = \dfrac{175 \text{ mi}}{9 \text{ gal}}$

b. 18 trees to 2,000 lb $= \dfrac{18 \text{ trees}}{2,000 \text{ lb}} = \dfrac{9 \text{ trees}}{1,000 \text{ lb}}$

PRACTICE 4

Express each rate in simplest form.

a. 12 nurses for 40 patients $\dfrac{3 \text{ nurses}}{10 \text{ patients}}$

b. 15 lab stations for every 60 students $\dfrac{1 \text{ lab station}}{4 \text{ students}}$

Frequently, we want to find a particular kind of rate—namely, the *rate per unit*. In the rate $\dfrac{\$13}{2 \text{ hr}}$, for instance, it would be useful to know what is earned for each hour (that is, the hourly wage). We need to rewrite $\dfrac{\$13}{2 \text{ hr}}$ so that the denominator is 1 hr.

$$\frac{\$13}{2 \text{ hr}} = \frac{\$13 \div 2}{2 \text{ hr} \div 2} = \frac{\$6.50}{1 \text{ hr}} = \$6.50 \text{ per hr, or } \$6.50/\text{hr}$$

Recall that
"per" means
"divided by."

Here, we divided the numbers in both the numerator and denominator by the number in the denominator. This rate is called a **unit rate**.

Definition

A **unit rate** is a rate in which the number in the denominator is 1.

EXAMPLE 5

Write as a unit rate.

a. $3,450 for 6 weeks   **b.** 275 mi in 5 hr

Solution First, we write each rate as a fraction. Then we divide numbers in the numerator and denominator by the number in the denominator, getting 1 in the denominator.

a. $3,450 for 6 weeks $= \dfrac{\$3,450}{6 \text{ weeks}} = \dfrac{\$3,450 \div 6}{6 \text{ weeks} \div 6} = \dfrac{\$575}{1 \text{ week}}$, or $\$575/\text{week}$

b. 275 mi in 5 hr $= \dfrac{275 \text{ mi}}{5 \text{ hr}} = \dfrac{275 \text{ mi} \div 5}{5 \text{ hr} \div 5} = \dfrac{55 \text{ mi}}{1 \text{ hr}}$, or 55 mph

PRACTICE 5

Express as a unit rate.

a. A fall of 192 ft in 4 sec   48 fps

b. 15 hits in 40 times at bat   0.375 hit per time at bat

EXAMPLE 6

An airline makes 270 flights in 18 mo to a certain city. How many flights per month does the airline make?

Solution $\dfrac{270 \text{ flights}}{18 \text{ mo}} = \dfrac{15 \text{ flights}}{1 \text{ mo}}$

So the airline makes 15 flights/month to the city.

PRACTICE 6

In a keyboarding class, you type 375 words in 5 min. How many words per minute do you type?   75 wpm

In order to get the better buy, we sometimes compare prices by computing the unit price of a single item. This **unit price** is a type of unit rate.

> **Definition**
>
> A **unit price** is the price of one item.

To find the unit price, we write the ratio of the price to the number of units and then simplify. Let's consider some examples of unit pricing.

EXAMPLE 7

Find the unit price.

a. $84,000 to educate 30 students

b. 6 credits for $234

c. A 10-oz box of wheat flakes for $2.76

Solution

a. $\dfrac{\$84,000}{30 \text{ students}} = \$2,800/\text{student}$

b. $\dfrac{\$234}{6 \text{ credits}} = \$39/\text{credit}$

c. $\dfrac{\$2.76}{10 \text{ oz}} \approx \$0.28/\text{oz}$ rounded to the nearest cent

PRACTICE 7

Determine the unit price.

a. 4 supersaver flights for $696 $174/flight

b. 5 fax machines for $4,475
$895/fax machine

EXAMPLE 8

A customer shopping for 150 greeting cards saw the following signs in two stores.

Which is the better buy?

PRACTICE 8

Which can of corn has the lower unit price? The 32-oz can at 2.875¢/oz

Solution CARDS FOR EVERYONE: $\dfrac{\$19.95}{25 \text{ cards}} = \$0.798/\text{card}$

CARDS GALORE: $\dfrac{\$24}{30} = \$0.80/\text{card}$

Since $\$0.798 < \$0.80$, the better buy is at CARDS FOR EVERYONE.

Have students explain how unit pricing is helpful in determining the relative savings on buying various sizes of packages.

HISTORICAL NOTE

The shape of a grand piano is dictated by the length of its strings. When a stretched string vibrates, it produces a particular pitch, say C. A second string of comparable tension will produce another pitch, which depends on the ratio of the string lengths. For instance, if the ratio of the new string to the original string were 16 to 15, then plucking the new string would produce the pitch B.

Around 500 B.C., the followers of the mathematician Pythagoras learned to adjust string lengths in various ratios so as to produce an entire scale. Thus the concept of ratio is central to the construction of pianos, violins, and many other musical instruments.

Sources:
John R. Pierce, *The Science of Musical Sound* (New York: Scientific American Library, 1983).

David Bergamini, *Mathematics* (New York: Time-Life Books, 1971).

Exercises 5.1

For Extra Help

 Tape 6

 InterAct Math
Tutorial Software

 AWL Tutor Center

 Student's Solutions
Manual

 www.awlonline.com/
akstbragg

Write each ratio in simplest form.

1. 6 to 9
$$\frac{2}{3}$$

2. 9 to 12
$$\frac{3}{4}$$

3. 10 to 15
$$\frac{2}{3}$$

4. 21 to 27
$$\frac{7}{9}$$

5. 55 to 35
$$\frac{11}{7}$$

6. 8 to 10
$$\frac{4}{5}$$

7. 12 to 8
$$\frac{3}{2}$$

8. 30 to 20
$$\frac{3}{2}$$

9. 2.5 to 10
$$\frac{1}{4}$$

10. 1.25 to 100
$$\frac{1}{80}$$

11. 60 min to 45 min
$$\frac{4}{3}$$

12. $40 to $25
$$\frac{8}{5}$$

13. 10 ft to 10 ft
$$\frac{1}{1}$$

14. 75 tons to 75 tons
$$\frac{1}{1}$$

15. 30¢ to 18¢
$$\frac{5}{3}$$

16. 66 yr to 32 yr
$$\frac{33}{16}$$

17. 7 mph to 24 mph
$$\frac{7}{24}$$

18. 21 gal to 20 gal
$$\frac{21}{20}$$

19. 1,000 acres to 50 acres
$$\frac{20}{1}$$

20. 2,000 mi to 25 mi
$$\frac{80}{1}$$

21. 8 g to 7 g
$$\frac{8}{7}$$

22. 19 oz to 51 oz
$$\frac{19}{51}$$

23. 24 sec to 30 sec
$$\frac{4}{5}$$

24. 28 ml to 42 ml
$$\frac{2}{3}$$

Write each rate in simplest form.

25. 25 telephone calls in 10 days
$$\frac{5 \text{ calls}}{2 \text{ days}}$$

26. 42 gal in 12 min
$$\frac{7 \text{ gal}}{2 \text{ min}}$$

27. 20 children in 6 families
$$\frac{10 \text{ children}}{3 \text{ families}}$$

28. $85 for 15 records
$$\frac{\$17}{3 \text{ records}}$$

29. 3,000 students for every 160 faculty
$$\frac{75 \text{ students}}{4 \text{ faculty}}$$

30. 2,000 m in 6 min
$$\frac{1000 \text{ m}}{3 \text{ min}}$$

31. 68 baskets in 120 attempts
$$\frac{17 \text{ baskets}}{30 \text{ attempts}}$$

32. 12 hits in 66 times at bat
$$\frac{2 \text{ hits}}{11 \text{ times at bat}}$$

33. 10 cars for every 10 people

$$\frac{1 \text{ car}}{1 \text{ person}}$$

34. 56 Cal in 4 oz of orange juice

$$\frac{14 \text{ Cal}}{1 \text{ oz}}$$

35. 400 mi to 18 gal of gas

$$\frac{200 \text{ mi}}{9 \text{ gal}}$$

36. 300 full-time students to 200 part-time students

$$\frac{3 \text{ full-time}}{2 \text{ part-time}}$$

37. 48 males for every 9 females

$$\frac{16 \text{ males}}{3 \text{ females}}$$

38. 3 lawyers for every 800 clients

$$\frac{3 \text{ lawyers}}{800 \text{ clients}}$$

39. 40 Democrats for every 35 Republicans

$$\frac{8 \text{ Democrats}}{7 \text{ Republicans}}$$

40. $12,500 in 6 mo

$$\frac{\$6,250}{3 \text{ mo}}$$

41. 2 lb of zucchini for 16 servings

$$\frac{1 \text{ lb}}{8 \text{ servings}}$$

42. 820 mi in 6 days

$$\frac{410 \text{ mi}}{3 \text{ days}}$$

43. 100 pages in 120 min

$$\frac{5 \text{ pages}}{6 \text{ min}}$$

44. 50 mm in 4 hr

$$\frac{25 \text{ mm}}{2 \text{ hr}}$$

45. 3 lb of grass seeds for 600 ft² of lawn

$$\frac{1 \text{ lb}}{200 \text{ ft}^2}$$

46. 908 g of a substance for every 2 lb of a certain mixture

$$\frac{454 \text{ g}}{1 \text{ lb}}$$

Determine the unit rate.

47. 110 mi in 2 hr   55 mph

48. 3,000 people to 1,500 acres of land   2 people/acre

49. 120 gal of heating oil for 15 days   8 gal/day

50. 48 yd in 8 carries   6 yd/carry

51. 3 tanks of gas to cut 10 acres of lawn   0.3 tank/acre

52. 80,000 library books for 5,000 students   16 books/student

53. 8 yd of material for 5 dresses   1.6 yd/dress

54. 30 hits in 80 times at bat   0.375 hit/time at bat

55. 20 hr of homework in 10 days   2 hr/day

56. $200 for 8 hr of work $25/hr

57. A run of 5 km in 20 min
0.25 km/min

58. 12 knockouts in 16 fights $\quad$ 0.75 knock-out/fight

59. 396 mi on 22 gal of fuel $\quad$ 18 mpg

60. 60 children for every 5 adults
12 children/adult

Find the unit price.

61. 12 bars of soap for $5.40 $\quad$ $0.45/bar

62. 4 credit hr for $200 $\quad$ $50/credit hr

63. 6 rolls of film that cost $17.70
$2.95/roll

64. 2 notebooks that cost $6.90
$3.45/notebook

65. 3 plants for $200 $\quad$ $66.67/plant

66. 3 lemons for $1.00 $\quad$ $0.33/lemon

67. 5 nights in a hotel for $495 $\quad$ $99/night

68. 60 min of Internet access for $12
$0.20/min

In each case which is the better buy?

69. Audio cassettes $\quad$ a

a. 4 cassettes for $11.96

b. 5 cassettes for $16.25

70. Video tapes $\quad$ b

a. 5 tapes for $17.50

b. 6 tapes for $19.50

71. Copies $\quad$ b

a. 100 copies for $5.75

b. 120 copies for $6.60

72. Soda $\quad$ a

a. a 12-pack for $5.00

b. a 6-pack for $2.52

73. Notebooks with the college logo $\quad$ b

a. 3 notebooks for $6.50

b. 4 notebooks for $7.50

74. Bus pass $\quad$ b

a. 11 rides for $15.00

b. 5 rides for $6.00

Applications

Solve. Simplify if possible.

75. In the U.S. House of Representatives, 228 members voted against the gun control bill, and 182 voted for it. What was the ratio of those who voted against the bill to those who voted in favor of it?

$$\frac{114}{91}$$

76. In a major city hospital, 88 patients were in the intensive care unit. If 22 nurses were on duty in the unit, what was the ratio of nurses to patients?

$$\frac{1}{4}$$

77. The number line shown is marked off in equal units. Find the ratio of the length of the distance x to the distance y. $\dfrac{2}{3}$

78. In the following rectangle, what is the ratio of the width to the length? $\dfrac{1}{2}$

Length = 6 ft

Width = 3 ft

79. In 10 oz of cashew nuts, there are 1,700 Cal. How many Calories are there per ounce? 170 Cal/oz

80. For a building valued at $200,000 the property tax is $4,000. Find the ratio of the tax to the building's value. $\dfrac{1}{50}$

81. On average, a person blinks 100 times in 4 min. How many times does a person blink in 1 min? 25 times/min

82. A bathtub contains 20 gal of water. If the tub empties in 4 min, what is the rate of flow of the water per minute? 5 gal/min

83. In a chemistry class, suppose that you had 400 cm³ of a substance with a mass of 7 g. To the nearest tenth of a cubic centimeter/gram, what is the unit rate? $57.1 \text{ cm}^3/\text{g}$

84. In a race, a sprinter completed 200 m in 22 sec. Find his speed, rounded to the nearest tenth of a meter per second. 9.1 m/sec

85. In a student government election, 1,000 students cast a vote for Alvin, 900 voted for Richard, and 100 cast a protest vote. What was the ratio of Alvin's vote to the total number of votes? $\dfrac{1}{2}$

86. In a sociology class, there are 100 females and 25 males. What is the ratio of males to the total number of students in the class? $\dfrac{1}{5}$

▦ *Using a calculator, solve the following problems, giving (a) the operation(s) carried out in the solution, (b) the exact answer, and (c) an estimate of the answer.*

87. In the insurance industry, a **loss ratio** is the ratio of total losses paid out by an insurance company to total premiums collected for a given time period.

$$\text{Loss ratio} = \frac{\text{Losses paid}}{\text{Premiums collected}}$$

In 2 months, a certain insurance company paid losses of $6,400,000 and collected premiums of $12,472,000. What is the loss ratio, rounded to the nearest hundredth? (a) Division; (b) 0.51; (c) possible answer: 0.5

88. Analysts for a brokerage firm prepare research reports on companies with stocks traded in various stock markets. One statistic that an analyst uses is the **price–earnings (P.E.) ratio**.

$$\text{P.E. ratio} = \frac{\text{Market price per share}}{\text{Earnings per share}}$$

Find the P.E. ratio to the nearest tenth for a stock that had a per-share market price of $70.75 and earnings of $5.37/share. (a) Division; (b) 13.18; (c) possible answer: 14

■ *Check your answers on page A-8.*

MINDSTRETCHERS

HISTORY

1. A rectangle is called a **golden rectangle** if the ratio of its length to its width is about 1.618 to 1 (the **golden ratio**).

1.618

1

To the ancient Egyptians and Greeks, a golden rectangle was considered to be the ratio most pleasing to the eye. Show that index cards in either of the two standard sizes (3×5 and 5×8) are close approximations to the golden rectangle.

$\dfrac{5}{3} = \dfrac{1.67}{1}, \dfrac{8}{5} = \dfrac{1.60}{1}$; both are close to $\dfrac{1.618}{1}$

INVESTIGATION

Tell students that in every circle the ratio $\dfrac{C}{d}$ is called π (pi) and is approximately 3.14.

2. The distance around a circle is called its **circumference** (C). The distance across the circle through its center is called its **diameter** (d).

a. Use a string and ruler to measure C and d for both circles shown.   Answers will vary.

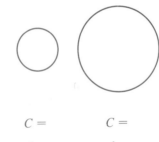

$C =$ $C =$

$d =$ $d =$

b. Compute the ratio of C to d for each circle. Are the ratios approximately equal?
Approximately 3.14; yes

$\dfrac{C}{d} =$ $\dfrac{C}{d} =$

WRITING

3. Sometimes we use *differences* rather than *quotients* to compare two quantities. Give an example of each kind of comparison and any advantages and disadvantages of each approach.   Answers may vary.

5.2 Solving Proportions

OBJECTIVES

- *To write and solve proportions*
- *To solve word problems involving proportions*

Writing Proportions

When two ratios—say, 1 to 2 and 4 to 8—are equal, we say that they are *in proportion*. We can write "1 is to 2 as 4 is to 8" as $\frac{1}{2} = \frac{4}{8}$. This equation is called a **proportion**.

Definition

A **proportion** is a statement that two ratios $\frac{a}{b}$ and $\frac{c}{d}$ are equal, written $\frac{a}{b} = \frac{c}{d}$.

One way to see if a proportion is true is to determine whether the *cross products* of the ratios are equal. For example, we see that the proportion

 Cross multiply.

is true because $2 \cdot 4 = 1 \cdot 8$, or $8 = 8$. However, the proportion $\frac{3}{5} = \frac{9}{10}$ is not true, since $5 \cdot 9 \neq 3 \cdot 10$.

EXAMPLE 1	**PRACTICE 1**

Determine whether the proportion 4 is to 3 as 16 is to 12 is true.

Solution First, we write the ratios in fractional form: $\frac{4}{3} = \frac{16}{12}$.

 Cross multiply.

$$3 \cdot 16 \overset{?}{=} 4 \cdot 12$$
$$48 \overset{\checkmark}{=} 48$$

The proportion 4 is to 3 as 16 is to 12 is true.

Determine whether the ratios 10 to 4 and 15 to 6 are in proportion.   Yes

---

EXAMPLE 2	**PRACTICE 2**

Is $\frac{15}{9} = \frac{8}{5}$ a true proportion?

Solution $\frac{15}{9} \overset{?}{=} \frac{8}{5}$

$$9 \cdot 8 \overset{?}{=} 15 \cdot 5 \quad \text{Cross multiply.}$$

$$72 \neq 75$$

The cross products are not equal, so the proportion is not true.

Determine whether $\frac{15}{6} = \frac{8}{3}$ is a true proportion.   Not a true proportion

EXAMPLE 3

A subcompact car travels 196 mi on 7 gal of gasoline, and a compact car travels 336 mi on 12 gal. Is the fuel economy (miles per gallon, or mpg) of both cars the same?

Solution The subcompact's fuel economy is $\dfrac{196 \text{ mi}}{7 \text{ gal}}$, and the compact's fuel economy is $\dfrac{336 \text{ mi}}{12 \text{ gal}}$. We want to know whether the two rates are equal.

$$\frac{196 \text{ mi}}{7 \text{ gal}} \overset{?}{=} \frac{336 \text{ mi}}{12 \text{ gal}}$$

$$7 \cdot 336 \overset{?}{=} 196 \cdot 12 \qquad \text{Cross multiply.}$$

$$2{,}352 \overset{\checkmark}{=} 2{,}352 \qquad \text{The cross products are equal.}$$

Both cars have the same fuel economy.

PRACTICE 3

Two candidates took a typing test for a certain job. Bob typed 300 words in 5 min, whereas Tamika typed 350 words in 6 min. Determine whether the two candidates had equal typing rates.   Different rates; Bob typed 60 wpm, and Tamika typed $58\frac{1}{3}$ wpm.

> Some students may decide to reduce each rate, getting
> 28 mpg $\overset{\checkmark}{=}$ 28 mpg.

Solving Proportions

Suppose that you make $840 for working 4 weeks in a book shop. At this rate of pay, how much money will you make in 10 weeks? To solve this problem, we can write a proportion in which the rates compare the amount of pay to the time worked.

$$\begin{array}{l} \text{Pay} \;\rightarrow \\ \text{Time} \;\rightarrow \end{array} \frac{840}{4} = \frac{x}{10} \begin{array}{l} \leftarrow \text{Pay} \\ \leftarrow \text{Time} \end{array}$$

We want to find the amount of pay corresponding to 10 weeks, so we call this missing value x. After setting the cross products equal, we find the missing value.

> Have students review Section 4.3, if necessary, before beginning these problems.

$$\frac{840}{4} = \frac{x}{10}$$

$$4 \cdot x = 840 \cdot 10 \qquad \text{Cross multiply.}$$

$$4x = 8{,}400$$

$$x = \frac{8{,}400}{4} \qquad \text{Divide each side of the equation by 4.}$$

$$x = 2{,}100$$

So you will make $2,100 in 10 weeks.

We can check our solution by substituting 2,100 for x in the original proportion.

$$\frac{840}{4} \overset{?}{=} \frac{2{,}100}{10} \qquad \text{Cross multiply.}$$

$$8{,}400 \overset{\checkmark}{=} 8{,}400$$

Our solution checks.

This example suggests the following rule.

To Solve a Proportion

■ find the cross products and set them equal,
■ solve the resulting equation, and
■ check the solution by substituting the value of the unknown in the original equation to be sure that the resulting proportion is true.

EXAMPLE 4

Solve and check: $\dfrac{2}{3} = \dfrac{x}{15}$

Solution

$\dfrac{2}{3} = \dfrac{x}{15}$

$3 \cdot x = 2 \cdot 15$ Cross multiply.

$3x = 30$ Divide each side by 3.

$x = 10$

Check:

$\dfrac{2}{3} = \dfrac{x}{15}$

$\dfrac{2}{3} \overset{?}{=} \dfrac{10}{15}$ Substitute 10 for x.

$2(15) \overset{?}{=} 3(10)$ Cross multiply.

$30 \overset{\checkmark}{=} 30$

PRACTICE 4

Solve and check: $\dfrac{x}{6} = \dfrac{12}{9}$ $x = 8$

EXAMPLE 5

Solve and check: $\dfrac{\frac{1}{4}}{12} = \dfrac{x}{96}$

Solution

$\dfrac{\frac{1}{4}}{12} = \dfrac{x}{96}$

$12 \cdot x = \dfrac{1}{4} \cdot 96$ Cross multiply.

$12x = 24$

$x = 2$ Divide each side by 12.

Check:

$\dfrac{\frac{1}{4}}{12} = \dfrac{x}{96}$

$\dfrac{\frac{1}{4}}{12} \overset{?}{=} \dfrac{2}{96}$ Substitute 2 for x.

$\dfrac{1}{4} \cdot (96) \overset{?}{=} 12(2)$ Cross multiply.

$24 \overset{\checkmark}{=} 24$

PRACTICE 5

Solve and check: $\dfrac{\frac{1}{2}}{2} = \dfrac{3}{x}$ $x = 12$

EXAMPLE 6

Forty pounds of sodium hydroxide are needed to neutralize 49 lb of sulfuric acid. At this rate, how many pounds of sodium hydroxide are needed to neutralize 98 lb of sulfuric acid? (*Source:* Peter Atkins and Loretta Jones, *Chemistry*)

Solution Let *n* represent the number of pounds of sodium hydroxide. We set up a proportion to compare the amount of sodium hydroxide to the amount of sulfuric acid.

$$\frac{40}{49} = \frac{n}{98}$$ ⟵ Pounds of sodium hydroxide
⟵ Pounds of sulfuric acid

$49n = 40 \cdot 98$ Cross multiply.

$49n = 3{,}920$

$n = 80$ Divide each side by 49.

Check: $\dfrac{40}{49} = \dfrac{n}{98}$

$\dfrac{40}{49} \overset{?}{=} \dfrac{80}{98}$ Substitute 80 for *n*.

$49 \cdot 80 \overset{?}{=} 40 \cdot 98$ Cross multiply.

$3{,}920 \overset{\checkmark}{=} 3{,}920$

So 80 lb of sodium hydroxide are needed to neutralize 98 lb of sulfuric acid.

PRACTICE 6

Saffron is a powder made from crocus flowers and is used in the manufacture of perfume. Some 8,000 crocus flowers are required to make 2 oz of saffron. How many flowers are needed to make 16 oz of saffron? (*Source: World Book Encyclopedia*)
64,000

> Point out to students the significance of where the unknown is placed when they are setting up a proportion.

TIP The ratios used in a proportion must always compare the same quantities. For example, if $\dfrac{\text{inches}}{\text{miles}}$ appears on one side of a proportion, then $\dfrac{\text{inches}}{\text{miles}}$ must appear on the other side as well, in that order.

EXAMPLE 7

The scale on a map is $\frac{1}{2}$ in. to 50 mi. How many miles correspond to 2.5 in.?

Solution We know that $\frac{1}{2}$ in. corresponds to 50 mi. Thus we can set up a proportion that compares inches to miles, letting *m* represent the unknown number of miles.

$$\frac{\frac{1}{2}\text{ in.}}{50\text{ mi}} = \frac{2.5\text{ in.}}{m\text{ mi}}$$

$$\frac{\frac{1}{2}}{50} = \frac{2.5}{m}$$

$$\frac{1}{2}m = (50)(2.5)$$ Cross multiply.

PRACTICE 7

A truck goes 60 mi on 10 gal of gas. At the same rate, how many gallons of gas will the truck need to go 90 mi?   15 gal

$$\frac{1}{2}m = 125$$

$$\frac{1}{2}m \div \frac{1}{2} = 125 \div \frac{1}{2} \qquad \text{Divide each side by } \tfrac{1}{2}.$$

$$\frac{1}{2}m \times \frac{2}{1} = 125 \times \frac{2}{1}$$

$$m = 250$$

Check: $\dfrac{\frac{1}{2}}{50} = \dfrac{2.5}{m}$

$$\dfrac{\frac{1}{2}}{50} \overset{?}{=} \dfrac{2.5}{250}$$

$$125 \overset{\checkmark}{=} 125$$

So 2.5 in. correspond to 250 mi.

EXAMPLE 8

In the following diagram, the heights and shadows of the two objects shown are in proportion. Find the height of the tree, h, in meters.

1.6 m

2.4 m   10.8 m

Solution The heights and shadows are in proportion, so we write the following.

$$\frac{h \text{ m}}{10.8 \text{ m}} = \frac{1.6 \text{ m}}{2.4 \text{ m}} \begin{array}{l} \leftarrow \text{height} \\ \leftarrow \text{shadow} \end{array}$$

$$\frac{h}{10.8} = \frac{1.6}{2.4}$$

$$2.4h = (10.8)(1.6)$$

$$\frac{2.4h}{2.4} = \frac{17.28}{2.4}$$

$$h = 7.2$$

The height of the tree is therefore 7.2 m.

Check: $\dfrac{h}{10.8} = \dfrac{1.6}{2.4}$

$$\frac{7.2}{10.8} \overset{?}{=} \frac{1.6}{2.4}$$

$$17.28 \overset{\checkmark}{=} 17.28$$

PRACTICE 8

In the following diagram, the ratio of the height to the base in each of the triangles is the same. Find the height, h, of the smaller triangle.   1.5 ft

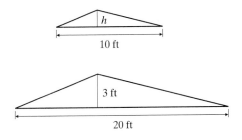

h

10 ft

3 ft

20 ft

Exercises 5.2

Indicate whether each statement is true or false.

1. Thirty is to 9 as 40 is to 12.
True

2. Nine is to 12 as 12 is to 16.
True

3. Two is to 3 as 7 is to 16.
False

4. Three is to 8 as 10 is to 27.
False

5. One and one-tenth is to 0.3
as 44 is to 12.   True

6. One and one-half is to 2 as 0.6 is to 0.8.
True

7. $\dfrac{3}{6} = \dfrac{2}{5}$
False

8. $\dfrac{4}{7} = \dfrac{5}{8}$
False

9. $\dfrac{12}{28} = \dfrac{18}{42}$
True

10. $\dfrac{28}{24} = \dfrac{7}{6}$
True

11. $\dfrac{6}{1} = \dfrac{3}{\frac{1}{2}}$
True

12. $\dfrac{5}{30} = \dfrac{\frac{1}{3}}{2}$
True

Solve and check.

13. $\dfrac{4}{8} = \dfrac{10}{x}$
$x = 20$

14. $\dfrac{2}{3} = \dfrac{x}{42}$
$x = 28$

15. $\dfrac{x}{19} = \dfrac{10}{5}$
$x = 38$

16. $\dfrac{1}{6} = \dfrac{x}{78}$
$x = 13$

17. $\dfrac{5}{x} = \dfrac{15}{12}$
$x = 4$

18. $\dfrac{15}{x} = \dfrac{6}{10}$
$x = 25$

19. $\dfrac{4}{1} = \dfrac{52}{x}$
$x = 13$

20. $\dfrac{1}{17} = \dfrac{x}{51}$
$x = 3$

21. $\dfrac{7}{4} = \dfrac{14}{x}$
$x = 8$

22. $\dfrac{x}{6} = \dfrac{15}{18}$
$x = 5$

23. $\dfrac{x}{8} = \dfrac{3}{6}$
$x = 4$

24. $\dfrac{7}{5} = \dfrac{35}{x}$
$x = 25$

25. $\dfrac{6}{21} = \dfrac{x}{70}$
$x = 20$

26. $\dfrac{4}{x} = \dfrac{92}{23}$
$x = 1$

27. $\dfrac{x}{12} = \dfrac{25}{20}$
$x = 15$

28. $\dfrac{20}{25} = \dfrac{x}{45}$
$x = 36$

29. $\dfrac{28}{x} = \dfrac{36}{27}$
$x = 21$

30. $\dfrac{27}{63} = \dfrac{24}{x}$
$x = 56$

31. $\dfrac{x}{10} = \dfrac{4}{3}$
$x = 13\frac{1}{3}$

32. $\dfrac{5}{6} = \dfrac{2}{x}$
$x = 2\frac{2}{5}$

33. $\dfrac{15}{2} = \dfrac{x}{2\frac{2}{3}}$
$x = 20$

34. $\dfrac{x}{11} = \dfrac{6}{5\frac{1}{2}}$
$x = 12$

35. $\dfrac{x}{27} = \dfrac{1.6}{24}$
$x = 1.8$

36. $\dfrac{2.4}{28} = \dfrac{18}{x}$
$x = 210$

37. $\dfrac{10.5}{x} = \dfrac{5}{10}$
$x = 21$

38. $\dfrac{32}{7.2} = \dfrac{x}{9}$
$x = 40$

39. $\dfrac{7}{0.9} = \dfrac{x}{36}$
$x = 280$

40. $\dfrac{18}{x} = \dfrac{4.8}{56}$
$x = 210$

41. $\dfrac{600}{x} = \dfrac{3}{\frac{1}{2}}$
$x = 100$

42. $\dfrac{2\frac{1}{3}}{5} = \dfrac{x}{12}$
$x = 5\frac{3}{5}$

43. $\dfrac{4}{x} = \dfrac{\frac{2}{5}}{10}$
$x = 100$

44. $\dfrac{\frac{3}{4}}{6} = \dfrac{3}{x}$
$x = 24$

45. $\dfrac{32}{x} = \dfrac{5.6}{49}$
$x = 280$

46. $\dfrac{x}{15} = \dfrac{1.6}{0.5}$
$x = 48$

47. $\dfrac{9}{6\frac{1}{2}} = \dfrac{4}{x}$
$x = 2\frac{8}{9}$

48. $\dfrac{32}{7\frac{1}{5}} = \dfrac{x}{9}$
$x = 40$

49. $\dfrac{x}{0.16} = \dfrac{0.15}{4.8}$
$x = 0.005$

50. $\dfrac{1.5}{1.25} = \dfrac{x}{0.5}$
$x = 0.6$

Applications

Solve, using proportions. Then check.

51. An average adult's heart beats 8 times every 6 sec, whereas a newborn baby's heart beats 7 times every 3 sec. Determine whether these rates are the same. (**Source:** *Mosby's Medical, Nursing, and Allied Health Dictionary*)   Not the same

52. In 1 week, you made $175 by working 25 hr. During the same week, your friend earned $35 by working 5 hr. Determine whether the pay rates are equal. Equal

53. A dripping faucet wastes about 15 gal of water daily. About how much water is wasted in 3 hr? (*Hint*: 1 day = 24 hr.)
$1\frac{7}{8}$ gal

54. A certain iceberg in Antarctica moves about 2 in./yr. At this rate, how long will the iceberg take to move 1 ft? (*Hint*: 1 ft = 12 in.)   6 yr

2. Work with a partner on the following.

a. Complete the following table.

x	0	1	2	3	4
5x	0	5	10	15	20

b. Write as many true proportions as you can, based on the values in the table.

$$\frac{1}{5} = \frac{2}{10} \qquad \frac{1}{5} = \frac{4}{20} \qquad \frac{3}{15} = \frac{4}{20}$$

$$\frac{1}{5} = \frac{3}{15} \qquad \frac{2}{10} = \frac{4}{20} \qquad \frac{2}{10} = \frac{3}{15}$$

c. Compare your answers with those of another student.

WRITING

3. The following two methods can be used to solve the proportion $\frac{x}{75} = \frac{29}{15}$.

Method 1	Method 2

$$\frac{x}{75} = \frac{29}{15} \qquad\qquad \frac{x}{75} = \frac{29}{15}$$

$$15x = 75 \cdot 29 \qquad\qquad 15x = 75 \cdot 29$$

$$\frac{\cancel{15}x}{\cancel{15}} = \frac{2{,}175}{15} \qquad\qquad \frac{\cancel{15}x}{\cancel{15}} = \frac{\overset{5}{\cancel{75}} \cdot 29}{\underset{1}{\cancel{15}}}$$

$$x = 145 \qquad\qquad x = 145$$

Which method do you prefer? Why? Explain.   Answers may vary.

_____

_____

KEY CONCEPTS and SKILLS

☐ = CONCEPT ☐ = SKILL

CONCEPT/SKILL	DESCRIPTION	EXAMPLE
[5.1] **Ratio**	A comparison of two numbers in terms of a quotient.	3 to 4, $\dfrac{3}{4}$, or 3:4
[5.1] **To simplify a ratio**	■ Write the ratio as a fraction. ■ Reduce the fraction to simplest form, dropping the units if given.	9:27 is the same as 1:3 because $\dfrac{9}{27} = \dfrac{1}{3}$ 21 hr to 56 hr $= \dfrac{21 \text{ hr}}{56 \text{ hr}} = \dfrac{21}{56} = \dfrac{3}{8}$
[5.1] **Rate**	A ratio of unlike quantities.	$\dfrac{10 \text{ students}}{3 \text{ tutors}}$
[5.1] **To simplify a rate**	■ Write the rate as a fraction. ■ Reduce the fraction to simplest form *keeping the units*.	175 miles per 7 gallons, $\dfrac{175 \text{ mi}}{7 \text{ gal}} = \dfrac{25 \text{ mi}}{1 \text{ gal}}$, or 25 mpg
[5.1] **Unit rate**	A rate in which the number in the denominator is 1.	180 Calories per ounce, or 180 Cal/oz
[5.1] **Unit price**	The price of one item.	$0.69 per can, or $0.69/can
[5.2] **Proportion**	A statement that two ratios are equal.	$\dfrac{5}{8} = \dfrac{15}{24}$
[5.2] **To solve a proportion**	■ Find the cross products. ■ Solve the resulting equation. ■ Check the solution by substituting the value of the unknown in the original equation to verify that the resulting proportion is true.	Solve: $\dfrac{6}{9} = \dfrac{2}{x}$ $6x = 18$ $x = 3$ Check: $\dfrac{6}{9} = \dfrac{2}{x}$ $\dfrac{6}{9} \overset{?}{=} \dfrac{2}{3}$ $6 \cdot 3 \overset{?}{=} 9 \cdot 2$ $18 \overset{\checkmark}{=} 18$

Chapter 5   Review Exercises

To help you review this chapter, solve these problems.

[5.1]

Write each ratio or rate in simplest form.

1. 10 to 15
$$\frac{2}{3}$$

2. 28 to 56
$$\frac{1}{2}$$

3. 3 to 4
$$\frac{3}{4}$$

4. 50 to 16
$$\frac{25}{8}$$

5. 10,400 votes to 6,500 votes
$$\frac{8}{5}$$

6. 9 cups to 12 cups
$$\frac{3}{4}$$

7. 88 ft in 10 sec
$$\frac{44 \text{ ft}}{5 \text{ sec}}$$

8. 16 applicants for 45 positions
$$\frac{16 \text{ applicants}}{45 \text{ positions}}$$

Write each ratio as a unit rate.

9. 4 lb of fertilizer to cover 1,600 ft² of lawn
    0.0025 lb/sq ft

10. 75 billion telephone calls in 150 days
    500,000,000 calls/day

11. 600 mi in 3 hr   200 mph

12. 750 VCRs in 500 households   1.5 VCRs/household

13. 21,000,000 vehicles produced in 2 yr
    10,500,000 vehicles/yr

14. 532,000 commuters traveled in 7 days
    76,000 commuters/day

Find the unit price for each item.

15. $475 for 4 nights $118.75/night

16. $1,785 for 105 tickets $17/ticket

17. $80,000 for 32 computer stations $2,500/station

18. $9,364 for 100 shares of stock $93.64/share

In each case which is the better buy?

19. Magazine subscription   a

a. 12 issues for $35.00

b. 36 issues for $107.50

20. Printing   b

a. 2 million posters for $1,400,000

b. 1.5 million posters for $1,000,000

21. Visits to a health club   b

a. 10 visits for $55

b. 20 visits for $100

22. College concert tickets   a

a. 12 tickets for $84.00

b. 15 tickets for $112.50

[5.2]

Indicate whether each proportion is true or false.

23. $\dfrac{15}{25} = \dfrac{3}{5}$
True

24. $\dfrac{3}{1} = \dfrac{1}{3}$
False

25. $\dfrac{50}{45} = \dfrac{10}{8}$
False

26. $\dfrac{15}{6} = \dfrac{5}{2}$
True

Solve for x. Then check.

27. $\dfrac{1}{2} = \dfrac{x}{12}$
$x = 6$

28. $\dfrac{9}{12} = \dfrac{x}{4}$
$x = 3$

29. $\dfrac{12}{x} = \dfrac{3}{8}$
$x = 32$

30. $\dfrac{x}{72} = \dfrac{5}{12}$
$x = 30$

31. $\dfrac{1.6}{7.2} = \dfrac{x}{9}$
$x = 2$

32. $\dfrac{x}{12} = \dfrac{1.2}{1.8}$
$x = 8$

33. $\dfrac{5}{37} = \dfrac{7}{x}$
$x = 51\dfrac{4}{5}$

34. $\dfrac{3}{5} = \dfrac{x}{2}$
$x = 1\dfrac{1}{5}$

35. $\dfrac{2\dfrac{1}{4}}{x} = \dfrac{1}{30}$
$x = 67\dfrac{1}{2}$

36. $\dfrac{3}{1\dfrac{3}{5}} = \dfrac{x}{24}$
$x = 45$

37. $\dfrac{0.36}{4.2} = \dfrac{2.4}{x}$
$x = 28$

38. $\dfrac{x}{0.21} = \dfrac{0.12}{0.18}$
$x = 0.14$

Mixed Applications

Solve.

39. An airplane has 12 first-class seats and 180 seats in coach. What is the ratio of first-class seats to coach seats? $\dfrac{1}{15}$

40. A computer store sells $23,000 worth of Dill computers and $45,000 worth of Orange computers in a given month. What is the ratio of Dill to Orange computer sales? $\dfrac{23}{45}$

41. If you earn $540 for a 6-day workweek, how much do you earn per day? $90/day

42. A glacier in Alaska moves about 2 in. in 16 mo. How far does the glacier move per month?   0.125 in./mo

43. You plan to divide your estate between your two children. If the older child receives $40,000 and the younger child $100,000, what is the ratio of the older child's inheritance to that of the entire estate? $\dfrac{2}{7}$

44. A city's public libraries spend about $9.50 in operating expenses for every book they circulate. If their operating expenses amount to $475,000, how many books circulate?   50,000 books

45. In your child's day-care center, the required staff-to-child ratio is 2 to 5. If there are 60 children and 12 staff in the day-care center, is the center in compliance with the requirement?   No

46. A sports car engine has an 8-to-1 compression ratio. Before compression, the fuel mixture in a cylinder takes up 440 cc of space. How much space does the fuel mixture occupy when fully compressed?   55 cc

47. Despite the director's protests, the 1924 silent film *Greed* was edited down from about 42 reels of film to 10 reels. If the original version was about 9 hr long, about how long was the edited version? (*Source: The Film Encyclopedia*)   2 hr

48. On a certain drawing, a measurement of 25 ft is represented by 2 in. If two houses are actually 62.5 ft apart, what is the distance between them on the drawing?   5 in.

49. The density of a substance is the ratio of its mass to its volume. To the nearest hundredth, find the density of gasoline if a volume of 317.45 cc has a mass of 216.21 g.   0.68 g/cc

50. An automobile uses 5.25 gal of gasoline to travel 176.56 mi. To the nearest hundredth, how many gallons of gasoline are needed to travel 254 mi?   7.55 gal

■ *Check your answers on page A-8.*

Chapter 5 Posttest

To see if you have mastered the topics in this chapter, take this test.

Write each ratio or rate in simplest form.

1. 8 to 12 $\dfrac{2}{3}$

2. 15 to 42 $\dfrac{5}{14}$

3. 55 oz to 31 oz $\dfrac{55}{31}$

4. 180 mi to 15 mi $\dfrac{12}{1}$

5. 65 revolutions in 60 sec $\dfrac{13 \text{ revolutions}}{12 \text{ sec}}$

6. 3 cm for every 75 km $\dfrac{1 \text{ cm}}{25 \text{ km}}$

Find the unit rate.

7. 340 mi in 5 hr

68 mph

8. $4,080 for 30 days

$136/day

9. 200-m dash in 30 sec

$6\dfrac{2}{3}$ m/sec

10. 302 mi on 17 gal of gasoline

$17\dfrac{13}{17}$ mpg

Determine whether each proportion is true.

11. $\dfrac{8}{21} = \dfrac{16}{40}$ No

12. $\dfrac{7}{3} = \dfrac{63}{27}$ Yes

Solve for x. Then check.

13. $\dfrac{15}{x} = \dfrac{6}{10}$

$x = 25$

14. $\dfrac{102}{17} = \dfrac{36}{x}$

$x = 6$

15. $\dfrac{0.9}{36} = \dfrac{0.7}{x}$

$x = 28$

16. $\dfrac{\frac{1}{3}}{4} = \dfrac{x}{12}$

$x = 1$

Solve.

17. To advertise your business, you can purchase 3 million e-mail addresses for $400 or 5 million e-mail addresses for $600. Which is the better buy?   5 million for $600

18. A house had an original value of $95,000 but increased in value to $110,000 in 5 years. What is the ratio of the increase to the original value?

$\dfrac{3}{19}$

19. A man $6\dfrac{1}{4}$ ft tall casts a 5-ft shadow. A nearby tree casts a 20-ft shadow. If the heights and shadows of the man and tree are proportional, how tall is the tree?

25 ft

20. You are a nurse taking your patient's pulse. If the patient's pulse beats 12 times in 15 sec, what is her pulse per minute?   48 beats/min

■ *Check your answers on page A-8.*

Cumulative Review Exercises

To help you review, solve the following:

1. Find the difference: $3\dfrac{1}{10} - 2\dfrac{7}{10}$ $\dfrac{2}{5}$

2. Multiply: $8.2 \times 1{,}000$ 8,200

3. Solve and check: $\dfrac{2}{9} = \dfrac{x}{45}$ $x = 10$

4. Simplify the ratio: 2.5 to 10 $\dfrac{1}{4}$

5. Find the unit price: 3 yd for \$12   \$4/yd

6. Estimate: $12\dfrac{1}{7} \div 3\dfrac{9}{10}$ Possible answer: 3

7. Suppose that coffee sells for \$$D$/lb. At this rate, how much will $\dfrac{11}{16}$ of a pound cost? $\$\dfrac{11}{16}D$

8. The time of the winning swimmer was 54.44 sec. If the swimmer in second place took 0.1 sec longer, what was her time?   54.54 sec

9. A rule-of-thumb for growing lily bulbs is to plant them 3 times as deep as they are wide. How deep should you plant a lily bulb that is 2.5 in. wide?   7.5 in.

10. At your company, a part-time employee works 10 hr a week and makes \$120. At this rate of pay, how much would the employee make for working 15 hr a week? \$180

■ *Check your answers on page A-8.*

CHAPTER **6**

Percents

6.1 Introduction to Percents

6.2 Solving Percent Problems

6.3 More on Percents

NUTRIENTS IN CEREAL

Researchers from the National Cancer Institute and the U.S. Department of Agriculture have reported that many children in the United States get a significant portion of their vitamins and other nutrients from fortified breakfast cereals. The listing of nutrition facts typically displayed on the side of a cereal box shows the amount of nutrients contained in a serving of the cereal. This amount is expressed as a percent of the recommended daily allowance (RDA).

For instance the following list copied from the box of a popular cereal indicates that a serving of the cereal contains 25% of the RDA of folic acid—a substance found in green leafy vegetables, liver, kidney, and yeast.

Vitamin A	4%
Vitamin C	2%
Calcium	25%
Iron	25%
Folic Acid	25%
Thiamin	25%
Riboflavin	35%
Zinc	30%

Chapter 6 Pretest

To see if you have already mastered the topics in this chapter, take this test.

Rewrite.

1. 5% as a fraction $\dfrac{1}{20}$

2. $37\dfrac{1}{2}$% as a fraction $\dfrac{3}{8}$

3. 250% as a decimal   2.5

4. 3% as a decimal   0.03

5. 0.007 as a percent   0.7%

6. 8 as a percent   800%

7. $\dfrac{2}{3}$ as a percent, rounded to the nearest whole percent
67%

8. $1\dfrac{1}{10}$ as a percent   110%

Solve.

9. What is 75% of 50 ft? $37\frac{1}{2}$ ft

10. Find 110% of 50.   55

11. Estimate 84% of $61.77.   Possible answer: $48

12. 2% of what number is 5?   250

13. What percent of 10 is 4?   40%

14. What percent of 4 is 10?   250%

15. In a savings account, you earn 4% interest on $350. How much money do you earn in interest? $14

16. What is the percent change for a quantity that increases from 6 to 8? $33\frac{1}{3}$%

17. In the depths of the Great Depression, 24% of the U.S. civilian labor force was unemployed. Write this percent as a simplified fraction. (*Source:* Bureau of Census) $\dfrac{6}{25}$

18. In a chemistry lab, you dissolved 10 ml of acid in 30 ml of water. What percent of the solution was acid?   25%

19. Your regular salary is $25,000. If you get a 10% bonus, how much is this bonus in dollars? $2,500

20. Health insurance covered 80% of the cost of your operation. You paid the remainder, which came to $2,000. Find the total cost of the operation. $10,000

■ *Check your answers on page A-8.*

6.1 Introduction to Percents

OBJECTIVES
- *To read and write percents*
- *To find the fraction and the decimal equivalent of a given percent*
- *To find the percent equivalent of a given fraction or decimal*
- *To solve word problems involving percent conversions*

What Percents Are and Why They Are Important

Percent means divided by 100. So 50% (read "fifty percent") means 50 divided by 100 (or 50 out of 100).

A percent can also be thought of as a ratio or a fraction. For example, we can look at 50% either as the ratio of 50 parts to 100 parts or as the fraction $\frac{50}{100}$, or $\frac{1}{2}$.

In the diagram at the right, the shaded portion represents 50%.

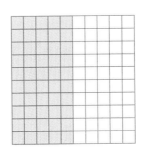

We can use diagrams to represent other percents.

> Point out that a **cent**ury has 100 years, a **cent**ennial is a 100th anniversary, and a **cent** is a hundredth of a dollar. Tell students that "percent" literally means "per one hundred."

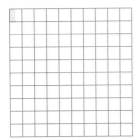

In this diagram, $\frac{1}{2}$ % is equivalent to the shaded portion,

$$\frac{\frac{1}{2}}{100}, \quad \text{or} \quad \frac{1}{200}.$$

> Point out that $\frac{\frac{1}{2}}{100} = \frac{1}{2} \div 100 = \frac{1}{2} \times \frac{1}{100} = \frac{1}{200}.$

The entire diagram at the right is shaded, so 100% means $\frac{100}{100}$, or 1.

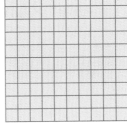

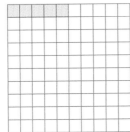

We can express 105% as $\frac{105}{100}$, or $1\frac{1}{20}$, as shown by the shaded portions of the diagrams.

Percents are in common use, as the following statements taken from a single page of a newspaper illustrate.

- About 10% of the city's budget goes to sanitation.
- Blanket Sale—30% to 40% off!
- The number of victims of the epidemic increased by 125% in just 6 months.

A key reason for using percents so frequently is that they are easy to compare. For instance, we can tell right away that a discount of 30% is larger than a discount of 22%, simply by comparing the whole numbers 30 and 22.

To see how percents relate to fractions and decimals, let's consider finding equivalent fractions, decimals, and percents. We have already discussed two of the six types of conversions:

■ changing a decimal to a fraction, and

■ changing a fraction to a decimal.

The remaining four types of conversion are

■ changing a percent to a fraction,

■ changing a percent to a decimal,

■ changing a decimal to a percent, and

■ changing a fraction to a percent.

Note that each type of conversion changes the way the number is written—but not the number itself.

Before moving on to the four new conversions, review with students the two conversions previously discussed: changing a decimal to a fraction and changing a fraction to a decimal.

Changing a Percent to a Fraction

Suppose that we want to rewrite a percent—say, 30%—as a fraction. Because percent means divided by 100, we simply drop the % sign, place 30 over 100, and simplify.

$$30\% = \frac{30}{100} = \frac{3}{10}$$

Therefore the fraction $\frac{3}{10}$ is just another way of writing the percent 30%. This result suggests the following rule.

> **To Change a Percent to the Equivalent Fraction**
> ■ drop the % sign from the given percent and place the number over 100, and
> ■ simplify the resulting fraction, if possible.

EXAMPLE 1

Write 7% as a fraction.

Solution To change this percent to a fraction, we drop the percent sign and write the 7 over 100.

$$7\% = \frac{7}{100}$$

The fraction is already reduced to lowest terms.

PRACTICE 1

Find the fractional equivalent of 21%.

$\frac{21}{100}$

EXAMPLE 2

Express 150% as a fraction.

Solution $150\% = \frac{150}{100} = \frac{3}{2}$, or $1\frac{1}{2}$

Note that the answer is larger than 1 because the original percent was more than 100%.

PRACTICE 2

Find the fractional equivalent of 200%. 2

Have students review how fractions are simplified.

EXAMPLE 3

Express $33\frac{1}{3}\%$ as a fraction.

Solution To find the equivalent fraction, we begin in the usual way, dropping the % sign and putting the number over 100.

Here change the fraction line to division.

$$\frac{33\frac{1}{3}}{100} = 33\frac{1}{3} \div 100 = 33\frac{1}{3} \div \frac{100}{1} = \frac{100}{3} \div \frac{100}{1} = \frac{\overset{1}{\cancel{100}}}{3} \times \frac{1}{\underset{1}{\cancel{100}}} = \frac{1}{3}$$

So $33\frac{1}{3}\%$ expressed as a fraction is $\dfrac{1}{3}$.

PRACTICE 3

Change $12\frac{1}{2}\%$ to a fraction. $\dfrac{1}{8}$

Remind students that the fraction line means to divide.

EXAMPLE 4

About 2% of babies born in the United States are twins. Express this percent as a fraction.

Solution $2\% = \dfrac{2}{100} = \dfrac{1}{50}$

Therefore $\dfrac{1}{50}$ of babies born in the United States are twins.

PRACTICE 4

Scientists estimate that 90% of the beaches on the U.S. East and Gulf Coasts are eroding. Express this percent as a fraction. $\dfrac{9}{10}$

Changing a Percent to a Decimal

Now let's consider rewriting a percent as a decimal. For instance, take 75%. We begin by writing this percent as a fraction.

$$75\% = \frac{75}{100}$$

We convert this fraction to a decimal by dividing.

$$\begin{array}{r} 0.75 \\ 100\overline{)75.00} \end{array}$$

Remind students that a shortcut for dividing a decimal by 100 is to move the decimal point two places to the left.

Note that we could have gotten this answer simply by moving the decimal point two places to the left and dropping the % sign.

$$75\% = 75.\% = .75, \quad \text{or} \quad 0.75$$

To Change a Percent to the Equivalent Decimal

move the decimal point two places *to the left*, and drop the % sign.

EXAMPLE 5

Change 400% to a decimal or a whole number.

Solution A decimal point to the right of the whole number 400 is understood. When we move the decimal point two places to the left and drop the % sign, it ends up immediately to the right of the 4.

$$400\% = 400.\% = 4.00, \quad \text{or} \quad 4$$

To simplify our answer, we dropped the two 0s and the decimal point at the right end of the number. Note that the answer is greater than 1 whole unit.

PRACTICE 5

Express 300% as a decimal.   3

EXAMPLE 6

Find the decimal equivalent of 1%.

Solution The unwritten decimal point lies to the right of the 1. Moving the decimal point two places to the left and dropping the % sign, we get the following.

$$1\% = 01.\% = .01, \quad \text{or} \quad 0.01$$

Did you note that we had to insert a 0 as a placeholder, because there was only a single digit to the left of the 1?

PRACTICE 6

What is the decimal equivalent of 5%?
0.05

EXAMPLE 7

Convert 37.5% to a decimal.

Solution $37.5\% = .375, \quad \text{or} \quad 0.375$

Note that the given number in this problem is a percent even though it involves a decimal point.

PRACTICE 7

Rewrite 48.2% as a decimal.   0.482

EXAMPLE 8

Change $12\frac{1}{2}\%$ to a decimal.

Solution To find the decimal equivalent of $12\frac{1}{2}\%$, we begin by converting the fraction in the mixed number to its decimal equivalent.

$$\frac{1}{2} = 2\overline{)1.0} \quad \frac{0.5, \text{ or } .5}{}$$

Next, we replace the fraction $\frac{1}{2}$ by its decimal equivalent (.5) in the original percent. Then we move the decimal point to the left two places, dropping the % sign.

$$12\frac{1}{2}\% = 12.5\% = .125, \quad \text{or} \quad 0.125$$

PRACTICE 8

Express the following percent as a decimal:
$62\frac{1}{4}\%$ 0.6225

You may want to warn students that a common error here is to forget to move the decimal point after changing the fraction to a decimal.

EXAMPLE 9	PRACTICE 9
In 1945, the public debt of the United States was 602% of what it had been in 1940. Write this percent as a decimal. **Solution** 602% = 602.%, or 6.02 The 1945 U.S. debt was 6.02 times what it had been 5 years earlier.	On an investment, you made a profit of 11%. Express this percent as a decimal. 0.11

Changing a Decimal to a Percent

Suppose that you want to change 0.75 to a percent. Because 100% is the same as 1, you can multiply by 100%.

$$0.75 \times 100\% = 75\%$$

Note that you could have gotten this answer simply by moving the decimal point to the right two places and adding the % sign.

$$0.75 = 075.\%, \quad \text{or} \quad 75\%$$

Note that we dropped the decimal point in the answer because it was to the right of the units digit.

> **To Change a Decimal to the Equivalent Percent**
> ■ move the decimal point two places to the right and insert the % sign.

EXAMPLE 10	PRACTICE 10
Write 0.125 as a percent. **Solution** First move the decimal point two places to the right. Then add the % sign. $$0.125 = 012.5\% = 12.5\%, \quad \text{or} \quad 12\tfrac{1}{2}\%$$	What percent is equivalent to the decimal 0.025? 2.5%

EXAMPLE 11	PRACTICE 11
Convert 0.03 to a percent. **Solution** 0.03 = 003.%, or 3%	Change 0.01 to a percent. 1%

EXAMPLE 12	PRACTICE 12
Express 0.1 as a percent. **Solution** In the given number, only a single digit is to the right of the decimal point. So to move the decimal point two places to the right, we need to insert a 0 as a placeholder. $$0.1 = 0.10 = 10.\%, \quad \text{or} \quad 10\%$$	What percent is equivalent to 0.7? 70% Remind students that inserting 0s at the right end of a decimal does not change its value.

EXAMPLE 13

What percent is equivalent to 2?

Solution Recall that 2 has a decimal point understood to its right.

Here we add two 0s as placeholders.

$$2 = 2. = 2.00 = 200.\%, \quad \text{or} \quad 200\%$$

So the answer is 200%, which makes sense: 200% is double 100%, just as 2 is double 1.

PRACTICE 13

Rewrite the whole number 3 as a percent.
300%

EXAMPLE 14

Express 0.2483 as a percent, rounded to the nearest whole percent.

Solution First, we obtain the exact percent equivalent.

$$0.2483 = 24.83\%$$

To round this number to the nearest whole percent, we underline the digit 4. Then we check the critical digit immediately to its right. This digit is 8, so we round up.

$$24.83\% \approx 25.\%, \quad \text{or} \quad 25\%$$

PRACTICE 14

Convert 0.718 to a percent, rounded to the nearest whole percent. 72%

> Use this problem to review the rounding of decimals.

EXAMPLE 15

Would your taxes be less if they were 0.4 of your income or 35% of your income? Explain.

Solution You want to compare the decimal 0.4 and the percent 35%. One way is to change the decimal to a percent.

$$0.40 = 40.\%, \quad \text{or} \quad 40\%$$

Because 40% is greater than 35%, your taxes would be less if they were 35% of your income.

PRACTICE 15

Air is a mixture of many gases. For example, 0.78 of air is nitrogen, and 0.93% is argon. Is there more nitrogen or argon in air? Explain. nitrogen; 78% > 0.93%, or 0.78 > 0.0093.

> Some students may want to change 35% to a decimal, and then compare.
> $$35\% = 0.35$$
> $$0.35 < 0.4$$

Changing a Fraction to a Percent

> You may want to show students that another way to find the equivalent percent is to multiply the given fraction by the number 1, expressed as 100%.
>
> $$\frac{1}{5} \times 100\% = \frac{1}{\cancel{5}} \times \frac{\overset{20}{\cancel{100}}}{1}\%$$
>
> $$= \frac{20}{1}\%$$
>
> $$= 20\%$$

Now let's change a fraction to a percent. Consider, for instance, the fraction $\frac{1}{5}$. One way to convert this number to a percent is first to convert it to a decimal and then to convert this decimal to a percent.

$$\frac{1}{5} = 0.2 = 20.\%, \quad \text{or} \quad 20\%$$

To Change a Fraction to the Equivalent Percent
- change the fraction to a decimal, and
- change the decimal to a percent.

EXAMPLE 16

Rewrite $\dfrac{1}{8}$ as a percent.

Solution To change the given fraction to a percent, we first find the equivalent decimal, which we then express as a percent.

$$\frac{1}{8} = 8\overline{)1.000} \;\; \begin{array}{c} 0.125 \end{array} \quad \text{and} \quad 0.125 = 12.5\%$$

PRACTICE 16

Convert $\dfrac{4}{25}$ to a percent.   16%

EXAMPLE 17

Which is larger: 130% or $1\dfrac{3}{8}$?

Solution To compare, let's express $1\dfrac{3}{8}$ as a percent.

$$1\frac{3}{8} = 1.375, \;\; \text{or} \;\; 137.5\%$$

Because 137.5% is larger than 130%, so is $1\dfrac{3}{8}$.

PRACTICE 17

True or false: $\dfrac{2}{3} > 60\%$. Justify your answer.   True. $\dfrac{2}{3} \approx 67\% > 60\%$

EXAMPLE 18

A student got 28 of 30 questions correct on a test. If all the questions were equal in value, what was the student's grade, rounded to the nearest whole percent?

Solution The student answered $\dfrac{28}{30}$ of the questions right. The task is to change this fraction to a percent.

$$\frac{28}{30} = 0.9333\ldots = 93.3\ldots\% \approx 93\%$$

Note that the critical digit is 3, so we round down. The rounded grade was therefore 93%.

PRACTICE 18

On a key vote, 95 of 435 Congressmen supported the administration's bill. What percent was this, rounded to the nearest whole percent?   22%

Point out to students that we do not need to compute any digit to the right of the critical digit, since it will not affect the rounded answer.

Before going any further, study the chart of equivalent fractions, decimals, and percents in the Appendix at the back of this book. These numbers come up frequently and are useful reference points for estimating the answer to percent word problems, as we demonstrate in Section 6.2.

Exercises 6.1

Change each percent to a fraction or mixed number. Simplify.

1. 5%
$\frac{1}{20}$

2. 3%
$\frac{3}{100}$

3. 250%
$2\frac{1}{2}$

4. 110%
$1\frac{1}{10}$

5. 33%
$\frac{33}{100}$

6. 41%
$\frac{41}{100}$

7. 18%
$\frac{9}{50}$

8. 6%
$\frac{3}{50}$

9. 14%
$\frac{7}{50}$

10. 45%
$\frac{9}{20}$

11. 65%
$\frac{13}{20}$

12. 92%
$\frac{23}{25}$

13. $\frac{3}{4}$%
$\frac{3}{400}$

14. $\frac{1}{10}$%
$\frac{1}{1000}$

15. $\frac{3}{10}$%
$\frac{3}{1000}$

16. $\frac{1}{5}$%
$\frac{1}{500}$

17. $12\frac{1}{2}$%
$\frac{1}{8}$

18. $37\frac{1}{2}$%
$\frac{3}{8}$

19. $33\frac{1}{3}$%
$\frac{1}{3}$

20. $66\frac{2}{3}$%
$\frac{2}{3}$

Change each percent to a decimal.

21. 6%
0.06

22. 9%
0.09

23. 72%
0.72

24. 25%
0.25

25. 0.1%
0.001

26. 0.2%
0.002

27. 102%
1.02

28. 113%
1.13

29. 87.5%
0.875

30. 12.5%
0.125

31. 18.2%
0.182

32. 82.4%
0.824

33. $6\frac{9}{10}$%
0.069

34. $1\frac{1}{10}$%
0.011

35. $3\frac{1}{2}$%
0.035

36. $2\frac{4}{5}$%
0.028

37. $\frac{9}{10}$%
0.009

38. $\frac{7}{10}$%
0.007

39. $\frac{3}{4}$%
0.0075

40. $\frac{1}{4}$%
0.0025

Change each decimal to a percent.

41. 0.31
31%

42. 0.05
5%

43. 0.875
87.5%

44. 0.125
12.5%

45. 0.3
30%

46. 0.4
40%

47. 0.04
4%

48. 0.17
17%

49. 0.18
18%

50. 0.27
27%

51. 1.29
129%

52. 1.07
107%

53. 2.9
290%

54. 3.5
350%

55. 2.87
287%

56. 12.91
1291%

57. 1.016
101.6%

58. 1.003
100.3%

59. 5
500%

60. 2
200%

Change each fraction to a percent.

61. $\dfrac{3}{4}$
75%

62. $\dfrac{1}{2}$
50%

63. $\dfrac{1}{10}$
10%

64. $\dfrac{7}{8}$
87.5%

65. 3
300%

66. 4
400%

67. $\dfrac{4}{5}$
80%

68. $\dfrac{5}{8}$
62.5%

69. $\dfrac{9}{10}$
90%

70. $\dfrac{7}{10}$
70%

71. $\dfrac{1}{8}$
12.5%

72. $\dfrac{2}{5}$
40%

73. $\dfrac{5}{9}$
$55\dfrac{5}{9}\%$

74. $\dfrac{2}{9}$
$22\dfrac{2}{9}\%$

75. $\dfrac{2}{3}$
$66\dfrac{2}{3}\%$

76. $\dfrac{5}{6}$
$83\dfrac{1}{3}\%$

77. $1\dfrac{5}{8}$
162.5%

78. $2\dfrac{3}{5}$
260%

79. $2\dfrac{1}{6}$
$216\dfrac{1}{6}\%$

80. $1\dfrac{1}{3}$
$133\dfrac{1}{3}\%$

Complete the following tables.

You may want to have students memorize these charts.

81.

Fraction	Decimal	Percent
$\dfrac{1}{2}$	0.5	50%
$\dfrac{1}{4}$	0.25	25%
$\dfrac{3}{4}$	0.75	75%
$\dfrac{1}{5}$	0.2	20%
$\dfrac{2}{5}$	0.4	40%
$\dfrac{3}{5}$	0.6	60%
$\dfrac{4}{5}$	0.8	80%

82.

Fraction	Decimal	Percent
$\dfrac{1}{6}$	$0.16\dfrac{1}{6}$	$16\dfrac{1}{6}\%$
$\dfrac{5}{6}$	$0.83\dfrac{1}{3}$	$83\dfrac{1}{3}\%$
$\dfrac{1}{8}$	0.125	12.5%
$\dfrac{3}{8}$	0.375	37.5%
$\dfrac{5}{8}$	0.625	$62\dfrac{1}{2}\%$
$\dfrac{7}{8}$	0.875	87.5%
$\dfrac{1}{10}$	0.1	10%

Applications

Solve.

83. Snow weighs 0.1 as much as water. Express this decimal as a percent.   10%

84. About 95% of all animal species on Earth are insects. Express this percent as a fraction. (*Source: Encyclopedia Americana*) $\dfrac{19}{20}$

85. Tuition accounted for 24% of the college's income. Convert this percent to a fraction. $\frac{6}{25}$

86. In a recent year, a single parasite reduced the U.S. corn crop by 15%. Express this percent as a decimal. 0.15

87. New York has the largest population of any U.S. city. But Juneau, Alaska, has the greatest area—9.7 times that of New York. Write this decimal as a percent. (*Source: Webster's New Geographical Dictionary*) 970%

88. A medical school accepted $\frac{2}{5}$ of its applicants. What percent of the applicants did the school accept? 40%

89. When the recession ended, the factory's output grew by 135%. Write this percent as a simplified mixed number. $1\frac{7}{20}$

90. According to a survey, 78% of the arguments that couples have are about money. Express this percent as a decimal. 0.78

91. In Nevada, the federal government controls about 80% of the land. Convert this percent to a decimal. 0.8

92. After an oil spill, 15% of the wildlife survived. Express this percent as a fraction. $\frac{3}{20}$

93. By age 75, about $\frac{1}{3}$ of women and about 40% of men have chronic hearing loss. Is this condition more common among men or among women? Explain. The condition is more common among men. $\frac{1}{3} = 33\frac{1}{3}\% < 40\%$

94. On a survey, 120 of 500 people who were sent questionnaires responded. What is the response rate expressed as a percent? 24%

▦ *Use a calculator to solve the following problems, giving (a) the operation(s) you used, (b) the exact answer and (c) an estimate of the answer.*

95. A senator reported that he had only missed 12 of 1,753 votes in the U.S. Senate, for an overall voting rate of about 99.32%. Was he correct?
(a) Subtraction, division, multiplication; (b) Yes; (c) Possible estimate: 100%

96. The first Social Security retirement benefits were paid in 1940 to Ida May Fuller of Vermont. She had paid in a total of $24.85 and got back $20,897 before her death in 1975. Express to the nearest percent the ratio of what she got back to what she put in. (*Source:* James Trager, *The People's Chronology*)   (a) Division, multiplication; (b) 84,093%; (c) Possible estimate: 84,000%

■ *Check your answers on page A-8.*

MINDSTRETCHERS

MATHEMATICAL REASONING

1. By mistake, you move the decimal point to the right instead of to the left when changing a percent to a decimal. Your answer is how many times as large as the correct answer?
10,000 times as large.

WRITING

2. A study of the salt content of seawater showed that the average salt content varies from 33‰ to 37‰, where the symbol ‰ (read "per mil") means "for every thousand." Explain why you think the scientist who wrote this study did not use the % symbol.
The salt content was low, so to avoid decimals, the scientist used mils.

CRITICAL THINKING

3. What percent of the region shown is shaded in? $31\frac{1}{4}\%$

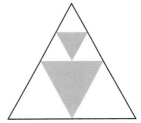

HISTORICAL NOTE

Throughout history, the concepts of percent and taxation have been interrelated. At the peak of the Roman Empire, the Emperor Augustus instituted an inheritance tax of 5% to provide retirement funds for the military. Another emperor, Julius Caesar, imposed a 1% sales tax on the population. And in Roman Asia, tax collectors exacted a tithe of 10% on crops. If landowners could not pay, the collectors offered to loan them funds at interest rates that ranged from 12% up to 48%.

Roman taxation served as a model for modern countries when these countries developed their own systems of taxation many centuries later.

Sources:

Frank J. Swetz, *Capitalism and Arithmetic, The New Math of the 15th Century* (La Salle, Illinois: Open Court, 1987).

Carolyn Webber and Aaron Wildavsky, *A History of Taxation and Expenditure in the Western World* (New York: Simon and Schuster, 1986).

6.2 Solving Percent Problems

OBJECTIVES

- *To identify the amount, the base, and the percent in a percent problem*
- *To find the amount, the base, or the percent in a percent problem*
- *To estimate the amount in a percent problem*
- *To solve word problems involving a percent*

The Three Basic Types of Percent Problems

Frequently, we think of a percent not in isolation but rather in connection with another number. In other words, we take *a percent of a number*.

Consider, for example, the problem of taking 50% of 8. This problem is equivalent to finding $\frac{1}{2}$ of 8, which gives us 4.

Note that this percent problem, like all others, involves three numbers.

$$50\% \quad \text{of} \quad 8 \quad \text{is} \quad 4.$$

Percent   Base   Amount

- The 50% is called the *percent* (or the *rate*). The percent always contains the % sign.

- The 8 is called the *base*. The base of a percent—the number that we are taking the percent of—always follows the word "of" in the statement of the problem.

- The remaining number 4 is called the *amount* (or the *part*).

Percent problems involve finding one of the three numbers. For example, if we omit the 4 in "50% of 8 is 4," we ask the question: What is 50% of 8? Omitting the 8, we ask: 50% of what number is 4? And omitting the 50%, we ask: What percent of 8 is 4?

There are several ways to solve these three basic percent questions. In this section, we discuss two ways—the translation method and the proportion method.

The Translation Method

In the translation method, a percent problem has the following form.

The percent of the base is the amount.

The percent problem gives two of the three quantities. To find the third, we translate to a simple equation that we then solve.

This method depends on translating the words in the given problem to the appropriate mathematical symbols.

Have students review how to solve simple equations, such as the following.

$$x = (0.2) \cdot 5$$

and

$$0.3x = 12$$

The proportion method is discussed on p. 308.

Word(s)	Math Symbol
What, what number, what percent	*x* (or some other letter)
is	=
of	$\times$ or $\cdot$
percent, %	Percent value expressed as a decimal or fraction

Let's translate several percent problems to equations.

EXAMPLE 1

Translate each question to an equation, using the translation method.

a. What is 10% of 2? **b.** 20% of what number is 5?

c. What percent of 8 is 4?

Solution

a. In this problem, we are looking for the amount.

What is 10% of 2?
$$x = 0.1 \cdot 2$$

Note that we chose to translate 10% to a decimal, but we could have used the fractional equivalent $\frac{1}{10}$. Also note that the constant 2 is left unchanged in the translation.

b. 20% of what number is 5?
$$0.2 \cdot x = 5$$

Do you see that this problem asks us to find the base—the number after the word "of"?

c. What percent of 8 is 4?
$$x \cdot 8 = 4$$

This problem asks "what percent?" So we are looking for the percent.

PRACTICE 1

Use the translation method to set up an equation.

a. What percent of 40 is 20? $x \cdot 40 = 20$

b. 50% of what number is 10? $0.5 \cdot x = 10$

c. What is 70% of 80? $x = 0.7 \cdot 80$

Finding an Amount

Now let's apply the translation method to solve the type of percent problem in which we are given both the percent and the base and are looking for the amount.

EXAMPLE 2

What is 25% of 8?

Solution First, we translate the question to an equation.

What is 25% of 8?
$$x = \frac{1}{4} \cdot 8$$

Then we solve this equation:
$$x = \frac{1}{4} \cdot 8 = \frac{1}{4} \cdot \frac{\overset{2}{\cancel{8}}}{1} = \frac{2}{1}$$
$$= 2$$

So 2 is 25% of 8. Note that we could have translated 25% to 0.25. Would we have gotten the same answer? Do you think that working with fractions or working with decimals is easier?

PRACTICE 2

What is 20% of 40? 8

Have your students write their responses in a journal.

EXAMPLE 3

Find 200% of 30.

Solution We can reword the problem as a question.

What is 200% of 30?
↓ ↓ ↓ ↓ ↓
x = 2 · 30

Solving this equation, we get $x = 60$. So 200% of 30 is 60.

PRACTICE 3

150% of 8 is what number?   12

| TIP | When the percent is less than 100%, the amount is *less* than the base. When the percent is more than 100%, the amount is *more* than the base. |

EXAMPLE 4

You win $400,000 in a lottery. What is the yearly payment if it amounts to $3\frac{1}{2}\%$ of this sum?

Solution We are looking for the amount of the yearly payment to you, which is $3\frac{1}{2}\%$ of $400,000.

What is $3\frac{1}{2}$ of $400,000?
↓ ↓ ↓ ↓ ↓
x = (0.035) · (400,000)

= 14,000

So the yearly payment is $14,000. Note that this amount has the same unit (dollars) as the base.

PRACTICE 4

Of the 600 workers at a factory, $8\frac{1}{2}\%$ belong to a union. How many workers are in the union?   51 workers

Estimating an Amount

In a percent problem, we frequently want to *estimate* the amount. Sometimes an approximate answer is good enough, and other times we want to check an exact answer we have already computed.

> Have students review the chart of equivalent fractions, decimals, and percents in the Appendix at the end of the book.

EXAMPLE 5

Approximately how much is 67% of 14.8?

Solution Here is one way to estimate the answer. We note that 67% is close to $66\frac{2}{3}\%$, which is equivalent to the fraction $\frac{2}{3}$. Also we see that 14.8 rounds to 15. So the answer to the given question is close to the answer to the following question.

What is $\frac{2}{3}$ of 15?
↓ ↓ ↓ ↓ ↓
x = $\frac{2}{3}$ · 15

We multiply mentally. $x = \frac{2}{\cancel{3}} \cdot \frac{\cancel{15}^{5}}{1} = \frac{10}{1} = 10$

So 67% of 14.8 is approximately 10. (The exact answer is 9.916— reasonably close to our estimate.)

PRACTICE 5

Estimate 49.3% of 401.6.   200

EXAMPLE 6

Fourteen-carat gold is an alloy that is 58% pure gold. How much gold, to the nearest gram, is needed to make 37 g of this alloy? Check by estimation.

Solution We write the question.

What is 58% of 37?

$$x = 0.58 \cdot 37$$

Multiplying, we get 21.46. So we need 21.46 g of gold to make 37 g of 14-carat gold.

To check by estimating the amount of gold, we mentally compute 60% of 40: $0.6 \times 40 = 24$, which is fairly close to our exact answer.

PRACTICE 6

The top 15% of a college's graduating class made the Dean's List. If 380 students graduated, compute the number of students who made the list. Then check by estimation.   57 students

> Discuss with students several strategies for estimating this answer, including $\frac{1}{8}$ of 400 and 0.1 of 380.

Finding a Base

Now let's consider some examples in which we use the translation method to find the base when we know the percent and the amount.

EXAMPLE 7

4% of what number is 8?

Solution We begin by writing the appropriate equation.

4% of what number is 8?

$$0.04 \cdot x = 8$$

Next, we solve this equation.

$$0.04x = 8$$

$$\frac{0.04}{0.04}x = \frac{8}{0.04} \qquad \text{Divide each side by 0.04 to isolate } x.$$

$$x = \frac{8}{0.04} = 200 \qquad 0.04\,\overline{)8.00} = 4\overline{)800.}$$

So 4% of 200 is 8.

PRACTICE 7

6 is 12% of what number?   50

EXAMPLE 8

A city's population this year is 156,000, which is 120% of the city's population 10 years ago. What was the population then?

Solution We must answer the following question.

120% of what number is 156,000?

$$1.2 \cdot x = 156,000$$

PRACTICE 8

There is a glut of office space in your city, with 400,000 sq ft, or 16% of the total office space, vacant. How much office space does your city have?   2,500,000 sq ft

> Students may prefer to solve this problem by the proportion method rather than by the translation method.

We solve the equation. $1.2x = 156{,}000$

$$\frac{1.2x}{1.2} = \frac{156{,}000}{1.2}$$

$$x = 130{,}000$$

The population 10 years ago was 130,000.

EXAMPLE 9

In a company, 55% of the employees are female. If 765 males work for the company, what is the total number of employees?

Solution Because 55% of the employees are female, 45% are male. So 765 males is the same as 45% of the employees.

55% of the employees (female)	45% of the employees (male) = 765

100% of the employees

We rephrase the question.   45% of what number is 765?

$$0.45 \cdot x = 765$$

Now we solve the equation. $0.45x = 765$

$$\frac{0.45}{0.45}x = \frac{765}{0.45}$$

$$x = 1{,}700$$

The company has a total of 1,700 employees.

PRACTICE 9

A quarterback completed 15 passes, or 20% of his attempted passes. How many of his attempted passes did he *not* complete?
60 passes

Students may prefer to solve this problem by the proportion method rather than by the translation method.

Finding a Percent

Finally, let's look at the third type of percent problem, where we are given the base and the amount and are looking for the percent.

EXAMPLE 10

What percent of 80 is 60?

Solution We begin by writing the appropriate equation.

What percent of 80 is 60?

$$x \cdot 80 = 60$$

$80x = 60$ Write the equation in standard form.

$\dfrac{80}{80}x = \dfrac{60}{80}$ Divide each side by 80.

$x = \dfrac{60}{80}$ or $\dfrac{3}{4}$ Simplify, reducing by 20.

PRACTICE 10

What percent of 6 is 5? $83\frac{1}{3}\%$

continued

We change $\dfrac{3}{4}$ to a percent. $x = \dfrac{3}{4} = 0.75,$ or   75%

So 75% of 80 is 60.

EXAMPLE 11

What percent of 60 is 80?

Solution We begin by writing the appropriate equation as shown to the right.

$$
\begin{array}{ccccc}
\text{What percent} & \text{of} & 60 & \text{is} & 80? \\
\downarrow & & \downarrow & \downarrow & \downarrow \\
x & \cdot & 60 & = & 80
\end{array}
$$

$60x = 80$ Rewrite the equation in standard form.

$\dfrac{60}{60}x = \dfrac{80}{60}$ Divide each side by 60.

$x = \dfrac{\overset{4}{80}}{\underset{3}{60}},$ or $\dfrac{4}{3}$ Simplify.

Finally, we want to change $\dfrac{4}{3}$ to a percent.

$$x = \dfrac{4}{3} = 1\dfrac{1}{3} = 1.33\dfrac{1}{3},\quad\text{or}\quad 133\dfrac{1}{3}\%$$

So $133\dfrac{1}{3}\%$ of 60 is 80.

PRACTICE 11

What percent of 5 is 6?   120%

> Have students compare the answers to Examples 10 and 11. Point out that changing the order of the numbers changes the answer. Ask students to explain why.

EXAMPLE 12

You buy a house for $125,000, making a down payment of $25,000 and paying the difference over time with a mortgage. What percent of the cost of the house was the down payment?

Solution We write the question as shown to the right.

$$
\begin{array}{ccccc}
\text{What percent} & \text{of} & \$125{,}000 & \text{is} & \$25{,}000? \\
\downarrow & & \downarrow & \downarrow & \downarrow \\
x & \cdot & 125{,}000 & = & 25{,}000
\end{array}
$$

$125{,}000x = 25{,}000$ Rewrite the equation in standard form.

$\dfrac{125{,}000}{125{,}000}x = \dfrac{25{,}000}{125{,}000}$ Divide each side by 125,000.

$x = \dfrac{25}{125} = \dfrac{1}{5}$ Simplify.

We change $\dfrac{1}{5}$ to a percent. $x = \dfrac{1}{5} = 0.2,$ or   20%

The down payment was 20% of the total cost of the house.

PRACTICE 12

In a 105-member state assembly, 45 members voted for a resolution. About what percent of the members voted for the resolution?   43%

The Proportion Method

So far, we have used the translation method to solve percent problems. Now let's consider an alternative approach, the proportion method.

Using the proportion method, we view a percent relationship in the following way.

$$\frac{\textbf{Amount}}{\textbf{Base}} = \frac{\textbf{Percent}}{\textbf{100}}$$

If we are given two of the three quantities, we set up this proportion and then solve it to find the third quantity.

> Have students review how to solve proportions, such as:
> $$\frac{4}{5} = \frac{x}{100}$$
> and
> $$\frac{5}{x} = \frac{4}{100}$$

EXAMPLE 13

What is 60% of 35?

Solution The base (the number after the word "of ") is 35. The percent (the number followed by the % sign) is 60%. The amount is unknown. We set up the proportion, substitute into it, and solve.

$$\frac{\text{Amount}}{\text{Base}} = \frac{\text{Percent}}{100}$$

$$\frac{x}{35} = \frac{60}{100}$$

$$100x = 60 \cdot 35 \qquad \text{Cross multiply.}$$

$$\frac{\cancel{100}}{\cancel{100}}x = \frac{2{,}100}{100} \qquad \text{Divide both sides by 100.}$$

$$x = 21$$

So 60% of 35 is 21.

PRACTICE 13

Find 108% of 250.   270

EXAMPLE 14

15% of what number is 21?

Solution Here the number after the word "of " is missing, so we are looking for the base. The amount is 21, and the percent is 15. We set up the proportion, substitute into it, and solve.

$$\frac{\text{Amount}}{\text{Base}} = \frac{\text{Percent}}{100}$$

$$\frac{21}{x} = \frac{15}{100}$$

$$15x = 2{,}100 \qquad \text{Cross multiply.}$$

$$\frac{\cancel{15}}{\cancel{15}}x = \frac{2{,}100}{15} \qquad \text{Divide both sides by 15.}$$

$$x = 140$$

So 15% of 140 is 21.

PRACTICE 14

2% of what number is 21.6?   1,080

EXAMPLE 15

What percent of $45 is $30?

Solution We are looking for the percent because we know that the base is 45 and that the amount is 30.

$$\frac{30}{45} = \frac{x}{100}$$

$$45x = 3,000 \quad \text{Cross multiply.}$$

$$\frac{45}{45}x = \frac{3,000}{45} \quad \text{Divide both sides by 45.}$$

$$x = 66\frac{2}{3}$$

We are looking for a percent, so we conclude that $66\frac{2}{3}\%$ of 45 is 30.

PRACTICE 15

What percent of 45 is 105? $233\frac{1}{3}\%$

> Point out to students that, in solving for a percent with the proportion method, the answer should not be multiplied by 100, since it is already in the form of a percent.

EXAMPLE 16

A car depreciated, that is, dropped in value, by 20% during its first year. By how much did the value of the car decline if it cost $12,500 new?

Solution The question here is: What is 20% of $12,500? So the percent is 20, the base is $12,500 and we are looking for the amount. We set up the proportion and solve.

$$\frac{x}{12,500} = \frac{20}{100}$$

$$100x = 250,000 \quad \text{Cross multiply.}$$

$$\frac{100}{100}x = \frac{250,000}{100} \quad \text{Divide both sides by 100.}$$

$$x = 2,500$$

The value of the car depreciated by $2,500.

PRACTICE 16

A company reduced its work force of 373,000 by 3%. How many employees were let go?   11,190 employees

EXAMPLE 17

Each day, you take tablets containing 24 mg of zinc. If this amount is 160% of the recommended daily allowance, how many milligrams are recommended?

Solution Here we are looking for the base. The question is: 160% of what amount is 24 mg? We set up the proportion and solve.

$$\frac{24}{x} = \frac{160}{100}$$

$$160x = 2,400 \quad \text{Cross multiply.}$$

$$\frac{160}{160}x = \frac{2,400}{160} \quad \text{Divide both sides by 160.}$$

$$x = 15$$

PRACTICE 17

A Nobel prize winner had to pay the Internal Revenue Service $129,200—or 38% of his prize—in taxes. How much was his Nobel prize worth? $340,000

Therefore the recommended daily allowance of zinc is 15 mg. Note that this base is less than the amount (24 mg). Why must that be true?

EXAMPLE 18

Of the 50 people who started a training program, 9 dropped out. What was the dropout rate, expressed as a percent?

Solution The question is: What percent of 50 is 9?

$$\frac{9}{50} = \frac{x}{100}$$

$$50x = 900 \qquad \text{Cross multiply.}$$

$$\frac{50}{50}x = \frac{900}{50} \qquad \text{Divide both sides by 50.}$$

$$x = 18$$

The dropout rate was 18%.

PRACTICE 18

Your computer can store 20MB of information. You have used 17.4MB. What percent of the storage capacity have you used? 87%

Percents on a Calculator

Some calculators have a percent key (%). However, the percent key functions differently on different models. Check to see if the following approaches work on your machine. If they don't, experiment to find an approach that does.

EXAMPLE

Use your calculator to find 50% of 8.

Solution

Input	8	×	50	%
Display	8.	8.	50.	4.

In computing the amount in this percent problem, we first entered the base, and then hit the × key. Next, we entered the percent and then hit the % key. Note that hitting the % key after *the second factor in a product* both changes the percent to a decimal and replaces the = key. However, on most models, entering the 50% before the 8 gives the wrong answer.

PRACTICE

What is 8.25% of $72.37, to the nearest cent? $5.97

continued

EXAMPLE

Calculate 5% more than 20.

Solution

Input	20	+	5	%
Display	20.	20.	5.	21.

Note that, in adding a percent of the base, we need not repeat the base.

PRACTICE

What is $299.95 reduced by 15%? $254.96

EXAMPLE

8% of what number is 7.25, to the nearest hundredth?

Solution To find the missing base, we could divide 7.25 by the decimal equivalent of 8%. But if your calculator has a percent key, try the following shortcut.

Input	7.25	÷	8	%
Display	7.25	8.	8.	90.625

So rounding 90.625 to the nearest hundredth, we get 90.63. Note that hitting the % key *in a quotient after the divisor* both changes the percent to a decimal and replaces the = key.

PRACTICE

To the nearest hundredth, 49.88 is 103% of what number? 48.43

EXAMPLE

18 is what percent of 25?

Solution Because the % key changes a percent to a decimal (and not a decimal to a percent), it won't help us find the missing percent. Instead, we divide the amount by the base, multiplying this quotient by 100.

Input	18	÷	25	=
Display	18.	18.	25.	0.72

Moving the decimal point two places to the right, we conclude that 18 is 72% of 25.

PRACTICE

To the nearest whole percent, $7.99 is what percent of $35.66?
22%

Exercises 6.2

Find the amount. Check by estimating.

1. What is 75% of 8?   6

2. Find 50% of 48.   24

3. Compute 100% of 23.   23

4. What is 200% of 6?   12

5. Find 41% of 7.   2.87

6. Calculate 6% of 9.   0.54

7. What is 35% of $400? $140

8. 40% of 10 mi is what?   4 mi

9. Compute 13% of 5 L.   0.65 L

10. How much is 19% of $10,000? $1,900

11. What is 3.1% of 20?   0.62

12. Find 0.5% of 7.   0.035

13. Compute $\frac{1}{2}$% of 20.   0.1

14. $\frac{1}{10}$% of 35 is what number?   0.035

15. What is 8% of $500? $40

16. 6% of $200 is how much money? $12

17. What is $12\frac{1}{2}$% of 32?   4

18. Compute $37\frac{1}{2}$% of 40.   15

19. What is $7\frac{1}{8}$% of $257.13, rounded to the nearest cent? $18.32

20. Calculate 8.9% of 7,325 mi, rounded to the nearest mile.   652 mi

Applications (Finding the Amount)

21. You are willing to spend up to 25% of your income on housing. What is the most you can spend if your annual income is $24,000? $6,000

22. In a dormitory, 40% of the rooms are especially equipped for disabled students. How many rooms are so equipped if the dorm has 80 rooms?   32 rooms

23. In a 35-hour-per-week job, you spend 70% of the time on data entry. How many hours a week do you enter data?   24.5 hr

24. In 1862, the U.S. Congress enacted the nation's first income tax, at the rate of 3%. How much in income tax would you have paid if you made $2,500?   (*Source:* Bureau of Census) $75

25. In a restaurant, 60% of the tables are in the no-smoking section. If the restaurant has 90 tables, how many tables are in the no-smoking section?   54 tables

26. You live in a town where the sales tax is 5%. Across the river, the tax is 4%. If it costs you $6 to make the round trip across the river, on which side of the river should you buy a $250 television set?   You should buy the TV set in your town.

27. The following graph shows the breakdown by gender of the 160 employees at a business. How many more women than men work there?   16 women

Men 45%

Women 55%

28. A classified ad describes a used car on sale for $7,500. You are interested but do not want to pay more than $6,000. If the owner is willing to reduce her price by at most 20%, can the two of you agree on terms?   Yes

In Exercises 29–48, find the base.

29. 25% of what number is 8?   32

30. 30% of what number is 120?   400

31. $12 is 10% of how much money? $120

32. 1% of what salary is $195? $19,500

33. 30 is 40% of what number?   75

34. 8 is 50% of what number?   16

35. 7% of what weight is 14 lb?   200 lb

36. 4% of what length is 36 ft?   900 ft

37. 5 is 200% of what number?   2.5

38. 70% of what amount is 14?   20

39. $20 is 10% of how much money? $200

40. 2% of what length is 3 m?   150 m

41. 2% of what amount of money is $5? $250

42. 8 is 20% of what number?   40

43. 3.5 is 200% of what number?   1.75

44. 150% of what number is 8.1?   5.4

45. 0.5% of what number is 23?   4,600

46. 0.75% of what is 24?   3,200

47. $3\frac{1}{2}$% of what is 98?   2,800

48. $\frac{1}{3}$% of what number is 15?   4,500

Applications (Finding the Base)

49. Yesterday, 8 employees in an office were absent. If these absentees constitute 25% of the employees, how many employees are there?   32 employees

50. Payroll deductions comprise 40% of your gross income. If your deductions total $240, what is your gross income? $600

51. According to the report on a country's economic conditions, 1.5 million people, or 8 percent of the workforce, were unemployed. How large was the workforce?   18,750,000 people

52. In what was described as a "bloodbath," the Dow Jones stock average in one day lost, in round numbers, a record 620 points, or 6 percent of its previous value. What was the previous value, to the nearest hundred points?
(***Source:*** *Deutsche Presse-Agentur*)   10,300

53. In the first quarter of last year, a steel mill produced 300 tons of steel. If this was 20% of the year's output, find that output.   1,500 tons

54. A company's profits amounted to 10% of its sales. If the profits were $3 million, compute the company's sales.
$30,000,000

In Exercises 55–72, find the percent.

55. 50 is what percent of 100?   50%

56. What percent of 13 is 13?   100%

57. What percent of 8 is 6?   75%

58. 5 is what percent of 15? $33\frac{1}{3}\%$

59. What percent of 12 is 10? $83\frac{1}{3}\%$

60. 7 is what percent of 21? $33\frac{1}{3}\%$

61. What percent of 50 is 20?   40%

62. 10 is what percent of 8?   125%

63. 2 mi is what percent of 8 mi?   25%

64. $15 is what percent of $20?   75%

65. $30 is what percent of $20?   150%

66. 35¢ is what percent of 21¢? $166\frac{2}{3}\%$

67. 2.5 is what percent of 4?   62.5%

68. 0.1 is what percent of 8?   1.25%

69. $8 is what percent of $240? $3\frac{1}{3}\%$

70. What percent of 60 trucks is 4 trucks?
$6\frac{2}{3}\%$

71. What percent of 5 is $\frac{1}{2}$?   10%

72. $\frac{2}{3}$ is what percent of 6? $11\frac{1}{9}\%$

Applications (Finding the Percent)

73. Your goal is to consume 2,000 Cal daily. What percent of this goal do you reach with a 300-Cal breakfast?   15%

74. In a recent year, the U.S. Supreme Court reviewed about 100 of 7,000 submitted cases. What percent is this, to the nearest whole percent?   1%

75. You enlisted in the Army for 36 mo. So far, you have served 9 mo. What percent of your enlistment has passed?   25%

76. You mailed a questionnaire to 5,000 people. Of these recipients, 3,000 responded. What was your response rate, expressed as a percent?   60%

77. You buy a house for $150,000. Of this amount, you are able to put down $30,000. The down payment is what percent of the purchase price?   20%

78. Last year, your income was $25,000. You donated $900 to charity. What percent of your income went to charity?   3.6%

79. A baseball player got 12 hits in 30 at-bats. What percent of the at-bats were not hits?   60%

80. In purchasing a condo for $80,000, you put down $10,000 and financed the remainder of the purchase price with a mortgage. What percent of the purchase price was the mortgage? $87\frac{1}{2}\%$

■ *Check your answers on page A-9.*

MINDSTRETCHERS

WRITING

1. Write a word problem whose answer is 40%, in which you give the base and the amount.
Answers will vary.

CRITICAL THINKING

2. At a college, 20% of the women commute, in contrast to 30% of the men. Yet more women than men commute. Explain how this result is possible.   More women than men attend the college.

MATHEMATICAL REASONING

3. What is *a*% of *b*% of *c*% of 1,000,000?   *abc*

6.3 More on Percents

OBJECTIVES

■ *To solve percent increase and decrease problems*
■ *To solve percent problems involving taxes, commissions, markups, and discounts*
■ *To solve simple and compound interest problems*
■ *To solve word problems of these types*

Finding a Percent Increase or Decrease

Next, let's consider a type of "what percent" problem that deals with a *changing quantity*. If the quantity is increasing, we speak of a *percent increase*; if it is decreasing, of a *percent decrease*.

Here is an example: In the presidential election of 1792, George Clinton earned 50 electoral votes. Sixteen years later, Clinton earned only 6 electoral votes. By what percent did Clinton's electoral vote count drop?

Note that this problem states the value of a quantity at two points in time. We are asked to find the percent decrease between these two values.

To solve, we first compute the difference between the values.

$$50 - 6 = 44$$

Change in value

The question posed is expressed as follows.

What percent of 50 is 44?

$$x \cdot 50 = 44$$

It is important to note that the *base* here—as in all percent change problems—is the original value of the quantity.

Next, we solve the equation.

$$50x = 44$$

$$\frac{50}{50}x = \frac{44}{50} = \frac{22}{25}$$

$$x = \frac{22}{25} = 0.88, \quad \text{or} \quad 88\%$$

So we conclude that the number of electoral votes *decreased* by 88%.

To Find a Percent Increase or Decrease
■ compute the difference between the two given values, and
■ compute what percent this difference is of *the original value*.

EXAMPLE 1	PRACTICE 1
The cost of a marriage license had been $10. Later it rose to $15. What percent increase was this?	To accommodate a flood of tourists, businesses in town boosted the number of hotel beds from 25 to 100. What percent increase is this? 300%

continued

317

Solution The earlier value of the cost of the license was $10, and the later value was $15. The change in value is therefore $15 − $10, or $5. So the question is as follows.

$$\text{What percent} \quad \text{of} \quad \$10 \quad \text{is} \quad \$5?$$

$$x \quad \cdot \quad 10 \quad = \quad 5$$

$$10x = 5$$

$$\frac{10}{10}x = \frac{5}{10}$$

$$x = \frac{1}{2}$$

We change $\frac{1}{2}$ to a percent. $\quad x = \frac{1}{2} = 0.5, \quad \text{or} \quad 50\%$

So the cost of the license *rose* by 50%.

EXAMPLE 2

Suppose that an animal species is considered to be endangered if its population drops by more than 60%. If a species' population fell from 40 to 18, should we consider the animal endangered?

Solution The population *dropped from* 40 to 18, that is, by 22. The question is how the percent decrease compares with 60%. We compute.

$$\text{What percent} \quad \text{of} \quad 40 \quad \text{is} \quad 22?$$

$$x \quad \cdot \quad 40 \quad = \quad 22$$

$$40x = 22$$

$$x = \frac{22}{40}, \quad \text{or} \quad \frac{11}{20}$$

We convert this fraction to a percent. $\quad x = \frac{11}{20} = 0.55, \quad \text{or} \quad 55\%$

The population *fell* by less than 60%, so we do not consider the species to be endangered.

PRACTICE 2

Major financial crashes took place on both Tuesday, October 29, 1929, and Monday, October 19, 1987. On the earlier date, the stock index dropped from 300 to 230. On the later date, it dropped from 2,250 to 1,750. As a percent, did the stock index drop more in 1929 or in 1987? (*Source: Wall Street Journal*) 1929

Ask students if they can solve this example another way—say, as an amount problem.

Ask students to discuss which is a more significant measure of a decline in stock value—the percent change or the difference in values.

Business Applications of Percent

The idea of percent is fundamental to business and finance. Percent applications are part of our lives whenever we buy or sell merchandise, pay taxes, and borrow or invest money.

Taxes

Governments levy taxes to pay for a variety of services, from supporting schools to paving roads. There are many kinds of taxes, including sales, income, property, and import taxes.

In general, the amount of a tax that we pay is a percent of a related value. For instance, sales tax is usually computed as a percent of the price of merchandise sold. Thus, in a town where the sales tax rate is 7%, we could compute the tax on any item sold by computing 7% of the price of that item.

Similarly, property tax is commonly computed by taking a given percent (the tax rate) of the property's assessed value. And an import tax is calculated by taking a specified percent of the market value of the imported item.

EXAMPLE 3

The sales tax on a $950 camcorder is $71.25. What is the sales tax rate, expressed as a percent?

Solution We must consider the following question.

71.25   is   what percent   of   950?

$$71.25 = x \cdot 950$$

$$950x = 71.25$$

$$\frac{950x}{950} = \frac{71.25}{950}$$

$$x = 0.075, \quad \text{or} \quad 7.5\%$$

The rate of the sales tax is 7.5%, or $7\frac{1}{2}\%$.

PRACTICE 3

What is the total cost of merchandise selling at $53.50 plus a sales tax of 7%, rounded to the nearest cent? $57.25

Commission

Many salespeople, instead of receiving a fixed salary, work on **commission** to encourage them to make more sales. Working on commission means that the amount of money that they earn is a specified percent—say, 10%—of the total sales for which they are responsible.

On some jobs, salespeople make a flat fee in addition to a commission based on sales. On other jobs, a salesperson may earn a higher rate of commission on sales over an amount previously agreed upon, as an extra incentive.

EXAMPLE 4

A real estate agent sold your condo in Miami for $122,000. On this amount, she received a commission of 6%. How much money will you make from the sale after paying the agent's fee?

Solution Let's first compute how much money the agent made. Here the question is as follows.

What   is   6%   of $122,000?

$$x = 0.06 \cdot 122,000$$

$$x = 7,320$$

The agent's fee was $7,320, so you cleared $122,000 − $7,320, or $114,680.

PRACTICE 4

In a retail shop, you are paid a flat salary of $225 a week. In addition, you receive a commission of $7\frac{1}{2}\%$ on that part of the shop's weekly sales in excess of $2,000. How much do you earn at this job during a week in which the shop's total sales are $4,500? $412.50

Discount

When buying or selling merchandise, we use the term **discount** to refer to a reduction on the merchandise's original price. The rate of discount is usually expressed as a percent of the selling price.

EXAMPLE 5

A drugstore gives senior citizens a 10% discount. If some pills normally sell for $16 a bottle, how much will a senior citizen pay?

Solution Note that, because senior citizens get a discount of 10%, they pay 100% − 10%, or 90%, of the normal price.

Let's draw a diagram to illustrate the problem.

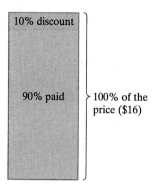

10% discount

90% paid   } 100% of the price ($16)

The question then becomes the one at the right.

What is 90% of $16?
  ↓   ↓   ↓   ↓   ↓
  x = 0.9 ·   16

We multiply.

$x = 0.9 \cdot 16 = 14.4$

So a senior citizen will pay $14.40 for a bottle of the pills.

Another way to solve this problem is first to compute the amount of the discount (10% of $16) and then to subtract this discount from the original price. With this approach, do you get the same answer?

PRACTICE 5

Find the sale price. $69.60

Famous Designer Jeans
Regularly $87

20% off
Today only

Have students explain their responses in a journal.

Markup

A retail firm must sell goods at a higher price (the selling price) than it pays for the merchandise (the cost price) to stay in business. The **markup** on an item is the difference between the selling price and the cost price. Often the markup rate on merchandise is a fixed percent of the selling price.

EXAMPLE 6

A clothing store sells a sweater for $35 at a markup rate of 55%. How much money is the markup on the sweater?

Solution We write the question shown at the right.

What is 55% of $35?
  ↓   ↓   ↓   ↓   ↓
  x = 0.55 ·   35 = 19.25

So the markup on the sweater is $19.25.

PRACTICE 6

A department store buyer purchases trousers at $480 per dozen and sells them for $80 each. What percent markup, based on the selling price, is the store making?
100%

Simple Interest

Anyone who has been late in paying a credit card bill or who has deposited money in a savings account knows about **interest**. When you loan or deposit money, you make interest. When you borrow money, you pay interest.

Interest depends on the amount of money borrowed (the **principal**), the annual rate of interest (usually expressed as a percent), and the length of time the money is borrowed (usually expressed in years). We can compute the amount of interest by multiplying the principal by the rate of interest and the number of years. This type of interest is called *simple interest* to distinguish it from *compound interest* (which we discuss later).

EXAMPLE 7	PRACTICE 7

EXAMPLE 7

You deposited $900 in a savings account that each year pays 5% in interest, which is credited to your account. What is your account balance after 1 yr?

Solution One way to solve this problem is to add percents, that is, to note that the total amount of money in the account after 1 yr is 100% + 5%, or 105%, of the original deposit.

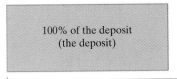

100% of the deposit (the deposit)

5% of the deposit (the interest)

105% of the deposit

From this point of view, the question becomes the one shown at the right.

What is 105% of $900?
$x = 1.05 \cdot 900$

$x = 945$

So you conclude that the account balance is $945.

Another way to solve this problem is to break it into two questions:

■ How much interest did you make?
■ What is the sum of that interest and the original deposit?

We reword the first question.

What is 5% of $900?
$x = 0.05 \cdot 900$

$x = 45$

Because the interest is $45, the account balance is $900 + $45, or $945.

PRACTICE 7

An account pays $8\frac{1}{2}$% interest on $1,600 for 1 yr. Compute the account balance after 1 yr. $1,736

EXAMPLE 8

How much interest do you make in 2 yr on principal of $825 and an annual interest rate of 5%?

Solution To compute the interest, we multiply the principal by the rate of interest and the number of years.

Principal	Rate of Interest	Number of Years

Interest = 825 × 0.05 × 2

= 82.5

You make $82.50 in interest.

PRACTICE 8

What is the interest on an investment of $20,000 for 3 yr at an annual interest rate of 6%? $3,600

Compound Interest

As we have seen, simple interest is paid on the principal. Most banks, however, pay their customers *compound interest*, which is paid on both the principal and the previous interest generated.

For instance, suppose that you have $1,000 deposited in a savings account that pays 5% interest compounded annually. Let's compute the balance in your account at the end of the third year.

The following table shows the balance in your account after you have left the money in the account for 3 yr. After 1 yr, the account will contain $1,050 (that is, the original $1,000 added to 5% of $1,000). Note that we could have computed this amount by multiplying the principal, $1,000, by 1.05 or 105%.

Year	Balance at the End of the Year
0	$1,000
1	$1,000 + 0.05 × $1,000 = $1,050.00
2	$1,050 + 0.05 × $1,050 = $1,102.50
3	$1,102.50 + 0.05 × $1,102.50 = $1,157.63

The balance in your account after the third year is $1,157.63, rounded to the nearest cent.

The following table shows another and perhaps easier way to get this answer.

Year	Balance at the End of the Year
0	$1,000
1	1.05 × $1,000 = $1,050.00
2	$(1.05)^2 \times \$1,000 = \$1,102.50$
3	$(1.05)^3 \times \$1,000 = \$1,157.63$

So the balance at the end of the third year is $(1.05)^3 \times \$1,000$, or again is $1,157.63. What would the balance be at the end of the fourth year? What is the relationship between the number of years the money has been invested and the power of 1.05?

In computing the preceding answer, we needed to raise the number 1.05 to a power. Before scientific calculators became available, compound interest problems were commonly solved by use of a compound interest table that contained information such as the following.

Number of Years	4%	5%	6%	7%
1	1.04000	1.05000	1.06000	1.07000
2	1.08160	1.10250	1.12360	1.14490
3	1.124864	1.15763	1.19102	1.22504

When you use such a table to calculate a balance, simply multiply the principal by the number in the table corresponding to the rate of interest and the number of years for which the principal is invested. For instance, after 3 yr a principal of $1,000 compounded at 5% per year results in a balance of 1.15763 × 1,000, or $1,157.63, as we previously noted.

> Have students write their responses in a journal.

Today, problems of this type are generally solved on a calculator.

EXAMPLE 9

You deposit $7,000 in a bank account and do not make any withdrawals or deposits in the account for 3 yr. The interest is compounded annually at a rate of 3.5%. What will be the amount in your account at the end of this period?

Solution Each year, the amount in the account is 100% + 3.5%, or 1.035 times the previous year's balance. So at the end of 3 yr, the number of dollars in the account is calculated as follows.

$$7,000 \ \times 1.035 \times 1.035 \times 1.035$$

It makes sense to use a calculator to carry out this computation.

One way to key in this computation on a calculator is as follows.

Input	1.035	✕	=	=	✕	7000	=
Display	1.035	1.035	1.071225	1.108717875	1.108717875	7000.	7761.025125

So at the end of 3 yr, you have $7,761.03 in your account, rounded to the nearest cent.

PRACTICE 9

Find the balance after 4 yr on a principal amount of $2,000 invested at a rate of 6% compounded annually. $2,524.95

Point out that different calculators will display a different number of decimal places. Also check if any student has a calculator with a special key (x^y) for raising a number to a power.

Exercises 6.3

Find the percent increase or decrease. Round to the nearest percent.

1.

Original Value	New Value	Percent Increase or Decrease
10	12	20% increase
10	8	20% decrease
6	18	200% increase
35	70	100% increase
14	21	50% increase
10	1	90% decrease
$8	$6.50	$18\frac{3}{4}$% decrease
$6	$5.25	$12\frac{1}{2}$% decrease

2.

Original Value	New Value	Percent Increase or Decrease
5	6	20% increase
12	10	$16\frac{2}{3}$% decrease
4	9	125% increase
25	45	80% increase
10	36	260% increase
100	20	80% decrease
4 ft	3 ft	25% decrease
8 lb	4.5 lb	$43\frac{3}{4}$% decrease

Compute the sales tax. Round to the nearest cent.

3.

Selling Price	Rate of Sales Tax	Sales Tax
$30.00	5%	$1.50
$24.88	3%	$0.75
$51.00	$7\frac{1}{2}$%	$3.83
$196.23	4.5%	$8.83

4.

Selling Price	Rate of Sales Tax	Sales Tax
$40.00	6%	$2.40
$16.98	4%	$0.68
$85.00	$5\frac{1}{2}$%	$4.68
$286.38	5%	$14.32

Compute the commission. Round to the nearest cent.

5.

Sales	Rate of Commission	Commission
$700	10%	$70.00
$450	2%	$9.00
$870	$4\frac{1}{2}$%	$39.15
$922	7.5%	$69.15

6.

Sales	Rate of Commission	Commission
$400	1%	$4.00
$670	3%	$20.10
$610	$6\frac{1}{2}$%	$39.65
$2,500	8.25%	$206.25

Compute the discount and sale price. Round to the nearest cent.

7.

Original Price	Rate of Discount	Discount	Sale Price
$700.00	25%	$175.00	$525.00
$18.00	10%	$1.80	$16.20
$43.50	20%	$8.70	$34.80
$16.99	5%	$0.85	$16.14

8.

Original Price	Rate of Discount	Discount	Sale Price
$200.00	30%	$60.00	$140.00
$21.00	50%	$10.50	$10.50
$88.88	10%	$8.89	$79.99
$72.50	40%	$29.00	$43.50

Compute the markup and selling price. Round to the nearest cent.

9.

Selling Price	Rate of Markup	Markup	Original Price
$10.00	50%	$5.00	$5.00
$23.00	70%	$16.10	$6.90
$18.40	10%	$1.84	$16.56
$13.55	60%	$8.13	$5.42

10.

Selling Price	Rate of Markup	Markup	Original Price
$20.00	40%	$8.00	$12.00
$81.00	25%	$20.25	$60.75
$74.20	30%	$22.26	$51.94
$300.00	8.5%	$25.50	$274.50

Calculate the simple interest and the final balance. Round to the nearest cent.

11.

Principal	Interest Rate	Time (in years)	Interest	Final Balance
$300	4%	2	$24.00	$324.00
$600	4%	2	$48.00	$648.00
$300	8%	2	$48.00	$348.00
$300	4%	4	$48.00	$348.00
$375	10%	3	$112.50	$487.50
$70,000	6%	30	$126,000.00	$196,000.00

12.

Principal	Interest Rate	Time (in years)	Interest	Final Balance
$100	6%	5	$30.00	$130.00
$800	3%	5	$120.00	$920.00
$100	3%	10	$30.00	$130.00
$800	6%	10	$480.00	$1,280.00
$250	1.5%	2	$7.50	$257.50
$300,000	5%	20	$300,000.00	$600,000.00

Calculate the final balance after compounding the interest. Round to the nearest cent.

13.

Principal	Interest Rate	Time (in years)	Final Balance
$500	4%	2	$540.80
$6,200	3%	5	$7,187.50
$300	5%	8	$443.24
$20,000	4%	2	$21,632.00
$145	3.8%	3	$162.17
$810	2.9%	10	$1,078.05

14.

Principal	Interest Rate	Time (in years)	Final Balance
$300	6%	1	$318.00
$2,900	5%	4	$3,524.97
$800	3%	5	$927.42
$10,000	3%	4	$11,255.09
$250	4.1%	2	$270.92
$200	3.3%	5	$235.25

Applications

Solve.

15. The value of a condominium apartment fell from $150,000 to $90,000. By what percent did it drop?   40%

16. Last year, a local team won 20 games. This year, it won 15 games. What was the percent decrease of games won?
25%

17. In 9 yr, the number of elderly nursing home residents rose from 200,000 to 1.3 million. By what percent did the number of residents increase?   550%

18. Between the 1970s and the 1990s, the price of a four-function hand-held calculator dropped from about $200 to about $5. Find the percent change.
$97\frac{1}{2}\%$ decrease

19. The first commercial telephone exchange was set up in New Haven, Connecticut, in 1878. Between 1880 and 1890, the number of telephones in the United States increased from 50 thousand to 200 thousand, in round numbers. What percent increase was this?
(*Source:* Bureau of Census)   300%

20. It used to take you 6 min to type a typical page. Since you got a word processor, however, it takes you only 4 min. By what percent did the time it takes drop? $33\frac{1}{3}\%$

21. You pay 8% sales tax on a coat which sells for $80 (the price before tax is added). How much sales tax will you pay? $6.40

22. You consult a table to figure out the taxes on your income of $46,900. According to the table, you owe $3,315 plus 28% of your income over $36,900. How much money do you owe?
$6,115

23. The sales tax on a $700 diamond ring is $53. What is the sales tax rate?   7.6%, to the nearest tenth of a percent

24. In your town, the sales tax rate is 6%. It costs you $5 to travel to a nearby town and back where the sales tax is only 5.8%. Is it worthwhile for you to make this trip to purchase an item that sells in both towns for $120?   No

25. Your commission amounts to $2\frac{1}{2}\%$ of the price of the merchandise that you sell. Yesterday you sold $20,000 worth of merchandise. Compute the commission that you earned. $500

26. A salesperson earns a flat salary of $150 plus a 10% commission on sales of $3,000. What was the salesperson's total earnings? $450

27. On a restaurant table, you see $1.35 left as a tip by the previous diner. Assuming that the party left a 15% tip, how much was the bill before the tip? $9

28. You earn a 5% commission on the first $2,000 in sales and 7% commission on sales above $2,000. How much commission do you earn on sales of $3,500?
$205

29. An article that cost $20 sells for $25. Calculate both the markup and the percent of markup based on the selling price. $5 and 20%

30. An antique store bought a table for $90 and resold it for $120. What is the markup percent, based on the selling price.   25%

31. What is the markup percent, based on selling price, on an item when the markup is $8 and the selling price is $15? $53\frac{1}{3}\%$

32. Managing a store, you mark up the cost of merchandise by $5. If the cost was $3, what percent of the selling price is the markup? $62\frac{1}{2}\%$

33. A store sells a television that lists for $399 at a 35% discount rate. What is the sale price? $259.35

34. An article that regularly sells for $64 is sold at a discount of $12. Calculate the discount rate. $18\frac{3}{4}\%$

35. You borrowed $3,000 for 1 yr at 5% simple interest to buy a computer. How much interest did you pay? $150

36. How much simple interest is earned on $600 at an 8% annual interest rate for 2 yr? $96

37. You deposit $5,000 in a bank. How much interest will you have earned after 1 yr if the interest rate is 5%? $250

38. You borrow $2,000 from a friend, agreeing to pay her 4% simple interest. If you promise to repay her the entire amount at the end of 3 yr, how much money must you pay her? $2,240

39. You invest $3,000 in an account that pays 6% interest, compounded annually. Find the amount in the account after 2 yr, rounded to the nearest cent. $3,370.80

40. A bank pays 5.5% interest, compounded annually on a 2-yr certificate of deposit (CD), which initially costs $500. What is the value of the CD at the end of the 2 yr? $556.51

41. A city had a population of 4,000. If the city's population increased by 10% per year, what was the population 4 yr later?   5,856

42. You bought a painting for $10,000. If the value of the painting increased by 50% per year, what was its value 4 yr later? $50,625

■ *Check your answers on page A-9.*

KEY CONCEPTS and SKILLS

□ = **CONCEPT**   □ = **SKILL**

CONCEPT/SKILL	DESCRIPTION	EXAMPLE
[6.1] Percent	A ratio or fraction with denominator 100. Written with the % sign, which means divided by 100.	$50\% = \dfrac{50}{100}$ ↑ Percent
[6.1] To change a percent to the equivalent fraction	■ Drop the % sign from the given percent, and place the number over 100. ■ Simplify the resulting fraction, if possible.	$25\% = \dfrac{25}{100} = \dfrac{1}{4}$
[6.1] To change a percent to the equivalent decimal	■ Move the decimal point two places to the left and drop the % sign.	$23.5\% = .235,\ \text{ or }\ 0.235$
[6.1] To change a decimal to the equivalent percent	■ Move the decimal point two places to the right and insert the % sign.	$0.125 = 12.5\%$
[6.1] To change a fraction to the equivalent percent	■ Change the fraction to a decimal. ■ Change the decimal to a percent.	$\dfrac{1}{5} = 0.2 = 20\%$
[6.2] Base	The number that we are taking the percent of. It always follows the word "of" in the statement of a percent problem.	50% of 8 is 4. ↑ Base
[6.2] Amount	The result of taking the percent of the base.	50% of 8 is 4. ↑ Amount
[6.2] To solve a percent problem using the translation method	■ Translate as follows: what → x is → = of → × or · % → decimal or fraction ■ Set up the equation. **The percent of the base is the amount.** ■ Solve.	50% of 8 is what? ↓ ↓ ↓ ↓ ↓ 0.5 · 8 = x $x = 4$ 30% of what is 6? ↓ ↓ ↓ ↓ ↓ 0.3 · x = 6 $\dfrac{0.3x}{0.3} = \dfrac{6}{0.3}$ $x = \dfrac{6}{0.3} = 20$ What percent of 8 is 2? ↓ ↓ ↓ ↓ x · 8 = 2 $x = \dfrac{2}{8} = \dfrac{1}{4} = 25\%$

continued

CONCEPT/SKILL	DESCRIPTION	EXAMPLE
[6.2] **To solve a percent problem using the proportion method**	■ Identify the amount, the base, and the percent. ■ Set up and substitute into the proportion. $$\frac{\text{Amount}}{\text{Base}} = \frac{\text{Percent}}{100}$$ ■ Solve.	50% of 8 is what? $$\frac{x}{8} = \frac{50}{100}$$ $$100x = 400$$ $$x = 4$$ 30% of what is 6? $$\frac{6}{x} = \frac{30}{100}$$ $$30x = 600$$ $$x = 20$$ What percent of 8 is 2? $$\frac{2}{8} = \frac{x}{100}$$ $$8x = 200$$ $$x = 25$$ So the answer is 25%.
[6.3] **Percent increase or decrease**	The percent increase or decrease between two values of a quantity, using the original value of the quantity as the base.	For a quantity that increases from 2 to 3, the percent change is 50%.
[6.3] **To find a percent increase or decrease**	■ Compute the difference between the two given values. ■ Determine what percent this difference is of the *original value*.	Find the percent change for a quantity that increases from 4 to 5. Difference: $5 - 4 = 1$ What percent of 4 is 1? $\downarrow$ $\downarrow$ $\downarrow$ $\downarrow$ x · $4 = 1$ $$x = \frac{1}{4} = 0.25, \quad \text{or} \quad 25\%$$

Chapter 6 Review Exercises

To help you review this chapter, solve these problems.

[6.1]

Complete the following tables.

1.

Fraction	Decimal	Percent
$\frac{1}{4}$	0.25	25%
$\frac{7}{10}$	0.7	70%
$\frac{3}{400}$	0.0075	$\frac{3}{4}$%
$\frac{5}{8}$	0.625	62.5%
$1\frac{1}{5}$	1.2	120%
$1\frac{1}{100}$	1.01	101%
$2\frac{3}{5}$	2.6	260%
$3\frac{3}{10}$	3.3	330%
$\frac{3}{25}$	0.12	12%
$\frac{7}{8}$	0.875	$87\frac{1}{2}$%
$\frac{1}{6}$	$0.16\frac{2}{3}$	$16\frac{2}{3}$%

2.

Fraction	Decimal	Percent
$\frac{3}{8}$	0.375	37.5%
$\frac{2}{5}$	0.4	40%
$\frac{1}{1,000}$	0.001	$\frac{1}{10}$%
$1\frac{1}{2}$	1.5	150%
$\frac{7}{8}$	0.875	87.5%
$\frac{5}{6}$	$.83\frac{1}{3}$	$83\frac{1}{3}$%
$2\frac{3}{4}$	2.75	275%
$1\frac{1}{5}$	1.2	120%
$\frac{3}{4}$	0.75	75%
$\frac{1}{10}$	0.1	10%
$\frac{1}{3}$	$0.33\frac{1}{3}$	$33\frac{1}{3}$%

Calculate.

3. Express $\frac{113}{758}$ as a percent, rounded to the nearest whole percent.   15%

4. Write $\frac{1}{32}$ as a percent.   3.125%

[6.2]

Solve.

5. What is 40% of 30?
12

6. What percent of 5 is 6?
120%

7. 2 feet is what percent of 4 ft?
50%

8. 30% of what number is 6?
20

9. What percent of 8 is 3.5?
43.75%

10. Find 55% of 10.
5.5

11. $12 is 200% of what amount of money?
$6

12. 2 is what percent of 10?
20%

13. What is 1.2% of 25?
0.3

14. Estimate 113% of 15.73.
16

15. 35% of $200 is what?
$70

16. $\frac{1}{2}$% of what number is 5?
1,000

17. 15 is what percent of 0.75?
2,000%

18. 4.5 is what percent of 18?
25%

19. Calculate $33\frac{1}{3}$% of $600.
$200

20. What percent of $9 is $4?
$44\frac{4}{9}$%

21. Estimate 59% of $19.99.
$12

22. 2.5% of how much money is $40?
$1,600

23. What percent of $7.99 is $1.35, to the nearest whole percent? 17%

24. 3.5 is $8\frac{1}{4}$% of what number, to the nearest hundredth?
42.42

[6.3]

Insert each missing entry.

25.

Original Value	New Value	Percent Decrease
24	16	$33\frac{1}{3}$%

26.

Selling Price	Rate of Sales Tax	Sales Tax
$50	6%	$3.00

27.

Sales	Rate of Commission	Commission
$600	4%	$24

28.

Original Price	Rate of Discount	Discount	Sale Price
$200	15%	$30	$170

29.

Selling Price	Rate of Markup	Markup	Original Price
$51	50%	$25.50	$25.50

30.

Principal	Interest Rate	Time (in years)	Simple Interest	Final Balance
$200	4%	2	$16	$216

Mixed Applications

Solve.

31. According to a recent survey, 85% of U.S. banks charge customers a fee for depositing a check that bounces. Express this percent as a fraction, reduced to lowest terms. $\dfrac{17}{20}$

32. Jonas Salk discovered the polio vaccine in 1954. The number of reported polio cases in the United States dropped from 29,000 to 15,000 between 1955 and 1956. What was the percent drop, to the nearest whole percent? (*Source:* Bureau of Census) 48%

33. For their fees, one real estate agent charges 11% of a year's rent and another charges the first month's rent. Which agent charges more? The agent that charges 11%.

34. A poll found that $\dfrac{3}{5}$ of respondents regularly recycle cans or bottles. What percent is this? 60%

35. According to a city survey, 49% of respondents approve of how the mayor is handling his job and 31% disapprove. What percent neither approved nor disapproved? 20%

36. In the TV ratings, the XYZ network got $\dfrac{1}{4}$ of U.S. households, and the TRS network got 18.8%. Which network did better? Network XYZ

37. If 75% of the jury found the defendant innocent, what fraction of the jury found him innocent? $\dfrac{3}{4}$

38. The rate of interest on a bank account was $6\dfrac{1}{4}\%$. Express this rate as a decimal. 0.0625

39. It takes you 50 min to commute to work. If you have been traveling for 20 min, what percent of your trip has been completed? 40%

40. Approximately 1 of every 10 Americans is left-handed. What percent is this? 10%

41. The standing of a baseball team was 0.632. Express the team's standing as a percent, rounded to the nearest whole percent. 63%

42. A company earned $6.3 million in profit on sales of $90 million. Express its profits as a percent of sales. 7%

43. In a recent year, 360 million of the 2 billion bushels of wheat grown in the United States came from Kansas. What percent is this, to the nearest whole percent? 18%

44. The length of a person's thigh bone is usually about 27% of his or her height. Estimate someone's height whose thigh bone is 20 inches long. (*Source: American Journal of Physical Anthropology*) Possible estimate: 80 in.

45. The winner of a men's U.S. Open tennis match got $87\dfrac{1}{2}\%$ of his first serves in. If he had 72 first serves, how many went in? 63 first serves

46. In a scientific study that relates weight to health, people are considered overweight if their actual weight is at least 20% above their ideal weight. If you weigh 160 lb and have an ideal weight of 130 lb, are you considered overweight? Yes

47. In purchasing a house, you made a 20% down payment. If this payment came to $20,000, how much did you pay for the house? $100,000

48. When you became editor, the magazine's weekly circulation increased from 50,000 to 60,000. By what percent did the circulation increase? 20%

49. Your salary had been $30,000 before you got a raise of $1,000. If the rate of inflation is 5%, has your salary kept pace with inflation?   No

50. The cat is the most popular pet in America. According to recent estimates, there are 53 million dogs. If there are 109% as many cats, how many cats are there? 57,770,000 cats

51. At an auction, you bought a table for $150. The auction house also charged a "buyer's premium"—an extra fee—of 10%. How much did you pay in all? $165

52. According to the news report, 80 tons of food met only 20% of the food needs in the refugee camp. How much additional food was needed?   320 tons

53. You need 14,000 more frequent flier miles to earn a free trip to Hawaii, which is 20% of the total number needed. How many frequent flier miles in all does this award require?   70,000 mi

54. A brand X light bulb shines for 8,600 hours, whereas brand Y's bulb lasts 10% longer. How long does a brand Y bulb last?   9,460 hr

55. How much commission does a salesperson make on sales totaling $5,000 at a 20% rate of commission? $1,000

56. At the end of the year, the receipts of a retail store amounted to $200,000. Of these receipts, 85% went for expenses; the rest was profit. How much profit did the store make? $30,000

57. If you deposit $7,000 in a bank account that pays a 6.5% rate of interest compounded annually, what will be the balance after 2 yr? $7,939.58

58. Suppose that a country's economy expands by 2% per year. By what percent will it expand in 10 yr, to the nearest whole percent?   122%

59. Complete the following table, which describes a company's income for the four quarters of last year:

Quarter	Income	Percent of Total Income
1	$375,129	27%
2	289,402	21%
3	318,225	23%
4	402,077	29%
Total	$1,384,833	100%

60. The following graph shows the sources from which the federal government received income in a recent year.

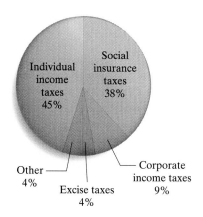

If the total amount of money taken in was $1,169 billion, compute how much money was received from each source, to the nearest $billion. (*Source:* U.S. Office of Management and Budget)   Individual income taxes: $526,050,000,000; social security taxes: $444,220,000,000; corporate income taxes: $105,210,000,000; excise taxes: $46,760,000,000; other: $46,760,000,000

■ *Check your answers on page A-10.*

Chapter 6 Posttest

To see if you have mastered the topics in this chapter, take this test.

Rewrite.

1. 4% as a fraction $\frac{1}{25}$

2. $62\frac{1}{2}$% as a fraction $\frac{5}{8}$

3. 150% as a decimal   1.5

4. 8% as a decimal   0.08

5. 0.009 as a percent   0.9%

6. 3 as a percent   300%

7. $\frac{5}{6}$ as a percent, rounded to the nearest whole percent
83%

8. $2\frac{1}{5}$ as a percent   220%

Solve.

9. What is 25% of 30 mi?   7.5 mi

10. Find 120% of 40.   48

11. Estimate 32% of $20.77.   Possible estimate: $6

12. 8% of what number is 16?   200

13. What percent of 10 is 6?   60%

14. What percent of 4 is 10?   250%

15. You make 5% simple interest per year on an initial balance of $300 for 2 yr. How much interest do you make? $30

16. 6 is what percent of $1\frac{1}{2}$?   400%

17. A basketball player made 45 of 50 attempted free throws. Express this success rate as a percent.   90%

18. Milk is approximately 50% cream. How much milk is needed to produce 2 pints of cream?   4 pt

19. After traveling abroad, you owe duty equal to 10% of the amount by which your purchases exceed $400. If your purchases cost $550, how much duty do you owe? $15

20. A college ended 6 straight years of double-digit tuition increases by raising its tuition from $3,000 to $3,100. Find the percent increase. $3\frac{1}{3}$%

■ *Check your answers on page A-11.*

Cumulative Review Exercises

To help you review, solve the following:

1. Divide: $1,962 \div 18$ 109

2. Express $\dfrac{5}{6}$ as a decimal, rounded to the nearest hundredth.   0.83

3. Multiply: 0.2×3.5 0.7

4. Find the sum of $3\dfrac{4}{5}$ and $1\dfrac{9}{10}$. $5\dfrac{7}{10}$

5. Solve for x: $\dfrac{x}{3} = 2.5$ 7.5

6. 20% of what amount is \$200?   \$1,000

Solve:

7. The government withdrew $\dfrac{1}{4}$ million of its 2 million troops. What fraction of the total is this? $\dfrac{1}{8}$

8. You need 3 lb of grass seed to plant a lawn 600 ft² in area. At this rate, how much grass seed is needed on a lawn with area 800 ft²?   4 lb

9. Three FM stations are highlighted on the radio dial shown. These stations have frequencies 99.5 (WBAI), 96.3 (WQXR), and 104.3 (WAXQ). Label the three stations on the dial.

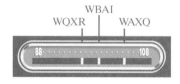

10. In a recent year in the United States, 21% of 760 thousand doctors were female. How many female doctors were there, to the nearest 10 thousand? (*Source:* The American Medical Association)   160 thousand

■ *Check your answers on page A-11.*

Signed Numbers

SIGNED NUMBERS IN CHEMISTRY

In chemistry, a valence is assigned to each element in a compound. Valences help us study the ways in which the elements combine to form the compound.

The valence is a positive or negative whole number that expresses the combining capacity of the element. For example in the compound H_2O (water), the element hydrogen (H) has a valence of $+1$ whereas the element oxygen (O) has a valence of -2.

The valences in any chemical compound add up to zero. So if you know how to perform signed number computations, you can predict the chemical formula of any compound.

Chapter 7 Pretest

To see if you have already mastered the topics in this chapter, take this test.

1. Which is larger, -23 or $+7$? $+7$

2. A negative number has an absolute value of 5. What is the number? -5

Compute.

3. $-8 + (-9)$ -17

4. $-20 + 20$ 0

5. $34 - 41$ -7

6. $-9 - (-9)$ 0

7. -5×15 -75

8. $-\dfrac{3}{4} \times \dfrac{2}{3}$ $-\dfrac{1}{2}$

9. $(-8)^2$ 64

10. $\left(-\dfrac{1}{2}\right)^2$ $\dfrac{1}{4}$

11. $-18 \div (-9)$ 2

12. $\dfrac{1}{2} \div (-4)$ $-\dfrac{1}{8}$

13. $-2 + 5 + (-3) + 8$ 8

14. $10 + (-3) - (-1)$ 8

15. $-9 - 3^2 \times (-5)$ 36

16. $8 \cdot (-2) + 3 \cdot (-1)$ -19

Solve.

17. The temperature outside is $-3\,°\mathrm{F}$. If it gets $4°$ colder, how cold will it be? $-7\,°\mathrm{F}$

18. Which is warmer: a temperature of $-3\,°\mathrm{F}$ or a temperature of $-1\,°\mathrm{F}$? $-1\,°\mathrm{F}$

19. You invested in a summer house, buying it for $25,000. Later, you sold it for $17,000. How much money, expressed as a signed number, did you make or lose? $-$8,000$ (a loss)

20. What is the average of -5, 2, and -6? -3

■ Check your answers on page A-11.

7.1 Introduction to Signed Numbers

OBJECTIVES

■ *To find the opposite and the absolute value of a signed number*
■ *To compare signed numbers*
■ *To solve word problems involving the comparison of signed numbers*

What Signed Numbers Are and Why They Are Important

Suppose that you visit your accountant who says, "I've got good news and bad news for you: Your company is earning $100,000 a year. But you owe the government $12,000 in back taxes." You're depressed, driving home from the accountant's office. As the temperature outside drops to 2° below 0, you begin to shiver. Then you remember that you are traveling to Hawaii next week where the temperature is 80°.

This story involves several applications of positive and negative numbers—the gain of $100,000, the debt of $12,000, and the temperatures of 2° below 0 and 80° above 0. The debt and the temperature below 0 are negative because each is less than 0. In this chapter we discuss negative numbers and show how they relate to positive numbers—the numbers greater than 0. Negative and positive numbers together are referred to as **signed numbers**.

Here are a few more applications of signed numbers.

■ In football, a positive number represents yards gained; a negative number, yards lost.

■ In terms of time, positive applies to a time after an event took place; negative, to a time before that event.

■ In the study of electricity, positive represents one kind of electric charge; negative, the opposite kind of electric charge.

These applications can help you develop intuition in working with negative numbers and understand what they represent.

The Number Line

Our previous discussion of the number line on page 3 included only positive numbers. However, the number line can be extended to represent the negative numbers also. If we label the numbers to the right of 0 as positive and extend the line leftward past 0, then we label the numbers to the left of 0 as negative.

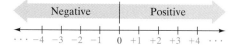

Note that we write "negative two" as −2 and "positive three" as +3, or just 3. However, we write no sign before 0 because 0 is neither negative nor positive.

Definitions

A **positive number** is a number greater than 0.

A **negative number** is a number less than 0.

A **signed number** is a number with a sign that is either positive or negative.

In drawing the number line, we usually label only the integers, that is, the whole numbers and the corresponding negatives: $\ldots, -4, -3, -2, -1, 0, +1, +2, +3, +4, \ldots$

Definitions

The **integers** are the numbers $\ldots, -4, -3, -2, -1, 0, +1, +2, +3, +4, \ldots,$ continuing indefinitely in both directions.

Fractions and decimals and their corresponding negatives can also be represented on the number line. Do you know how to locate the points on the number line that correspond to the numbers $\frac{1}{2}$, 3.8, and -2.4? The following number line shows these locations.

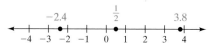

EXAMPLE 1

Locate $\frac{1}{2}$, -2.8, $-\frac{1}{8}$ and 1.2 on the number line.

Solution

PRACTICE 1

Locate $1\frac{9}{10}$, -1, -3.1 and 0 on the number line.

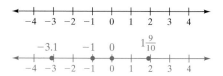

On the number line, note that -1 and $+1$ (or 1) are the opposite of each other. Similarly -50 and $+50$ (or 50) are opposites. What is the opposite of 0?

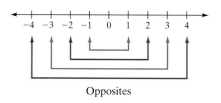

Opposites

Definition

Two numbers that are the same distance from 0 on the number line but on opposite sides of 0 are called **opposites**.

EXAMPLE 2

Find the opposite of each number in the table.

Solution

Number	Opposite
a. 5	−5
b. $-\dfrac{1}{2}$	$\dfrac{1}{2}$
c. 1.5	−1.5
d. −100	100

PRACTICE 2

Find the opposite of each number.

Number	Opposite
a. 9	−9
b. $-4\dfrac{9}{10}$	$4\dfrac{9}{10}$
c. −2.9	2.9
d. 31	−31

Because the number −2 is negative, it lies 2 units to the left of 0. The number 2, which is positive, lies in the opposite direction: 2 units to the right of 0.

When you locate a number on the number line, the *distance* of that number from 0 is called its **absolute value**. Thus the absolute value of +2 is 2, and the absolute value of −2 is 2.

Definition

The **absolute value** of a number is its distance from 0 on the number line. The absolute value of a number is represented by the symbol | |.

For example. we write the absolute value of −2 as $|-2|$.

Several properties of absolute value follow from this definition.

■ The absolute value of a positive number is the number itself.

■ The absolute value of a negative number is its opposite.

■ The absolute value of 0 is 0.

■ The absolute value of a number is always positive or 0.

These properties help us compute the absolute value of any number.

EXAMPLE 3

Compute.

a. $|-8|$ **b.** $|0|$ **c.** $\left|-\dfrac{1}{2}\right|$ **d.** $|5.3|$ **e.** $|2.5|$

Solution

a. Because −8 is negative, its absolute value is its opposite, or 8.

b. The absolute value of 0 is 0.

c. $\dfrac{1}{2}$ **d.** 5.3   **e.** 2.5

PRACTICE 3

Compute.

a. $|9|$ **b.** $\left|1\dfrac{3}{4}\right|$ **c.** $|-4.1|$ **d.** $|-5|$

9    $1\frac{3}{4}$ 4.1   5

EXAMPLE 4

Determine the sign and the absolute value.

a. 25   **b.** −1.9

Solution

a. Sign: +   absolute value: 25

b. Sign: −   absolute value: 1.9

PRACTICE 4

Determine the sign and the absolute value.

	Sign	Absolute value
a. −4	−	4
b. $6\frac{1}{2}$	+	$6\frac{1}{2}$

Comparing Signed Numbers

The number line helps us compare two signed numbers, that is, to decide which number is larger and which is smaller. On the number line, a number to the right of another number is the larger number.

Smaller | Larger

−4 −3 −2 −1 0 1 2 3 4

To Compare Signed Numbers

■ locate the points that you want to compare on the number line; the number to the right is larger than the number to the left.

When comparing signed numbers, you should remember the following.

■ Zero is greater than any negative number because all negative numbers lie to the left of 0.

■ Zero is less than any positive number because all positive numbers lie to the right of 0.

■ Any positive number is greater than any negative number because all positive numbers lie to the right of all negative numbers.

EXAMPLE 5

Which is larger, 2 or 0?

Solution Because 2 (or +2) is to the right of 0 on the number line, 2 is greater than 0.

PRACTICE 5

Which is larger, 0 or $\frac{1}{2}$? $\frac{1}{2}$

EXAMPLE 6

Which is smaller, −1 or −3?

Solution Because −1 is to the right of −3, −3 < −1, and −3 is smaller.

PRACTICE 6

Which is smaller, −5 or −2?   −5

EXAMPLE 7

Which is larger, 1 or −3?

Solution Because 1 is to the right of −3, 1 > −3; that is, 1 is the larger of the two numbers.

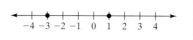

PRACTICE 7

Which is larger, 2 or −4?   2

Now let's try some practical applications of comparing signed numbers. The key is to be able to identify the signs of numbers. Becoming familiar with the following words that indicate the sign of a number will help you do so.

Negative	Positive
Loss	Gain
Below	Above
Decrease	Increase
Down	Up
Withdrawal	Deposit
Past	Future
Before	After

EXAMPLE 8

Express as a signed number: The Dow Jones Industrial Average on the stock market lost 3 points today.

Solution The number in question represents a loss (or a decline), so we write it as a negative number: −3

PRACTICE 8

Represent as a signed number: The New York Giants gained 2 yards on a play.
+2

EXAMPLE 9

Your house is located in a valley 15 ft below sea level. Your friend's house is in the same valley 10 ft below sea level. Whose house is higher?

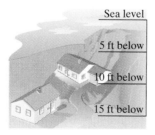

Sea level

5 ft below

10 ft below

15 ft below

Solution We need to compare −15 and −10. Because −10 > −15, your friend's house at 10 ft below sea level is higher. The diagram confirms this conclusion.

PRACTICE 9

The following table shows the temperature below which various plants freeze and die.

Plant	Asters	Carnations	Mums
Hardy to	−20 °F	−5 °F	−30 °F

In a very cold climate, which of these plants would you plant? (*Source: The American Horticultural Society A–Z Encyclopedia of Garden Plants*)   Mums

Exercises 7.1

Find the opposite of each number.

1. 8   **2.** −3   **3.** 10.2   **4.** −25   **5.** −5   **6.** −5$\frac{1}{2}$

   −8   3   −10.2   25   5   5$\frac{1}{2}$

7. 2$\frac{1}{3}$ **8.** $\frac{3}{4}$ **9.** −4.1   **10.** 0.5   **11.** −1.2   **12.** −2.1

   −2$\frac{1}{3}$ −$\frac{3}{4}$ 4.1   −0.5   1.2   2.1

Mark the corresponding point for each number on the number line.

13. −2   **14.** 1.1   **15.** 0   **16.** −3$\frac{9}{10}$

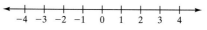

Express each quantity as a signed number.

17. A withdrawal of $150 from an   **18.** 6 km below sea level   −6 km
account   −$150

19. A rise in temperature of 14.5 °C   **20.** A loss of 3$\frac{1}{4}$ lb while on a diet
+14.5 °C

   −3$\frac{1}{4}$ lb

Determine the sign and the absolute value of each number.

	Sign	Absolute Value			Sign	Absolute Value
21. 9	+	9		**22.** 12	+	12
23. −4.3	−	4.3		**24.** 9.2	+	9.2
25. −7	−	7		**26.** −30	−	30
27. $\frac{1}{5}$	+	$\frac{1}{5}$		**28.** −$\frac{3}{4}$	−	$\frac{3}{4}$

Evaluate.

29. $|-6|$

6

30. $|39|$

39

31. $\left|-\dfrac{4}{5}\right|$

$\dfrac{4}{5}$

32. $|-5.8|$

5.8

33. $|2|$

2

34. $|8|$

8

35. $|-0.6|$

0.6

36. $\left|-1\dfrac{2}{3}\right|$

$1\dfrac{2}{3}$

Solve.

37. How many numbers have an absolute value of 5?   Two; -5 and 5

38. How many numbers have an absolute value of 0.5?   Two; -0.5 and 0.5

39. Can you name a number whose absolute value is -1?   No; absolute value is always positive, or 0.

40. Can you name 3 different numbers that have the same absolute value?   No

Circle the larger number in each pair.

41. -4 and -7
-4

42. 4 and -7
4

43. 12 and 0
12

44. 0 and -87
0

45. -3 and 2
2

46. -3 and -14
-3

47. -4 and $-2\dfrac{1}{3}$
$-2\dfrac{1}{3}$

48. 5.1 and 8
8

49. -29 and -2
-2

50. -4 and 7
7

51. 9 and -22
9

52. 0 and -5
0

53. -8 and -2
-2

54. $+3$ and -14
$+3$

55. -7 and $7\dfrac{1}{4}$
$7\dfrac{1}{4}$

56. 3.888 and 4
4

Indicate whether each inequality is true or false.

57. $-5 > -7$
T

58. $3 < -1$
F

59. $-1 < 3.4$
T

60. $-5 > 0$
F

61. $0 > -2\dfrac{3}{4}$
T

62. $-6 < -1$
T

63. $2 > -2$
T

64. $-100 < 0$
T

65. $-3.5 > -3.4$
F

66. $-1.6 < -1.7$
F

67. $-4\dfrac{1}{3} < 0$
T

68. $-\dfrac{5}{2} > -\dfrac{7}{2}$
T

Arrange the numbers in each group from smallest to largest.

69. $3, -3, 0$ **70.** $3.5, -3.1, -3, 0, 4$ **71.** $-9, 9, -4.5$ **72.** $-2\frac{1}{2}, -2, 3, -2.7$

$-3, 0, 3$ $-3.1, -3, 0, 3.5, 4$ $-9, -4.5, 9$ $-2.7, -2\frac{1}{2}, -2, 3$

Applications

Solve.

73. Today you owe $100. Last week, you owed $1,000. Were you better off financially last week, or are you better off today?   Better off today

74. Would your company be doing better if it lost $1,000,000 or if it had a profit of $1,000?   If it had a profit of $1,000

75. Will you weigh more if you lose 5 lb or gain 1 lb?   Gain 1 lb

76. Which is later: 10 sec before blastoff or 3 sec after blastoff?   3 sec after blastoff

77. Which has a better record: the Trendsetters, which won 2 games, or the Yellowjackets, which lost 23 games?   The Trendsetters

78. Would you be higher if you took the elevator down 2 floors or if you took it down 5 floors?   Down 2 floors

79. Which is warmest: a temperature of -2 °F, a temperature of -1 °F, or a temperature of -3 °F? -1 °F

80. Three of the coldest temperature readings ever recorded on Earth were -64.8 °C, -64.3 °C, and -54.5 °C. Of these 3 temperatures, which was the coldest? -64.8 °C

81. Each liquid has its own boiling point — the temperature at which it changes to a gas. Liquid chlorine, for example, boils at -35 °C, whereas liquid fluorine boils at -188 °C. Which liquid has the higher boiling point? (*Source: Handbook of Chemistry & Physics*)   Liquid chlorine

82. The temperature to which a plant is hardy is the coldest temperature at which it will survive. If daylilies are hardy to -25 °F, will they survive at -20 °F?   Yes

■ *Check your answers on page A-11.*

MINDSTRETCHERS

GROUPWORK

1. a. List several numbers between -2 and -3.

    Possible answers: $-2.1, -2\frac{1}{2}, -2.2, -2\frac{1}{3}$, etc.

b. How many numbers are there between -2 and -3?
Infinitely many

MATHEMATICAL REASONING

2. Suppose that you will win $1 million if you can guess the temperature outside to within $2°$. If the outside temperature is $-1°$, highlight the winning temperatures on the thermometer at the right.   Winning temperatures:
$-3, -2, -1, 0, 1$

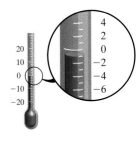

WRITING

3. Consider the following statement.

The negative part of the number line is the mirror image of its positive part.

What does this statement mean? Is it true?   The negative part of the number line and the positive part are similar except that the signs of the numbers and the order of the numbers are reversed. Yes, it is true.

HISTORICAL NOTE

Up until the work of sixteenth-century Italian physicists, no one was able to measure temperature. Liquid-in-glass thermometers were invented around 1650, when glass blowers in Florence were able to create the intricate shapes that thermometers require. Thermometers from the seventeenth and eighteenth centuries provided a model for working with negative numbers that led to their wider acceptance in the mathematical and scientific communities. Numbers above and below 0 represented temperatures above and below the freezing point of water, just as they do on the Centigrade scale today. Before the introduction of these thermometers, a number such as −1 was difficult to interpret for those who believed that the purpose of numbers is to count or to measure.

By contrast, the early Greek mathematicians had rejected negative numbers, calling them absurd. A thousand years later in the seventh century A.D., the Indian mathematician Brahmagupta argued for accepting negative numbers and put down the first comprehensive rules for computing with them.

Sources:

Lancelot Hogben, *Mathematics in the Making* (London: Galahad Books, 1960).

Henri Michel, *Scientific Instruments in Art and History* (New York: Viking Press, 1966).

Calvin C. Clawson, *The Mathematical Traveler* (New York and London: Plenum Press, 1994).

7.2 Adding Signed Numbers

OBJECTIVES

■ *To add signed numbers*
■ *To solve word problems involving the addition of signed numbers*

Our previous work in addition, subtraction, multiplication, and division was restricted to positive numbers—whether those positive numbers happened to be whole numbers, fractions, or decimals. Now we consider computations involving *any* signed numbers. Let's first consider the operation of addition.

Suppose that we want to add two negative numbers—say, -1 and -2. It's helpful to look at this problem in terms of money. If you have a debt of \$1 and another debt of \$2, altogether you *owe* \$3.

$$-1 + (-2) = -3$$

We can also look at this problem on the number line. To add -1 and -2, we start at the point corresponding to the first number, -1. The second number, -2, is *negative*, so we move 2 units to the *left*. We end at -3, which is the answer we expected.

Now let's consider adding the signed numbers -1 and $+3$. Thinking of this problem in terms of money may make it clearer. Suppose that you pay a debt of \$1 and earn \$3. What happens then? You still have \$2.

$$-1 + 3 = 2$$

We can picture this problem by starting at -1 on the number line. The second number, 3, is *positive*, so we move 3 units to the *right*. We end at 2, which is the answer.

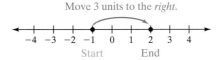

The following method provides a shortcut for adding signed numbers:

To Add Two Signed Numbers

■ if they have the same sign, add the absolute values and keep the sign, but
■ if they have different signs, subtract the smaller absolute value from the larger and take the sign of the number with the larger absolute value.

EXAMPLE 1

Add -1 and -2.

Solution The sum of the absolute values is 3.

$$|-1| + |-2| = 1 + 2 = 3$$

Both -1 and -2 are negative, so their sum is negative.

$$(-1) + (-2) = -3$$

Visual Check: Move 2 units to the *left*.

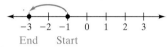

End   Start

PRACTICE 1

Add -8 and -17. -25

EXAMPLE 2

Add: $(-3.9) + (-0.5)$

Solution $|-3.9| = 3.9$ and $|-0.5| = 0.5$

$$3.9 + 0.5 = 4.4$$

The sum of two negative numbers is negative, so

$$(-3.9) + (-0.5) = -4.4.$$

Visual Check: Move 0.5 units to the *left*.

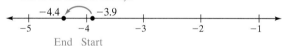

End   Start

PRACTICE 2

Find the sum: $-3 + \left(-1\frac{1}{2}\right)$ $-4\frac{1}{2}$

EXAMPLE 3

Add: $3 + (-1)$

Solution Here we are adding numbers with different signs. First, we find the absolute values.

$$|3| = 3 \quad \text{and} \quad |-1| = 1$$

Next, we subtract the smaller absolute value from the larger.

$$3 - 1 = 2$$

Because 3 has the larger absolute value and its sign is positive, the sum is also positive. Our answer is 2, or $+2$.

$$3 + (-1) = 2$$

Visual Check: Move 1 unit to the *left*.

End Start

PRACTICE 3

Find the sum of -2 and 9.   7

Note in Example 3 that, when we added a negative number to 3, we got a smaller result—namely, 2.

EXAMPLE 4

Combine: $(-2) + (+2)$

Solution $|-2| = 2$ and $|+2| = 2$ Find the absolute values.

$\qquad 2 - 2 = 0$ Subtract the absolute values.

Zero is neither positive nor negative, for it has no sign.

$$(-2) + (+2) = 0$$

Visual Check: Move 2 units to the *right*.

Do you see why the sum of -44 and $+44$ is 0? How about $2.77 + (-2.77)$ or $2\frac{5}{8} + \left(-2\frac{5}{8}\right)$?

Have students explain their responses in a journal.

PRACTICE 4

Combine: $-35 + 35$ 0

EXAMPLE 5

Add -2.1 to 0.8.

Solution $|-2.1| = 2.1$ and $|0.8| = 0.8$ Find the absolute values.

$$\begin{array}{r} 2.1 \\ -0.8 \\ \hline 1.3 \end{array}$$ Subtract the absolute values.

Because $|-2.1|$ is greater than $|0.8|$, the answer is negative. So

$$(-2.1) + 0.8 = -1.3.$$

PRACTICE 5

Add: $3\frac{4}{5} + \left(-1\frac{1}{5}\right)$ $2\frac{3}{5}$

Some addition problems involve the sum of three or more signed numbers. Rearranging the signed numbers to add the positives separately from negatives can make the addition easier. Note that this rearrangement does not affect the sum because addition is a commutative and associative operation.

EXAMPLE 6

Find the sum: $3 + (-1) + (-8) + 2 + (-11)$

Solution We are adding two positive and three negative numbers. Rearranging the numbers by sign, we get the following.

$$\underbrace{3 + 2}_{\text{Positives}} + \underbrace{(-1) + (-8) + (-11)}_{\text{Negatives}}$$

First, we add the positives. $3 + 2 = 5$

Then, we add the negatives. $(-1) + (-8) + (-11) = -20$

Finally, we combine the positive and the negative subtotals.

$$5 + (-20) = -15$$

Thus $3 + (-1) + (-8) + 2 + (-11) = -15$.

PRACTICE 6

$-3 + 1 + 8 + (-6) = ?$ 0

EXAMPLE 7	PRACTICE 7

The most famous of all comets is Halley's comet, which flies by Earth every 76 yr. For other comets, however, the length of time between visits is much longer. The Great Comet, for example, comes near Earth only once every 3,000 yr. If this comet visited Earth about 1200 B.C., approximately when was its next visit? (*Source:* Mark R. Kidger, "Some Thoughts on Comet Hale-Bopp")

Solution   To help you understand this problem, we draw the number line. Any number line involving time is called a *time line*. On a time line, positive years are A.D. and negative years are B.C.

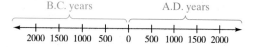

So 1200 B.C. is represented by -1200. We must add -1200 and $+3000$.

$$|-1200| = 1200 \quad \text{and} \quad |3000| = 3000 \qquad \text{Find the absolute values.}$$

$$3000 - 1200 = 1800 \qquad \text{Subtract the absolute values.}$$

The absolute value of 3000 is greater than the absolute value of -1200, so the answer is positive.

$$(-1200) + 3000 = 1800$$

Thus the Great Comet came near the Earth in about 1800 A.D.

Visual Check:

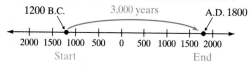

The Roman emperor Julius Caesar was born in 100 B.C. and died 56 yr later. In what year did he die? (*Source: Encyclopedia Americana*)   44 B.C.

> Point out to students that just as we often write positive numbers with no sign, so A.D. years are usually written without the A.D.

  Signed Numbers on a Calculator

The numbers that we have entered so far on a calculator have been positive numbers. To enter a negative number, we need to hit a special key that indicates that the sign of the number is negative. On many machines, this key is the *change of sign key*, $\boxed{+/-}$.

Recall that some calculators have a negative sign key, $\boxed{(-)}$, and others have a change of sign key. Do not confuse either of these keys with the subtraction key, $\boxed{-}$.

EXAMPLE

On a calculator, enter −5.

Solution

Input	5	+/−
Display	5.	−5.

By looking at the display on your calculator, check to see whether the negative sign must be entered *after* the absolute value of the number.

PRACTICE

Enter the number −12.

Now let's use a calculator to carry out computations involving signed numbers.

EXAMPLE

Compute: $(-2) + 3$

Solution This problem requires us to add −2 and +3.

Input	2	+/−	+	3	=
Display	2.	−2.	−2.	3.	1.

The answer is +1, or 1.

PRACTICE

Find the sum: $(-1) + 7$ 6

EXAMPLE

Calculate: $-1.3 + (-5.891) + 4.713$

Solution

Input	1.3	+/−	+	5.891	+/−	+	4.713	=
Display	1.3	−1.3	−1.3	5.891	−5.891	−7.191	4.713	−2.478

So $-1.3 + (-5.891) + 4.713 = -2.478$. Note that in this example the calculator displays a running total, −7.191, when the first two addends are entered. Check to see whether your calculator does the same.

PRACTICE

Add: $-6.002 + (-9.37) + (-0.22)$ −15.592

Exercises 7.2

Find the sum of each pair of numbers. Use the number line as a visual check.

1. $6 + (-5)$ 1

2. $3 + (-9)$ -6

3. $-2 + 5$ 3

4. $-9 + (-2)$ -11

5. $7 + 0$ 7

6. $3 + (-2)$ 1

7. $7 + (-7)$ 0

8. $-4 + 9$ 5

Find the sum.

9. $67 + (-67)$
0

10. $23 + (-2)$
21

11. $-10 + 5$
-5

12. $0 + (-12)$
-12

13. $-100 + 300$
200

14. $-60 + (-20)$
-80

15. $8 + (-2)$
6

16. $5,000 + (-3,000)$
2,000

17. $60 + (-90)$
-30

18. $-5 + (-4)$
-9

19. $-7 + 2$
-5

20. $2 + (-7)$
-5

21. $-7 + 0$
-7

22. $-3 + 3$
0

23. $-2 + 2$
0

24. $-2 + (-2)$
-4

25. $5.2 + (-0.3)$
4.9

26. $-0.6 + 1$
0.4

27. $-0.2 + 0.3$
0.1

28. $-5.5 + 0$
-5.5

29. $60 + (-0.5)$
59.5

30. $-0.7 + 0.7$
0

31. $-9.8 + 3.9$
-5.9

32. $6.1 + (-5.9)$
0.2

33. $(-5.6) + (-8.9)$
-14.5

34. $(-0.8) + (-0.5)$
-1.3

35. $\left(-\dfrac{1}{2}\right) + \left(-5\dfrac{1}{2}\right)$
-6

36. $-1\dfrac{1}{3} + \left(-2\dfrac{2}{3}\right)$
-4

37. $-1\dfrac{1}{5} + \dfrac{3}{5}$

$-\dfrac{3}{5}$

38. $2\dfrac{1}{6} + \left(-\dfrac{5}{6}\right)$

$1\dfrac{1}{3}$

39. $-\dfrac{2}{5} + 2$

$1\dfrac{3}{5}$

40. $-14 + \dfrac{1}{3}$

$-13\dfrac{2}{3}$

41. $\dfrac{1}{2} + \left(-1\dfrac{3}{5}\right)$

$-1\dfrac{1}{10}$

42. $1\dfrac{3}{8} + \left(-\dfrac{1}{4}\right)$

$1\dfrac{1}{8}$

43. $(-24) + 20 + (-98)$
-102

44. $35 + (-17) + (-18)$
0

45. $12 + (-7) + (-12) + 7$
0

46. $-8 + (-4) + (-8) + 3$
-17

47. $(-7) + 12 + 0 + (-7)$
-2

48. $(-3) + 8 + (-9) + 3 + (-4)$
-5

49. $-0.3 + (-2.6) + (-4)$
-6.9

50. $-5.25 + (-0.4) + 3$
-2.65

51. $-12 + 7.58 + 12$
7.58

52. $-3.7 + (-1.88) + 5$
-0.58

▦ 53. $8.756 + (-9.08) + (-4.59)$
-4.914

▦ 54. $-5.405 + 6 + (-6.89)$
-6.295

▦ 55. $-3.001 + (-0.59) + 8$
4.409

▦ 56. $-10 + 5.17 + (-10.002)$
-14.832

Applications

Solve. Express each answer as a signed number.

57. The temperature on the top of a mountain was 3° below 0. If the temperature then rose 10°, what was the temperature? $+7°$

58. An elevator goes up 3 floors, down 2 floors, up 5 floors, and finally up 1 floor. What's the overall change in position of the elevator?   +7 floors (up)

59. At the beginning of this week, you had $197 in your savings account. During the week, you made a deposit of $50 and then withdrew $70. How much money was left in your account? $177

60. You are president of a corporation that last year took in $132,000 and had expenses of $148,000. How much money did the corporation make or lose?   −$16,000 (a loss)

61. In a physics class, you study the properties of atomic particles, including protons and electrons. You learn that a proton has an electric charge of +1, whereas an electron has an electric charge of −1. What is the total charge of a collection of 3 protons and 4 electrons?   −1

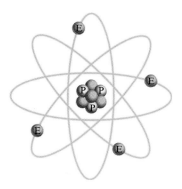

62. You are a chemist. In order to conduct an experiment, you cooled a substance to −10°C. In the course of this experiment, a chemical reaction took place that raised the temperature of the substance by 15°. What was the final temperature?   5°C

63. Ten years ago, you got married. Four years later, you got divorced. How many years ago was your divorce?   −6 (6 years ago)

64. Cleopatra became queen of Egypt in 51 B.C. She left the throne 20 yr later. In what year was that?   −31 (31 B.C.)

65. A football team gained 5 yd on its first play and lost 7 yd on its second play. What was the overall change in position as a result of these plays?   −2 (loss of 2 yd)

66. In the last 4 mo, you lost 5 lb, gained 2 lb, lost 1 lb, and maintained your weight, respectively. What was your overall change in weight?   −4 (you lost 4 lb)

67. You have two bank accounts with balances of $1,050.75 and $2,177.85. If you withdraw $1,000 from each account, will you have enough money left to buy a $1,000 used car?   Yes; you will have $1,228.60.

68. With $227.50 in your checking account, you write checks for $87.25 and $92.50. You also deposit $25. After the two checks clear and your deposit is credited, will you still have enough money to cover a $100 check?   No; you will only have $72.75 left.

■ *Check your answers on page A-11.*

MINDSTRETCHERS

GROUPWORK

1. Work with a partner on the following.

a. Fill in the following addition table.

+	+3	−2	−1
+4	7	2	3
−3	0	−5	−4
−1	2	−3	−2

b. Why do the nine numbers that you entered sum to 0?   +3, −2, and −1 sum to 0; so do +4, −3, and −1. In adding the nine numbers in the table, 0 is added repeatedly.

continued

WRITING

2. For signed numbers, does *adding* mean *increasing*? Explain.   <u>No, adding a negative</u>

<u>number results in a sum smaller than the other addend.</u>

MATHEMATICAL REASONING

3. In the diagram shown, + stands for a positive electrical charge and − for a negative charge.

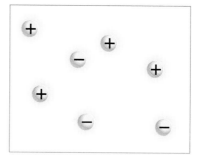

If a positive and a negative charge cancel each other, what is the result of combining all these charges?   +1

7.3 Subtracting Signed Numbers

OBJECTIVES

■ *To subtract signed numbers*
■ *To solve word problems involving the subtraction of signed numbers*

The subtraction of signed numbers is based on two topics previously covered — adding signed numbers and finding the opposite of a signed number.

Let's first consider a subtraction problem involving money. Suppose that you have $10 in your bank account and withdraw $3. Then $7 will be left in the account.

$$10 - (+3) = 7$$

Now suppose that you had started with $10 in the account but that the bank imposed a monthly service charge of $3. The balance in your account once more would be $7.

$$10 + (-3) = 7$$

The answers to the two problems are the same;

$$10 - (+3) = 7 \quad \text{and} \quad 10 + (-3) = 7$$

are equivalent problems.

In other words, we can change a problem in subtracting signed numbers to an equivalent problem in adding signed numbers by adding the *opposite* of the number that we want to subtract.

Show students the following alternative method of subtracting:
$10 - (+3) = 7$ because
$7 + 3 = 10$.

To Subtract Two Signed Numbers

■ change the operation of subtraction to addition, and change the number being subtracted to its opposite, and then
■ follow the method for addition.

To see if this method works when the number we are subtracting is negative, consider $4 - (-1)$. Recall that every subtraction problem has a related addition problem.

$$\underbrace{5 - 3 = 2}_{\text{Subtraction}} \quad \text{because} \quad \underbrace{2 + 3 = 5}_{\text{Related addition}}$$

Stress the importance of using parentheses in writing signed number problems to distinguish "subtraction" from "negative."

Therefore $4 - (-1) = 5$ because $5 + (-1) = 4$. Note that we get the same result using the method previously mentioned.

$$4 \underset{\substack{\uparrow \\ \text{Subtract}}}{-} \underset{\substack{\uparrow \\ \text{Negative 1}}}{(-1)} = 4 \underset{\substack{\uparrow \\ \text{Add}}}{+} \underset{\substack{\uparrow \\ \text{Positive 1}}}{(+1)} = 5$$

EXAMPLE 1

Find the difference: $-2 - (-4)$

Solution We leave the first number, -2, as is but change the operation of subtraction to addition and change the second number from -4 to $+4$.

$$-2 - (-4)$$
$$= -2 + (+4)$$

We already know how to add a negative and a positive number.

$$-2 + 4 = +2, \quad \text{or} \quad 2$$

PRACTICE 1

Find the difference: $-4 - (-2)$ -2

Have students explain their answers in a journal.

Note that in Example 1 and Practice 1 the numbers in the differences are the same except for their order. But the answers are quite different: $-2 - (-4) = 2$, whereas $-4 - (-2) = -2$. Why do you think this is so?

EXAMPLE 2

Compute: $3 - (-9)$

Solution

$$3 - (-9) = 3 + (+9) = +12, \quad \text{or} \quad 12$$

Negative 9   Positive 9

Subtract   Add

Note that when we subtracted -9 from 3, we got an answer larger than 3.

PRACTICE 2

Subtract: $9 - (-9)$ 18

EXAMPLE 3

Subtract: $-2 - 8\frac{1}{3}$

Solution $-2 - 8\frac{1}{3} = -2 + \left(-8\frac{1}{3}\right) = -10\frac{1}{3}$

PRACTICE 3

Find the difference: $-9 - 12.1$ -21.1

EXAMPLE 4

Calculate: $5 + (-6) - (-11)$

PRACTICE 4

$-2 - 3 + (-5) = ?$ -10

Solution This problem involves addition and subtraction. According to the rule for order of operations, we work from left to right.

$$5 + (-6) - (-11) \quad \text{Add } 5 + (-6).$$

$$= \quad -1 \quad - (-11) \quad \text{Subtract } -11.$$

$$= \quad -1 \quad + \ 11 \quad \text{Add } 11.$$

$$= \quad 10$$

EXAMPLE 5	PRACTICE 5

Normally we think of oxygen as a gas. However, when cooled to $-183\ °C$ (its boiling point), oxygen becomes a liquid. If it is cooled further to $-218\ °C$ (its melting point), oxygen becomes a solid. How much higher is the boiling point of oxygen than its melting point? (*Source: Handbook of Chemistry & Physics*)

Solution We need to compute how much greater -183 is than -218.

$$(-183) - (-218) = (-183) + (+218)$$

$$= +35$$

The boiling point of oxygen is 35 °C higher than its melting point.

One of the lowest temperatures ever recorded in this country was $-80\ °F$, in Alaska. One of the highest ever recorded was $134\ °F$, in California. How much warmer is the second temperature than the first? (*Source: The Weather Almanac*)

214 °F

Exercises 7.3

Find the difference.

1. 23 − 8
15

2. 8 − 23
−15

3. −34 − 7
−41

4. 5 − 9
−4

5. −9 − 5
−14

6. −44 − 2
−46

7. 42 − (−2)
44

8. 22 − 35
−13

9. 50 − 75
−25

10. −44 − (−2)
−42

11. 20 − (−1)
21

12. 85 − (−85)
170

13. 3 − (−3)
6

14. −3 − 3
−6

15. 0 − 38
−38

16. 38 − 0
38

17. −13 − 13
−26

18. 13 − 13
0

19. 13 − (−13)
26

20. 5 − (−2)
7

21. 0 − 1
−1

22. 1 − 0
1

23. 800 − (−200)
1,000

24. 30 − (−10)
40

25. 7 − 8.52
−1.52

26. 9.1 − 10.84
−1.74

27. 9.2 − (−0.5)
9.7

28. (−3) − (−0.2)
−2.8

29. −5.2 − (−5.2)
0

30. 0.5 − (−0.5)
1

31. 8.6 − (−1.9)
10.5

32. −1.9 − 8.6
−10.5

33. −10 − (−9.5)
−0.5

34. −6 − 8.7
−14.7

35. $4\frac{1}{2} - 9\frac{1}{2}$
−5

36. $9\frac{1}{2} - 4\frac{1}{2}$
5

37. $10 - 2\frac{1}{4}$

$7\frac{3}{4}$

38. $-10 - \left(-2\frac{1}{4}\right)$

$-7\frac{3}{4}$

39. $-7 - \frac{1}{4}$

$-7\frac{1}{4}$

40. $-9 - \frac{1}{8}$

$-9\frac{1}{8}$

41. $5\frac{3}{4} - \left(-1\frac{1}{2}\right)$

$7\frac{1}{4}$

42. $-6\frac{1}{2} - \left(-1\frac{1}{3}\right)$

$-5\frac{1}{6}$

Combine.

43. $4 + (-6) - (-9)$
7

44. $-10 - (-6) + 8$
4

45. $7 - 7 + (-5)$
-5

46. $10 + (-10) - (-5)$
5

47. $-8 + (-4) - 9 + 7$
-14

48. $-5 - (-1) + 6 + (-3)$
-1

49. $7.043 - 9.002 - 1.883$
-3.842

50. $-6.192 - 0.337 - (-23.94)$
17.411

51. $-8.722 + (-3.913) - 3.86$
-16.495

52. $2.884 - 0.883 + (-6.125)$
-4.124

Applications

Solve. Express each answer as a signed number.

53. Two airplanes take off from the same airport. One flies north and the other south, as shown. How far apart are they?   1,000 mi

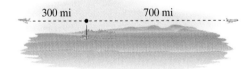

300 mi   700 mi

54. You and a friend get on different elevators at the same floor. She goes up 2 floors, and you go down 4 floors. How many floors are you and your friend apart?   6 floors

55. Ethiopia was founded around 1,000 B.C., the United States in 1789 A.D. How much older is Ethiopia than the United States? (*Source: The Concise Columbia Encyclopedia*)   About 2,789 yr

56. Paper was invented in China in about 100 B.C. About how many years ago was that? (*Source: World of Invention*)
About 2,100 yrs

57. A company reported that last year it lost $367,120 and that this year it lost $7,301. By how much money were the losses reduced? $359,819

58. Your checking account had a balance of $92.75. You then wrote a check for $102, which cleared. What was your new balance?   −$9.25

59. The value of a stock rose by $\frac{1}{8}$ of a dollar and then dropped by $\frac{1}{2}$ of a dollar. What was the overall change in value?
$-\$\frac{3}{8}$ (dropped)

60. Last year, your company suffered a loss of $20,000. This year, it showed a profit of $50,000. How big an improvement was this? $70,000

■ *Check your answers on page A-11.*

MINDSTRETCHERS

GROUPWORK

1. Working with a partner, rearrange the numbers in the magic square on the left so that the sum of every row, column, and diagonal is −6.

−3	+2	−2
−5	−4	−6
0	−1	+1

−3	+2	−5
−4	−2	0
+1	−6	−1

MATHEMATICAL REASONING

2. The following two columns of numbers add up to the same sum.

3	7
7	?
1	7
9	8
5	9

What number does the question mark represent?   −6

WRITING

3. Consider the following two problems:

$$8 - (-2) = 8 + 2 = 10$$

$$8 \div \frac{4}{7} = 8 \times \frac{7}{4} = 14$$

Explain in what way the two problems are similar.   In both problems there is a conversion from one operation to its opposite; from subtraction to addition and from division to multiplication.

7.4 Multiplying Signed Numbers

OBJECTIVES

■ *To multiply signed numbers*
■ *To solve word problems involving the multiplication of signed numbers*

We now turn to multiplication of signed numbers. Consider, for example, the problem of finding the product of $+4$ and -2, or $4(-2)$. Multiplication is repeated addition, so we know that multiplying a number by 4 means the same as adding the number to itself four times. Using the method for adding signed numbers, we get the following.

$$4(-2) = -2 + (-2) + (-2) + (-2)$$

$$= -8$$

Note that, when we multiply a positive number by a negative number, we get a negative answer.

Let's take another look at this same problem in practical terms. Suppose that you are on a diet and you *lose* 2 lb/mo. Compared to your current weight, how much will you weigh 4 mo from now? The answer is that you will weigh 8 lb *less*.

$$4(-2) = -8$$

Now we examine a different question. Again assume that you lose 2 lb/mo by dieting. Four months ago, you were heavier than you are now. How much heavier were you? To answer this question, note that each month you lost 2 lb but that you are going back in time 4 mo. So you weighed 8 lb *more than* you do now.

$$-4(-2) = +8, \quad \text{or} \quad 8$$

Note that when we multiply two negative numbers, we get a positive number.

We can use the following method to multiply signed numbers.

To Multiply Two Signed Numbers

■ multiply their absolute values, and
■ if the numbers have the same sign, their product is positive; if the numbers have different signs, their product is negative.

EXAMPLE 1

Find the product of -2 and -1.

Solution First, we find the absolute values.

$$|-2| = 2 \quad \text{and} \quad |-1| = 1$$

Next, we multiply the absolute values.

$$2 \cdot 1 = 2$$

The numbers have the same sign, so the product is positive. The answer is $+2$, or 2.

$$(-2)(-1) = 2$$

PRACTICE 1

Compute: $-8(-4)$ 32

EXAMPLE 2

Calculate: $(5)(-10)$

Solution $|5| = 5$ and $|-10| = 10$ Find the absolute values of each factor.

$$5 \cdot 10 = 50 \quad \text{Multiply the absolute values.}$$

The factors have different signs, so the product is negative.

$$(5) \cdot (-10) = -50$$

PRACTICE 2

Multiply: $(-5)(2)$ -10

EXAMPLE 3

Evaluate $(-7)^2$.

Solution Recall that $(-7)^2$ means $(-7)(-7)$. Because the two factors have the same sign, their product is positive. So $(-7)^2 = 49$.

PRACTICE 3

$(-1)^2 = ?$ 1

EXAMPLE 4

Simplify: -7^2

Solution $-7^2 = -(7 \cdot 7) = -(49)$

$$= -49$$

Do you see the difference in the solutions to Examples 3 and 4?

PRACTICE 4

Evaluate: -1^2 -1

> Have students explain their responses in a journal.

EXAMPLE 5

Calculate: $8(-2)(-3)$

Solution We multiply from left to right:

$$\underbrace{8(-2)}(-3) \quad \text{positive} \cdot \text{negative} = \text{negative}$$

$$= \underbrace{-16 \cdot (-3)} \quad \text{negative} \cdot \text{negative} = \text{positive}$$

$$= \quad 48$$

PRACTICE 5

Multiply: $-8(-2)(-3)$ -48

> Have students write their responses in a journal.

Comparing Example 5 and Practice 5, we note that both problems have the same absolute values but different signs. The product in Example 5 is positive because there are two negative factors. By contrast, the answer to Practice 5 is negative because there are three negative factors. Can you explain why a product is positive if it has an even number of negative factors, whereas a product is negative if it has an odd number of negative factors?

EXAMPLE 6

Find the product of $2\frac{1}{5}$ and -5.

Solution $2\frac{1}{5} \cdot 5 = \frac{11}{5} \cdot \frac{5}{1} = 11$ Multiply the absolute values.

The factors $2\frac{1}{5}$ and -5 have different signs, so the product is negative.

$$2\frac{1}{5} \cdot (-5) = -11$$

PRACTICE 6

Multiply: $(-1.4)(-0.6)$ 0.84

EXAMPLE 7

Simplify: $8 - 10(-5)^2$

Solution Use the order of operations rule.

$$8 - 10 \cdot (-5)^2$$ Square first.

$$= 8 - 10 \cdot 25$$ Then multiply.

$$= 8 - 250$$ Subtract 250.

$$= 8 + (-250)$$ Add -250.

$$= -242$$

PRACTICE 7

Calculate: $-4 + (-2)^2 \cdot 3$ 8

EXAMPLE 8

You are drilling for oil, hoping to strike it rich. Each day you drill down 20 ft farther until you hit a pool of oil, as shown. Will you reach oil by the end of the fifth day?

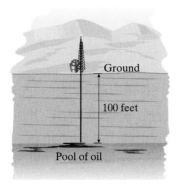

Solution Let's represent movement downward by a negative number. Each 5 days we drill 20 ft farther down, so we need to compute $5 \cdot (-20)$. Using the rule for multiplying signed numbers, we get as our answer -100. Therefore the drill will reach 100 ft *below* ground level by the fifth day—the depth of the pool of oil.

PRACTICE 8

Suppose that you are the captain of the submarine shown. While exploring the ocean, your submarine plunges 50 m/min from the surface. Will the submarine reach the ocean floor in 4 min?   No; the submarine will plunge 200 m in 4 min.

Exercises 7.4

Find the product.

1. $(2)(-5)$
-10

2. $-4 \cdot 9$
-36

3. $-80 \cdot 90$
$-7,200$

4. $(-7)(-100)$
700

5. $-5 \cdot (-5)$
25

6. $4 \cdot (-3)$
-12

7. $-34(-9)$
306

8. $8(-100)$
-800

9. $2 \cdot (-8)$
-16

10. $-1 \cdot 5$
-5

11. $907 \cdot (-9)$
$-8,163$

12. $-5 \cdot (-100)$
500

13. $5(-8)$
-40

14. $2 \cdot (-53)$
-106

15. $-88 \cdot 2$
-176

16. $20 \cdot (-30)$
-600

17. $(-200)(-4)$
800

18. $-4 \cdot (-200)$
800

19. $4 \cdot 3$
12

20. $-3 \cdot 0$
0

21. $-2 \cdot 5$
-10

22. $-1(-2)$
2

23. $(2.5)(-2)$
-5

24. $(0.3)(-0.2)$
-0.06

25. $(0.2)(-50)$
-10

26. $3 \cdot (-0.3)$
-0.9

27. $(-1.2)(-4.6)$
5.52

28. $(-0.7)(-1.8)$
1.26

29. $(5)(-1.6)$
-8

30. $(-40)(2.7)$
-108

31. $-\dfrac{1}{3} \cdot \dfrac{5}{9}$
$-\dfrac{5}{27}$

32. $\left(-\dfrac{5}{6}\right) \cdot \left(-\dfrac{2}{3}\right)$
$\dfrac{5}{9}$

33. $1\dfrac{1}{4}\left(-\dfrac{2}{3}\right)$
$-\dfrac{5}{6}$

34. $-\dfrac{1}{5} \cdot 2\dfrac{1}{2}$
$-\dfrac{1}{2}$

35. -5^2
-25

36. $(-5)^2$
25

37. $(-100)^2$
10,000

38. $(-300)^2$
90,000

39. $(-0.5)^2$
0.25

40. $(-0.4)^2$
0.16

41. $\left(-\dfrac{3}{4}\right)^2$
$\dfrac{9}{16}$

42. $\left(-\dfrac{1}{5}\right)^3$
$-\dfrac{1}{125}$

43. $(-1)^3$
-1

44. $(-4)^4$
256

45. $(9)(12)(-2)$
-216

46. $(2)(-3)(-200)$
1,200

47. $(5)(-2)(-1)(3)(-2)$
-60

48. $(-5)(-2)(-1)(3)(-2)$
60

49. $(-5)(-3)(0)$
0

50. $(-7)(0)(-10)$
0

51. $10 \cdot \left(-\dfrac{1}{2}\right) \cdot (-1)$
5

52. $\left(-\dfrac{1}{2}\right) \cdot (-4) \cdot \left(-\dfrac{1}{2}\right)$
-1

53. $\dfrac{4}{5} \cdot \left(-\dfrac{8}{9}\right) \cdot \dfrac{1}{3}$
$-\dfrac{32}{135}$

54. $-\dfrac{3}{4} \cdot \dfrac{1}{2} \cdot \left(-\dfrac{5}{7}\right)$
$\dfrac{15}{56}$

55. $(-0.308)^2$
0.094864

56. $(-7.96)^2$
63.3616

57. $(-2.64)(0.03)(-1.85)$
0.14652

58. $(5.24)(-0.18)(-2.4)$
2.26368

59. $(-3)^2 + (-4)$
5

60. $10^2 - (-5)$
105

61. $-7 + 3(-3) - 10$
-26

62. $5 - 2(-8) - (-2)$
23

63. $-3(4) + (-6)(-2)$
0

64. $8 \cdot (-2) + 3 \cdot (-1)$
-19

65. $2(-8) + 3(-4)$
-28

66. $-2 - 5(-7)$
33

67. $(-0.5)^2 + 1^2$
1.25

68. $(-0.3)^2 + 0.3^2$
0.18

69. $\dfrac{3}{5}(-10) - 6$
-12

70. $\dfrac{1}{5}(-15) + 32$
29

71. $-5 \cdot (-3 + 1.2)$
9

72. $(5 - 0.3) \cdot (-11)$
-51.7

Complete the following tables.

73.

Input	Output
a. -2	$(-3)(\mathbf{-2}) - 1 = 5$
b. -1	$(-3)(\mathbf{-1}) - 1 = 2$
c. 0	$(-3)(\mathbf{0}) - 1 = -1$
d. $+1$	$(-3)(\mathbf{+1}) - 1 = -4$
e. $+2$	$(-3)(\mathbf{+2}) - 1 = -7$

74.

Input	Output
a. -2	$(-5)(\mathbf{-2}) + 1 = 11$
b. -1	$(-5)(\mathbf{-1}) + 1 = 6$
c. 0	$(-5)(\mathbf{0}) + 1 = 1$
d. $+1$	$(-5)(\mathbf{+1}) + 1 = -4$
e. $+2$	$(-5)(\mathbf{+2}) + 1 = -9$

Applications

Solve.

75. The price of a stock that you own fell $\$\dfrac{1}{4}$ each week for 6 weeks in a row. What was the overall change in its price?   It fell $\$1\frac{1}{2}$ $\left(-\$1\frac{1}{2}\right)$.

76. On the stock market, the Dow Jones Industrial Average lost $1\dfrac{1}{8}$ points each of 5 consecutive days. What was the total loss?   It lost $5\frac{5}{8}$ points $\left(-5\frac{5}{8}\right)$.

77. On March 31, the temperature in Chicago was 40 °F. The temperature then dropped 3 °F per day for 3 days. If 32 °F is freezing on the Fahrenheit scale, was the temperature below freezing on April 3?   Yes; the actual temperature was 31 °F.

78. During a drought, the water level in a reservoir dropped 2 in./week for 6 straight weeks. How much less water was there in the reservoir at the end of this period?   12 in. less (-12 in.)

79. Two seconds after release, the height of an object is $\frac{1}{2}(-32)(2)^2$ ft below the point of release. What is this height? -64 ft; 64 ft below the point of release

80. You are taking a course in running a small business. In that course, you pretend to own a company that took out a loan to buy 12 microcomputers. If each computer cost $1,150, what was the company's debt?   The company owes $13,800 ($-$13,800).

81. In the 10 games you played this season, your team won 3 games by 2 points, won 2 games by 1 point, lost 4 games by 1 point, and tied in the final game. In these games, did your team score more or fewer points than the opposing teams?   Your team scored 4 more points than its opponents (+4 points).

82. Temperatures can be measured in both the Fahrenheit and Celsius scales. To find the Celsius equivalent of the temperature -4 °F, we need to compute $\frac{5}{9} \cdot (-4 - 32)$. Simplify this expression. -20 °C

■ *Check your answers on page A-11.*

MINDSTRETCHERS

GROUPWORK

1. Ask a partner to think of two negative numbers. Then you decide which is larger—the product of these numbers or their sum. Switch roles with your partner and repeat the exercise.   The product is larger.

CRITICAL THINKING

2. Fill in the following times table.

×	−1	3	−2
2	−2	6	−4
−3	3	−9	6
−2	2	−6	4

Verify that the nine entries sum to 0. Why is this so?   The column headings sum to 0.

WRITING

3. Describe in words two ways to evaluate −3(7 − 10).
- ■ The product of −3 and the difference between 7 and 10.

- ■ −3 multiplied by the result of subtracting 10 from 7.

7.5 Dividing Signed Numbers

OBJECTIVES

■ *To divide signed numbers*
■ *To solve word problems involving the division of signed numbers*

Now let's consider an example of the last of the four basic operations—namely, division. Suppose that you and a friend together owe $8 and you both agree to split the debt evenly. Then each of you will owe $4.

A debt is considered negative, so this problem requires us to calculate $-8 \div 2$. Recall that every division problem has a related multiplication problem.

$$\underbrace{10 \div 5 = 2}_{\text{Division}} \quad \text{because} \quad \underbrace{2 \cdot 5 = 10}_{\text{Related multiplication}}$$

We see that $-8 \div 2 = -4$ because $-4 \cdot 2 = -8$. Note that when we divide a negative number by a positive number, we get a negative quotient.

Let's look at an example in which we divide two negative numbers. Suppose that your friend owes you $8 and agrees to repay the debt in installments of $2 each. How many installments must your friend pay? The answer, of course, is 4.

This problem asks us to calculate $(-8) \div (-2)$. We know that $4 \cdot (-2) = -8$, so it follows that $(-8) \div (-2) = 4$. This example illustrates that dividing two negative numbers gives a positive quotient.

We can use the following method for dividing signed numbers.

To Divide Two Signed Numbers

■ divide their absolute values, and
■ if the numbers have the same sign, their quotient is positive; if the numbers have different signs, their quotient is negative.

EXAMPLE 1

Find the quotient: $-16 \div (-8)$

Solution First, find the absolute values.

$$|-16| = 16 \quad \text{and} \quad |-8| = 8$$

Next, divide the absolute values.

$$16 \div 8 = 2$$

The numbers have the same sign, so the quotient is positive.

$$-16 \div (-8) = 2$$

PRACTICE 1

Divide: $-24 \div (-2)$ 12

EXAMPLE 2

Simplify: $\dfrac{-8}{16}$

Solution $|-8| = 8$ and $|16| = 16$ Find the absolute values.

$\dfrac{8}{16} = \dfrac{1}{2}$ Express the quotient of the absolute values as a fraction.

Because the numbers have different signs, the answer is negative.

$\dfrac{-8}{16} = -\dfrac{1}{2}$

PRACTICE 2

Simplify: $\dfrac{9}{-15}$ $-\dfrac{3}{5}$

Remind students how to simplify fractions.

TIP   When a fraction has a negative sign in its numerator or denominator, we often rewrite the fraction so that the negative sign precedes it. For instance, we write $\dfrac{-1}{2}$ as $-\dfrac{1}{2}$ and $\dfrac{1}{-2}$ as $-\dfrac{1}{2}$.

EXAMPLE 3

$7.4 \div (-2) = ?$

Solution $|7.4| = 7.4$ and $|-2| = 2$ Find the absolute values.

$7.4 \div 2 = 3.7$ Divide the absolute values.

The numbers have different signs, so their quotient is negative.

$7.4 \div (-2) = -3.7$

PRACTICE 3

Divide: $-1.4 \div 5$ -0.28

EXAMPLE 4

Divide: $-8 \div \left(-1\dfrac{3}{5}\right)$

Solution We divide the absolute values of -8 and $-1\dfrac{3}{5}$.

$8 \div 1\dfrac{3}{5} = 8 \div \dfrac{8}{5} = 8 \times \dfrac{5}{8}$

$= 5$

The quotient of two negative numbers is positive.

$-8 \div \left(-1\dfrac{3}{5}\right) = 5$

PRACTICE 4

Find the quotient: $-\dfrac{1}{2} \div 3$ $-\dfrac{1}{6}$

EXAMPLE 5

The federal deficit in 1910 was about $20 million. Five years later, it was $60 million. How many times as great as the deficit of 1910 was that of 1915?

Solution The problem asks us to compute $-60 \div (-20)$. The quotient of numbers with the same sign is positive, so the answer is 3. That is, the 1915 deficit was three times as great as the deficit of 1910.

PRACTICE 5

Three partners in a company agree to split a debt of $16,480 evenly. How much money does each partner owe?   Each owes $5,493.33 ($-$5,493.33).

EXAMPLE 6

You invested in the stock market. During the past 4 weeks, the value of your stocks changed as follows.

Week	Change
1	$200 ↑
2	$300 ↓
3	$500 ↓
4	$100 ↑

On average, what was the weekly change in the value of your stocks?

Solution To compute the average change, we add the four changes and divide this sum by 4.

$$\frac{200 + (-300) + (-500) + 100}{4}$$

We must find the sum in the numerator before dividing by the denominator because parentheses around the numerator of a fraction are understood.

$$\frac{300 + (-800)}{4} = \frac{-500}{4} = -125$$

The average weekly change during the 4 weeks was therefore down $125.

PRACTICE 6

A young girl has a fever. The following chart shows how her temperature changed each day this week.

Monday	Up 2°
Tuesday	Up 1°
Wednesday	Down 1°
Thursday	Up 1°
Friday	Down 3°

What was the average daily change in her temperature?   0

You may want to review the order of operations rule on page 55.

Exercises 7.5

Find the quotient. Simplify.

1. $-20 \div (-4)$
  5

2. $-7 \div (-1)$
  7

3. $0 \div (-5)$
  0

4. $0 \div 3$
  0

5. $10 \div (-2)$
  -5

6. $-9 \div 3$
  -3

7. $16 \div (-8)$
  -2

8. $-12 \div 4$
  -3

9. $-250 \div (-10)$
  25

10. $-300 \div (-3)$
  100

11. $-200 \div 8$
  -25

12. $-20 \div 10$
  -2

13. $-35 \div (-5)$
  7

14. $8 \div (-4)$
  -2

15. $6 \div (-3)$
  -2

16. $-8 \div 2$
  -4

17. $-17 \div (-1)$
  17

18. $20 \div (-2)$
  -10

19. $30 \div (-6)$
  -5

20. $-40 \div (-10)$
  4

21. $\left(-\dfrac{2}{3}\right) \div \dfrac{4}{5}$
  $-\dfrac{5}{6}$

22. $\left(-\dfrac{5}{6}\right) \div \left(-\dfrac{5}{6}\right)$
  1

23. $7 \div \left(-\dfrac{1}{3}\right)$
  -21

24. $-7 \div \left(-\dfrac{1}{3}\right)$
  21

25. $-40 \div 2\dfrac{1}{2}$
  -16

26. $2\dfrac{1}{2} \div (-40)$
  $-\dfrac{1}{16}$

27. $-1.5 \div 5$
  -0.3

28. $-0.26 \div 2$
  -0.13

29. $-4 \div 0.2$
  -20

30. $9 \div (-0.6)$
  -15

31. $-4.8 \div 0.3$
  -16

32. $-2.6 \div 0.2$
  -13

33. $(-15.1214) \div (-2.45)$
  6.172

34. $-0.749 \div 0.214$
  -3.5

35. $-12.25 \div 3.5$
  -3.5

36. $50.8369 \div (-7.13)$
  -7.13

Simplify.

37. $\dfrac{-1}{5}$

$-\dfrac{1}{5}$

38. $\dfrac{-1}{-5}$

$\dfrac{1}{5}$

39. $\dfrac{-11}{-11}$

1

40. $\dfrac{-3}{-11}$

$\dfrac{3}{11}$

41. $\dfrac{2}{-5}$

$-\dfrac{2}{5}$

42. $\dfrac{1}{-2}$

$-\dfrac{1}{2}$

43. $\dfrac{-11}{-2}$

$5\dfrac{1}{2}$

44. $\dfrac{-2}{-11}$

$\dfrac{2}{11}$

45. $\dfrac{-17}{-4}$

$4\dfrac{1}{4}$

46. $\dfrac{-26}{-5}$

$5\dfrac{1}{5}$

47. $\dfrac{-3}{-4}$

$\dfrac{3}{4}$

48. $\dfrac{-7}{-8}$

$\dfrac{7}{8}$

49. $-8 \div (-2)(-2)$

-2

50. $-3(-4) \div (-2)$

-6

51. $(3-7)^2 \div (-4)$

-4

52. $(4-6)^2 \div (1-5)^2$

0.25

53. $\dfrac{2^2 - (-6)}{2}$

5

54. $\dfrac{3^2 \cdot (-4)}{-1}$

36

55. $\dfrac{2^2 + (-6)}{-2}$

1

56. $\dfrac{3^2 \cdot (-4)^2}{-1}$

-144

57. $\left(\dfrac{-8}{-2}\right)\left(\dfrac{8}{-2}\right)$

-16

58. $\dfrac{-10}{2} \cdot \dfrac{-6}{5}$

6

59. $\dfrac{-9 - (-3)}{2}$

-3

60. $\dfrac{-5 + (-7)}{2}$

-6

61. $\dfrac{3(-0.2)^2}{-2}$

-0.06

62. $\dfrac{(-16)(1.5^2)}{-1}$

36

63. $(-15) + (-3)^2 - 2 \cdot (-1)$

-4

64. $24 \div (-8) + (-5) \cdot 6$

-33

65. $(-13 - 3) \div (-2 - 6)$

2

66. $-12 \cdot 2 + (-2)^2 - (-5) \cdot 3$

-5

67. $-49 \div (-7)^2 - 4 \cdot (-3)$

11

68. $10 + (-8) \div (-4)(-5)$

0

Applications

Solve. Express each answer as a signed number.

69. The population of a certain city decreased by 60,989 in 10 yr. Find the average annual change in population.
A decrease of 6,098.9/yr (−6,098.9)

70. You lost 18 lb in 15 weeks. What was your average weekly change in weight?
You lost $1\frac{1}{5}$ lb/week ($-1\frac{1}{5}$ lb/week).

71. A company's business expenses for the year totaled $60,000. What were the average monthly expenses?
$5,000/mo in expenses (−$5,000)

72. A football running back lost 4 yd on each of several plays. His total yardage lost was 24 yd. How many plays were involved?   6 plays

73. In 3 days, the value of a share of stock dropped by $7\frac{1}{2}$ points. What was the average daily loss?   A loss of $2\frac{1}{2}$ points per day ($-2\frac{1}{2}$ points per day)

74. Over a 5-yr period, the height of a cliff eroded by 3.5 ft. By how many feet did it erode per year?   It eroded by 0.7 ft/yr (−0.7 ft).

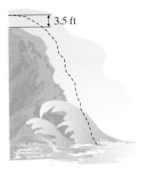
3.5 ft

75. As part of your job as a meteorologist, your boss expects you to accurately predict the average high temperature for the next 5 days. This week the high temperatures were 3°, 0°, −8°, −11°, and 1°. If your prediction for these days was −3°, was it correct?   Yes.

76. In a statistics course, you need to carry out the following computation:

$$\frac{(-0.5)^2 + (0.3)^2 + (0.2)^2}{3}$$

Find this number, rounded to the nearest hundredth.   0.13

■ *Check your answers on page A-12.*

MINDSTRETCHERS

PATTERNS

1. Find the missing numbers in the following sequence.

$$+1296, +648, -216, -108, +36, +18, -6, \underline{}, \underline{}, \underline{\phantom{\tfrac{1}{2}}}$$

GROUPWORK

2. Do the following with a partner.   Answers may vary.

- ■ Take your partner's age in years.
- ■ Square it.
- ■ Subtract 9.
- ■ Divide the result by 3 less than your partner's age.
- ■ Subtract 53.
- ■ Add your partner's age.
- ■ Divide by 2.
- ■ Add 5^2.

Verify that you wind up where you started—with your partner's age.

WRITING

3. When a certain number is divided by -2, the result is greater than 0. Was the number positive or negative? Explain.   The number was negative. Whenever the divisor is a negative number, the only possible way to get a positive quotient is with a negative dividend.

KEY CONCEPTS and SKILLS

▭ = CONCEPT   ▭ = SKILL

CONCEPT/SKILL	DESCRIPTION	EXAMPLE
[7.1] **Positive number**	A number greater than 0.	$5, \dfrac{1}{3}, 2.7$
[7.1] **Negative number**	A number less than 0.	$-5, -\dfrac{1}{3}, -2.7$
[7.1] **Signed number**	A number with a sign that is either positive or negative.	$5, -5, \dfrac{1}{3}, -\dfrac{1}{3}, 2.7, -2.7$
[7.1] **Integers**	The numbers ... , $-4, -3, -2, -1, 0, 1, 2, 3, 4, ...$ continuing indefinitely in both directions.	$+2, -2$
[7.1] **Opposites**	Two numbers that are the same distance from 0 on the number line but on opposite sides of 0.	$+2$ and -2
[7.1] **Absolute value**	The distance of a number from 0 on the number line, represented by the symbol $\mid \ \mid$.	2 units ┐ ┌ 2 units ◄─┼──┼──┼──●──┼──┼──┼──●──┼──┼─► $-4 \ -3 \ -2 \ -1 \quad 0 \quad 1 \quad 2 \quad 3 \quad 4$ $\mid-2\mid = 2, \quad \mid+2\mid = 2$
[7.1] **To compare signed numbers**	■ Locate the points that you want to compare on the number line. The number to the right is larger than the number to the left.	◄ Smaller │ Larger ► ◄─┼──┼──┼──┼──┼──┼──●──┼──┼─► $-4 \ -3 \ -2 \ -1 \quad 0 \quad 1 \quad 2 \quad 3 \quad 4$ $-2 < -1$ and $2 > -1$
[7.2] **To add two signed numbers**	■ If they have the same sign, add the absolute values and keep the sign. ■ If they have different signs, subtract the smaller absolute value from the larger and take the sign of the number with the larger absolute value.	$-0.5 + (-1.7) = -2.2$ because $\mid-0.5\mid + \mid-1.7\mid =$ $0.5 + 1.7 = 2.2$ $3\dfrac{1}{2} + (-9) = -5\dfrac{1}{2}$ because $\mid-9\mid > \left\mid +3\dfrac{1}{2}\right\mid$ and $9 - 3\dfrac{1}{2} = 5\dfrac{1}{2}$
[7.3] **To subtract two signed numbers**	■ Change the operation of subtraction to addition, and change the number being subtracted to its opposite. ■ Follow the rule for addition.	$-2 - (-5) = -2 + 5 = +3,$ or 3
[7.4] **To multiply two signed numbers**	■ Multiply their absolute values. ■ If the numbers have the same sign, their product is positive; if the numbers have different signs, their product is negative.	$(-8)\left(-\dfrac{1}{2}\right) = +4, \quad$ or $\quad 4$ $-0.2 \times 4 = -0.8$
[7.5] **To divide two signed numbers**	■ Divide their absolute values. ■ If the numbers have the same sign, their quotient is positive; if the numbers have different signs, their quotient is negative.	$\dfrac{-8}{-4} = +2, \quad$ or $\quad 2$ $18 \div (-2) = -36$

Chapter 7 Review Exercises

[7.1]

Find the opposite signed number.

1. +5
−5

2. −4
4

3. −5$\frac{1}{2}$
5$\frac{1}{2}$

4. 10.1
−10.1

Find the absolute value.

5. $|-8|$
8

6. $|+2.5|$
2.5

7. $\left|-1\frac{1}{5}\right|$
1$\frac{1}{5}$

8. $|12|$
12

Circle the larger number.

9. −4 and −2$\frac{1}{4}$
−2$\frac{1}{4}$

10. 9 and −5$\frac{1}{3}$
9

11. −8 and −22.5
−8

12. −6.75 and 2
2

Arrange the numbers in each group from smallest to largest.

13. −8, 8, −3.5
−8, −3.5, 8

14. 9, −6, −9.7
−9.7, −6, 9

15. −2$\frac{1}{2}$, 0, −2.9
−2.9, −2$\frac{1}{2}$, 0

16. −4, −1$\frac{1}{4}$, 0
−4, −1$\frac{1}{4}$, 0

[7.2]

Find the sum.

17. −10 + (−10)
−20

18. 8 + (−10)
−2

19. −5$\frac{1}{2}$ + 12
6$\frac{1}{2}$

20. −$\frac{1}{4}$ + $\left(-\frac{3}{4}\right)$
−1

21. 0.9 + (−5)
−4.1

22. −1.2 + (−0.8)
−2

23. −8 + 5$\frac{1}{2}$ + (−4)
−6$\frac{1}{2}$

24. 12 + (−12) + $\left(-\frac{1}{4}\right)$
−$\frac{1}{4}$

[7.3]

Find the difference.

25. −10 − (−10)
0

26. 14 − (−14)
28

27. 5 − 15
−10

28. −2 − 9
−11

29. 2.5 − (−0.5)
3

30. −$\frac{1}{8}$ − 4
−4$\frac{1}{8}$

[7.4]

Find the product.

31. $-10 \times (-10)$
100

32. -15×3
-45

33. $\dfrac{-2}{-3} \times \left(\dfrac{+10}{-11}\right)$
$-\dfrac{20}{33}$

34. $3.5 \times (-2.1)$
-7.35

35. $\left(\dfrac{1}{4}\right)^2$
$\dfrac{1}{16}$

36. $(-3.1)^2$
9.61

[7.5]

Find the quotient.

37. $\dfrac{-14}{-14}$
1

38. $20 \div (-4)$
-5

39. $-2 \div 8$
-0.25

40. $-5 \div 2$
-2.5

41. $-\dfrac{1}{8} \div (-4)$
$\dfrac{1}{32}$

42. $15 \div (-0.3)$
-50

[7.2–7.5]

Simplify.

43. $-8 - (-3) + 20$
15

44. $12 \cdot (-3)^2 - (-6)$
114

45. $(-7 + 3) \cdot (-5)^2$
-100

46. $(20 - 30) \div (-10)$
1

47. $\dfrac{(-9.1)(-0.6)}{2}$
2.73

48. $\dfrac{-8 - 5.1}{5}$
-2.62

49. $10^2 + \dfrac{-8 - 2}{2}$
95

50. $\dfrac{10}{2} - (5 - 9)^2$
-11

Mixed Applications

Solve.

51. The Chou dynasty ruled China between 1027 B.C. and 256 B.C. The philosopher Confucius was born in about 551 B.C. and died in about 479 B.C. Was the Chou dynasty in power throughout Confucius's lifetime? (*Source: Asian History on File*)   Yes

52. Buddha was born in 563 B.C. and died in 483 B.C. Was he alive in 500 B.C.? (*Source: Compton's Encyclopedia*)
Yes

53. You invest in the stock market, buying 200 shares of a stock. Unfortunately, the value of the stock drops, and you lose $1.25 for every share that you own. What's your total loss?   A loss of $250 ($-$250)

54. In your last 4 months on a diet, you have lost 4 lb, lost 3 lb, gained 1 lb, and lost 1 lb, respectively. What was your total change in weight during that period of time?
A loss of 7 lb ($-$7 lb)

55. Last week, a cold wave hit Chicago, resulting in a temperature of -2 °F. At the same time, it was 83 °F in Honolulu. How much warmer was Honolulu than Chicago?   85 °F warmer

56. You pay off a debt of $1,000 in 5 equal installments. How much is each installment? $200 per installment

57. A submarine can descend 100 feet per minute. How many minutes will it take to descend 650 ft?   6.5 min

58. Two of the most influential math books in history were *The Elements*, which Euclid wrote in 323 B.C., and *The Principia*, which Isaac Newton wrote in 1687 A.D. How many years apart were these books written? (***Source***: *Notable Mathematicians from Ancient Times to the Present*) 2,010 years apart

59. For each of the last 2 yr, your company made $20,000. This year, however, it lost $10,000. What was your company's average profit or loss over the 3-yr period? An average profit of $10,000

60. On your first day of scuba diving, you dove to a depth of 30 ft below the surface of the sea. If on the next day you dove to a depth 3 times as great, how deep did you dive on that day?   90 feet below the surface (-90 ft)

61. Golf scores are commonly expressed as over or under *par*—the number of expected swings on each hole. To *birdie* a hole is to take 1 less hit than par, to *eagle* is to take 2 fewer hits than par, and to *bogie* is to take 1 more hit than par. With 3 birdies, 2 eagles, 1 par, and 2 bogies, how far over or under par are you altogether? 5 under par

62. Physicists have shown that, if an object is thrown upward at a speed of 100 feet per second, its location after 5 seconds will be

$$-16 \times 5^2 + 100 \times 5$$

feet above the point at which the object was thrown. How far above or below that point will the object be at that time?   100 ft above the point of release ($+100$ ft)

■ *Check your answers on page A-12.*

Chapter 7 Posttest

To see if you have mastered the topics in this chapter, take this test.

1. Which is smaller, -10 or -4?
-10

2. A number has an absolute value of $\frac{1}{2}$ and is negative. What is the number? $-\frac{1}{2}$

Evaluate.

3. $-8 + 8$ 0

4. $4.5 + (-5)$ -0.5

5. $42 - 91$ -49

6. $-12 - (-12)$ 0

7. -23×9 -207

8. -0.5×0.2 -0.1

9. $(-12)^2$ 144

10. $\left(-\dfrac{1}{4}\right)^2$ $\dfrac{1}{16}$

11. $-1.8 \div (-0.9)$ 2

12. $-4 \div \dfrac{1}{2}$ -8

13. $-4 + 6 + (-7) + 9$ 4

14. $15 - (-7) + (-1)$ 21

15. $-8 - 4^2 \cdot (-3)$ 40

16. $(2 - 8)^2 \div (-2)$ -18

Solve.

17. Which is colder: a temperature of $-2\,°C$ or a temperature of $0\,°C$? $-2\,°C$

18. You lost $88\frac{1}{2}$ lb on a diet in 15 mo. Find your average monthly loss, expressed as a signed number.   You lost $5\frac{9}{10}$ lb/mo $\left(-5\frac{9}{10}\text{ lb/mo}\right)$

19. In your savings account, you had a balance of $370. You then made a deposit of $100 and two withdrawals of $75 and $80. What was your new balance, expressed as a signed number? $+\$315$

20. Chlorine boils at $-34.6\,°C$ and melts at $-100.98\,°C$. How much higher is the boiling point than the melting point? (**Source:** *Handbook of Chemistry & Physics*) $66.38\,°C$

■ *Check your answers on page A-12.*

Cumulative Review Exercises

To help you review, solve the following.

1. Round 2,891 to the nearest thousand.   3,000

2. Estimate: $86.66 \div 1.7$ Possible answer: 45

3. Multiply: $(4)\left(2\dfrac{1}{2}\right)$ 10

4. Add: $8 + 2.1 + 3.9$ 14

5. Express $16\dfrac{2}{3}\%$ as a fraction. $\dfrac{1}{6}$

6. What percent of 2.5 is 0.5?   20%

7. Solve for *n*: $\dfrac{1.4}{7} = \dfrac{13}{n}$ $n = 65$

Solve.

8. Last year State University awarded twenty-five $11,000 scholarships, and City University awarded twenty $15,000 scholarships. Which institution spent more money on scholarships? How much more?   City University spent $25,000 more.

9. You can type a 5-page sociology paper in 55 min. At that rate, how long will it take you to type a 15-page paper? Express your answer in hours.   2.75 hr

10. When mortgage rates dropped, the number of housing starts rose from 4,000 to 5,000. What percent increase is this?   25%

■ *Check your answers on page A-12.*

C H A P T E R **8**

Basic Statistics

STATISTICS AND THE LAW

Increasingly lawyers use statistical evidence to win their cases.

Statistics on the distribution of blood and hair types in the general population are commonly used as evidence in physical assault and robbery trials. Where plaintiffs claim that they are suffering from exposure to a toxic agent, their lawyers often present statistical evidence about the general incidence of their illness. And cases of race and sex discrimination frequently focus on such statistics as the proportion of people who are admitted or hired, or the average length of time that employees have spent in positions before being promoted.

This use of statistics in the U.S. legal system goes back to a landmark nineteenth century trial, wherein the claim was made that the signature of Sylvia Howland on her will was forged. The turning point of the case was the testimony of an expert witness — a Harvard mathematician — who developed a system of statistically analyzing the degree of similarity among 42 signatures of the deceased. On the basis of these statistics, he testified that the signature on the will was unreasonably similar to another of Ms. Howland's from which it had probably been traced.

Chapter 8 Pretest

To see if you have already mastered the topics in this chapter, take this test.

1. Find the range: 10, 2, 2, 5, 11   9

2. In an accounting office, three employees made an annual income of $19,000, two made $27,000, and one made $55,000. What was the median income of these six employees? $23,000

3. The director of a personnel office kept records on the number of absences for the 20 employees who work for the company. During the past 2 mo each was absent the following number of times.

$$2 \quad 0 \quad 0 \quad 4 \quad 1 \quad 3 \quad 1 \quad 0 \quad 1 \quad 0$$
$$0 \quad 1 \quad 1 \quad 3 \quad 9 \quad 2 \quad 1 \quad 0 \quad 3 \quad 0$$

What was the mean number of absences?   1.6

4. In a year that is not a leap year, what is the mode of the number of days in a month? (*Reminder*: February has 28 days; April, June, September, and November have 30 days; and January, March, May, July, August, October, and December have 31 days.)   31

5. Late in the spring term, your grades were: Spanish I (3 credits) — A; Music (2 credits) — A; Social Science (4 credits) — C; and Physical Education (1 credit) — B. The grades are assigned the following points: A = 4, B = 3, C = 2, D = 1, and F = 0. Calculate your GPA.   3.1

In each case use the given table or graph to answer the question.

6. The following mortality table gives estimates for the life spans of individuals (in years), taking into account such factors as year of birth and gender. (*Source:* U.S. National Center for Health Statistics)

Year of Birth	1920	1930	1940	1950	1960	1970	1980	1990
Male	53.6	58.1	60.8	65.6	66.6	67.1	70.0	71.8
Female	54.6	61.6	65.2	71.1	73.1	74.7	77.5	78.8

A female born in 1950 is expected to live how much longer than a male born in 1950?
5.5 yr longer

7. The following bar graph shows the distance (in feet) that various cars need in order to stop when they are traveling at 70 mph.

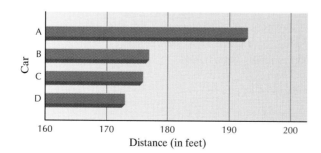

About how many feet does car D need in order to stop?   Possible answer: 173 ft

8. The following line graph shows the number of males for every 100 females among various age groups of Americans in a recent year.

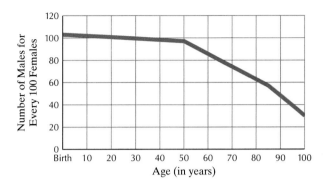

At age 50, are there more males or females?   There are more females (100) than males (approximately 97).

9. The following graph shows the opening values of two stocks, A and B, for each work day last year.

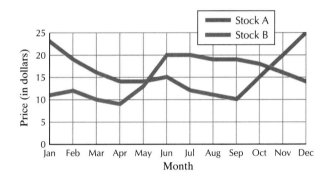

In September, which stock was worth more?   Stock B

10. The federal government conducted a survey to find out the sources of income for older Americans. The results are depicted in the pie chart.

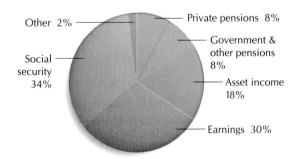

Does a larger percent of the respondents' income come from earnings or from pensions? (*Source: Business Week*)   Earnings account for a larger percent of their income (30%) than pensions (16%).

■ *Check your answers on page A-12.*

HISTORICAL NOTE

A seventeenth-century English clothing salesman named John Graunt had the insight to apply a numerical approach to major social problems. In 1662, he published a book entitled *Natural and Political Observations upon the Bills of Mortality*, and so founded the science of statistics.

Graunt was curious about the periodic outbreaks of the bubonic plague in London, and his book analyzed the number of deaths in London each week due to various causes. He was the first to discover that, at least in London, the number of male births exceeded the number of female births. He also found that there was a higher death rate in urban areas than in rural areas, and that more men than women died violent deaths. Graunt summarized large amounts of information to make it understandable, and made conjectures about large populations based on small samples. Graunt was also a pioneer in examining expected life span — a statistic which became vital to the insurance companies formed at the end of the seventeenth century.

Sources:

Morris Kline, *Mathematics, A Cultural Approach* (Reading, Massachusetts: Addison-Wesley Publishing Company, 1962), p. 614.

F.N. David, *Games, Gods and Gambling* (NY: Hafner Publishing Company, 1962).

8.1 Introduction to Basic Statistics

OBJECTIVES

- *To find the mean, median, and mode(s) of a list of numbers*
- *To find the range of a list of numbers*
- *To interpret these statistics and use them to solve word problems*

What Basic Statistics Is and Why It Is Important

Statistics is the branch of mathematics that deals with ways of handling large quantities of information. The goal is to make this information easier to interpret.

With unorganized data, spotting trends and making comparisons is difficult. The study of statistics teaches you how to organize data in various ways in order to make the data more understandable.

One approach is to calculate special numbers, also called statistics, which describe the data. In this section, we consider four statistics: the mean, the median, the mode, and the range.

Another approach to organizing data is to display this information in the form of a table or graph. We discuss tables and graphs in the other section of this chapter.

Many situations lend themselves to the application of statistical techniques. Wherever there are large quantities of information — from sports to business — statistics can help us to find meaning where, at first glance, there seems to be none.

Averages

We begin our introduction to statistics by considering averages again. Previously, (see page 56), we discussed the meaning of average. There we defined the average of a list of numbers to be the sum of the numbers divided by however many numbers are on the list. This statistic, which is more precisely called the *arithmetic mean*, or just the **mean**, is what most people think of as the average.

A second average, the **median**, may describe the numbers better than the mean when there is an unusually large or unusually small number on the list to be averaged. The third average, the **mode**, has a special property — unlike the mean and the median, it is always on the list of numbers being averaged.

In this section, we first look at more examples of the mean.

Mean

EXAMPLE 1	PRACTICE 1
In his eight games against Cincinnati, a baseball pitcher threw the following number of strikeouts. 0 1 3 1 0 2 2 1 In his five games against Los Angeles, his record of strikeouts was as follows. 2 0 1 2 1 On average, did he throw more strikeouts against Cincinnati or against Los Angeles?	The five textbooks that you bought cost the following amounts. $25.70 $31.70 $12.50 $42.50 $31.75 Was the cost of the $31.70 book above the average cost for the books? Yes, the cost was $2.87 above average.

continued

Solution We need to compare the mean for the number of strikeouts in the games against Cincinnati with the corresponding mean for the number of strikeouts against Los Angeles. Against Cincinnati, the mean is:

$$\frac{0 + 1 + 3 + 1 + 0 + 2 + 2 + 1}{8} = \frac{10}{8} = \frac{5}{4} = 1.25$$

Against Los Angeles, the mean is:

$$\frac{2 + 0 + 1 + 2 + 1}{5} = \frac{6}{5} = 1.2$$

Because $1.25 > 1.2$, he threw more strikeouts on the average against Cincinnati than against Los Angeles.

Another kind of mean, called the **weighted average**, is used when some numbers on the list count more heavily than others. For instance, suppose that you want to compute the average of your test scores in a class and that the final exam counts twice as much as any of the other tests. Or suppose that you are computing your grade point average (GPA), and some courses carry more credits than others.

EXAMPLE 2

Last term, your grades were as follows.

Course	Credits	Grade	Grade Equivalent
Psychology	4	A	4
English	4	C	2
Art	3	B	3
Physical Education	1	B	3

Compute your GPA for the term.

Solution To calculate your GPA, you first multiply the number of credits each course carries by the numerical grade equivalent received. You then add these products to find the total number of grade points. Finally, you divide this sum by the total number of credits.

┌ Number of credits for the first course
│ ┌ Grade equivalent of the first course

$$\text{GPA} = \frac{4(4) + 4(2) + 3(3) + 1(3)}{12}$$

└ Total number of credits

$$= \frac{16 + 8 + 9 + 3}{12} = \frac{36}{12} = 3$$

Your GPA of 3 is exactly equivalent to a B.

PRACTICE 2

Suppose that you scored 90 on a final exam that counts as two quizzes. On each of the three remaining quizzes, you got an 80. If you wanted an average of at least 85, did you achieve your goal? No, the average was 84.

Median

Definition

On a list of numbers arranged in numerical order, the **median** is the number in the middle. If there are two numbers in the middle, the median is the mean of the two middle numbers.

EXAMPLE 3	PRACTICE 3

Find the median.

a. 6, 8, 2, 1, and 5   **b.** 6, 8, 2, and 5

Solution

a. We arrange the numbers from smallest to largest.

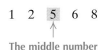

1   2   **5**   6   8
↑
The middle number

So the median is 5. If we arrange the numbers from largest to smallest, we get the same answer.

The middle number
↓
8   6   **5**   2   1

The median is still 5.

b. We order the numbers from smallest to largest.

2   5   6   8

Since four numbers are on the list and four is even, no single number is in the middle. In this case, the median is the mean of the two middle numbers.

Two middle numbers
↓
2   **5   6**   8
↑
$$\frac{5 + 6}{2} = 5.5$$

So 5.5 is the median.

Compute the median.

a. 7, 2, 8, 5, 10, 7, 9, 10, 2, 5, 8, and 6   7

b. 0, 4, 1, 5, 7, 2, 5, 9, and 3   4

EXAMPLE 4

In a recent year, the ages of the nine Supreme Court justices were 54, 85, 73, 63, 69, 57, 57, 45, and 60. (*Source: World Almanac*)

a. What was the median age of the nine justices?

b. Suppose that a year later, the same justices, each 1 yr older, were on the Court. Show whether the median age increased by 1 yr.

Solution

a. To compute the median age of the justices, let's first arrange their ages in increasing order.

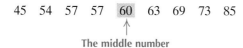

45   54   57   57   **60**   63   69   73   85

↑
The middle number

Because 60 is the middle number, the median age was 60.

b. A year later, the ages of the justices were as follows, again in increasing order.

46   55   58   58   **61**   64   70   74   86

↑
The middle number

The median is 61, so the median age did increase by 1 yr.

PRACTICE 4

At the college bookstore, you buy six textbooks at the following prices.

$20.99 $57.49 $68.75 $125
$35.98 $145.50

a. Find the median price of these textbooks.
$63.12

b. Later you buy a seventh textbook at $69.95. Is the cost of this textbook above or below the median of the other six textbooks?   Above the median

Mode

> **Definition**
>
> The **mode** is the number (or numbers) occurring most frequently on a list of numbers.

EXAMPLE 5

Compute the mode(s).

a. 8, 6, 10, 8, 10, 8, 9, and 6

b. 2, 9, 3, 5, 7, 12, 3, 2, 18, 12, 2, and 3

Solution

a. When we count how often each number occurs on this list, we see that 6 occurs twice, 8 occurs three times, 9 once, and 10 twice. Because there are more 8s than any other number, 8 is the mode.

b. Here 2 occurs three times, 3 occurs three times, 5 once, 7 once, 9 once, 12 twice, and 18 once. So both 2 and 3, occurring most frequently, are modes.

PRACTICE 5

Find the mode(s).

a. 7, 2, 5, 1, 2, 5, and 2   2

b. 9, 1, 0, 4, 9, 4, 1, 5, 9, and 4   4 and 9

> If you want to discuss bimodal data, use this example as an illustration.

EXAMPLE 6

The minimum age for a regular driver's license varies among the 50 states: one state has age 15, seventeen states have age 16, one state has age $16\frac{1}{2}$, four states have 17, twenty-five states have age 18, and two states have age 21. What is the mode of these minimum ages?
(*Source:* Federal Highway Administration)

Solution The most common minimum age is 18 because it occurs more frequently (25 times) than any other age. So 18 is the mode of the minimum ages.

PRACTICE 6

The marriage laws of the states set the minimum age for getting married. With parental consent, the minimum age for women is as follows: one state has age 12, one state has age 13, five states have age 14, two states have age 15, thirty states have age 16, four states have 17, and five states have age 18. (Two states have no minimum age.) Find the mode of these ages. (*Source: World Almanac 2000*)   16

Range

The last statistic that we consider is called the **range**. The range is not an average because it does not represent a typical number on a list. Instead, the range is a measure of the spread of the numbers on the list.

Definition

The **range** of a list of numbers is the difference between the largest and the smallest number on the list.

EXAMPLE 7

Find the range of the numbers 3, 13, 2, 5, 9, and 2.

Solution The largest number on the list is 13, and the smallest is 2. So the range is $13 - 2$, or 11.

PRACTICE 7

What is the range of 8, 10, 3, and 8?   7

EXAMPLE 8

In History your test scores were 80, 82, 80, 86, and 100. In French your scores were 72, 95, and 79. In which class was the spread of your test scores greater?

Solution The highest History score was 100, and the lowest score was 80. The range of History scores was $100 - 80$, or 20. Similarly, the range of French scores was $95 - 72$, or 23. The greater range of the French scores indicates a larger spread.

PRACTICE 8

Among the guests at a party, there were three 21-year-olds, two 19-year-olds, one 20-year-old, and one 25-year-old. If the range of the ages of the guests at your previous party was 7, at which party did the ages of your guests have a greater spread?
The greater spread was at the previous party; the range at the later party was 6.

Exercises 8.1

Compute the indicated statistics. Round to the nearest tenth, where necessary.

1.

Numbers	Mean	Median	Mode(s)	Range
a. 8, 2, 9, 4, 8	6.2	8	8	7
b. 3, 0, 0, 3, 10	3.2	3	0 and 3	10
c. 6.5, 9, 8.5, 6.5, 8.1	7.7	8.1	6.5	2.5
d. $3\frac{1}{2}, 3\frac{3}{4}, 4, 3\frac{1}{2}, 3\frac{1}{4}$	$3\frac{3}{5}$	$3\frac{1}{2}$	$3\frac{1}{2}$	$\frac{3}{4}$
e. 4, −2, −1, 0, −1	0	−1	−1	6

2.

Numbers	Mean	Median	Mode(s)	Range
a. 5, 3, 5, 5, 3	4.2	5	5	2
b. 8, 0, 7, 5, 0	4	5	0	8
c. 2.1, 2.6, 2.4, 2.5, 2.4	2.4	2.4	2.4	0.5
d. $4\frac{1}{2}, 3\frac{3}{4}, 4, 4\frac{1}{2}, 4\frac{1}{4}$	$4\frac{1}{5}$	$4\frac{1}{4}$	$4\frac{1}{2}$	$\frac{3}{4}$
e. −1, −3, −3, −2, −2	−2.2	−2	−3 and −2	2

3. Calculate the mean, rounded to the nearest cent.
   $9,125.88 $11,724.87 $12,705 $11,839.75 $13,500.79 $14,703.71
   $12,266.67

4. Find the mean, rounded to the nearest foot, of the following measurements.
   3,725 ft   3,719 ft   3,740 ft   3,726 ft   3,729 ft   3,734 ft   3,725 ft
   3,728 ft

Applications

Solve and check.

5. Here were your grades last term: A in College Skills (2 credits), B in World History (4 credits), C in Music (2 credits), A in History (3 credits), and B in Physical Education (1 credit). Did you make the Dean's List, which requires a GPA of 3.5? Explain. (*Reminder*: A = 4, B = 3, C = 2, and D = 1.) No; your GPA was 3.25

6. You are in a class of 20 students. On a test, 9 students earned 80, 10 students earned 70, and you earned 75. Was your grade below the class average (mean), exactly average, or above the class average? Explain. The grade was below the class average of 74.75.

7. A woman leaves $1,000,000 to her 10 heirs. What is the mean amount left to each heir? Can you compute the median amount with the given information? **Explain.** The mean amount is $100,000. We can't compute the median amount because we don't know the actual amounts given to each heir.

8. An airline advertised that the median time for one of its daily flights is 45 min. For the past 7 days, the flight took 70 min, 38 min, 42 min, 45 min, 52 min, 71 min, and 40 min. Does this support the airline's claim? Explain. Yes it does; 45 is the middle number.

9. In the U.S. House of Representatives, 435 members of Congress represent the 50 states. If 10 members represent a state, is its representation below average, average, or above average? The number of this state's representatives is above average; the average is 8.7.

10. You and four other people are riding in an elevator. Two are taller than you, and two are shorter. Who has the median height of the people in the elevator? You do.

11. In a recent year, the numbers of over-the-air TV stations in the dozen biggest U.S. cities were as follows.

7 14 11 8 11 7 19 9 7
14 10 9

a. Find the mode of these numbers. 7

b. What is their range? 12

12. A used car dealer sold the following number of cars in the past 10 days.

1 5 3 2 1 0 4 2 6 1

a. What is the mode of daily car sales? 1

b. Find the range. 6

13. The diameters for the nine planets of the solar system, rounded to the nearest 1,000 mi, are as follows.

Planet	Miles (in thousands)
Mercury	3
Venus	8
Earth	8
Mars	4
Jupiter	89
Saturn	75
Uranus	32
Neptune	31
Pluto	1

(*Source: Encyclopedia Americana*)

Find each of the following distances, rounded to the nearest 1,000 mi.

a. Mean diameter   28,000 mi   **b.** Median diameter   8,000 mi

c. Mode(s) of the diameters   8,000 mi   **d.** Range of the diameters   88,000 mi

14. Suppose that the following were your utility bills for the past 10 mo.

Jan	Feb	Mar	Apr	May	Jun	Jul	Aug	Sep	Oct
$90	$80	$90	$70	$100	$110	$140	$140	$100	$90

Find each of the following amounts.

a. Mean bill $101   **b.** Median bill $95

c. Mode(s) of the bills $90   **d.** Range of the bills $70

■ *Using a calculator, solve each problem, giving (a) the operation(s) carried out in your solution, (b) the exact answer, and (c) an estimate of the answer.*

15. In 1980—when the U.S. population was 226 million—the U.S. Postal Service handled some 106 billion pieces of mail. Ten years later, when the population was 248 million, the U.S. Postal Service handled about 166 billion pieces of mail. On average, how many more pieces of mail did a resident of the United States receive in 1990 than in 1980? (*Source:* Bureau of Census)   (a) Division, subtraction; (b) 200.32; (c) possible answer: 250

16. Your rent went up last year. For each of the first 4 mo, it was $635.88. For the remaining 8 mo, it was $699.47/mo. Find the mean amount of your monthly rent.
(a) Multiplication, addition, division; (b) $678.27; (c) possible answer: $700

■ *Check your answers on page A-12.*

MINDSTRETCHERS

CRITICAL THINKING

1. Suppose that a billionaire walks into your classroom. Which will increase more, the mean income or the median income of people in the room? Explain.   Assuming that the other people in the room are not billionaires, the mean income will increase more than the median will. There will be a large increase in the total income and therefore a large increase in the mean income. The median is the middle income or the mean of the two incomes in the middle after ranking. The new median income will be either one of the incomes of the nonbillionaires or the mean of two of their incomes. So the change in median will be less than the change in mean.

Graphs

Now let's discuss displaying data in the form of graphs. We deal with three kinds of graphs: bar graphs, line graphs, and circle graphs.

Bar Graphs

On a **bar graph**, quantities are represented by thin, parallel rectangles called bars. The length of each bar is proportional to the quantity that it represents.

On some graphs, the bars extend to the right. On others, they extend upward or downward. Sometimes, bar lengths are labeled. Other times, bar lengths are read against an *axis* — a straight line parallel to the bars and similar to a number line.

Bar graphs are especially useful for making comparisons or contrasts among a few quantities, as the following example illustrates.

EXAMPLE 3	PRACTICE 3
The following graph shows a company's net income for the 4 quarters of last year.	The following graph shows the number of people who died as a result of hurricanes in the United States during various years.

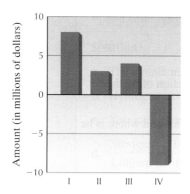

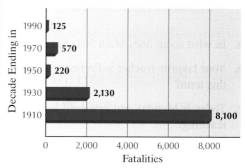

(*Source:* U.S. National Oceanic and Atmospheric Administration)

EXAMPLE 3

a. What was the net income for the company in the first quarter of last year?

b. How much money did the company lose in the fourth quarter?

c. Estimate the total net income of the company for the year.

Solution

a. About $8 million.

b. About $9 million.

c. About $8 + 3 + 4 + (-9)$, or $6 million. Your answer may differ somewhat from this answer because reading this graph involves estimating bar lengths.

PRACTICE 3

a. In which decade were there the fewest fatalities from hurricanes?   The decade ending in 1990

b. Approximately what was the ratio of fatalities in the decade ending in 1910 to the fatalities in the decade ending in 1930?   4 to 1

c. In a sentence, describe the story that you think this graph is telling. Possible answer: The number of fatalities as a result of hurricanes in the United States declined by 7,975 from 1910 to 1990.

A **pictograph** is a variation on the bar graph. Instead of bars, pictographs use images of people, books, coins, and so on, to represent quantities.

Pictographs are visually appealing. However, they make it difficult to distinguish between small differences — say, between a half and a third of a quantity.

EXAMPLE 4

The following pictograph shows the number of robots installed in four car factories.

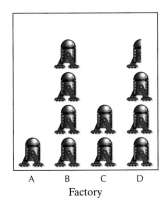

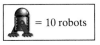 = 10 robots

Factory

a. What does the symbol mean in the *legend* at the right of the graph?

b. How many more robots are in factory B than in factory A?

c. About how many robots have been installed in factory D?

Solution

a. According to the legend, the symbol represents 10 robots.

b. Factory B has 40 robots, and factory A has 10 robots. So factory B has 30 more robots than factory A.

c. In factory D about $30 + \frac{1}{2} \cdot 10$, or 35, robots have been installed.

PRACTICE 4

The following pictograph shows takeoffs and landings at four busy airports in a recent year.

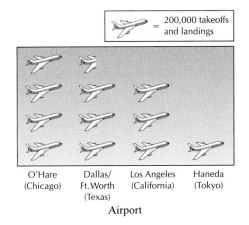

= 200,000 takeoffs and landings

Airport

a. What does each 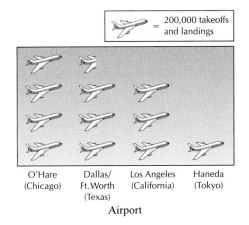 represent?
200,000 takeoffs and landings

b. About how many more takeoffs and landings were there at Dallas/Ft. Worth than at Haneda?

c. Los Angeles's takeoffs and landings were about what percent of Chicago's takeoffs and landings?   Los Angeles had 600,000 and Chicago had 800,000. Therefore Los Angeles had about 75% as many takeoffs and landings as Chicago.

b. Dallas/Ft. Worth had about 700,000, and Haneda had 200,000. Therefore Dallas/Ft. Worth had about 500,000 more takeoffs and landings than Haneda.

EXAMPLE 5

The following graph describes the living arrangements of elderly American men and women (aged 65 or older) in a recent year.

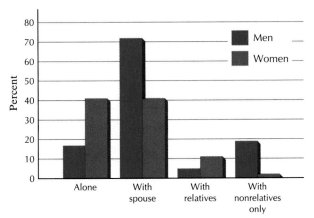

(*Source:* U.S. Census Bureau)

a. Is the percent of elderly men who live alone more or less than the percent of elderly women who live alone? About what percent difference?

PRACTICE 5

The following bar graph shows the number of vehicles that a company sold in the months of January, February, and March of one year.

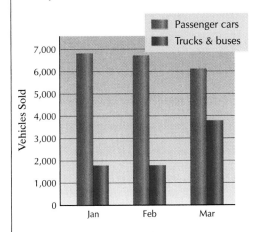

continued

b. Suppose that there are 20 million elderly men. To the nearest million, about how many of these men live with their spouse?

Solution

a. A larger percent of elderly women live alone (about 41%) than men (about 17%). The difference is 41% − 17%, or 24%.

b. The percent of men living with their spouse is about 72%, and 72% of 20,000,000 is 14,400,000. To the nearest million, about 14 million men live with their spouse.

a. About how many more trucks and buses were sold in March than in February? Approximately 2,000 more trucks and buses were sold in March than in February.

b. In January, about how many more passenger cars than trucks and buses were sold? About 5,000 more passenger cars than trucks and buses were sold in January.

Line Graphs

Line graphs represent quantities as points connected by straight line segments. The height of any point on a line is read against the vertical axis.

Line graphs, also called **broken-line graphs**, are commonly used to highlight changes and trends over a period of time. Especially when we have data for many points in time, we are more likely to use a line graph than a bar graph.

EXAMPLE 6

The following line graph shows a measles epidemic.

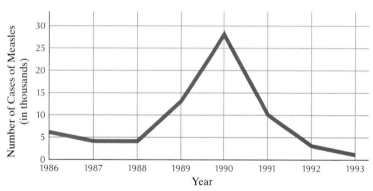

(*Source:* U.S. Center for Disease Control and Prevention)

a. In which year did the epidemic peak?

b. About how many cases of measles were there in 1989?

c. Between 1990 and 1993, the line drops to the right. Explain what this means in terms of the number of cases of measles.

Solution

a. The epidemic peaked in 1990.

b. In 1989, there were about 13,000 cases of measles.

c. The dropping line represents a decline each year in the number of cases of measles.

PRACTICE 6

The following graph shows the value of a share of stock for the past 5 days.

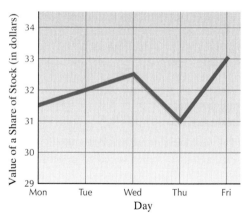

a. About what was a share worth on Wednesday?   About $32.50

b. By about how much did the value of a share decline between Wednesday and Thursday?   The value declined by about $1.50.

c. By about how much did the value of a share increase from Monday to Friday?   The value increased by about $1.50.

Some line graphs compare two or more changing quantities, as Example 7 illustrates.

| **EXAMPLE 7** | **PRACTICE 7** |

A law school surveyed its graduates to determine how old they were when they received their law degrees. The ages of the students at the time of graduation are shown in the graph.

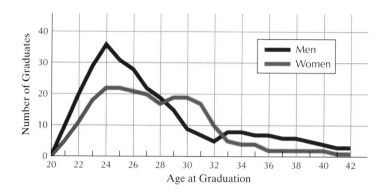

a. About how many women graduated at age 30?

b. At what ages did more women than men graduate?

c. Among 20-year-olds, did more men or women graduate?

Solution

a. At age 30, about 19 women graduated.

b. For ages 29 through 32, more women than men graduated.

c. Neither — no one aged 20, either man or woman, graduated.

Two candidates — Cardoso and da Silva — competed in an election campaign. The following graph shows their changing popularity.

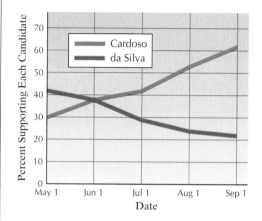

a. About what percent of the public supported Cardoso on May 1?   About 30% supported Cardoso.

b. When was the difference in popularity between Cardoso and da Silva greatest?   September 1

c. What happened in the campaign from June 1 through July 1?   During the period from June 1 to July 1 support shifted from da Silva to Cardoso.

Circle Graphs

Circle graphs are commonly used to show how a whole amount — say, an entire budget or population — is broken into its parts. The graph resembles a pie, representing the whole, that has been cut into slices, representing the parts.

Each slice (or *sector*) is proportional in size to the part of the whole that it represents. Each slice is appropriately labeled with either its actual count or the percent of the whole that it represents.

Example 8 on the next page illustrates how to read and interpret the information given by a circle graph.

EXAMPLE 8

The following graph shows the sources of energy for a particular state.

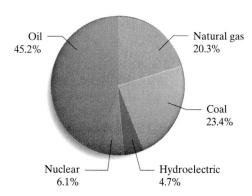

a. What percent of the energy is hydroelectric, to the nearest whole percent?

b. Nearly half the energy comes from which source?

c. Verify that the sources of energy sum to 100%, aside from any difference caused by rounding.

Solution

a. 4.7% of the energy is hydroelectric; rounding gives 5%.

b. Oil provides 45.2%, or nearly half, the energy.

c. Adding the percents corresponding to the graph sectors, we get 45.2% + 20.3% + 23.4% + 4.7% + 6.1% = 99.7%, which is close to 100%. The difference presumably is due to rounding.

PRACTICE 8

The following graph shows the amounts (in $US billions) that U.S. citizens invested in different parts of the world during a recent year.

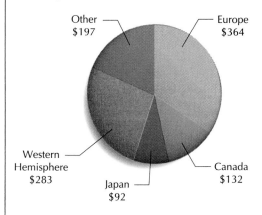

a. How much money did U.S. citizens invest in Europe? $364,000,000,000

b. By how much did investments in the Western Hemisphere exceed those in Other? $86,000,000,000

c. What was the ratio of investments in Canada to investments in Japan? $\frac{33}{23}$

Exercises 8.2

Solve.

1. The following table gives the area (in thousands of square miles) of the five largest islands in the world.

Island	Area
Australia	2,940
Greenland	840
New Guinea	300
Borneo	290
Madagascar	230

(*Source:* U.S. Census Bureau)

a. Is the area of Australia more or less than 1 million sq mi?   More than; the actual area is 2,940,000 sq mi

b. About how many times as large as Borneo is Greenland?   Possible answer: 2.8 times

2. In a recent Winter Olympics, the times for some competitors in the men's downhill alpine skiing were as follows.

Skier	Time
Norwegian	1 min, 45.79 sec
U.S.	1 min, 45.75 sec
French	1 min, 46.09 sec
Austrian	1 min, 46.01 sec
Canadian	1 min, 45.87 sec

(*Source: The New York Times*)

a. What was the nationality of the skier who took the least time?   U.S.

b. What was the nationality of the skier who took the most time?   French

3. The following table shows in various years the average age of U.S. citizens who married for the first time.

Year	Average Age of Bride	Average Age of Groom
1900	21.9	25.9
1940	21.5	24.3
1980	22.0	24.7

(*Source:* U.S. National Center for Health Statistics)

a. In 1980, what was the average age of a groom?   24.7 years old

b. In 1900, what was the difference between the average age of a groom and the average age of a bride?   4 yr

c. What was the difference between the average age of a bride in 1900 and in 1940?   0.4 yr

d. Were grooms, on average, older than brides in all three years?   Yes, they were.

4. An atlas contains a table of road distances (in miles) between various U.S. cities.

	Los Angeles	Chicago	Houston
Los Angeles	—	2,112	1,556
Chicago	2,112	—	1,092
Houston	1,556	1,092	—

(*Source: Rand McNally Road Atlas*)

a. According to this table, how far is Los Angeles from Houston?   1,556 mi

b. Chicago is how much closer to Houston than to Los Angeles?   1,020 mi

c. What do the blanks in the table mean?   The blanks mean zero (there is no distance between a city and itself).

5. The following table shows how to determine a stockbroker's commission in any stock transaction. The commission depends on both the number of shares sold and the price per share.

Price per Share	**Number of Shares**				
	1–100	200	300	400	500
$1–$20	$40	$50	$60	$70	$80
>$20	$40	$60	$80	$90	$100

a. What is the broker's commission on a sale of 300 shares of stock at $15.75 a share? $60

b. What is the commission on a sale of 500 shares of stock at $30 a share? $100

c. Will you pay your broker a lower commission if you sell 400 shares of stock in a single deal or 200 shares of stock in each of two deals?   You will pay a lower commission if you sell 400 shares in a single deal; 400 shares in a single deal will cost you $70 or $90, whereas 2 deals of 200 shares will cost you $100 or $120.

6. At a college, the math courses that students take depend on their scores on the Arithmetic and Algebra Placement Test. Test scores get translated into course placements as follows.

		Arithmetic Score		
		0–10	11–15	16–20
Algebra Score	0–10	Math 1 and then Math 3	Math 2 and then Math 3	Math 3
	11–15	Math 1 and then Math 3	Math 4	Math 3
	16–20	Math 1	Math 2	Math 5

a. If you score 12 in arithmetic and 8 in algebra, what should you take?   Math 2 and then Math 3

b. Suppose that you score 8 in arithmetic and 12 in algebra. What is your placement?   Math 1 and then Math 3

c. What must you score to take Math 5?   You must score 16–20 in both algebra and arithmetic.

7. The following bar graph, reflecting data from a recent year, shows the percent of government officials in various positions who were women.

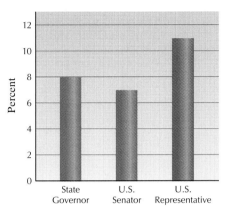

(*Source: Statistical Handbook on Women in America*)

a. What percent of the state governors were women?   8%

b. In the U.S. House of Representatives, about what fraction of the members were women? $\dfrac{11}{100}$

c. Of the 100 U.S. Senators, about how many were women?   7

8. The following graph shows the number of daily servings of various food groups recommended for women.

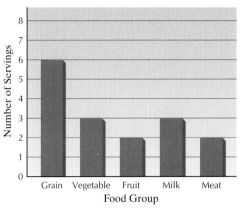

(*Source:* U.S. Department of Agriculture)

a. For which food group is the largest number of servings recommended?   Grain

b. How many servings of fruit are recommended?   Two servings of fruit are recommended.

c. How many more servings of grain than of meat are recommended?   Four more servings

9. The following pictograph shows the number of households in a particular city with telephones, in various years.

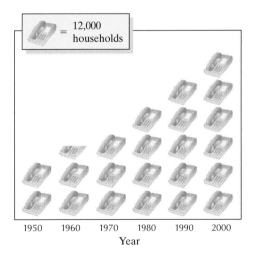

a. How many households in the city had telephones in 1950?   24,000 households

b. About how many households had telephones in 1960?   Possible answer: 30,000

c. Estimate in what year 48,000 households had telephones.   1980

10. The pictograph shows the number of physician's assistants who worked in a clinic in selected years.

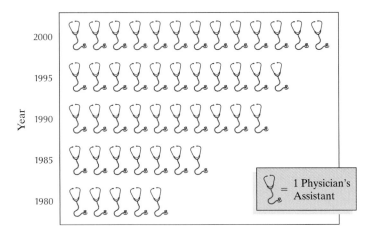

a. How many physician's assistants worked in the clinic in 1985?   7

b. Suppose that 5 doctors worked in the clinic in 1990. How many more physician's assistants than doctors worked in the clinic then?   There were 5 more physician's assistants than doctors in 1990.

c. Describe in words the trend that this graph shows.   Every 5 yr, the number of physician's assistants working in the clinic increased.

11. The following graph shows the amount of water that a reservoir contained last year.

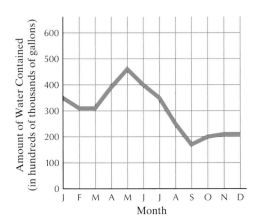

a. Throughout January, was there less than 400,000 gallons of water in the reservoir?
Yes; there was between 300,000 and 350,000 gallons of water in the reservoir throughout January.

b. In which month did the reservoir contain the least water?   September

c. If the capacity of the reservoir is 600,000 gal, in which month was the reservoir closest to overflowing?   May

12. The following graph shows the number (in thousands) of U.S. engineers in various specialties during a recent year.

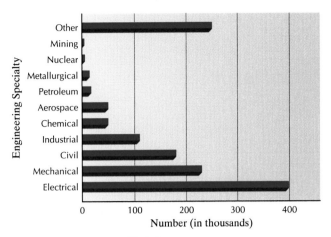

(*Source:* U.S. Department of Labor)

a. Which engineering specialty is the most popular?   Electrical engineering

b. Approximately how many engineers were there altogether?   Possible answer: 1,300,000

13. A human child and a chimp were raised together. Scientists graphed the number of words that the child and the chimp understood at different ages.

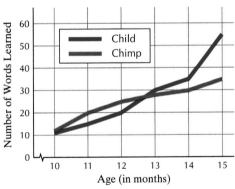

(*Sources:* A.H. Kritz, *Problem Solving in the Sciences;* W.N. Kellogg and L.A. Kellogg, *The Ape and the Child*)

a. At about what age was the child's vocabulary first better than that of the chimp?   Approximately 13 months

b. At age 15 months, about how many more words did the child know than the chimp?   Possible answer: 20 more words

14. The following graph shows how often a hospital used three kinds of painkillers — aspirin, ibuprofen, and acetaminophen.

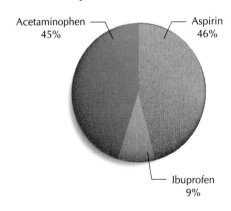

a. What percent of the time was aspirin used?   46%

b. How many times as often was acetaminophen used as ibuprofen?   5 times

15. The following graph shows how a typical dollar was spent by one city government in a recent year.

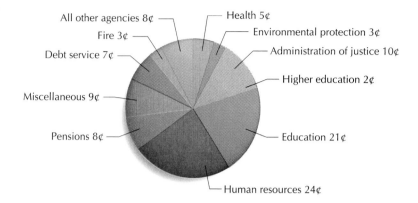

a. How much of each dollar spent by the city went for environmental protection?   3¢

b. Did education and higher education together account for nearly 25% of the city's expenses?   Yes; education and higher education accounted for 23¢ of every dollar (23%).

■ *Check your answers on page A-13.*

MINDSTRETCHERS

WRITING

1. Consider the following table, which shows the number of hurricanes by month for the period 1886–1991.

Month	Jan–Apr	May	Jun	Jul	Aug	Sept	Oct	Nov	Dec
Number	1	3	23	36	150	189	96	22	3

(*Source:* J.B. Elsner and A.B. Karon, *Hurricanes of the North Atlantic*)

Write the steps that must have led to creating this table.   1. Tally for each year the number of hurricanes in each month. 2. Add the tallies for all Januaries, for all Februaries, etc. 3. Make January through April an interval because there was only one hurricane for those months. Then create a table and enter the data.

CRITICAL THINKING

2. The following graph shows the number of games in which a baseball player got various numbers of hits.

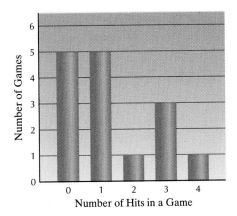

Identify the mode(s) of the number of hits in a game for this player.   0 and 1

MATHEMATICAL REASONING

3. Consider the following three formats for representing the same data—the number of Democrats and Republicans in the Senate during a recent year.

Party	Number of Senators
Democrats	55
Republicans	45

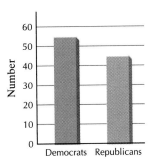

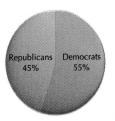

Which format do you prefer? Explain why.   Answers may vary.

KEY CONCEPTS

$\boxed{}$ = CONCEPT

CONCEPT	DESCRIPTION	EXAMPLE
[8.1] **Mean**	Given a list, the sum of the numbers divided by however many numbers are on the list.	For 0, 0, 1, 3, and 5 the mean is $$\frac{0 + 0 + 1 + 3 + 5}{5} = \frac{9}{5} = 1.8$$
[8.1] **Median**	Given a list of numbers arranged in numerical order, the number in the middle. If there are two numbers in the middle, the mean of the two middle numbers.	For 0, 0, 1, 3, and 5, the median is 1.
[8.1] **Mode**	The number (or numbers) occurring most frequently on a list.	For 0, 0, 1, 3, and 5, the mode is 0.
[8.1] **Range**	The difference between the largest number and the smallest number on a list of numbers.	For 0, 0, 1, 3, and 5, the range is $5 - 0$, or 5.
[8.1] **Table**	A rectangular display of data consisting of rows and columns.	Column $\downarrow$ Row → <table><tr><td></td><td>Dems.</td><td>Reps.</td></tr><tr><td>Men</td><td>37</td><td>45</td></tr><tr><td>Women</td><td>14</td><td>15</td></tr></table>
[8.2] **Bar graph**	A graph in which quantities are represented by thin, parallel rectangles called bars. The length of each bar is proportional to the quantity that it represents.	*Decade Ending in:* 1990: 125; 1970: 570; 1950: 220; 1930: 2,130; 1910: 8,100 (horizontal axis: Fatalities, 0 to 8,000)
[8.2] **Pictograph**	A variation on the bar graph in which images of people, books, coins, and so on represent the quantities.	= 10 robots (Factory A, B, C, D)

continued

CONCEPT	DESCRIPTION	EXAMPLE
[8.2] **Line graph**	A graph in which quantities are represented as points connected by straight line segments. The height of any point on a line is read against the vertical axis.	
[8.2] **Circle graph**	A graph that resembles a pie, representing a whole amount, that has been cut into slices, representing the parts of the whole.	

Chapter 8 Review Exercises

To help you review this chapter, solve these problems.

[8.1]

Compute the desired statistic for each list of numbers.

1.

List of numbers	Mean	Median	Mode	Range
a. 6, 7, 4, 10, 4, 5, 6, 8, 7, 4, 5	6	6	4	6
b. 1, 3, 4, 4, 2, 3, 1, 4, 5, 1	2.8	3	1 and 4	4

Mixed Applications

2. The following table shows how long the first five American presidents and their wives lived.

President	Age	President's Wife	Age
George Washington	67	Martha Washington	70
John Adams	90	Abigail Adams	74
Thomas Jefferson	83	Martha Jefferson	34
James Madison	85	Dolley Madison	81
James Monroe	73	Eliza Monroe	62

(*Source: Presidents, First Ladies, and Vice Presidents*)

a. Using the median as the average, did the husbands or the wives live longer?
The husbands (83 yr) lived longer than the wives (70 yr).

b. By how many years?   13 yr

3. According to the census of 1990, the median age in the United States was 32.9. (*Source:* U.S. Census Bureau)

a. Explain what this statement means.   Half of the people were younger than 32.9 and half were older.

b. In 1990, by how many years was your age above or below average?   Answers will vary.

4. A soda machine is considered reliable if the range of the amounts of soda that it dispenses is less than 2 fluid ounces (fl oz). In 10 tries, a particular machine dispensed the following amounts (in fl oz).

8.1   7.8   8.6   8.1   8.4   7.8   8   7.7   6.9   8.4

Is the machine reliable? Explain.   This machine is reliable because the range is 1.7 fl oz.

5. According to its producer, a Broadway show would make a profit if it averaged at least 1,000 paying customers a night. For the past 10 nights, the number of paying customers, rounded to the nearest hundred, was as follows.

900   700   1,500   800   1,100   800   700   1,600   800   1,100

Using the mean as the average, determine whether the show was making a profit.
The average was 1,000; therefore the show was making a profit.

6. A newspaper sports section contained the following baseball table.

Team	This Season			In the Last 10 Games		At Home This Season		Away This Season	
	Won	**Lost**	**Pct.**	**W**	**L**	**W**	**L**	**W**	**L**
Detroit	75	51	.595	6	4	39	23	36	28
Toronto	76	53	.589	6	4	38	23	38	30
New York	71	56	.594	6	4	38	20	33	36
Milwaukee	70	58	.547	8	2	38	27	32	31
Boston	61	66	.480	6	4	41	24	20	42
Baltimore	59	70	.457	6	4	26	36	33	34
Cleveland	49	80	.380	3	7	29	38	20	40

a. How many games has Milwaukee played so far this season?   128 games

b. Which team has the highest season winning percentage?   Detroit

c. Which team has the highest winning percentage for the last 10 games?   Milwaukee

d. How can you tell that there is a mistake in the information on the Cleveland team?   The total of losses at home and away (78) is not equal to the total losses for the season (80).

7. In a recent year, the amount of cargo (in tons) that the five busiest U.S. ports handled was as follows.

Port	Cargo (in tons)		
	Total	**Domestic**	**Foreign**
Port of South Louisiana	189,374,451	100,252,774	89,121,677
Houston, Texas	131,233,876	62,943,241	68,290,635
New York, New York	126,860,940	83,082,893	43,778,047
Valdez, Alaska	99,616,158	99,576,371	39,787
Baton Rouge, Louisiana	87,630,227	49,341,472	38,288,755

(*Source: World Almanac 2000*)

a. To the nearest 10,000 tons, how much domestic cargo did Baton Rouge handle?   49,340,000 tons

b. Which of these five ports handled the least foreign cargo?   Valdez, Alaska (39,787 tons)

c. To the nearest 10,000 tons, how much more domestic cargo did the Port of South Louisiana handle than New York?   17,170,000 tons

d. Which of these ports handled more domestic than foreign cargo?   All except Houston, Texas

8. The following graph shows the number of motion picture theaters in a city in various years.

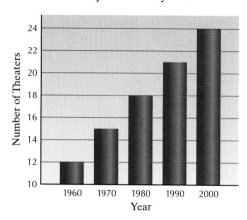

a. About how many motion picture theaters were there in the city in 1990? 21 theaters

b. Describe the trend that this graph illustrates. Every 10 yr the number of theaters has increased. Since 1970, the number has increased by 3 every 10 yr.

c. On the basis of this trend, about how many theaters do you predict there will be in the year 2010? 27

10. Consider the following learning curve that shows how long a rat running through a maze takes on each run.

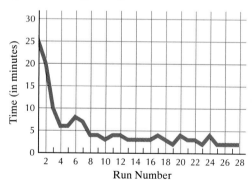

a. On which run does the rat run through the maze in 10 min? Run number 3

b. How long does the rat take on the tenth run? Approximately 3 min

c. What general conclusion can you draw from this learning curve? With practice, the rat ran through the maze more quickly.

9. The following pictograph shows the number of successful launches from a missile range in various months.

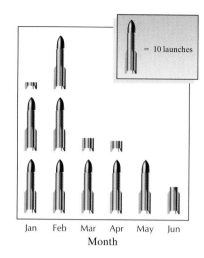

a. In January, February, and March, about how many total launches were there? Possible answer: 65 launches

b. In which month were there the most launches? February

c. Describe the trend that this pictograph shows. The number of launches peaked in February and then declined.

11. The following graph shows where immigrants to the United States came from during the 1980s.

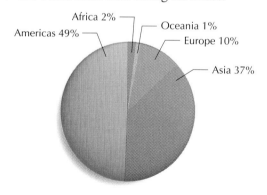

(*Source:* U.S. Immigration and Naturalization Service)

a. What percentage of immigrants came from Europe? 10%

b. Did more than $\frac{1}{3}$ of the immigrants come from Asia? Yes; 37% of the immigrants came from Asia, which is more than $\frac{1}{3}$ (33.3%).

c. If there were about 7 million immigrants altogether, approximately how many came from the Americas? Approximately 3.5 million immigrants

12. The following graph shows the population of New York City and the rest of New York State for the 50 yr between 1940 and 1990.

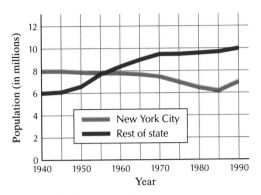

(*Source:* U.S. Census Bureau)

a. What was the approximate population of New York City in 1950?   8 million

b. In 1990, what was the approximate ratio of the population of the rest of the state to that of New York City? Possible answer: $\frac{10}{7}$

c. Express in words the trend that this graph is illustrating. Since 1940, the population of New York City had stayed almost constant (8 million) until the period between 1970 and 1980 when it fell by approximately 1 million and then rebounded. Over the same time period (1940–1990) the population for the rest of the state had increased from 6 million to 10 million.

13. Recently a national study concluded that trash was composed of the following types of materials.

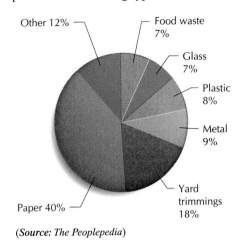

(*Source: The Peoplepedia*)

a. Which single type of trash is the largest?   Paper

b. Glass constitutes what fraction of trash? $\frac{7}{100}$

c. What percent of trash is either plastic or metal?   17%

■ *Check your answers on page A-13.*

Chapter 8 Posttest

To see whether you have mastered the topics in this chapter, take this test.

1. Find the mode: 10 8 6 9 5 3 4 8 5 1 3 0 6 5 5

2. To test the exhaust fumes from a car, a government inspector took 10 samples of its exhaust. The samples contained the following amounts of toxic gas in parts per million (ppm).

$$7 \quad 4 \quad 6 \quad 5 \quad 8 \quad 4 \quad 6 \quad 7 \quad 5 \quad 7$$

If the maximum allowable mean is 6 ppm, did the car pass the test? Explain.   Yes it did, because the mean was 5.9 ppm.

3. Your annual salary during each of the past 10 years was the following.

$9K $12K $15K $18K $25K $16K $13K $12K $17K $17K

(K represents 1,000.) What was the range of these salaries? $16,000

4. The following table shows the number of people unemployed in the United States in selected years.

Year	Unemployed (in millions)
1940	8
1950	3
1960	4
1970	4
1980	8
1990	7

(*Source:* U.S. Bureau of Labor Statistics)

In these years, what was the median number of unemployed?   5,500,000

5. Late in a term, your grades were as follows: English Composition I (4 credits) — A; Freshman Orientation (1 credit) — A; College Skills (2 credits) — C; and Physical Education (1 credit) — B. If A = 4, B = 3, C = 2, D = 1, and F = 0, calculate your GPA.   3.375

Use the given table or graph to answer each question.

6. To compare the hourly wages for various job titles, members of two railroad unions (Metro-North and L.I.R.R.) prepared the following table.

Job Title	Metro-North	L.I.R.R.
Foreman	$20.67	$26.33
Inspector	21.00	20.17
Engineer	23.89	22.37
Conductor	21.57	22.37
Bartender	15.33	13.46

To the nearest cent, how much more was the hourly wage of an L.I.R.R. conductor than a Metro-North bartender? $7.04

7. The following graph pictures the enrollment at a college for various quarters.

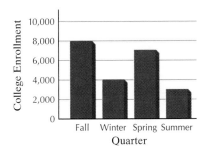

About how many students were enrolled during the summer quarter?   Approximately 3,000 students

8. The following line graph is based on the data from a recent year, showing the number of U.S. children between the ages of 18 and 40 living with their parents.

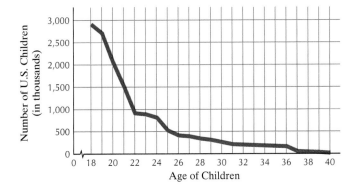

About how many more 20-year-olds than 30-year-olds lived with their parents? Approximately 1,750,000; there were roughly 2,000,000 20-year-olds and 250,000 30-year-olds.

9. You made a circle graph showing costs associated with a recent vacation.

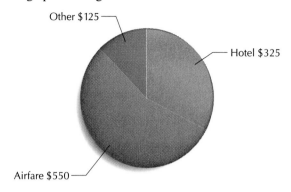

Other $125

Hotel $325

Airfare $550

What fraction of your vacation expenses went to paying the airfare? $\frac{11}{20}$

10. The following graph shows the number of Nobel Prizes awarded in the sciences for research conducted in the United States and in the United Kingdom.

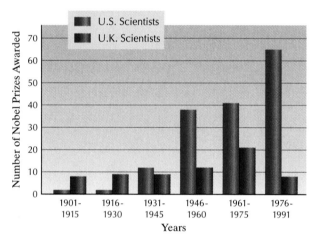

(*Source: Who's Who of Nobel Prize Winners*)

In which period of time did scientists in the United States receive about twice as many prizes as scientists in the United Kingdom? 1961–1975; U.S. scientists received approximately 40 and U.K. scientists received approximately 20 Nobel Prizes.

■ *Check your answers on page A-13.*

Cumulative Review Exercises

To help you review, solve the following.

1. Round to the nearest hundred dollars: $3,384 $3,400

2. Simplify: $\dfrac{20}{25}$ $\dfrac{4}{5}$

3. Estimate the following product: $\left(8\dfrac{1}{10}\right)\left(4\dfrac{9}{10}\right)$ 40

4. $(3.01)(1,000) = ?$ 3,010

5. What is $12\dfrac{1}{2}\%$ of 16?   2

6. Solve for x: $x - 7\dfrac{1}{4} = 10$ $x = 17\dfrac{1}{4}$

7. Solve for n: $\dfrac{7}{10} = \dfrac{n}{30}$ $n = 21$

Solve.

8. The Great Pyramid of Khufu — the last surviving wonder of the ancient world — was built around 2680 B.C. To the nearest thousand years, how long ago was this pyramid built? (*Source: The Concise Columbia Encyclopedia*)   5,000 yr

9. In a recent year, $2\dfrac{1}{2}$ million students graduated from high school. By contrast, 1 million students graduated from college. The number of high school graduates was how many times as large as the number of college graduates, expressed as a percent?   250%

10. A newspaper surveyed 400 scientists, asking which of various types of locations they would prefer to live near. The graph below shows their responses. About how many more scientists chose to live near an airport than near a chemical plant?
64 more scientists

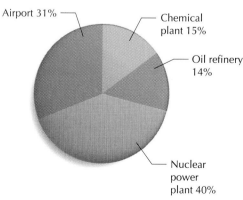

Airport 31%

Chemical plant 15%

Oil refinery 14%

Nuclear power plant 40%

■ *Check your answers on page A-13.*

More on Algebra

EQUATIONS IN ACCOUNTING

Accountants commonly compute financial quantities — say tax or profit — using a computer program called a *spreadsheet*.

Like a piece of accounting paper, a spreadsheet has the appearance of a grid, with rows and columns and numbers in the cells where the rows and columns intersect. For example, A3 is the cell at the intersection of the A column and the third row.

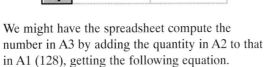

	A	B
1	128	
2	A2	
3	A3	
4		

We might have the spreadsheet compute the number in A3 by adding the quantity in A2 to that in A1 (128), getting the following equation.

$$A3 = 128 + A2$$

This equation shows the relationship between the variables A2 and A3.

Chapter 9 Pretest

To see if you have already mastered the topics in this chapter, take this test.

Solve and check.

1. $y + 8 = 2$ $y = -6$

2. $x - 6 = -8$ $x = -2$

3. $-6x = 24$ $x = -4$

4. $-1 = \dfrac{a}{5}$ $a = -5$

5. $3x + 1 = 10$ $x = 3$

6. $4 = 5y - 1$ $y = 1$

7. $-2c + 3 = 7$ $c = -2$

8. $\dfrac{a}{2} - 7 = -12$ $a = -10$

9. $\dfrac{w}{3} + 1 = -4$ $w = -15$

10. $2x + 3x = 10$ $x = 2$

11. $2x - 4 = x - 5$ $x = -1$

12. $4x = 2(x - 4)$ $x = -4$

13. The length, l, of a woman's radius bone — the longest bone in the arm — is about $\dfrac{1}{7}$ of her height, h. Translate this relationship to a formula. (*Source:* Colin Evans, *Casebook of Forensic Detection*) $l = \frac{1}{7}h$

14. Given the formula $S = \dfrac{n(n + 1)}{2}$ for the sum S of the first n whole numbers, find the value of S when $n = 13$. $S = 91$

Write an equation for each word problem. Solve and check.

15. You worked 4 hr last Saturday. Your boss paid you $45 for the day, which included a $5 tip. How much money did you earn per hour? $45 = 4x + 5$, where x is your hourly rate; $x = \$10/\text{hr}$

16. A disc jockey charges $250 to provide 3 hr of music for a party. For each additional hour, he charges $75. If he received $475 for working at a graduation party, how many hours did he work? $475 = 250 + 75x$, where $x = $ additional hours worked; $x = 3$, so he worked a total of 6 hr

17. Each used book at a library sale sold for the same amount. You bought 6 short-story and 8 mystery books. Then you bought a college poster for $5. If you spent $19 in all, how much did each book cost? $14x + 5 = 19$, where $x = $ cost of 1 book; $x = \$1$

18. Three times the sum of a number and 7 is 30. Find the number. $3(n + 7) = 30$; $n = 3$

19. Lumber is frequently sold by the board foot, where the number of board feet, B, for a piece of lumber is $\dfrac{1}{12}$ the product of the lumber's thickness, t, width, w, and length, l. (The thickness and width are given in inches, and the length is given in feet.)

  a. Express this relationship as a formula. $B = \dfrac{1}{12}twl$

  b. How many board feet would a 2-in. $\times$ 4-in. $\times$ 20 ft piece of lumber contain? $B = 13\frac{1}{3}$ board ft

20. In marketing, the sale price, s, equals the regular price, r, minus the discount price, d.

  a. Write a formula for this relationship. $s = r - d$

  b. Find the sale price of an item when $r = \$400$ and $d = \$99$. $s = \$301$

■ *Check your answers on page A-13.*

9.1 Solving Two-Step Equations

OBJECTIVES

- *To solve simple equations involving signed numbers*
- *To solve equations involving two steps*
- *To solve word problems involving equations with signed numbers or two steps*

Solving Simple Equations Involving Signed Numbers

In Chapter 4, we solved simple equations such as the following.

$$x + 3 = 5, \quad x - 3 = 7, \quad 3x = 6, \quad \text{and} \quad \frac{x}{4} = 3$$

In this chapter, we extend the discussion to include equations that involve

- both positive and negative numbers or

- more than a single operation.

Recall that in solving simple equations the key is to isolate the variable, that is, to get the variable alone on one side of the equation.

Definition

A **solution** to an equation is a value of the variable that makes the equation a true statement. To **solve** an equation means to find all solutions of the equation.

We have already discussed rules for solving addition, subtraction, multiplication, and division equations involving positive numbers. Let's now apply these rules to simple equations involving both positive and negative numbers.

EXAMPLE 1

Solve and check: $y + 5 = 3$

Solution $y + 5 = 3$

$y + 5 - 5 = 3 - 5$ Subtract 5 from each side of the equation.

$y + 0 = -2$ $\quad 5 - 5 = 0$ and
$\quad\quad\quad\quad\quad 3 - 5 = 3 - (+5) = 3 + (-5) = -2$

$y = -2$ $\quad y + 0 = y$

Check: $y + 5 = 3$

$-2 + 5 \overset{?}{=} 3$ Substitute -2 for y in the original equation.

$3 \overset{\checkmark}{=} 3$

The solution is -2. Note that, because 5 was added to y in the original equation, we solved the equation by subtracting 5 from each side of the equation in order to "undo" the addition.

PRACTICE 1

Solve and check: $x + 7 = 4 \quad x = -3$

Have students review subtracting signed numbers on page 359.

Have students review adding signed numbers on page 349.

EXAMPLE 2

Solve and check: $n - 5 = -11$

Solution $n - 5 = -11$

$n - 5 + 5 = -11 + 5$ Add 5 to each side of the equation.

$n + 0 = -6$

$n = -6$

Check: $n - 5 = -11$

$-6 - 5 \overset{?}{=} -11$ Substitute -6 for n in the original equation.

$-11 \overset{\checkmark}{=} -11$

The solution is -6.

PRACTICE 2

Solve and check: $m - 7 = -19$ $m = -12$

EXAMPLE 3

Solve and check: $-7x = 21$

Solution $-7x = 21$

$\dfrac{-7x}{-7} = \dfrac{21}{-7}$ Divide each side of the equation by -7.

$x = -3$ $\dfrac{-7x}{-7} = 1x = x$ and $\dfrac{21}{-7} = -3$

Check: $-7x = 21$

$-7(-3) \overset{?}{=} 21$ Substitute -3 for x in the original equation.

$21 \overset{\checkmark}{=} 21$

The solution is -3.

PRACTICE 3

Solve and check: $9y = -18$ $y = -2$

> Have students review how to divide signed numbers on page 373.

> Have students review how to multiply signed numbers on page 365.

EXAMPLE 4

Solve and check: $-1 = \dfrac{x}{6}$

Solution $-1 = \dfrac{x}{6}$

Multiply each side of the equation by 6:

$-1 \cdot 6 = \dfrac{x}{6} \cdot 6$ $\dfrac{x}{6} \cdot 6 = \dfrac{x}{6} \cdot \dfrac{6}{1} = \dfrac{x}{1} = x$

$-6 = x,$ or $x = -6$

Check: $-1 = \dfrac{x}{6}$

$-1 \overset{?}{=} \dfrac{-6}{6}$ Substitute -6 for x in the original equation.

$-1 \overset{\checkmark}{=} -1$

The solution is -6.

PRACTICE 4

Solve and check: $\dfrac{y}{-3} = -2$ $y = 6$

EXAMPLE 5

Suppose that a contractor lost $80 an hour on a job because of bad weather. How many hours of work did the contractor lose if the loss totaled $1,600?

Solution We can represent a loss of $80 an hour as $-\$80$ and a loss of $1,600 as $-\$1,600$. To find the number of hours lost, we write the following equation, letting h represent the number of hours lost.

$$-80h = -1,600$$

We solve this equation for h.
$$-80h = -1,600$$

$$\frac{-80h}{-80} = \frac{-1,600}{-80}$$

$$h = 20$$

Check: $-80h = -1,600$

$$-80(20) \overset{?}{=} -1,600$$

$$-1,600 \overset{\checkmark}{=} -1,600$$

Thus the contractor lost 20 hr of work.

PRACTICE 5

A company lost 2% of its income to store theft. What was the company's income if these losses amounted to $50,000?

$2,500,000

Solving Equations Involving Two Steps

We now turn our attention to solving equations that involve more than one operation. Solving equations such as the following requires two steps.

$$3y + 5 = 26 \qquad \text{and} \qquad \frac{c}{3} - 4 = 7$$

To solve a two-step equation
- first use the rule for solving addition or subtraction equations and
- then use the rule for solving multiplication or division equations.

> Point out to students that when solving equations they reverse (or "undo") the order of operations rule. When evaluating expressions use multiplication/division first and then addition/subtraction. When solving equations use addition/subtraction first and then multiplication/division.

EXAMPLE 6

Solve and check: $3y + 5 = 26$

Solution $3y + 5 = 26$

$3y + 5 - 5 = 26 - 5$ Subtract 5 from each side of the equation.

$3y = 21$

$\dfrac{3y}{3} = \dfrac{21}{3}$ Divide each side of the equation by 3.

$y = 7$

PRACTICE 6

Solve and check: $2x + 8 = -6$ $x = -7$

continued

Check: $3y + 5 = 26$

$3(7) + 5 \stackrel{?}{=} 26$ Substitute 7 for y in the original equation.

$21 + 5 \stackrel{?}{=} 26$

$26 \stackrel{\checkmark}{=} 26$

The solution is 7.

EXAMPLE 7	**PRACTICE 7**

Solve and check: $\dfrac{c}{3} - 4 = 7$

Solve and check: $\dfrac{k}{5} - 5 = -3$ *k = 10*

Solution $\dfrac{c}{3} - 4 = 7$

$\dfrac{c}{3} - 4 + 4 = 7 + 4$ Add 4 to each side of the equation.

$\dfrac{c}{3} = 11$

$\cancel{3} \cdot \dfrac{c}{\cancel{3}} = 3 \cdot 11$ Multiply each side of the equation by 3.

$c = 33$

Check: $\dfrac{c}{3} - 4 = 7$

$\dfrac{33}{3} - 4 \stackrel{?}{=} 7$ Substitute 33 for c in the original equation.

$11 - 4 \stackrel{?}{=} 7$

$7 \stackrel{\checkmark}{=} 7$

The solution is 33.

> Have students explain their answers in a journal.

     Note that in Example 6 we solved by subtracting before dividing, and in Example 7, we added before multiplying. Can you think of any other way to solve these equations?

EXAMPLE 8	**PRACTICE 8**

Solve and check: $1 - 2x = 5$

Solve and check: $8 - 3d = -4$ *d = 4*

Solution $1 - 2x = 5$

$1 - 1 - 2x = 5 - 1$ Subtract 1 from each side of the equation.

$-2x = 4$

$\dfrac{-2x}{-2} = \dfrac{4}{-2}$ Divide each side of the equation by -2.

$x = -2$

Check: $1 - 2x = 5$

$1 - 2(-2) \overset{?}{=} 5$ Substitute −2 for x in the original equation.

$1 + 4 \overset{?}{=} 5$

$5 \overset{\checkmark}{=} 5$

The solution is −2.

EXAMPLE 9

Solve and check: $\dfrac{a}{2} + 7 = 5$

Solution $\dfrac{a}{2} + 7 = 5$

$\dfrac{a}{2} + 7 - 7 = 5 - 7$ Subtract 7 from each side of the equation.

$\dfrac{a}{2} = -2$

$\dfrac{a}{2}(2) = -2(2)$ Multiply each side of the equation by 2.

$a = -4$

Check: $\dfrac{a}{2} + 7 = 5$

$\dfrac{-4}{2} + 7 \overset{?}{=} 5$ Substitute −4 for a in the original equation.

$-2 + 7 \overset{?}{=} 5$

$5 \overset{\checkmark}{=} 5$

The solution is −4.

PRACTICE 9

Solve and check: $\dfrac{x}{8} + 7 = -9$
$x = -128$

EXAMPLE 10

You spent $35 shopping, buying a magazine for $3 and cassette tapes at $8 apiece.

a. Write an equation to determine how many tapes you bought.

b. Solve this equation.

continued

PRACTICE 10

Suppose that an empty box for oranges weighs 2 kg. If a typical orange weighs about 0.2 kg and the total weight of the box with oranges is 10 kg, how many oranges does the box contain? $2 + 0.2x = 10$, where x equals the number of oranges in the box; $x = 40$ oranges

Solution

a. If you let x represent the number of tapes bought, then $8x$ stands for the total amount you spent for the tapes.

    Write a word sentence for the problem and then translate it to an algebraic equation.

Number of
dollars spent for tapes   plus   cost of
magazine   equals   total
amount spent.

↓   ↓   ↓   ↓   ↓

$8x$ $+$ 3 $=$ 35

b. Now solve the equation $8x + 3 = 35$.

$$8x + 3 - 3 = 35 - 3$$

$$8x = 32$$

$$x = 4$$

Check:

$$8x + 3 = 35$$

$$8(4) + 3 \overset{?}{=} 35$$

$$32 + 3 \overset{?}{=} 35$$

$$35 \overset{\checkmark}{=} 35$$

You bought 4 cassette tapes.

Exercises 9.1

Solve and check.

1. $a - 7 = -21$
$a = -14$

2. $x - 6 = -9$
$x = -3$

3. $b + 4 = -7$
$b = -11$

4. $y + 12 = -12$
$y = -24$

5. $-11 = z - 4$
$z = -7$

6. $-15 = m - 20$
$m = 5$

7. $x + 21 = 19$
$x = -2$

8. $a + 12 = 10$
$a = -2$

9. $c + 33 = 14$
$c = -19$

10. $d + 27 = 15$
$d = -12$

11. $z + 2.4 = -5.3$
$z = -7.7$

12. $t + 2.3 = -6.7$
$t = -9$

13. $2.3 = x + 5.9$
$x = -3.6$

14. $4.1 = d + 6.9$
$d = -2.8$

15. $y - 2\frac{1}{3} = -3$
$y = -\frac{2}{3}$

16. $s - 4\frac{1}{2} = -8$
$s = -3\frac{1}{2}$

17. $-5 = t + 1\frac{1}{4}$
$t = -6\frac{1}{4}$

18. $-3 = 1\frac{2}{3} + c$
$c = -4\frac{2}{3}$

19. $39 = z + 51$
$z = -12$

20. $33 = c + 49$
$c = -16$

21. $-5x = 30$
$x = -6$

22. $-8y = 8$
$y = -1$

23. $-36 = -9n$
$n = 4$

24. $-125 = -25x$
$x = 5$

25. $\dfrac{m}{-15} = 1$
$m = -15$

26. $\dfrac{x}{-7} = 1.3$
$x = -9.1$

27. $\dfrac{w}{10} = -24$
$w = -240$

28. $\dfrac{a}{5} = -40$
$a = -200$

29. $-6 = \dfrac{x}{-2}$
$x = 12$

30. $-8 = \dfrac{y}{-4}$
$y = 32$

31. $1.7t = -51$
$t = -30$

32. $-1.5x = 45$
$x = -30$

33. $\dfrac{y}{9} = -\dfrac{5}{3}$
$y = -15$

34. $\dfrac{z}{3} = \dfrac{-4}{3}$
$z = -4$

35. $-10y = 4$
$y = -\dfrac{2}{5}$

36. $-15a = 3$
$a = -\dfrac{1}{5}$

▦ *Solve. Round each solution to the nearest tenth. Check.*

37. $895 = -624n$
$n = -1.4$

38. $-145 = 209p$
$p = -0.7$

39. $-2.5 = \dfrac{x}{5.91}$
$x = -14.8$

40. $-4.6 = \dfrac{z}{2.78}$
$z = -12.8$

Solve and check.

41. $4n - 20 = 36$
$n = 14$

42. $3a - 1 = 7$
$a = 2\frac{2}{3}$

43. $3x + 1 = 7$
$x = 2$

44. $7m + 1 = 22$
$m = 3$

45. $6k + 23 = 5$
$k = -3$

46. $2x + 21 = 7$
$x = -7$

47. $3x + 20 = 20$
$x = 0$

48. $4x + 28 = 28$
$x = 0$

49. $31 = 3 - 4h$
$h = -7$

50. $-68 = 10 - 3x$
$x = 26$

51. $34 = 13 - 4p$
$p = -5\frac{1}{4}$

52. $36 = 25 - 3c$
$c = -3\frac{2}{3}$

53. $-7b + 8 = -6$
$b = 2$

54. $-2x + 15 = -9$
$x = 12$

55. $21 + \dfrac{a}{3} = 10$
$a = -33$

56. $25 + \dfrac{w}{5} = 15$
$w = -50$

57. $\dfrac{1}{2}y + 5 = -13$
$y = -36$

58. $\dfrac{x}{5} + 15 = 0$
$x = -75$

59. $5 - \dfrac{x}{12} = 1$
$x = 48$

60. $16 - \dfrac{a}{2} = 15$
$a = 2$

61. $\dfrac{c}{3} + 3 = -4$
$c = -21$

62. $\dfrac{m}{4} + 1 = -5$
$m = -24$

63. $\dfrac{4}{9}x - 13 = -5$
$x = 18$

64. $\dfrac{5}{4}y - 19 = 26$
$y = 36$

65. $-8 - x = 11$
$x = -19$

66. $-24 = 2 - x$
$x = 26$

Solve. Round each solution to the nearest tenth. Check.

67. $58.3r + 23.58 = 2.79$
$r = -0.356603774$, or -0.4

68. $-51.5 = 29m - 4.06$
$m = -1.635862069$, or -1.6

69. $\dfrac{x}{0.24} - 0.03 = -0.14$
$x = -0.0264$, or -0.0

70. $\dfrac{a}{2.5} + 11.9 = 0.02$
$a = -29.7$

Write an equation for each word problem. Then solve and check.

71. While dieting, you lose 3 lb/mo. How many months will it take you to lose 21 lb? $-3x = -21$, where x represents the number of months; $x = 7$ mo

72. The temperature at mid-afternoon was 12°C. By early evening, the temperature had dropped to −7°C. What was the temperature change? $-7 = 12 + x$, where x is the temperature change; $x = -19$°C

73. Three business partners agree to split a debt of $6,000 equally. How much money does each partner owe? $3x = -6,000$, where x represents how much each one owes; $x = -2,000$; each partner owes $2,000.

74. If you were to write a check for $350, your account would be overdrawn by $200. What is the present balance of your account? $x - 350 = -200$; $x = \$150$

75. Fifteen members of the Drama Club at a college were given a special group discount of $25 off the total price of their tickets. If they paid $155 with the discount, what was the original price of each ticket? $15x - 25 = 155$, where x is the original price of each ticket; $x = \$12$

76. A car rents for $35/day plus 30¢/mi. How many miles were driven if the rental fee for a day was $65?
$35 + 0.3x = 65; x = 100$ mi

77. At a graduation dinner, an equal number of guests were seated at each of 8 large tables, and 3 late-arriving guests were seated at a small table. If there were 43 guests in all, how many guests were seated at the large tables?
$8x + 3 = 43$, where x represents the number of people at the large tables; $x = 5$ people

78. You spent $58.50 shopping, buying a CD holder for $19.50 and CDs at $13 apiece. How many CDs did you buy?
$19.50 + 13x = 58.50; x = 3$ CDs

🔲 *Use a calculator to solve the following problems, giving (a) the equation, (b) the exact answer, and (c) an estimate of the answer.*

79. Suppose that a computer service charges $8.95 per month plus $4.75 per hour for word processing. If a client paid $251.20 one month, how many hours did she use the service?
(a) $251.20 = 8.95 + 4.75x$; (b) exact answer: 51 hr; (c) possible estimate: 48 hr

80. At a football game, you sold bottles of juice for $1.25 each. At the end of the game, you had $86.25, including the $7.50 that you had started with. How many bottles of juice had you sold?

(a) $86.25 = 7.50 + 1.25x$; (b) exact answer: 63 bottles; (c) possible estimate: 80 bottles

■ *Check your answers on page A-13.*

MINDSTRETCHERS

GROUPWORK

1. In the following magic square, the sum of every row, column, and diagonal is 15. Working with a partner, solve for *a*, *b*, and *c*. $a = 1; b = 2; c = 24$

6	1	$\frac{c}{4} + 2$
7	$2b + 1$	3
$3a - 1$	9	4

MATHEMATICAL REASONING

2. Suppose that you have 3 objects to weigh: **a**, **b**, and **c**. You are told that 2 of the 3 objects are equal in weight but that the third object weighs more than either of the other two. On the balance scale shown, indicate how you can identify which object is the heaviest one by *only one weighing.*

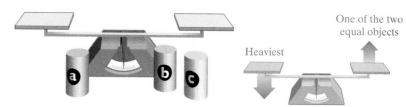

Place one object on each end of the scale. If the scale remains horizontal, the object that's *not* on the scale is the heaviest. If the scale moves, the end that moves *downward* contains the heaviest object.

WRITING

3. Write two different word problems that are applications of each of the following equations.

a. $12x + 500 = 24{,}500$

- Possible solution: Including a $500 bonus, your gross income at the end of the year was $24,500. What was your monthly income?

- Renting a booth at the trade expo required a $500 deposit plus a weekly fee. At the end of 12 weeks you paid $24,500. What was your weekly fee?

b. $\frac{x}{2} - 40 = 60$

- You decide to go shopping with half your savings. After buying a pair of shoes for $40, you have $60 left. What was the amount of your savings?

- Half the riders on a train got off at the first stop. At the second stop, 40 got off, leaving 60 riders. If no one got on at either stop, how many riders were originally on the train?

9.2 Solving Multistep Equations

OBJECTIVES
- *To combine like terms*
- *To solve equations involving like terms*
- *To solve equations containing parentheses*
- *To solve word problems with like terms or parentheses*

Solving Equations Involving Like Terms

In this section we extend the discussion of simple equations involving two steps to include equations that involve more than two steps. These multistep equations sometimes involve *like terms* or contain parentheses.

Definition

Like terms are terms that have the same variables and exponents.

In Chapter 4, we considered algebraic expressions such as $2x$ or $x - 1$ that consist of one or two terms. Some algebraic expressions, such as $2x + x + 1$, have three terms.

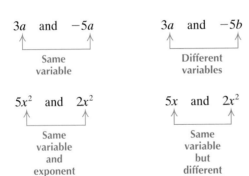

The terms $2x$ and x are like terms.

Like terms	**Unlike terms**
$3a$ and $-5a$	$3a$ and $-5b$
Same variable	Different variables
$5x^2$ and $2x^2$	$5x$ and $2x^2$
Same variable and exponent	Same variable but different exponents

We can combine only like terms. We can't combine unlike terms, such as $3a$ and $-5b$.

To combine like terms

- use the distributive property in reverse, and
- add or subtract.

EXAMPLE 1

Combine like terms.

a. $7x + 3x$ **b.** $6y - 8y$ **c.** $4a + a + 9$

Solution

a. $7x + 3x = (7 + 3)x$ Use the distributive property in reverse.

　　　　$= 10x$ Add 7 and 3.

PRACTICE 1

Simplify.

a. $5x + 7x$ $12x$

b. $7y - y$ $6y$

c. $4z - 5z + 6$ $-z + 6$

continued

b. $6y - 8y = (6 - 8)y$ Use the distributive property in reverse.

$\quad\quad = -2y$ Subtract 8 from 6.

c. $4a + a + 9 = (4 + 1)a + 9$ Use the distributive property in reverse. Recall that $a = 1a$.

$\quad\quad = 5a + 9$ Add 4 and 1.

Some equations have like terms on one side. To solve this type of equation, we collect or combine all like terms before using the rules for solving multiplication and division equations.

EXAMPLE 2

Solve and check: $3x - 5x = 10$

Solution $3x - 5x = 10$

$\quad\quad -2x = 10$ Combine like terms: $3x - 5x = (3 - 5)x = -2x$

$\quad\quad \dfrac{-2x}{-2} = \dfrac{10}{-2}$ Divide each side of the equation by -2.

$\quad\quad x = -5$

Check: $\quad\quad 3x - 5x = 10$

$\quad\quad 3(-5) - 5(-5) \overset{?}{=} 10$ Substitute -5 for x in the original equation.

$\quad\quad -15 - (-25) \overset{?}{=} 10$

$\quad\quad 10 \overset{\checkmark}{=} 10$

The solution is -5.

PRACTICE 2

Solve and check: $2y - 3y = 8$ $y = -8$

Some equations have like terms on both sides. To solve this type of equation, we use the addition or subtraction rule to get like terms on the same side so that they can be combined.

EXAMPLE 3

Solve and check: $8y - 11 = 5y - 2$

Solution $\quad\quad 8y - 11 = 5y - 2$

$\quad\quad 8y - 5y - 11 = 5y - 5y - 2$ Subtract $5y$ from each side of the equation.

$\quad\quad 3y - 11 = -2$ Combine like terms.

$\quad\quad 3y - 11 + 11 = -2 + 11$ Add 11 to each side of the equation.

$\quad\quad 3y = 9$

$\quad\quad y = 3$

PRACTICE 3

Solve and check: $10x - 1 = 2x + 3$
$x = \frac{1}{2}$

Check: $8y - 11 = 5y - 2$

$8(3) - 11 \overset{?}{=} 5(3) - 2$ Substitute 3 for y in the original equation.

$24 - 11 \overset{?}{=} 15 - 2$

$13 \overset{\checkmark}{=} 13$

The solution is 3.

Solving Equations Containing Parentheses

Some equations contain parentheses. To solve this type of equation, we first remove the parentheses, using the distributive property.

EXAMPLE 4

Solve and check: $6x = 4(x + 5)$

Solution $6x = 4(x + 5)$

$6x = 4x + 20$ Use the distributive property.

$6x - 4x = 4x - 4x + 20$ Subtract 4x from each side of the equation.

$2x = 20$ Combine like terms.

$x = 10$

Check: $6x = 4(x + 5)$

$6(10) \overset{?}{=} 4(10 + 5)$ Substitute 10 for x in the original equation.

$60 \overset{?}{=} 4(15)$

$60 \overset{\checkmark}{=} 60$

The solution is 10.

PRACTICE 4

Solve and check: $4(x + 2) = 3x$
$x = -8$

EXAMPLE 5

The product of 5 and a number is 24 less than twice that number.

a. Write an equation to find the number.

b. Solve this equation.

Solution

a. If we let x represent the number, $5x$ represents the product of 5 and that number. Then we represent 24 less than twice that number by $2x - 24$.

We can now write a word sentence for this problem, which we can then translate to an algebraic equation: $5x = 2x - 24$.

continued

PRACTICE 5

The sum of a number and 3 times that number is 9 more than the number.

a. Write an equation to find the number.
$x + 3x = x + 9$

b. Solve this equation. $x = 3$

b. Solve the equation $5x = 2x - 24$.

$$5x = 2x - 24$$
$$5x - 2x = 2x - 2x - 24$$
$$3x = -24$$
$$x = -8$$

Check:
$$5x = 2x - 24$$
$$5(-8) \stackrel{?}{=} 2(-8) - 24$$
$$-40 \stackrel{?}{=} -16 - 24$$
$$-40 \stackrel{\checkmark}{=} -40$$

The number is -8.

EXAMPLE 6

You buy a ticket for a concert. One day before the concert, two of your friends ask to join you. When you order their tickets, each costs $5 more than your ticket. The total cost of the three tickets is $88. What is the cost of each ticket?

Solution Let c represent the cost of your ticket. Then $c + 5$ represents the cost of a ticket for each friend.

Cost of your ticket		Cost of tickets for both friends		Total cost for the three tickets
↓		↓		↓
c	$+$	$2(c + 5)$	$=$	88

$$c + 2(c + 5) = 88$$
$$c + 2c + 10 = 88$$
$$3c + 10 = 88$$
$$3c = 78$$
$$c = 26$$

Check:
$$c + 2(c + 5) = 88$$
$$26 + 2(26 + 5) \stackrel{?}{=} 88$$
$$26 + 2(31) \stackrel{?}{=} 88$$
$$26 + 62 \stackrel{?}{=} 88$$
$$88 \stackrel{\checkmark}{=} 88$$

So your ticket costs $26, and each ticket for your friends costs $26 + $5, or $31.

PRACTICE 6

A taxi fare is $1.50 for the first mile and $1.25 for each additional mile. If your total cost was $4.25, how far did you travel in the taxi?   2.2 mi

Exercises 9.2

Simplify.

1. $4x + 3x$
$7x$

2. $y + 5y$
$6y$

3. $4a - a$
$3a$

4. $3d - 2d$
d

5. $6y - 9y$
$-3y$

6. $2x - 7x$
$-5x$

7. $2n - 3n$
$-n$

8. $4z - 7z$
$-3z$

9. $2c + c + 12$
$3c + 12$

10. $5a + a - 1$
$6a - 1$

11. $8 + x - 7x$
$8 - 6x$

12. $5 - a + 4a$
$5 + 3a$

13. $-y + 5 + 3y$
$2y + 5$

14. $2x - 4 - x$
$x - 4$

Solve and check.

15. $5m + 4m = 36$
$m = 4$

16. $3y + 8y = 22$
$y = 2$

17. $18 = 4y - 2y$
$y = 9$

18. $24 = 3x - x$
$x = 12$

19. $2a - 3a = 7$
$a = -7$

20. $x - 7x = 12$
$x = -2$

21. $7 = -5b + b$
$b = -1\frac{3}{4}$

22. $5 = -y + 3y$
$y = 2\frac{1}{2}$

23. $n + n - 13 = 13$
$n = 13$

24. $y + 2y - 3 = 18$
$y = 7$

25. $6 = 7x - 3x - 6$
$x = 3$

26. $11 = s - 4s + 5$
$s = -2$

27. $n + 3n - 7 = 29$
$n = 9$

28. $6m + m - 9 = 30$
$m = 5\frac{4}{7}$

29. $2 + 3y - y = -8$
$y = -5$

30. $6 - a + 4a = -6$
$a = -4$

▦ *Solve. Round each solution to the nearest tenth. Check.*

31. $60r - 17r + 23.58 = 2.79$
$r = -0.5$

32. $-51.5 = 20m + 9m - 4.06$
$m = -1.6$

33. $1.02m + 3.007m = 50.1$
$m = 12.4$

34. $5.8y + 2.05y = 0.929$
$y = 0.1$

Solve and check.

35. $5x = 2x + 12$
$x = 4$

36. $3t + 8 = t$
$t = -4$

37. $4p + 1 = 3p - 1$
$p = -2$

38. $10w + 1 = 3w + 1$
$w = 0$

39. $8x + 1 = x - 6$
$x = -1$

40. $5x - 7 = 2x + 2$
$x = 3$

41. $3p - 2 = -p + 4$
$p = 1\frac{1}{2}$

42. $2y - 3 = y + 1$
$y = 4$

43. $4n - 6 = 3n + 6$
$n = 12$

44. $2x + 1 = -x - 4$
$x = -1\frac{2}{3}$

45. $3 - 6t = 5t - 19$
$t = 2$

46. $5 - 8r = 2r - 25$
$r = 3$

47. $-7y + 2 = 3y - 8$
$y = 1$

48. $-6s - 2 = -8s - 4$
$s = -1$

49. $3x - 2 = 8 - 5x$
$x = 1\frac{1}{4}$

50. $6y - 8 = 4 - 3y$
$y = 1\frac{1}{3}$

🔲 *Solve. Round each solution to the nearest tenth. Check.*

51. $0.0125x + 0.07x = 3{,}625$
$x = 43{,}939.4$

52. $3.61n = 4 + 2.135n$
$n = 2.7$

53. $5.31y + 2.83 = -6.91 - 3.77y$
$y = -1.1$

54. $0.138a = -4.667a + 2.931a - 4.625$
$a = -2.5$

Solve and check.

55. $2(n - 3) = 12$
$n = 9$

56. $3(b + 4) = 24$
$b = 4$

57. $8(x - 1) = -24$
$x = -2$

58. $3(5 - r) = 18$
$r = -1$

59. $6n = 5(n + 7)$
$n = 35$

60. $3x = 2(x + 4)$
$x = 8$

61. $3(y - 5) = 2y$
$y = 15$

62. $5(m - 3) = 2m$
$m = 5$

63. $5y = 3(y + 1)$
$y = 1\frac{1}{2}$

64. $6x = 4(x - 1)$
$x = -2$

65. $4n = 7(n - 9)$
$n = 21$

66. $2(3 - x) = x$
$x = 2$

67. $6r + 2(r - 1) = 14$
$r = 2$

68. $3t + 2(t + 4) = 13$
$t = 1$

69. $x + 3(2 - x) = 12$
$x = -3$

70. $2(1 - m) + 3m = 5$
$m = 3$

🖩 *Solve. Round each solution to the nearest tenth. Check.*

71. $0.06x + 0.10(10,000 - x) = 900$
$x = 2,500$

72. $0.53(n + 2.6) = 8.31$
$n = 13.1$

73. $1.72y = 3.16(y - 8.72)$
$y = 19.1$

74. $6.19t + 3.81(1 - t) = 2.72$
$t = -0.5$

Write an equation for each word problem. Then solve and check.

75. You worked 4 hr last Saturday and 3 hr on Sunday. You were paid $63 for the 2 days. If you earned the same hourly rate each day, how much money did you earn per hour? $4x + 3x = 63$;
$x = \$9/\text{hr}$

76. The difference between three times a number and that number is -8. What is the number? $3x - x = -8; x = -4$

77. The perimeter of a rectangle 70 ft long is 200 ft. Find the width, w.

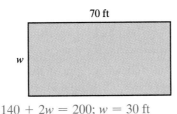

70 ft

w

$140 + 2w = 200; w = 30$ ft

78. In the triangle shown, what is the measure of the two equal angles, x?

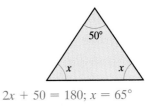

50°

x x

$2x + 50 = 180; x = 65°$

79. On an average day, you spend $\frac{1}{5}$ of your study time on mathematics and $\frac{1}{2}$ of your study time on English. The time for these two subjects totals $1\frac{3}{4}$ hr. How much time do you spend studying?
$\frac{1}{5}x + \frac{1}{2}x = 1\frac{3}{4}; x = 2\frac{1}{2}$ hr

80. Your family budget allows $\frac{1}{3}$ of the family's monthly income for housing and $\frac{1}{4}$ of its monthly income for food. If a total of $1,050 a month is budgeted for housing and food, what is your family's monthly income?
$\frac{1}{3}x + \frac{1}{4}x = 1,050; x = \$1,800$

81. At a publisher's "break-even" point, the publisher's income is equal to its production costs. Suppose that the cost of producing n books is given by $15n + 3,000$ and that the resulting income is given by $20n$. To break even, how many books must the publisher produce? $15n + 3,000 = 20n$;
$n = 600$ books

82. In the spreadsheet shown, C1 = A1 · B1, B3 = B1 + B2, C2 = A2 · B2, and C3 = C1 + C2. Find B1.

	A	B	C
1	2		
2	3		
3		20	40

Let x = B1; $2x + 3(20 - x) = 40$;
$x = 20$

83. In any polygon with n sides, the sum of the measures of the interior angles is $180(n - 2)$. If the measures of the angles of a polygon sum to 540°, how many sides does the polygon have?

$180(n - 2) = 540$; $n = 5$ sides

84. A company buys a copier for $10,000. The Internal Revenue Service values the copier at $10,000\left(1 - \dfrac{n}{20}\right)$ after n yr. After how many years will the copier be valued at $5,500?

$10,000\left(1 - \dfrac{n}{20}\right) = 5,500$; $n = 9$ yr

■ *Check your answers on page A-14.*

MINDSTRETCHERS

MATHEMATICAL REASONING

1. The *algebra tiles* pictured represent $3x + 1$ and $2x - 3$ respectively.

Represent each of the following expressions by algebra tiles.

a. $2x - 1$ a.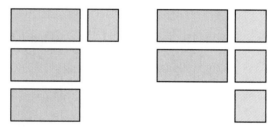

b. $x + 4$

GROUPWORK

2. Work in a small group. Write a list of steps for solving equations with variables on both sides. Compare your list of steps with that of another group.   Possible answer: 1) Get the terms with the variables on the same side of the equals sign and combine them.

2) Place the terms without variables on the "other" side of the equals sign. 3) Solve for the variable by performing the appropriate operation.

CRITICAL THINKING

3. Suppose that you work through a two-step equation and obtain $6x = 5x$. Then you divide both sides by x, getting $6 = 5$. Is there a solution to the original two-step equation? Explain.   The solution is $x = 0$, which could be found by moving the variable terms to one side of the equation.

9.3 Using Formulas

OBJECTIVES
- To translate a rule to a formula
- To evaluate formulas
- To solve word problems involving formulas

Translating Rules to Formulas

In a formula, letters and mathematical symbols are used to represent words. We often use formulas as shorthand for expressing a stated rule or relationship.

EXAMPLE 1

To convert a temperature expressed in Celsius degrees, C, to Fahrenheit degrees, F, we multiply the Celsius temperature by $\frac{9}{5}$ and then add 32. Write a formula for this relationship.

Solution Stating the rule briefly in words, we get

Fahrenheit equals nine-fifths times Celsius plus 32.

Now we can easily translate the rule to mathematical symbols.

$$F = \frac{9}{5}C + 32$$

PRACTICE 1

To predict the temperature, t, at a particular altitude, a, scientists who study weather conditions subtract $\frac{1}{200}$ of the altitude from the temperature on the ground, g. Here t and g are in degrees Fahrenheit and a is in feet. Write this relationship as a formula.

$t = g - \dfrac{1}{200}a$

Evaluating Formulas

In business and the health and physical sciences, as well as many other areas of life, we often must evaluate a formula.

EXAMPLE 2

Given the distance formula $d = rt$, find the value of d if rate $r = 50$ miles per hour and time $t = 1.6$ hours.

Solution $d = rt$

$$= 50\frac{\text{mi}}{\text{hr}} \cdot 1.6\ \text{hr} \qquad \text{Replace } r \text{ with } 50\ \frac{\text{mi}}{\text{hr}}, \text{ and } t \text{ with } 1.6 \text{ hr.}$$

$$= 80\ \text{mi}$$

So d is 80 mi. Note that we "cancelled" units as if they were numbers, which results in the correct unit for the answer.

PRACTICE 2

The formula for finding simple interest is $I = Prt$ where I is the interest, P is the principal, r is the rate of interest, and t is the time that the principal has been on deposit. Evaluate $I = Prt$ when $P = \$3,000$, $r = 0.06$, and $t = 2$ yr. $I = \$360$

EXAMPLE 3

The conversion of a temperature from Celsius degrees, C, to Kelvin degrees, K, is given by the formula $K = C + 273$. Suppose that in a chemistry experiment, C equals -4. What is the value of K?

Solution We use the formula $K = C + 273$ and substitute for C.

$$K = C + 273$$
$$= -4 + 273$$
$$= 269$$

So when the temperature is -4 °C, it is 269 °K.

PRACTICE 3

When an article is sold, the net profit on the sale p, the income i, and the original cost c are related by the formula $p = i - c$, where all quantities are in dollars. As store manager, you must compute your company's net profit on the sale of a certain item at the end of the day. Find the net profit if the item cost $498 and the income received was $800. $p = \$302$

EXAMPLE 4

The top of a can of soda is a circle with radius 1 in. Find the area of the top of the can.

1 in.

Solution We use the formula $A = \pi r^2$ and substitute for r.

$$A = \pi r^2$$
$$\approx 3.14 \times (1 \text{ in.})^2$$
$$\approx 3.14 \times 1 \text{ in}^2$$
$$\approx 3.14 \text{ in}^2$$

The top of the can has area 3.14 in^2.

PRACTICE 4

The distance that a free-falling object drops, ignoring friction, is given by the following formula.

$$s = \frac{1}{2} g t^2$$

Here, s is the distance, g is the force of gravity, and t is time. Suppose that in a physics experiment you find that the time it takes an object to fall is equal to 2 sec. If g equals 32 ft/sec^2, what distance does the object fall? $s = 64$ ft

Exercises 9.3

Translate each stated rule or relationship to a formula.

1. To figure out how far away in kilometers a bolt of lightning hits the ground, *d*, we divide the number of seconds, *t*, between the flash of lightning and the associated sound of thunder by 3.

$d = \dfrac{t}{3}$

2. The length, *l*, of a certain spring in centimeters is 25 more than 0.4 times the weight, *w*, in grams of the object hanging from it. $l = 0.4w + 25$

3. The average, *A*, of two numbers, *a* and *b*, is the sum of the numbers divided by 2. $A = \dfrac{a + b}{2}$

4. In a particular state, the sales tax, *t*, can be computed by multiplying 0.0625 by the price of the item sold, *p*. $t = 0.0625p$

5. The total surface area, *A*, of a cube equals 6 times the square of the length of one of its edges, *e*. $A = 6e^2$

6. The equivalent energy, *E*, of a mass equals the product of the mass, *m*, and the square of the speed of light, *c*. $E = mc^2$

Evaluate each formula for the given quantity in Exercises 7–14.

	Formula	Given	Find
7.	$F = \dfrac{9}{5}C + 32$	$C = -5°$	F 23°
8.	$C = K - 273$	$K = 270°$	C −3°
9.	$i = prt$	$p = \$1,000, r = 7.5\%,$ and $t = 3$ yr	i \$225
10.	$d = rt$	$r = 88\dfrac{\text{km}}{\text{hr}}$ and $t = 2.5$ hr	d 220 km
11.	$C = \dfrac{5}{9}(F - 32)$	$F = 32°$	C 0°
12.	$A = \dfrac{a + b + c}{3}$	$a = -8, b = 6,$ and $c = -4$	A −2

	Formula	Given	Find
13.	$s = \dfrac{1}{2}gt^2$	$g = 32\ \dfrac{\text{ft}}{\text{sec}^2}$ and $t = 10$ sec	s 160 ft
14.	$R = \dfrac{s^2}{A}$	$s = 10$ ft and $A = 25$ ft^2	R 4

Solve.

15. The formula for finding the present value of an item that depreciates yearly is $v = c - crt$. In this formula, v is the present value, c is the original cost, r is the rate of depreciation per year, and t is the number of years that have passed. After 5 yr, what is the value of a car originally costing $27,000 that depreciated at a rate of 0.1 per year?
$v = \$13,500$

16. The markup, m, on an item is its selling price, s, minus its cost, c. If a CD player costs $199.95 and sells for $299, how much is the markup?
$m = \$99.05$

17. One of the formulas for calculating the correct dosage for a child is as follows.

$$C = \frac{a}{a + 12} \cdot A$$

Here, C is the child's dosage in milligrams, a is the age of the child, and A is the adult dosage in milligrams. What is the prescribed dosage of a certain medicine for a child 6 years old, if the adult dosage of the medicine is 180 mg? $C = 60$ mg

18. When a cricket chirps n times per minute, you can find the temperature in degrees Fahrenheit, F, by using the following formula.

$$F = \frac{n}{4} + 37$$

What is the temperature when a cricket chirps 10 times/min? $F = 39.5°$

🖩 *Use a calculator to solve the following problems, giving (a) the equation, (b) the exact answer, and (c) an estimate of the answer.*

19. Scientists studying the American buffalo use formulas to predict the number of male, m, and female, f, calves that will be born in a year.

$$m = 0.48A \qquad \text{and} \qquad f = 0.42A$$

This prediction is based on the number of adult females, A, in the herd. If there are 328 adult females in a particular herd, how many calves will be born during the coming year, to the nearest 10 calves?   (a) $T = 0.48A + 0.42A$, where T represents the total number of calves born; (b) exact answer: 295 calves; (c) possible estimate: 270 calves

20. A formula for the breaking distance, d, in feet of a car traveling on dry pavement is $d = \dfrac{s^2}{25}$, where s is the speed in miles per hour. The corresponding formula for wet pavements is $d = \dfrac{s^2}{15}$. At a speed of 74 mph, how much greater is the breaking distance on wet pavement than on dry pavement, to the nearest foot?   (a) $D = \dfrac{s^2}{15} - \dfrac{s^2}{25}$, where D is the difference in breaking distances; (b) exact answer: 146 ft; (c) possible estimate: 150 ft

■ *Check your answers on page A-14.*

MINDSTRETCHERS

WRITING

1. A formula used in mathematics is

$$S = \frac{a}{1 - r}$$

For which value of r is there no value of S? Explain.   When $r = 1$, there is no value for

S because the denominator would equal zero. We cannot divide by zero.

GROUPWORK

2. Working with a partner, give an example of a situation that the following formula might describe.   Possible answer: Calculating net pay

$$d = bc - a$$

Explain what each variable represents in your example. d: net pay; b: hourly rate;
c: total number of hours worked; a: deductions (taxes, etc.)

MATHEMATICAL REASONING

3. Consider the following table.

Counting number, C	1	2	3	4	...	1,000
Odd number, O	1	3	5	7	...	1,999

Write a formula expressing O in terms of C. $O = 2C - 1$

HISTORICAL NOTE

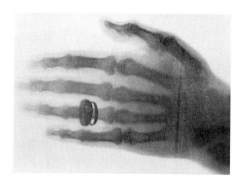

These two pictures both reflect the mathematical tradition of representing an unknown by the letter *x*. The painting of an unnamed woman, entitled *Madame X* (1884), is by the American artist John Singer Sargent. About a dozen films made throughout the twentieth century dealing with enigmatic women bear the same name.

The small photograph is one of the first taken by the German physicist Wilhelm Roentgen in 1895 after he discovered strange rays which he called *X-rays* since they were a mysterious phenomenon. Roentgen's discovery allowed doctors to see inside the body for the first time without surgery, and earned him the first Nobel prize awarded for physics.

The practice of using an *x* and other letters from the end of the alphabet to represent mathematical unknowns goes back to the seventeenth-century French mathematician René Descartes, who made major contributions to the development of algebra.

Florian Cajori, *A History of Mathematical Notations* (Chicago: The Open Court Publishing Company, 1929). (available from Dover)

KEY CONCEPTS and SKILLS

☐ = CONCEPT   ☐ = SKILL

CONCEPT/SKILL	DESCRIPTION	EXAMPLE
[9.1] **Solution**	A value of the variable that makes an equation a true statement	-2 is a solution of $x + 5 = 3$ because $-2 + 5 = 3$.
[9.1] **To solve a two-step equation**	■ First use the rule for solving addition or subtraction equations. ■ Then use the rule for solving multiplication or division equations.	$2y - 7 = 13$ $2y - 7 + 7 = 13 + 7$ $2y = 20$ $\dfrac{2y}{2} = \dfrac{20}{2}$ $y = 10$
[9.2] **Like terms**	Terms that have the same variables and exponents	$2x$ and $3x$
[9.2] **To combine like terms**	■ Use the distributive property in reverse. ■ Add or subtract.	$2x + 3x = (2 + 3)x$ $= 5x$

Chapter 9 Review Exercises

To help you review this chapter, solve these problems.

[9.1]

Solve and check.

1. $x + 2 = -4$
$x = -6$

2. $y + 3 = -6$
$y = -9$

3. $a + 19 = 19$
$a = 0$

4. $c + 25 = -25$
$c = -50$

5. $d + 9 = 0$
$d = -9$

6. $w + 11 = 0$
$w = -11$

7. $8 = x + 17$
$x = -9$

8. $4 = y + 20$
$y = -16$

9. $c + 9 = -9$
$c = -18$

10. $p + 11 = 11$
$p = 0$

11. $a - 7 = -9$
$a = -2$

12. $b - 9 = -11$
$b = -2$

13. $-10 = y - 4$
$y = -6$

14. $-12 = d - 5$
$d = -7$

15. $x - 9 = -9$
$x = 0$

16. $w - 14 = -4$
$w = 10$

17. $c - 4 = -12$
$c = -8$

18. $p - 11 = -3$
$p = 8$

19. $3x + 1 = 13$
$x = 4$

20. $2a + 3 = 7$
$a = 2$

21. $4y - 3 = 17$
$y = 5$

22. $5w - 1 = 9$
$w = 2$

23. $2y - 1 = 6$
$y = 3\frac{1}{2}$

24. $3y - 5 = 3$
$y = 2\frac{2}{3}$

25. $-x + 9 = 4$
$x = 5$

26. $-a + 7 = 6$
$a = 1$

27. $1 - y = 6$
$y = -5$

28. $2 - w = 4$
$w = -2$

29. $-c - 6 = 6$
$c = -12$

30. $-a - 9 = 9$
$a = -18$

31. $\frac{a}{3} + 1 = 9$
$a = 24$

32. $\frac{w}{4} + 3 = 7$
$w = 16$

33. $\frac{x}{5} + 4 = -1$
$x = -25$

34. $\frac{b}{2} + 2 = -3$
$b = -10$

35. $\frac{c}{7} - 1 = -1$
$c = 0$

36. $\frac{d}{8} - 2 = -2$
$d = 0$

[9.2]

Solve and check.

37. $4y - 2y = 18$
$y = 9$

38. $-2b + 7b = 30$
$b = 6$

39. $2c + c = -6$
$c = -2$

40. $-8x + 3x = -11$

$x = 2\dfrac{1}{5}$

41. $y + y + 2 = 18$

$y = 8$

42. $5 - t - t = -1$

$t = 3$

43. $3x - 4x + 6 = -2$

$x = 8$

44. $7 = 4m - 2m + 1$

$m = 3$

45. $0 = -7n + 4 - 5n$

$n = \dfrac{1}{3}$

46. $a - 5a - 6 = 30$

$a = -9$

47. $4r + 2 = 3r - 6$

$r = -8$

48. $-5y + 8 = -3y + 10$

$y = -1$

49. $2x - 8 = x + 1$

$x = 9$

50. $4y - 1 = 2y - 3$

$y = -1$

51. $7s + 4 = 5s + 8$

$s = 2$

52. $2 + 8x = 3x - 8$

$x = -2$

53. $5m - 9 = 9 - 4m$

$m = 2$

54. $x = -x + 12$

$x = 6$

55. $s + s - 6\dfrac{2}{3} = 4\dfrac{1}{3} + s$

$s = 11$

56. $5z - 6\dfrac{1}{2} = z - 4z + \dfrac{1}{2}$

$z = \dfrac{7}{8}$

57. $2(8 + w) = 22$

$w = 3$

58. $-3(z + 5) = -15$

$z = 0$

59. $6x + 12(10 - x) = 84$

$x = 6$

60. $3(y - 1) + y = 37$

$y = 10$

61. $6(y + 4) = 2y - 8$

$y = -8$

62. $5(m - 1) = 11 - m$

$m = 2\dfrac{2}{3}$

63. $0 = \dfrac{1}{3}(6b + 9) + b$

$b = -1$

64. $-\dfrac{1}{5}(10d - 5) = 9$

$d = -4$

[9.3]

Translate each stated rule or relationship to a formula.

65. Engineers use the Rankin temperature scale, where a Rankin temperature, R, is 460° more than the corresponding Fahrenheit temperature, F. $R = F + 460$

66. In medicine, a child's dosage, C, of a certain medication is equal to the adult dosage, A, times the quotient of the weight w of the child (in pounds) and 150.
(*Source:* Richard Burton, *Physiology by Numbers*)

$C = \dfrac{W}{150} \cdot A$

67. The average, A, of the numbers a, b, c, d, and e is the sum of these numbers divided by 5.

$A = \dfrac{a + b + c + d + e}{5}$

68. The midrange, m, of a collection of numbers is one-half the sum of the smallest number, s, and the largest number, l.

$m = \dfrac{1}{2}(s + l)$

Evaluate each formula for the given quantity.

	Formula	Given	Find
69.	$F = ma$	$m = 3.6$ and $a = 14$	F 50.4
70.	$d = 16t^2$	$t = 3$	d 144
71.	$P = n(p - c)$	$n = 8$, $p = 350$, and $c = 240$	P 880
72.	$C = \dfrac{5}{9}(F - 32)$	$F = -4$	C -20

Mixed Applications

Solve.

73. In marketing, the cost, c, of a certain number of units of an item can be found by using the formula $c = p \cdot n$, where p is the price per unit and n is the number of units. If the total cost of 50 units of an item is $269, what is the price per unit? $p = \$5.38/\text{unit}$

74. You plan to make a down payment of $45 on a camcorder and then make weekly payments of $50. How many weeks will it take you to pay for the camcorder if it sells for $895? 17 weeks

75. The perimeter of a rectangular rug is 42 ft. If the length of the rug is 12 ft, what is the width? $w = 9$ ft

76. The sale price, s, on an item equals the regular price, r, minus the discount, d. Find the sale price of a coat that was regularly priced at $325 and was later reduced by $53.95. $s = \$271.05$

■ **77.** The density, D, of an object can be found by using the formula $D = \dfrac{m}{V}$, where m is its mass and V is its volume. Find the density of a piece of ice if $m = 10$ g and $V = 10.9$ ml. Round your answer to the nearest tenth. $D = 0.9$ g/ml

■ **78.** When police discover a human thigh bone, police doctors need to be able to predict the height of the entire skeleton. A formula that they use is $h = 2.38t + 61.41$, where h is the height of the skeleton and t is the length of the thigh bone, both in centimeters. What length thigh bone would lead to a predicted height of 150 cm? (*Source:* Colin Evans, *Casebook of Forensic Detection*) $t = 37.2$ cm rounded to the nearest cm

■ *Check your answers on page A-14.*

Chapter 9 Posttest

To see whether you have mastered the topics in this chapter, take this test.

Solve and check.

1. $x + 5 = 0$
  $x = -5$

2. $y - 6 = -6$
  $y = 0$

3. $27 = -9a$
  $a = -3$

4. $-2 = \dfrac{b}{5}$
  $b = -10$

5. $2x + 3 = 4$
  $x = \dfrac{1}{2}$

6. $4y + 1 = -9$
  $y = -2\dfrac{1}{2}$

7. $-10 = 7w - 1$
  $w = -1\dfrac{2}{7}$

8. $-y + 2 = 11$
  $y = -9$

9. $\dfrac{a}{5} + 6 = 6$
  $a = 0$

10. $3x - 5x = 12$
  $x = -6$

11. $-4c + 3 = 7 - 2c$
  $c = -2$

12. $-2x = 4(x - 3)$
  $x = 2$

13. When you drop an object, the distance it falls, D, in meters is equal to 4.9 times the square of the length of time, t, in seconds that it has been falling. Translate this relationship to a formula. $D = 4.9t^2$

14. Given the formula $y = mx + b$, find the value of y when $m = -\dfrac{1}{2}$, $x = -3$, and $b = -4$. $y = -2\dfrac{1}{2}$

Write an equation for each word problem. Then solve and check.

15. The cost of a long distance telephone call to your friend is $0.35 for the first minute and $0.16 for each additional minute. What was the total length of a call that cost $2.75? $0.35 + 0.16x = 2.75$, where x represents the number of additional minutes; $x = 15$ min, so the total length of the call was 16 min.

16. An automobile dealer sells cars at a price that is $\dfrac{8}{9}$ the suggested retail price, plus a $100 handling fee. What is the suggested retail price of a car that the dealer sells for $12,906? $100 + \dfrac{8}{9}P = 12{,}906$, where P is the suggested retail price; $P = \$14{,}406.75$.

17. A computer system consisting of a monitor, a keyboard, and a CPU with a hard drive cost $1,800. The monitor costs $300 more than the keyboard, and the CPU costs $900 more than the keyboard. Find the cost of the keyboard. $(x + 300) + (x + 900) = 1{,}800$; $x = \$300$

18. Four times the sum of a number and 3 is 5 times that number. What is the number? $4(n + 3) = 5n$; $n = 12$

Solve.

19. The *reaction distance* is the distance that a car travels during the time that the driver is getting his or her foot on the brake. Studies have shown that the reaction distance, d, is approximately 2.2 times the rate, r, of the car. Here, the reaction distance is in feet, and the rate is in miles per hour. (*Source:* National Highway Traffic Safety Administration)

a. Write a formula for the reaction distance. $d = 2.2r$

b. Find the distance when $r = 55$ mph. $d = 121$ ft

20. To calculate the speed of an object in feet per second, f, we multiply its speed expressed in miles per hour, m, by $\dfrac{22}{15}$.

a. Write this relationship as a formula. $f = \dfrac{22}{15}m$

b. If an object is moving 45 mph, what is its speed in fps? $f = 66$ fps

■ *Check your answers on page A-14.*

Cumulative Review Exercises

To help you review, solve the following.

1. Round to the nearest hundred: 3,817   3,800

2. How is the number 3,000,400 read?   Three million, four hundred

3. Compute: $(0.1)^3$ 0.001

4. What percent of 30 is 25? $83\frac{1}{3}\%$

5. Calculate: $(-1)^2 - 3$ -2

6. You serve on a large military ship on which there are 75 officers and 1,275 enlisted personnel. What is the ratio of officers to enlisted personnel among the ship's crew? $\frac{1}{17}$

7. Solve: $3x - 5 = 13$ $x = 6$

8. A doctor changed a patient's dosage of thyroxine from 0.075 mg to 0.1 mg. Explain whether this change represented an increase or a decrease.   This change represented an increase because 0.1 is larger than 0.075 (by 0.025).

9. You are interested in buying stock in a particular company. The following table shows the selling price of its stock last week.

Day	M	Tu	W	Th	F
Price per Share	$2\frac{1}{4}$	$2\frac{5}{16}$	$3\frac{1}{16}$	$2\frac{7}{8}$	$2\frac{1}{2}$

On which day was the stock least expensive?
Monday

10. According to some medical research, women with a body mass index between 22.0 and 23.4 are likely to have the longest life span. The formula for body mass index (represented by the letter i) is as follows.

$$i = \frac{697.5w}{h^2}$$

In this formula the weight, w, is in pounds and the height, h, is in inches. Estimate i if $w = 100$ lb and $h = 60$ in. (*Source: The New England Journal of Medicine*)
Possible answer: $i = 20$

■ *Check your answers on page A-14.*

Measurement and Units

10.1 U.S. Customary Units

10.2 Metric Units and Metric/U.S. Customary Unit Conversions

UNITS AND INTERNATIONAL TRADE

Although consumers have been reluctant to change, the United States is increasingly "going metric." The federal government is replacing inches with centimeters and gallons with liters in its business dealings. The largest U.S. car manufacturer began building car parts in metric units (everything except the speedometer) in the early 1970s. Many packaged goods are now labeled in grams and liters as well as ounces and quarts.

Industry leaders may have to switch over to the metric system if they want to survive in the global economy. One major U.S. corporation had a shipment of appliances rejected by Saudi Arabia because the connecting cords were 6 ft long instead of the required standard of 2 m (approximately 6.6 ft).

However a complete switch to the metric system would be costly, entailing endless changes in machines, tools, dials, containers, signs, contracts, and laws.

Chapter 10 Pretest

To see if you have already mastered the topics in this chapter, take this test.

Solve.

1. 8 qt = _____2_____ gal

2. 5 tons = _10,000_ lb

3. Add: 7 ft 11 in. + 4 ft 7 in.   12 ft 6 in.

4. Which of the following units is a measure of length?

    a. a gram   **b.** a meter   **c.** a liter

    d. a second b.

5. The width of a dime is about

    a. a meter   **b.** a kilometer

    c. a millimeter   **d.** a centimeter
    d.

6. 3.5 kg = _3,500_ g

7. 2,100 mm = _2.1_ m

8. Which is larger: 2.3 m or 700 cm?   700 cm

9. A bottle contains 0.5 liters (L) of olive oil. Express this amount in milliliters.   500 ml

10. The doctor ordered 75 mg of the drug Demerol. How many grams is this?   0.075 g

■ *Check your answers on page A-14.*

10.1 U.S. Customary Units

OBJECTIVES

- *To identify units in the U.S. customary system*
- *To change a measurement from one U.S. customary unit to another*
- *To add and subtract measurements expressed in U.S. customary units*

What Measurement and Units Are and Why They Are Important

When taking measurements, we use devices to express quantities in standardized units. For instance, a scale allows us to express the weight of a package in pounds, just as a yardstick lets us express the length of a room in yards.

By expressing quantities in standardized units such as pounds and yards, we can compare characteristics of physical objects. For instance, we can decide which of two packages is heavier or how many times longer one room is than another.

Throughout this chapter, we focus on measuring four quantities: length, weight, capacity, and time. The units that we use to express these measurements come from two systems of measurement: the U.S. customary system and the metric system.

We begin with U.S. customary units because we use them most often in everyday situations.

U.S. Customary Units of Length, Weight, Capacity, and Time

Sometimes, we need to change the unit in which a measurement is expressed. To convert units, we must know how different units are related. We refer to these units as *U.S. customary units*, or simply *U.S. units*.

In measuring *length*, the main U.S. units are inches (in.), feet (ft), yards (yd), and miles (mi). The key conversion relationships among these units are as follows.

$$12 \text{ in.} = 1 \text{ ft}$$

$$3 \text{ ft} = 1 \text{ yd}$$

$$5{,}280 \text{ ft} = 1 \text{ mi}$$

The main U.S. units of *weight* are ounces (oz), pounds (lb), and tons. Note the following relationships.

$$16 \text{ oz} = 1 \text{ lb}$$

$$2{,}000 \text{ lb} = 1 \text{ ton}$$

Next, let's consider the main U.S. units of *capacity* (or *liquid volume*): fluid ounces (fl oz), pints (pt), quarts (qt), and gallons (gal). Here is how these units are related.

$$16 \text{ fl oz} = 1 \text{ pt}$$

$$2 \text{ pt} = 1 \text{ qt}$$

$$4 \text{ qt} = 1 \text{ gal}$$

Finally, there are the main U.S. units of *time*: seconds (sec), minutes (min), hours (hr), days, weeks (wk), months (mo), and years (yr). Some of the key relationships among these units of time are as follows.

$$60 \text{ sec} = 1 \text{ min}$$

$$60 \text{ min} = 1 \text{ hr}$$

$$24 \text{ hr} = 1 \text{ day}$$

$$7 \text{ days} = 1 \text{ wk}$$

$$52 \text{ wk} = 1 \text{ yr}$$

$$12 \text{ mo} = 1 \text{ yr}$$

$$365 \text{ days} = 1 \text{ yr}$$

Study these relationships so that you can use them to solve problems involving units.

TIP The abbreviations of the units are the same regardless of whether they are singular or plural. For example, the abbreviation for foot (ft) is the same as the abbreviation for feet (ft).

Changing Units

Suppose that you are ordering the rug pictured from a catalog. If you know that the length of the space to be covered is 2 ft, how long is it in inches? To change a length given in feet to inches, we need to know that there are 12 in. in 1 ft.

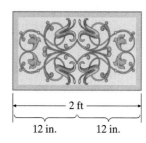

|← 2 ft →|
| 12 in. | 12 in. |

$$2 \text{ ft} = 2 \times (1 \text{ ft}) \qquad \text{Substitute 12 in. for 1 ft.}$$

$$= 2 \times 12 \text{ in.}$$

$$= 24 \text{ in.}$$

The rug is 24 in. long.

TIP When we change *a large unit to a small unit*, the numerical part of the answer is larger than the original number. But when we change *a small unit to a large unit*, the numerical part of the answer is smaller than the original number.

You also may want to discuss these conversion problems from the point of view of ratio and proportion, as in the following example.

$$\frac{x \text{ in.}}{2 \text{ ft}} = \frac{12 \text{ in.}}{1 \text{ ft}}$$

Another way of solving this problem is to multiply the original measurement by the *unit factor* $\frac{12 \text{ in.}}{1 \text{ ft}}$. The unit factor method, commonly used in the physical and health sciences, is particularly helpful in solving challenging conversion problems. Because the numerator 12 in. and the denominator 1 ft represent the same length, the unit factor $\dfrac{12 \text{ in.}}{1 \text{ ft}}$ is equivalent to 1.

$$2 \text{ ft} = 2 \text{ ft} \times \frac{12 \text{ in.}}{1 \text{ ft}}$$

$$= \frac{2 \times 12 \text{ in.}}{1} = 24 \text{ in.}$$

Note how we simplified the answer by canceling common units, as if the units were numbers.

Or suppose that we wanted to change 36 in. to feet. We can solve this problem by multiplying the original measurement by the unit factor $\dfrac{1 \text{ ft}}{12 \text{ in.}}$.

$$36 \text{ in.} = \overset{3}{36} \text{ in.} \times \frac{1 \text{ ft}}{\underset{1}{12 \text{ in.}}}$$

$$= 3 \text{ ft}$$

Let's look back at the two problems that we just solved. Both involved inches and feet. In the first problem, we multiplied by the unit factor $\dfrac{12 \text{ in.}}{1 \text{ ft}}$; in the second problem, we multiplied by its reciprocal, $\dfrac{1 \text{ ft}}{12 \text{ in.}}$. (Note that both unit factors are equivalent to 1.)

TIP   When converting from one unit to another unit, multiply the original measurement by the unit factor that has the desired unit in its numerator.

EXAMPLE 1

5 qt = _____ pt

Solution From the capacity chart on page 461 we know that 2 pt = 1 qt. We want to change from *quarts* to *pints*, so we use the unit factor $\dfrac{2 \text{ pt}}{1 \text{ qt}}$.

$$5 \text{ qt} = 5 \text{ qt} \times \frac{2 \text{ pt}}{1 \text{ qt}}$$

$$= 10 \text{ pt}$$

PRACTICE 1

5 pt = ___$2\frac{1}{2}$___ qt

EXAMPLE 2

A lecture lasted 30 min. How long was the lecture in hours?

Solution $30 \text{ min} = \overset{1}{30} \text{ min} \times \dfrac{1 \text{ hr}}{\underset{2}{60 \text{ min}}}$

$$= \frac{1}{2} \text{ hr}$$

The lecture lasted half an hour.

PRACTICE 2

The capacity of an oil barrel is 30 qt. How many gallons of oil are needed to fill the barrel?   7.5 gal

---

EXAMPLE 3

How many fluid ounces are equivalent to 3 gal?

Solution Note that the capacity chart on page 461 does not indicate the number of fluid ounces in a gallon. To solve this problem, we must use several steps.

Step 1. Change gallons to quarts.

Step 2. Change quarts to pints.

Step 3. Change pints to fluid ounces.

To combine these three steps, we multiply the original measurement by a chain of appropriate unit factors to get *fluid ounces* in the final answer.

$$3 \text{ gal} = 3 \text{ gal} \times \frac{4 \text{ qt}}{1 \text{ gal}} \times \frac{2 \text{ pt}}{1 \text{ qt}} \times \frac{16 \text{ fl oz}}{1 \text{ pt}}$$

$$= \frac{3 \times 4 \times 2 \times 16}{1} \text{ fl oz}$$

$$= 384 \text{ fl oz}$$

PRACTICE 3

How many seconds are there in a day?
86,400 sec

EXAMPLE 4

The sign on a bridge reads: "Warning! Maximum load — 2 tons." If your car weighs about 3,000 lb, can you cross the bridge safely?

Solution To compare the weight limit of the bridge and the weight of your car, you must express the two measurements in the same unit.

$$2 \text{ tons} = 2 \text{ tons} \times \frac{2,000 \text{ lb}}{1 \text{ ton}}$$

$$= 4,000 \text{ lb}$$

Because the bridge can support 4,000 lb — a lot more than your car weighs — you can safely cross the bridge.

PRACTICE 4

You buy 3 pt of milk. Can you completely fill a 2-qt container?   No; 3 pt = 1.5 qt

---

> Have students explain their responses in a journal.

In Example 4, we converted tons to pounds. Explain why.

Adding and Subtracting Measurements

There are many practical situations in which we need to add or subtract measurements—for instance, when we want to find out how much longer one car is than another or how much paint there will be if we combine the contents of several cans.

In addition or subtraction problems, only quantities having the same unit can be added or subtracted.

$$\begin{array}{r} 3 \text{ mi } 250 \text{ ft} \\ +5 \text{ mi } 400 \text{ ft} \\ \hline 8 \text{ mi } 650 \text{ ft} \end{array} \qquad \begin{array}{r} 2 \text{ hr } 10 \text{ min} \\ -1 \text{ hr } 3 \text{ min} \\ \hline 1 \text{ hr } 7 \text{ min} \end{array}$$

Often addition or subtraction problems involve changing units.

EXAMPLE 5

Find the sum: 7 ft 9 in.
                +2 ft 5 in.

Solution 7 ft 9 in.   Start with the inches column. (The smaller unit is
            +2 ft 5 in.   always the column on the right.) Add the inches.
             14 in.

   1 ft
   7 ft 9 in.   Because 14 in. = 1 ft 2 in., replace the 14 in. by
  +2 ft 5 in.   2 in. and carry the 1 ft to the feet column.
   14 in.
    2 in.

   1 ft
   7 ft 9 in.
  +2 ft 5 in.
   10 ft 2 in.   Add the feet.

The sum is 10 ft 2 in.

PRACTICE 5

Add 3 lb 10 oz and 1 lb 14 oz.   5 lb 8 oz

EXAMPLE 6

A local theater is showing a double feature of your two favorite science fiction films—*Aliens* and *Invasion of the Body Snatchers*. The first film runs 2 hr 18 min and the second 1 hr 57 min. How long is the double feature?

Solution
        1 hr
   2 hr 18 min   Start with the minutes column. Because 75 min =
  +1 hr 57 min   1 hr 15 min, replace the 75 min with 15 min, and
   4 hr 75 min   carry the 1 hr to the hours column. Add the hours.
        15 min

The double feature lasts 4 hr 15 min.

Now let's look at some examples of subtraction.

PRACTICE 6

You worked at three temporary jobs, one after the other. The first lasted 8 mo, the second 11 mo, and the third 1 yr 2 mo. How long did you work in all?   2 yr 9 mo

EXAMPLE 7

Subtract 1 yd 2 ft from 3 yd 1 ft.

Solution 3 yd 1 ft   Position the quantities so that like units are in the
            −1 yd 2 ft   same column.

   2   4
   3 yd 1 ft   Because 2 ft is larger than 1 ft, exchange 1 yd
  −1 yd 2 ft   from the 3 yd in the minuend, leaving 2 yd.
                Replace the borrowed yard by 3 ft, which when
                added to the 1 ft gives 4 ft.

   2   4
   3 yd 1 ft   Subtract within each column.
  −1 yd 2 ft
   1 yd 2 ft

PRACTICE 7

How much greater is 6 gal than 5 gal 1 qt?
3 qt

continued

Check: Remember that we can check a subtraction by using addition.

$$
\begin{array}{r}
1\text{ yd} \\
1\text{ yd } 2\text{ ft} \\
+1\text{ yd } 2\text{ ft} \\
\hline
3\text{ yd } \cancel{4}\text{ ft} \\
1\text{ ft}
\end{array}
$$

Adding, we get 3 yd 1 ft, so our answer checks.

EXAMPLE 8

On a 4-hr video cassette, you recorded a television show that ran 1 hr 50 min. How much recording time was left on the tape?

Solution

$$
\begin{array}{r}
4\text{ hr} \\
-1\text{ hr } 50\text{ min} \\
\end{array}
$$
Borrow 1 hr from 4 hr.

$$
\begin{array}{r}
\overset{3}{\cancel{4}}\text{ hr } 60\text{ min} \\
-1\text{ hr } 50\text{ min} \\
\hline
2\text{ hr } 10\text{ min}
\end{array}
$$
1 hr = 60 min

So 2 hr 10 min were left on the tape.

PRACTICE 8

Two of the most famous horses in history were Citation and Secretariat. In 1948, Citation won the Kentucky Derby with a time of 2 min 5 sec. Twenty-five years later, Secretariat won in 1 min 59 sec. What was the difference between the two times?
(*Source: Facts and Dates of American Sports*)

6 sec

Exercises 10.1

Change each quantity to the indicated unit.

1. 48 in. = _4_____ ft   **2.** 60 ft = _720_____ in.   **3.** 9 ft = _3_____ yd

4. 10 yd = _30_____ ft   **5.** 6 pt = _3_____ qt   **6.** 8 qt = _16_____ pt

7. 32 oz = _2_____ lb   **8.** 4 lb = _64_____ oz   **9.** 36 mo = _3_____ yr

10. 2 yr = _104_____ wk   **11.** 7 min = _420_____ sec   **12.** 5 hr = _300_____ min

13. 32 pints = _512_____ fl oz   **14.** 60 min = _1_____ hr   **15.** 30 min = $\frac{1}{2}$ hr

16. 36 hr = _1.5_____ day   **17.** 2 mi = _3,520_ yd   **18.** 5 gal = _40_____ pt

19. 5 wk = _35_____ day   **20.** 32 fl oz = _2_____ pt   **21.** $\frac{1}{2}$ hr = $\frac{1}{48}$ day

22. $\frac{1}{4}$ day = _6_____ hr   **23.** $\frac{1}{2}$ gal = _2_____ qt   **24.** $1\frac{1}{2}$ qt = $\frac{3}{8}$ gal

25. 7 pt = _3.5_____ qt   **26.** $2\frac{1}{2}$ qt = _5_____ pt

27. $1\frac{1}{2}$ tons = _3,000_ lb   **28.** 7,000 lb = $3\frac{1}{2}$ tons

29. 5 min 10 sec = _310_____ sec   **30.** 20 yr 2 mo = _242_____ mo

31. 117 mi = _617,760_ ft   **32.** 15,840 ft = _3_____ mi

Compute.

33.   4 lb 7oz
    −2 lb 9 oz
    1 lb 14 oz

34.   2 hr
    −1 hr 2 min
    58 min

35.   20 lb 5 oz
    + 9 lb 10 oz
    29 lb 15 oz

36.   5 lb 10 oz
    +1 lb 8 oz
    7 lb 2 oz

37. 5 yr 7 mo + 3 yr 11 mo
    9 yr 6 mo

38. 4 yr − 2 yr 3 mo
    1 yr 9 mo

39. 5 gal 1 qt − 2 gal 2 qt   **40.** 1 pt 10 fl oz + 3 pt 8 fl oz   **41.** 2 qt 1 pt + 1 qt 1 pt
    2 gal 3 qt   5pt 2 fl oz   4 qt

42. 2 ft − 5 in.   **43.**   6 min   2 sec   **44.**   20 ft   5 in.
    1 ft 7 in.   70 sec   5 ft   7 in.
                             +1 min   3 sec   + 9 ft 10 in.
                             8 min 15 sec   35 ft 10 in.

Applications

Solve.

45. Which weighs more, a $\frac{1}{4}$-lb hamburger or a 4-oz piece of cheese?   They both weigh the same.

46. Which is longer, a 2-mi race or a 10,000-ft race?   A 2-mi race is longer than a 10,000-ft race.

47. In the kitchen, there is a 35-in. opening in which you would like to fit a refrigerator. If the refrigerator is 3 ft wide, will it fit in the space?   No; the refrigerator is too wide.

48. A recipe calls for $\frac{1}{2}$ qt of milk. If you have $1\frac{1}{2}$ pt of milk on hand, is this enough?   Yes; $\frac{1}{2}$ qt = 1 pt

49. The record for a person holding his or her breath underwater without special equipment is 823 sec. Express this time in minutes and seconds. (*Source: Atlantic Monthly*)   13 min 43 sec

50. One of the tallest women who ever lived was an American named Sandy Allen, who, at age 22, was 91 in. tall. What was her height in feet and inches? (*Source: Guiness Book of Records*)   7 ft 7 in.

51. Abraham Lincoln spoke of "four score and seven years ago." If a score is 20 yr, how many months are there in four score and seven years?   1,044 mo

52. Suppose that you walk 3 mi. Express this distance in yards.   5,280 yd

53. Born in 1934, the Dionne sisters were Canadians who become world famous as the first quintuplets to survive beyond infancy. At birth, the tiniest of these babies weighed 1 lb 15 oz, and the largest weighed 3 lb 4 oz. What was the difference between their weights? (*Source: Encyclopedia Brittanica*)   1 lb 5 oz

54. In 1940, U.S. athlete Cornelius Warmerdam used a bamboo pole to vault 15 ft 8 in., setting a record. In 1962, U.S. athlete Dave Tork, using a fiberglass pole, vaulted 16 ft 2 in. How much higher was Tork's vault than Warmerdam's? (*Source: Facts and Dates of American Sports*)   6 in.

55. The lease on your apartment runs for 3 yr. If you have lived in the apartment already for 1 yr 2 mo, how much time is left on the lease?   1 yr 10 mo

56. You park your car next to a meter that will expire in 1 hr 55 min. If you pay for an additional $1\frac{1}{2}$ hr of parking, is there enough time on the meter for you to park for $3\frac{1}{4}$ hr?   Yes; your total meter time is 3 hr 25 min.

■ *Use a calculator to solve the following problems, giving (a) the operation(s) carried out in your solution, (b) the exact answer, and (c) an estimate of the answer.*

57. The highest mountain in the world is Mt. Everest. If its peak is 29,028 ft above sea level, find the height of the mountain to the nearest tenth of a mile. (a) Division; (b) 5.5 mi; (c) possible estimate: 6 mi

58. An airplane is cruising at an altitude of 33,000 feet. What is the plane's altitude to the nearest mile?   (a) Division; (b) 6 mi; (c) possible estimate: 6 mi

■ *Check your answers on page A-14.*

MINDSTRETCHERS

WRITING

1. Not all units are standardized. For example, the term *city block* varies in meaning from town to town. Explain the consequences of this lack of standardization.   The lack of standardization prevents you from comparing the sizes of city blocks in different towns.

MATHEMATICAL REASONING

2. In measuring, we often introduce errors. Suppose that each of two measurements could be as much as an inch off. If we then add the two measurements, how far from the truth could our sum be?   2 in.

INVESTIGATION

3. The foot is not the only body part used as a measure. For instance, the ancient Egyptians used the *mouthful* as a unit of measure of volume. In your college library or on the Web, investigate other examples.   Answers may vary.

HISTORICAL NOTE

Joseph Louis Lagrange (1736–1813) was a French mathematician who, as chairman of the French commission on weights and measures, was influential during the years following the French Revolution of 1789 in developing the metric system of measures based on decimals and powers of ten. Since then, the United States has been resistant to adopting the metric system, although in 1790, Thomas Jefferson — then secretary of state and later president — argued that the country should adopt a decimal system of weights and measures.

Source: Gullberg, *Mathematics, From the Birth of Numbers* (New York: W.W. Norton, 1997), p. 52.

10.2 Metric Units and Metric/U.S. Customary Unit Conversions

OBJECTIVES

- *To identify units in the metric system*
- *To change a measurement from one metric unit to another*
- *To change a measurement from a metric unit to a U.S. customary unit, and vice versa*

Units of Length, Weight, and Capacity

Now let's turn to metric units. Developed by French scientists some 200 years ago, the metric system (formally known as the International System of Units, or SI) has become standard in most countries of the world. Even in the United States, metric units predominate in many important fields, including scientific research, medicine, the film industry, food and drink packaging, sports, and the import–export industry.

As in Section 10.1, in the U.S. customary system, we consider measurements of length, weight, and capacity. Time units are identical in both systems, so we do not discuss them in this section. Again abbreviations of units in the singular and plural are the same. For example, the abbreviation for meter (m) is the same as the abbreviation for meters (m).

We begin this discussion of the metric system by considering the basic metric units:

- the **meter**, a unit of length (which gives the metric system its name);

- the **gram**, a unit of weight; and

- the **liter**, a unit of capacity (liquid volume).

There are quite a few other metric units as well. The name for each of the other units is formed by combining a basic unit with one of the metric prefixes. These prefixes are listed in the following table.

METRIC PREFIXES		
Prefix	**Symbol**	**Meaning**
Milli-	m	One thousandth $\left(\dfrac{1}{1{,}000}\right)$
Centi-	c	One hundredth $\left(\dfrac{1}{100}\right)$
Deci-	d	One tenth $\left(\dfrac{1}{10}\right)$
Deka-	da	Ten (10)
Hecto-	h	Hundred (100)
Kilo-	k	One thousand (1,000)

Next, let's see how the three basic metric units combine with the metric prefixes to form new units. We begin with units of length.

Length

The following table shows the four most commonly used metric units of length. Memorize the following table, noting what each unit means as well as its symbol.

METRIC UNITS OF LENGTH		
Unit	**Symbol**	**Meaning**
Millimeter	mm	$\frac{1}{1,000}$ meter
Centimeter	cm	$\frac{1}{100}$ meter
Meter	m	1 meter
Kilometer	km	1,000 meters

Point out to students that there is no unit of length in the U.S. customary system comparable to that of the millimeter in the metric system.

In this table, the first unit of length, the *millimeter*, is the smallest — about the thickness of a dime. We use the millimeter to measure short lengths — say, the dimensions of an insect.

1 mm

The next unit of length, the *centimeter*, is approximately the width of your little finger, or a bit less than half an inch. In the metric system, the width of an envelope is expressed in centimeters.

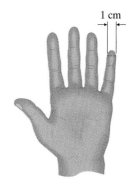
1 cm

The *meter*, the basic metric unit of length, is a bit longer than a yard, or about the width of a twin bed. In the metric system, we use meters to measure medium-size lengths — say, the length of a room.

1 m

The largest unit of length in the table is the *kilometer*. A kilometer is a little more than half a mile, or about three times the height of the Empire State Building. Great lengths, such as the distance between two cities, are expressed in kilometers.

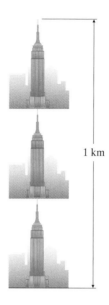

1 km

Weight

Now, we turn to the metric units of weight. Memorize the following table, which describes the three most commonly used metric units of weight.

METRIC UNITS OF WEIGHT		
Unit	**Symbol**	**Meaning**
Milligram	mg	$\frac{1}{1,000}$ gram
Gram	g	1 gram
Kilogram	kg	1,000 grams

The smallest unit, the *milligram*, is tiny — about the weight of a hair. It is therefore used to measure light weights — say, that of a small pill.

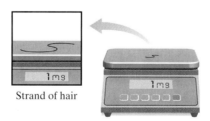

Strand of hair

The next unit, the *gram*, is larger but is still only about $\frac{1}{30}$ oz, or about the weight of a raisin.

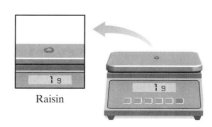

Raisin

The largest unit of weight in the table is the *kilogram*. A kilogram is approximately 2 lb, or about the weight of this textbook. Large weights — say, that of a car or of a person — are expressed in kilograms.

This textbook

Capacity (Liquid Volume)

Amounts of liquid are commonly measured in terms of liquid volume, or equivalently, the capacity of containers that hold the liquid. Memorize the following table, which describes the two primary metric units of capacity.

METRIC UNITS OF CAPACITY		
Unit	Symbol	Meaning
Milliliter	ml	$\frac{1}{1,000}$ liter
Liter	L	1 liter

In this table, the first unit, the *milliliter*, represents a very small amount of liquid — about as much as an eyedropper contains. Milliliters are used in measuring small volumes of liquid — say, the amount of perfume in a tiny bottle.

1 ml

The second unit of liquid volume, the *liter*, is much larger — slightly more than a quart. Liters are used in measuring larger quantities of liquid, such as the amount of water that a sink will hold.

1 L

Changing Units

Students should review here how to multiply and divide a decimal by a power of 10, covered on pages 191 and 203.

As we have already shown, sometimes we need to change the unit in which a measurement is expressed. Unit conversions in the metric system are much easier than those in the U.S. customary system because they involve multiplying or dividing by a power of 10, such as 100 or 1,000. That is why the metric system is so popular. The following table shows several metric conversion relationships.

Length	Weight	Capacity
1,000 mm = 1 m	1,000 mg = 1 g	1,000 ml = 1 L
100 cm = 1 m	1,000 g = 1 kg	
1,000 m = 1 km		

EXAMPLE 1

1.5 g = _____ mg

Solution Because 1,000 mg = 1 g and we want to convert to *milligrams*, we use the unit factor $\dfrac{1,000 \text{ mg}}{1 \text{ g}}$ to solve this problem.

$1.5 \text{ g} = 1.5 \text{ g} \times \dfrac{1,000 \text{ mg}}{1 \text{ g}}$

$= 1.5 \times 1,000 \text{ mg}$ **To multiply 1.5 by 1,000, move the decimal point in 1.5 three places to the right.**

$= 1,500 \text{ mg}$

Why is the number of milligrams (1,500) more than the number of grams (1.5) for an equivalent measurement?

PRACTICE 1

3,100 mg = ___3.1___ g

You may want to show students how to solve this problem from the point of view of ratio and proportion:

$$\frac{1.5 \text{ g}}{x \text{ mg}} = \frac{1 \text{ g}}{1,000 \text{ mg}}$$

Have students explain their responses in a journal.

Note in Example 1 that a part of a metric unit is usually expressed as a decimal, not as a fraction. That is, we write 1.5 g, not $1\frac{1}{2}$ g.

EXAMPLE 2

500 m = _____ km

Solution We want *kilometers*, so we solve this problem by multiplying the original measurement by the unit factor $\dfrac{1 \text{ km}}{1,000 \text{ m}}$.

$500 \text{ m} = 500 \text{ m} \times \dfrac{1 \text{ km}}{1,000 \text{ m}}$

$= \dfrac{500}{1,000} \text{ km}$ **To divide 500 by 1,000, move the decimal point in 500 three places to the left.**

$= 0.5 \text{ km}$

As a quick check, note that the number of the larger unit (0.5 km) is less than the number of the smaller unit (500 m) for an equivalent measurement.

PRACTICE 2

2,500 cm = ___25___ m

EXAMPLE 3

Express 3 km in millimeters.

Solution Because the table of lengths on page 472 does not indicate how many millimeters are equivalent to a kilometer, we need to solve the problem in steps.

Step 1. Change kilometers to meters.

Step 2. Change meters to millimeters.

To combine these two steps, we multiply the original measurement by a chain of appropriate unit factors to get *millimeters* for the final answer.

$$3 \text{ km} = 3 \text{ km} \times \frac{1{,}000 \text{ m}}{1 \text{ km}} \times \frac{1{,}000 \text{ mm}}{1 \text{ m}}$$

$$= 3 \times 1{,}000 \times 1{,}000 \text{ mm}$$

$$= 3{,}000{,}000 \text{ mm}$$

Do you see that, because a kilometer is larger than a millimeter, the number of kilometers (3) is less than the number of millimeters (3,000,000)?

PRACTICE 3

Change 5,000,000 mm to kilometers.

5 km

EXAMPLE 4

For a 20-year-old female, the U.S. Recommended Daily Allowance (RDA) for calcium and iron is 1 g and 10 mg, respectively. Which RDA is higher? (*Source: The World Almanac and Book of Facts, 2000*)

Solution When comparing quantities expressed in different units, we convert them to the same unit, usually the smaller unit. Here, we change 1 g to milligrams.

$$1 \text{ g} = 1 \text{ g} \times \frac{1{,}000 \text{ mg}}{1 \text{ g}}$$

$$= 1{,}000 \text{ mg}$$

The RDA for calcium is 1,000 mg, which is higher than the 10-mg RDA for iron.

PRACTICE 4

The water level in a beaker stands at 271 ml. Is this more or less than a liter?

Less than a liter

Metric/U.S. Customary Unit Conversions

In some situations, we need to change a measurement expressed in a U.S. unit to a metric unit, or vice versa. For example, if we were driving in Canada and saw a road sign giving the distance to the next town in kilometers, we might want to translate that distance to miles.

Or suppose that we had gone shopping to buy mouthwash and wondered how many pint bottles are equal in capacity to a 750-ml bottle. Here, we might want to change pints to milliliters.

To convert, we must either have memorized or have access to metric/U.S. unit conversion relationships. The following table contains *approximate* relationships for conversions that occur most often.

METRIC/U.S. UNIT CONVERSION RELATIONSHIPS

Length	Weight	Capacity
2.5 cm ≈ 1 in.	28 g ≈ 1 oz	470 ml ≈ 1 pt
30 cm ≈ 1 ft	450 g ≈ 1 lb	2.1 pt ≈ 1 L
39 in. ≈ 1 m	2.2 lb ≈ 1 kg	1.1 qt ≈ 1 L
3.3 ft ≈ 1 m	910 kg ≈ 1 ton	3.8 L ≈ 1 gal
3,300 ft ≈ 1 km		260 gal ≈ 1 kl
1,600 m ≈ 1 mi		
1.6 km ≈ 1 mi		

EXAMPLE 5

Express 2 oz in grams.

Solution According to the conversion table, 1 oz ≈ 28 g. Because we want *grams*, we multiply 2 oz by the unit factor $\dfrac{28 \text{ g}}{1 \text{ oz}}$.

$$2 \text{ oz} \approx 2 \text{ oz} \times \frac{28 \text{ g}}{1 \text{ oz}}$$

$$\approx 56 \text{ g}$$

So 2 ounces is about 56 grams.

PRACTICE 5

Express 10 gal in terms of liters.   38 L

Note that our answer in Example 5 is only an approximation because the unit factor is not exact. Also note that the number of grams is more than the number of ounces because an ounce is larger than a gram.

EXAMPLE 6

The typical length of a bed is about 1.9 m. Express this length in inches, rounded to the nearest inch.

Solution We know from the conversion table that 39 in. ≈ 1 m. Because we want to convert to inches, we use the unit factor $\dfrac{39 \text{ in.}}{1 \text{ m}}$.

$$1.9 \text{ m} \approx 1.9 \text{ m} \times \frac{39 \text{ in.}}{1 \text{ m}}$$

$$\approx 74.1 \text{ in.}$$

When we round 74.1 in. to the nearest inch, we get 74 in. So the typical length of a bed is about 74 in.

PRACTICE 6

The distance from Washington, D.C., to Atlanta is 973 km. Express this distance in miles, rounded to the nearest tenth of a mile.   608.1 mi

Exercises 10.2

For Extra Help

 Tape 12

 InterAct Math
Tutorial Software

 AWL Tutor Center

 Student's Solutions
Manual

 www.awlonline.com/
akstbragg

Choose the unit that you would most likely use to measure each quantity.

1. The volume of liquid in a test tube
a. millimeter **b.** milligram **c.** milliliter
c

2. The weight of a television set
a. milligram **b.** gram **c.** kilogram
c

3. The width of a room
a. millimeter **b.** meter **c.** kilometer
b

4. The distance between two cities
a. kilometer **b.** kilogram **c.** millimeter
a

Choose the best estimate in each case.

5. The capacity of a large bottle of soda
a. 1 ml **b.** 1 L **c.** 1 kg
b

6. The width of film for slides
a. 35 mm **b.** 35 cm **c.** 35 m
a

7. The height of the Washington Monument
a. 170 cm **b.** 170 m **c.** 170 km
b

8. The length of a mailing envelope
a. 20 mm **b.** 20 cm **c.** 20 m
b

9. The capacity of a bottle of hydrogen peroxide
a. 400 ml **b.** 400 L **c.** 400 g
a

10. The weight of an adult
a. 70 mg **b.** 70 g **c.** 70 kg
c

Change each quantity to the indicated unit.

11. 1,000 mg = <u>1</u> g

12. 253 mm = <u>0.253</u> m

13. 750 g = <u>0.75</u> kg

14. 2 L = <u>2,000</u> ml

15. 0.08 km = <u>80</u> m

16. 4.3 kg = <u>4,300</u> g

17. 3.5 m = <u>3,500</u> mm

18. 900 m = <u>0.9</u> km

19. 5 ml = <u>0.005</u> L

20. 250 mg = <u>0.25</u> g

21. 4,000 mm = <u>4</u> m

22. 5 L = <u>5,000</u> ml

23. 7,000 ml = <u>7</u> L

24. 2,500 mg = <u>2.5</u> g

25. 413 cm = <u>4.13</u> m

26. 2.8 m = <u>280</u> cm

27. 0.002 kg = <u>2,000</u> mg

28. 3,000 mm = <u>300</u> cm

Combine.

29. 3 km + 250 m
3,250 m

30. 5 L − 600 ml
4,400 ml

31. 98 kg + 25.6 g
98,025.6 g

32. 30 cm + 2 m
230 cm

Change each quantity to the indicated unit. If needed, round the answer to the nearest tenth of the unit.

33. 30 oz ≈ <u>840</u> g

34. 4 mi ≈ <u>6.4</u> km

35. 10 cm ≈ <u>4</u> in.

36. 900 g ≈ <u>2</u> lb

37. 48 in. ≈ <u>1.2</u> m

38. 6 qt ≈ <u>5.5</u> L

Applications

Solve. If needed, round the answer to the nearest tenth of the unit.

39. According to a medical journal, the average daily U.S. diet contains 6,000 mg of sodium. How many grams is this? (*Source:* Journal of the American Medical Association)   6 g

40. In the Summer Olympics, a major track-and-field event is the 100-m dash. How long is this race in kilometers?

0.1 km

41. You must administer a 4-ml dose of a drug daily to a patient. If there is 1 L of this drug on hand, will it last your patient 120 days?   Yes

42. Vitamin C commonly comes in pills with a strength of 500 mg. How many of these pills will you need to take if you want a dosage of half a gram? Take 1 pill.

43. In a physics lab, you measured the length of a pendulum string as 7.5 cm. Express this length in inches.   3 in.

44. The speed limit on many European highways is 100 km/hr. Express this speed in miles per hour.   62.5 mph

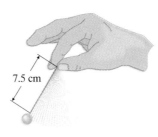

7.5 cm

45. A can of shaving cream weighs 312 g. How many kilograms is this?   0.3 kg

46. A mirror is 250 cm long. Round this length to the nearest meter.   3 m

47. The side of a square tile is 75 mm long. If 100 of these tiles are placed on the floor side by side, what is their total length in centimeters?   750 cm

48. The standard width of a bowling lane is 41 in. Express this width in feet.
$3\frac{5}{12}$ ft

49. Your chemistry professor mixes the contents of two beakers containing 2.5 L and 700 ml of a liquid. What is the combined amount?   3.2 L

50. A prehistoric bird had a wing span of 8 m. Express this wing span in feet. (*Source: Guiness Book of Records*)   26.4 ft

51. In 1989, an oil tanker in Alaska spilled 10,000,000 gal of oil in Prudhoe Bay. What is the quantity of this spillage expressed in liters?   38,000,000 L

52. A passenger car is generally considered small if the distance between its front and back wheels is less than 95 inches. What is this distance expressed in meters?   2.4 m

53. One of the heaviest babies ever born was an Italian boy who at birth weighed 360 oz. What was the baby's weight in kilograms? (*Source: Guiness Book of Records*) 10.1 kg

54. A train can climb the mountain at a speed of 55 kilometers per hour. Express this speed in miles per hour. 34.4 mph

■ *Check your answers on page A-14.*

MINDSTRETCHERS

INVESTIGATION

1. The metric system includes some extremely tiny units, including the angstrom. An angstrom (abbreviated Å) is $\dfrac{1}{10,000,000}$ millimeter. In your college library or on the Web, find situations in which you might want to use angstroms.   Answers may vary.

WRITING

2. Consider the following measurement expressed in the metric system.

$$\underbrace{37,}_{km}\underbrace{568.}_{m}\underbrace{251}_{mm} \text{ meters}$$

Note how we can split this measurement in meters into other metric units. Would this work with U.S. units? Explain.   No, because U.S. units are not based on the decimal system, that is, base ten.

GROUPWORK

3. A liter is about 5% more than a quart. Work with a partner to answer the following questions.

a. Do you think that, as the United States goes metric, containers will increase in size? Explain.   Answers may vary.

b. Suppose containers were to increase in size. What do you think the economic consequence of this increase would be?   Answers may vary.

KEY UNITS

QUANTITY	U.S. CUSTOMARY UNITS	RELATIONSHIPS
[10.1] **Length**	Inch (in.), foot (ft), yard (yd), and mile (mi)	12 in. = 1 ft 3 ft = 1 yd 5,280 ft = 1 mi
[10.1] **Weight**	Ounce (oz), pound (lb), and ton	16 oz = 1 lb 2,000 lb = 1 ton
[10.1] **Capacity (Liquid Volume)**	Fluid ounce (fl oz), pint (pt), quart (qt), and gallon (gal)	16 fl oz = 1 pt 2 pt = 1 qt 4 qt = 1 gal
[10.1] **Time**	Second (sec), minute (min), hour (hr), day, week (wk), month (mo), and year (yr)	60 sec = 1 min 60 min = 1 hr 24 hr = 1 day 7 days = 1 wk 52 wk = 1 yr 12 mo = 1 yr 365 days = 1 yr

QUANTITY	METRIC UNITS	RELATIONSHIPS
[10.2] **Length**	Millimeter (mm), centimeter (cm), meter (m), and kilometer (km)	$1 \text{ mm} = \dfrac{1}{1,000} \text{ m}$ $1 \text{ cm} = \dfrac{1}{100} \text{ m}$ $1 \text{ km} = 1,000 \text{ m}$
[10.2] **Weight**	Milligram (mg), gram (g), and kilogram (kg)	$1 \text{ mg} = \dfrac{1}{1,000} \text{ g}$ $1 \text{ kg} = 1,000 \text{ g}$
[10.2] **Capacity (Liquid Volume)**	Milliliter (ml) and liter (L)	$1 \text{ ml} = \dfrac{1}{1,000} \text{ L}$
[10.2] **Time**	Same as U.S. customary units.	

KEY METRIC/U.S. UNIT CONVERSION RELATIONSHIPS

[10.2] Length	2.5 cm ≈ 1 in. 30 cm ≈ 1 ft 1 m ≈ 39 in. 1 m ≈ 3.3 ft 1 km ≈ 3,300 ft 1,600 m ≈ 1 mi 1.6 km ≈ 1 mi
[10.2] Weight	28 g ≈ 1 oz 450 g ≈ 1 lb 1 kg ≈ 2.2 lb 910 kg ≈ 1 ton
[10.2] Capacity (Liquid Volume)	470 ml ≈ 1 pt 1 L ≈ 2.1 pt 1 L ≈ 1.1 qt 3.8 L ≈ 1 gal 1 kl ≈ 260 gal

Chapter 10 Review Exercises

To help you review this chapter, solve these problems.

[10.1]–[10.2]

Change each quantity to the indicated unit.

1. 5 yd = <u>15</u> ft

2. 20 mo = <u>$1\frac{2}{3}$</u> yr

3. 8 oz = <u>$\frac{1}{2}$</u> lb

4. 10 ft = <u>$3\frac{1}{3}$</u> yd

5. $3\frac{1}{2}$ tons = <u>7,000</u> lb

6. $8\frac{1}{2}$ lb = <u>136</u> oz

7. $3\frac{1}{2}$ gal = <u>14</u> qt

8. 2 pt = <u>32</u> fl oz

9. 3 yd = <u>9</u> ft

10. 150 sec = <u>$2\frac{1}{2}$</u> min

11. 4 oz ≈ <u>112</u> g, rounded to the nearest gram.

12. 5 cm ≈ <u>2</u> in., rounded to the nearest inch.

13. 32 km ≈ <u>20</u> mi, rounded to the nearest mile.

14. 4 gal ≈ <u>15</u> L, rounded to the nearest liter.

15. 10,000 ft = <u>2</u> mi, rounded to the nearest mile.

16. 2,000 oz = <u>125</u> lb, rounded to the nearest pound.

Compute the given sum or difference.

17.   4 hr 20 min
    +3 hr 50 min
    8 hr 10 min

18.   20 ft
    − 1 ft 3 in.
    18 ft 9 in.

19. 7 min 2 sec − 53 sec
    6 min 9 sec

20. 3 lb 6 oz + 2 lb 9 oz + 1 lb 3 oz
    7 lb 2 oz

21. 3 gal 2 qt − 1 gal 3 qt
    1 gal 3 qt

22. 4 pt − 2 pt 14 fl oz
    1 pt 2 fl oz

484

[10.2]

Choose the unit that you would most likely use to measure each quantity.

23. The weight of a car

a. milligrams   **b.** grams   **c.** kilograms

c.

24. The width of a pencil's point

a. millimeters   **b.** centimeters   **c.** meters

a.

25. The capacity of an oil barrel

a. milliliters   **b.** liters   **c.** meters

b.

26. The distance from Detroit to Montreal

a. millimeters   **b.** centimeters   **c.** kilometers

c.

Choose the best estimate in each case.

27. The width of a piece of typing paper

a. 16 mm   **b.** 16 cm   **c.** 16 km

b.

28. The capacity of a bottle of mouthwash

a. 100 ml   **b.** 100 L   **c.** 100 g

a.

29. The weight of an aspirin pill

a. 200 mg   **b.** 200 g   **c.** 200 kg

a.

30. The length of an athlete's long jump

a. 6.72 mm   **b.** 6.72 cm   **c.** 6.72 m

c.

Mixed Applications

Solve. If needed, round the answer to the nearest tenth of the unit.

31. Your favorite compact disc plays for 72 min. Express this playing time in hours.   1.2 hr

32. In a recent year, a typical U.S. resident used about 1,600 gal of water a day for residential, agricultural, and industrial purposes. How many pints is this? (*Source:* U.S. Geological Survey)   12,800 pt

33. *Frankenstein* (130 min) and *Dracula* (1 hr 15 min) are two classic horror films made in 1931. Which film is longer?   *Frankenstein*

34. Some doctors recommend that athletes drink about 600 ml of fluid each hour. Round this amount to the nearest liter.   1 L

35. A teaspoon of common table salt contains about 2,000 mg of sodium. How many grams of sodium is this?   2 g

36. In a factory, a chemical process produced 3 mg of a special compound each hour. How many grams were produced in 24 hr?   0.1 g

37. The average level of cholesterol in U.S. children is 160 mg/100 ml of blood. Express this ratio in terms of grams per liter. (*Source: The Peoplepedia*)   1.6 g/L

38. According to a journal article, 750 kg of pesticide is sprayed on a typical U.S. golf course each year. How many grams is this?   750,000 g

39. The following table shows how you spent your day. How long after leaving home did you return?   6 hr 30 min

Activity	Length of Time
Trip to Work	45 min
Work	$5\frac{1}{4}$ hr
Trip Home	30 min

40. The following diagram shows the heights of an average U.S. woman and an average 10-year-old U.S. girl.

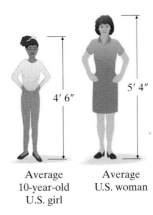

Average 10-year-old U.S. girl   Average U.S. woman

What is the difference in their heights? (*Source: Archives of Pediatrics and Adolescent Medicine*)   10 in.

41. One of the shortest dinosaurs that ever lived was only 60 cm long when fully grown. What was the length of this dinosaur in inches? (*Source: Encarta Learning Zone Encyclopedia*)   24 in.

42. There are about 6 qt of blood in an average-sized man. Express this amount in liters.   5.5 L

43. The time is now 12:51 P.M. What time will it be in $3\frac{1}{2}$ hours?   4:21 P.M.

44. How many 200-ml portions of soda can you serve from a 2.4-L bottle?   12 portions

■ *Check your answers on page A-14.*

Chapter 10 Posttest

To see if you have mastered the topics in this chapter, take this test.

1. 120 sec = 2 min

2. 7 yd = 21 ft

3. How many seconds are there in an hour? 3,600 sec

4. Which of the following units is a measure of length?
   a. a gram   **b.** a meter   **c.** a liter
   b.

5. The weight of a baby is measured in
   a. milligrams   **b.** grams   **c.** kilograms
   c.

6. 400 cm = 4 m

7. 500 ml = 0.5 L

8. Which is larger: 4 mm or 2 km? 2 km

9. The following diagram shows the radius of Earth. Express this radius in terms of meters, rounded to the nearest million meters. 6,000,000 m

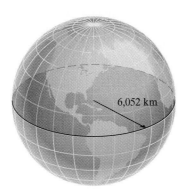

6,052 km

10. The Mona Lisa is one of the best known paintings in the world.

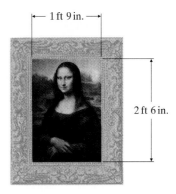

1 ft 9 in.

2 ft 6 in.

By how many inches does the painting's height exceed its width? 9 in.

■ *Check your answers on page A-14.*

Cumulative Review Exercises

To help you review, solve the following.

1. Compute: 9^3
729

2. Express as a mixed number: $\dfrac{11}{2}$
$5\dfrac{1}{2}$

3. Divide: $\dfrac{2.8}{0.2}$
14

4. What is 20% more than 30?
36

5. $(-3)^2 + 5 = ?$
14

6. Find the range: 8, 6, 2, 9, 1, 6
8

7. Fill in the blank: 7 ft = <u>84</u> in.

8. In a recent year, the average person in the United States ate about 44 pt of ice cream. How many quarts is this?
22 qt

9. A house cat spends about 80% of its time resting, or catnapping. Express this percent as a fraction.
$\dfrac{4}{5}$

10. According to an agricultural study, 5 kg of grain are required to produce 1 kg of steak. Express this statement in terms of grams.
According to an agricultural study, 5000 g of grain are required to produce 1,000 g of steak.

■ *Check your answers on page A-14.*

Basic Geometry

GEOMETRY AND ARCHITECTURE

Students of geometry study abstract figures in space, whereas architects design real structures in space. The two fields, geometry and architecture, are therefore closely related.

The simplest architectural structures have basic geometric shapes. An igloo in the far North and a dome that graces a church or state capital are shaped like hemispheres. A tepee is in the shape of a cone, and the peak of a roof is triangular.

The rectangle plays an especially important role in architectural design. Bricks, windows, doors, rooms, buildings, lots, city blocks, and street grids are all based on the rectangle — one of the most adaptable shapes for human needs.

Of all the rectangles with a given area, the square has the smallest perimeter. Because of this special property of the square, it is therefore the least expensive of these rectangles to build. So warehouses are often square-shaped, whereas houses, hotels, and hospitals — for which daylight and a long perimeter are more important — usually are not.

Chapter 11 Pretest

To see if you have already mastered the topics in this chapter, take this test.

1. Sketch and label an example of each figure.

    a. Angle   **b.** Right triangle

2. Find each indicated square root.

    a. $\sqrt{36}$ 6   **b.** $\sqrt{121}$ 11

3. Find the supplement of 100°.   80°

4. Find the complement of 36°.   54°

Find each perimeter or circumference. Use $\pi \approx 3.14$ when needed.

5.

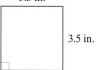

3.5 in.   14 in.

3.5 in.

6.

8 ft   20 ft

2 ft

7. A circle with a diameter of 4 in.   Approximately 12.56 in.

Find each area. Use $\pi \approx 3.14$ when needed.

9.

6 in.

10 in.

30 sq in.

10.

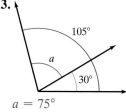

8 ft

50.24 sq ft

8. An equilateral triangle 2.6 m wide.   7.8 m

Find each volume. Use $\pi \approx 3.14$ when needed.

11.

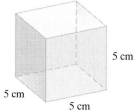

5 cm

5 cm   5 cm

125 cm³

12.

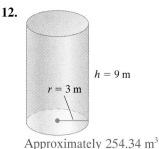

$h = 9$ m

$r = 3$ m

Approximately 254.34 m³

Find the unknown measure(s) in each figure in Exercises 13–116.

13.

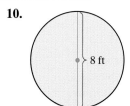

105°

a

30°

$a = 75°$

14.

$x = \sqrt{34}$ m

5 m x

3 m

15.

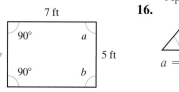

7 ft

90° a

y 5 ft

90° b

x

$a = b = 90°$; $x = 7$ ft; $y = 5$ ft

16.

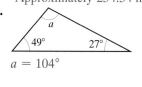

a

49°   27°

$a = 104°$

17. For the diagrams shown, $\triangle ABC$ is similar to $\triangle DEF$. Find y.

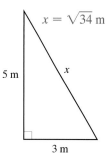

B

6 cm   9 cm

A 12 cm C

E

10 cm y

D 20 cm F

$y = 15$ cm

18. You are searching for survivors of a shipwreck that took place within a mile of a rock. What is the area of the region that would be most appropriate to search for survivors? Round to the nearest square mile.   3 mi²

19. What is the area of a square whose side measures 15 cm?   225 cm²

20. A swimming pool is 60 ft long, 5 ft deep, and 30 ft wide. How much water does it take to fill the pool so that the water is 1 ft below the top of the pool? 7,200 ft³ of water

■ *Check your answers on page A-15.*

11.1   Introduction to Basic Geometry

OBJECTIVES
- *To identify basic geometric concepts*
- *To identify basic geometric figures*
- *To solve word problems involving basic geometric concepts and figures*

What Geometry Is and Why It Is Important

The word **geometry**, which dates back thousands of years, means "measurement of the earth." Today, we use the term to mean the branch of mathematics that deals with concepts such as point, line, angle, perimeter, area, and volume.

Ancient peoples, including the Egyptians, used the principles of geometry in their construction projects. They understood these principles because of observations they made in their daily lives and their studies of the physical forms in nature.

Geometry also has many practical applications in such diverse fields as art and design, architecture, physics, and engineering. To give one example, city planners often make use of geometric concepts, relationships, and notation when designing the layout of a city. Note how the use of geometric thinking helps to transform the street plan on the left to the geometric diagram on the right, making it easier to focus on the street plan's key features.

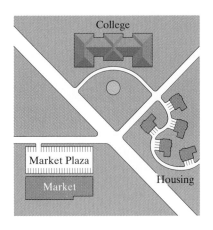
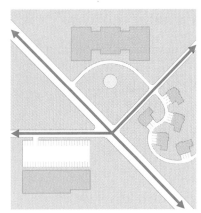

Basic Geometric Concepts

Let's first consider some of the basic concepts that underlie the study and application of geometry. The following table gives the definitions of some basic geometric terms illustrated in the preceding street plan. We use these terms throughout this chapter.

Definition	Example
A **point** is an exact location in space. A point has no dimension.	A • (read "point A")
A **line** is a collection of points along a straight path that extends endlessly in both directions.	$C \quad B$ ⟵————⟶ $\overleftrightarrow{CB}$ (read "line CB")
A **line segment** is a part of a line having two endpoints. A line segment has only one dimension — its length.	$A \qquad B$ •————• $\overline{AB}$ (read "line segment AB") The length of $\overline{AB}$ is denoted AB.
	continued

A **ray** is a part of a line having only one endpoint.

$\overrightarrow{CD}$ (read "ray CD")
(The endpoint is always the first letter.)

An **angle** consists of two rays that have a common endpoint called the **vertex** of the angle.

$\angle ABC$ (read "angle ABC")
(The vertex is always the middle letter.)
$\angle ABC$ can also be written as $\angle CBA$ or just $\angle B$.

A **plane** is a flat surface that extends endlessly in all directions.

Plane $ABCD$

Angles are measured in *degrees* (°). Angles are classified according to their measures.

Definition	Example
A **straight angle** is an angle that measures 180°.	$\angle ABC$ is a straight angle.
An **right angle** is an angle that measures 90°.	Symbol for right angle. $\angle DEF$ is a right angle.
An **acute angle** is an angle that measures less than 90°.	$\angle XYZ$ is an acute angle.
An **obtuse angle** is an angle that measures more than 90° and less than 180°.	$\angle CDE$ is an obtuse angle.

Complementary angles are two angles whose sum is 90°.

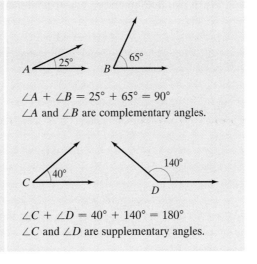

$\angle A + \angle B = 25° + 65° = 90°$
$\angle A$ and $\angle B$ are complementary angles.

Supplementary angles are two angles whose sum is 180°.

$\angle C + \angle D = 40° + 140° = 180°$
$\angle C$ and $\angle D$ are supplementary angles.

Lines in a plane are either intersecting or parallel.

Definition	Example
Intersecting lines are two lines that meet.	$\overleftrightarrow{AC}$ intersects $\overleftrightarrow{DE}$ at point B.
Parallel lines are two lines in the same plane that do not intersect.	$\overleftrightarrow{EF} \parallel \overleftrightarrow{GH}$ is read "$\overleftrightarrow{EF}$ is parallel to $\overleftrightarrow{GH}$."
Perpendicular lines are two lines that intersect to form right angles.	$\overleftrightarrow{RT} \perp \overleftrightarrow{PQ}$ is read "$\overleftrightarrow{RT}$ is perpendicular to $\overleftrightarrow{PQ}$." $\angle RSP$, $\angle RSQ$, $\angle PST$, and $\angle QST$ are all right angles.

When two lines intersect, two special pairs of angles are formed.

Definition	Example
Vertical angles are two equal angles formed by two intersecting lines.	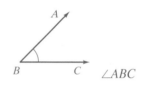 $\angle BAE$ and $\angle DAC$ are vertical angles. $\angle BAD$ and $\angle EAC$ are vertical angles.

Try drawing another pair of vertical angles. Do you think that they are equal? What is true about the pair of angles formed by intersecting lines that are not vertical angles?

Now let's consider more examples involving these basic geometric terms.

EXAMPLE 1

Sketch and label a line segment. Write the symbol for the line segment.

Solution First sketch a line segment.

Then label the line segment.

F

E

This line segment is written $\overline{EF}$, and is read "line segment EF."

PRACTICE 1

Draw an angle. Label the angle with three letters, and write the symbol for this angle.

A

B C $\angle ABC$

EXAMPLE 2

$\angle A$ and $\angle B$ are complementary angles. Find the measure of $\angle A$ if $\angle B = 69°$.

Solution Because $\angle A$ and $\angle B$ are complementary angles, $\angle A + \angle B = 90°$.

$$\angle A + \angle B = 90°$$

$$\angle A + 69° = 90°$$

$$\angle A + 69° - 69° = 90° - 69° \quad \text{Subtract 69° from each side.}$$

$$\angle A = 21°$$

PRACTICE 2

What is the measure of the angle complementary to 37°?   53°

EXAMPLE 3

Find the angle that is supplementary to 89°.

Solution To find the measure of the angle that is supplementary to 89°, we write the following equation.

$$89° + x = 180°$$

Here, x represents the measure of the supplementary angle.

$$89° + x = 180°$$

$$89° - 89° + x = 180° - 89° \quad \text{Subtract 89° from each side.}$$

$$x = 91°$$

So 91° is supplementary to 89°.

PRACTICE 3

What is the measure of the angle supplementary to 15°?   165°

EXAMPLE 4

In the following diagram, $\angle ABC$ is a straight angle. Find y.

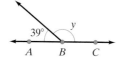

Solution Because $\angle ABC$ is a straight angle, $y + 39° = 180°$. We solve this equation for y.

$$y + 39° = 180°$$

$$y + 39° - 39° = 180° - 39° \quad \text{Subtract 39° from each side.}$$

$$y = 141°$$

PRACTICE 4

In the diagram shown, find x. $x = 82°$

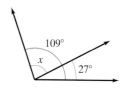

EXAMPLE 5

Find the values of x and y in the diagram shown at the right.

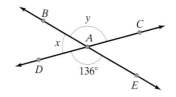

Solution Because $\angle BAC$ and $\angle DAE$ are vertical angles and $\angle DAE = 136°$, $\angle BAC = 136°$, or $y = 136°$. Because $\angle DAC$ is a straight angle, the sum of x and y is 180°.

$$x + y = 180°$$

$$x + 136° = 180°$$

$$x + 136° - 136° = 180° - 136°$$

$$x = 44°$$

So $x = 44°$ and $y = 136°$.

PRACTICE 5

In the following diagram, what are the values of a and b? $a = b = 153°$

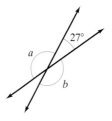

Basic Geometric Figures

Here we use the concepts just discussed to define some basic geometric figures: triangles, trapezoids, parallelograms, rectangles, squares, and circles. Except for circles, these figure are *polygons*.

Definition

A **polygon** is a closed plane figure made up of line segments.

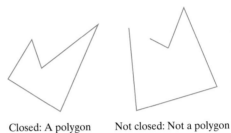

Closed: A polygon   Not closed: Not a polygon

Polygons are classified according to the number of their sides. Here we examine two types of polygons — triangles and quadrilaterals.

Definition	Example
A **triangle** is a polygon with three sides.	$\triangle DEF$ (read "triangle *DEF* ") $\triangle DEF$ has three *vertices* (plural of *vertex*) — points *D*, *E*, and *F*. $\triangle DEF$ has sides $\overline{DE}$, $\overline{EF}$, and $\overline{DF}$.
A **quadrilateral** is a polygon with four sides.	Quadrilateral *ABCD* has four vertices — points *A*, *B*, *C*, and *D*.

Triangles are classified according to the measures of either their sides or their angles.

Definition	Example
An **equilateral triangle** is a triangle with *three* sides equal in length.	$PQ = QR = PR$

An **isosceles triangle** is a triangle with *two* sides equal in length.

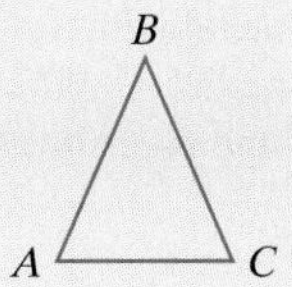

$AB = BC$

A **scalene triangle** is a triangle with *no* sides equal in length.

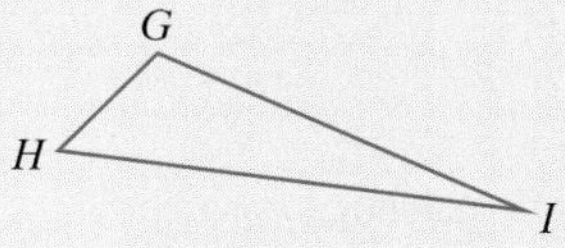

$GH \neq GI, GH \neq HI,$ and $GI \neq HI$

An **acute triangle** is a triangle with *three acute* angles.

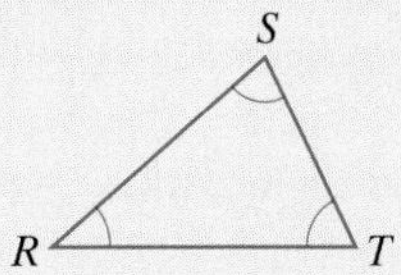

$\angle R$, $\angle S$, and $\angle T$ are acute angles.

A **right triangle** is a triangle with *one right* angle.

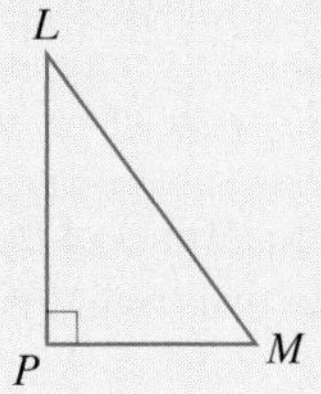

$\angle P = 90°$

An **obtuse triangle** is a triangle with *one obtuse* angle.

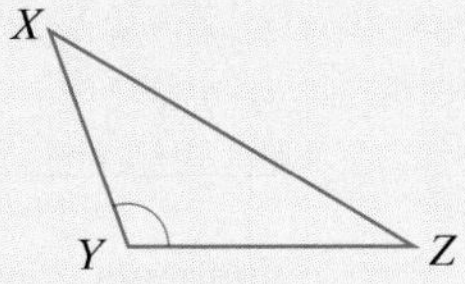

$\angle Y$ is an obtuse angle.

In any triangle, the measures of all three angles sum to 180°.

We have already shown that a polygon with four sides is called a *quadrilateral*. Let's consider special types of quadrilaterals.

Have students use a paper triangle to demonstrate that the measures of its angles sum to 180°.

Definition	Example
A **trapezoid** is a quadrilateral with only one pair of opposite sides parallel.	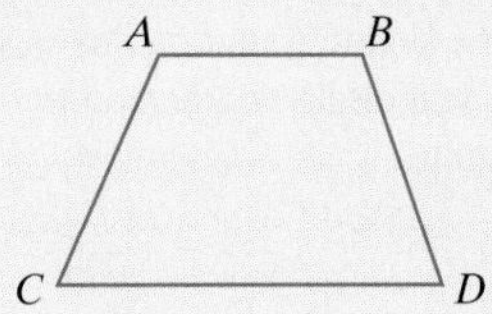 $\overline{AB} \parallel \overline{CD}$

continued

A **parallelogram** is a quadrilateral with both pairs of opposite sides parallel. Opposite sides are equal in length, and opposite angles have equal measures.

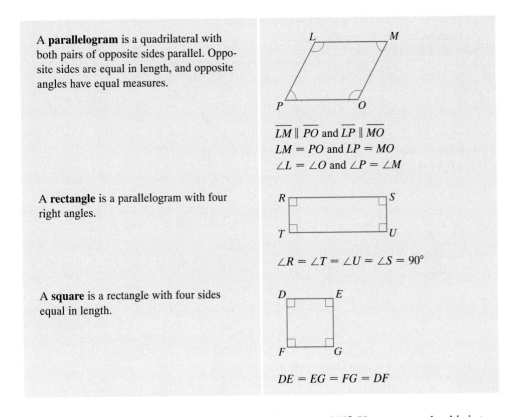

$\overline{LM} \parallel \overline{PO}$ and $\overline{LP} \parallel \overline{MO}$
$LM = PO$ and $LP = MO$
$\angle L = \angle O$ and $\angle P = \angle M$

A **rectangle** is a parallelogram with four right angles.

$\angle R = \angle T = \angle U = \angle S = 90°$

A **square** is a rectangle with four sides equal in length.

$DE = EG = FG = DF$

In any quadrilateral, the measures of the angles sum to 360°. You can see why this is true by cutting a quadrilateral into two triangles. Each triangle has three angles that sum to 180°.

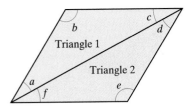

The last basic geometric figure we consider here is the circle.

Definition	Example
A **circle** is a closed plane figure made up of points that are all the same distance from a fixed point called the **center**.	Circle with center O
A **diameter** is a line segment that passes through the center of a circle and has both endpoints on the circle.	Diameter $\overline{AB}$

A **radius** is a line segment with one endpoint on the circle and the other at the center.

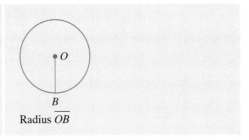

Radius $\overline{OB}$

Note that the diameter of a circle is twice the radius, or $d = 2r$.

EXAMPLE 6

Sketch and label an isosceles triangle. Name the equal sides.

Solution

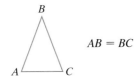

$AB = BC$

PRACTICE 6

Draw and label a quadrilateral that has at least one right angle with opposite sides equal and parallel. Name both pairs of parallel sides.

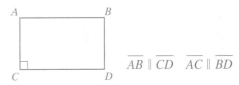

$\overline{AB} \parallel \overline{CD}$ $\overline{AC} \parallel \overline{BD}$

EXAMPLE 7

In $\triangle DEF$, $\angle D = 45°$ and $\angle E = 65°$. Find the measure of $\angle F$.

Solution First we draw a diagram.

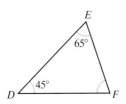

The sum of the measures of the angles is 180°, so we write the following.

$$\angle D + \angle E + \angle F = 180°$$

$$45° + 65° + \angle F = 180°$$

$$110° + \angle F = 180°$$

$$110° - 110° + \angle F = 180° - 110°$$

$$\angle F = 70°$$

PRACTICE 7

In triangle RST, where $\angle S$ is a right angle and $\angle T = 30°$, find the measure of $\angle R$.

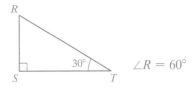

$\angle R = 60°$

EXAMPLE 8

In the quadrilateral shown, find the measure of ∠D.

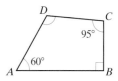

Solution   The sum of the measures of the four angles is 360°. Note that ∠B is a right angle, so ∠B = 90°. We write the following.

$$\angle A + \angle B + \angle C + \angle D = 360°$$

$$60° + 90° + 95° + \angle D = 360°$$

$$245° + \angle D = 360°$$

$$245° - 245° + \angle D = 360° - 245°$$

$$\angle D = 115°$$

PRACTICE 8

In the trapezoid shown, what is the measure of ∠U?   ∠U = 60°

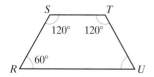

EXAMPLE 9

Over the years, one of the world's most popular cookies has changed in size. The current version is circular and $1\frac{3}{4}$ in. across. What is the distance from the center of the cookie to a point on the edge of the cookie?

Solution   The diameter of the cookie is $1\frac{3}{4}$ in. The distance from the center of the cookie to a point on the edge of the cookie is the radius. To find the radius, we divide the diameter by 2.

$$1\frac{3}{4} \div 2 = \frac{7}{4} \div \frac{2}{1} = \frac{7}{4} \times \frac{1}{2} = \frac{7}{8}$$

The cookie's radius is $\frac{7}{8}$ in.

PRACTICE 9

A certain tree trunk has a radius of about 0.67 ft. What is the diameter of the trunk?

1.34 ft

Exercises 11.1

Sketch and label each geometric object. Where appropriate, use symbols to express your answer.

1. Point

A
•

2. Line $\overleftrightarrow{AB}$

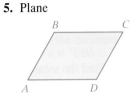

3. Line segment $\overline{BC}$

4. Angle $\angle ABC$

5. Plane

6. Ray

7. Parallel lines

8. Perpendicular lines

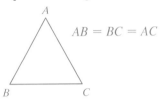

9. Right angle

10. Obtuse angle

11. Equilateral triangle

$AB = BC = AC$

12. Isosceles triangle

$PQ = PR$

13. Circle

○

14. Square

$AB = BD = CD = AC$

15. Rectangle

$KL = NM; LM = KN$

16. Trapezoid $\overline{QR} \parallel \overline{ST}$

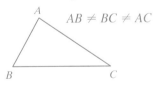

17. Scalene triangle

$AB \neq BC \neq AC$

18. Right triangle

$90°$

19. Acute angle

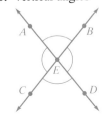

20. Vertical angles

$\angle AEB = \angle CED$

$\angle AEC = \angle BED$

Exercises 11.5

Find each missing side.

1. $x = 10\frac{2}{3}$ in.

$\triangle ABC \sim \triangle DEF$

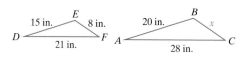

2. $y = 12$ ft

$\triangle LOM \sim \triangle RST$

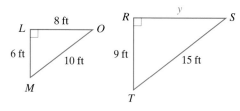

3. $x = 24$ m

$\triangle DOT \sim \triangle PAN$

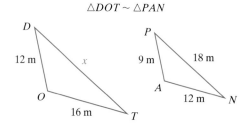

4. $y = 15$ cm

$\triangle ACT \sim \triangle MLK$

5. $x = 15$ ft, $y = 20$ ft

$\triangle DEF \sim \triangle ABC$

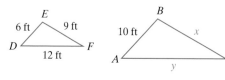

6. $x = 14.4$ m, $y = 14.4$ m

$\triangle DOT \sim \triangle PIN$

7. $x = 24$ yd, $y = 12$ yd

$\triangle TAP \sim \triangle RON$

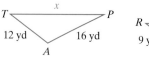

8. $x = 4$ cm, $y = 3.5$ cm

$\triangle FEG \sim \triangle CBD$

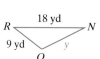
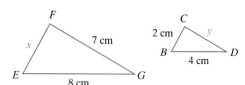

The number π equals $3.1415926\ldots$. It is an *irrational number*, because when π is expressed as a decimal, the digits go on indefinitely without any pattern being repeated. For convenience, we often use approximate values of π, such as 3.14 and $\dfrac{22}{7}$ when calculating circumferences by hand.

EXAMPLE 4	**PRACTICE 4**

Find the circumference of the circle shown. Use $\pi \approx 3.14$.

Solution   The radius of the circle is 4 m. We use the formula for the circumference of a circle in terms of the radius.

$$C = 2\pi r$$

$$\approx 2(3.14)(4) \quad \text{Substitute 4 for } r.$$

$$\approx 25.12$$

Therefore the circumference is approximately 25.12 m.

What is the circumference of the circle shown? Use $\pi \approx \dfrac{22}{7}$.   Approximately 132 in.

EXAMPLE 5	**PRACTICE 5**

Suppose that the diameter of a rolling wheel is 20 in. How far does it travel in one complete turn?

Solution   First let's draw a diagram.

The wheel makes one complete turn, so we know that it travels a distance equal to its circumference. We use the formula for the circumference in terms of the diameter.

A circular swimming pool has a radius of 18 ft. If a metal guardrail is to be placed around the edge of the pool, how many feet of railing are needed?   Approximately 113.0 ft

continued

$$C = \pi d$$

$$\approx 3.14(20) \quad \text{Substitute 20 for } d.$$

$$\approx 62.8$$

Thus the wheel travels approximately 62.8 in. in one turn. Explain how to solve this problem using the formula $C = 2\pi r$.

Have students write their responses in a journal.

Composite Figures

Two or more basic figures may be combined to form a **composite figure**.

EXAMPLE 6

Find the perimeter of the following figure, which consists of a semicircle and a rectangle.

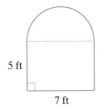

Solution The top of the figure is a semicircle with a diameter of 7 ft. The distance around the semircle is $\frac{1}{2}$ the circumference of the entire circle. So let's compute this circumference and then divide by 2.

$$C = \pi d$$

$$\approx \frac{22}{7}(7)$$

$$\approx 22$$

The circumference of the semicircle is approximately $\frac{22}{2}$, or 11 ft.

Now we find the perimeter of three sides of the rectangle at the bottom of the figure.

$$P = 5 + 7 + 5$$

$$= 17$$

The perimeter of the bottom is 17 ft.

Total perimeter = Circumference of top + Perimeter of bottom

= 11 ft + 17 ft

= 28 ft

Therefore the perimeter of the composite figure is approximately 28 ft.

PRACTICE 6

What is the perimeter of the figure shown, which is composed of a square and a semicircle?   Approximately 164.5 yd

36 yd

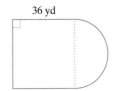

EXAMPLE 7

To prevent a cellar from being flooded, a plumber puts drainage pipes around the outside of a building complex. The outline of the building, consisting of two rectangles and a square, is shown.

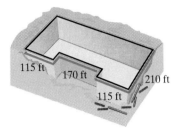

If you ignore the fact that the pipe is slightly away from the walls, how much pipe is required to go around the complex?

Solution We want to find the perimeter of the building. So we break the composite figure into three basic shapes — two rectangles and one square.

We find the length of the indented part of each rectangular shape by subtracting 170 ft from 210 ft to get 40 ft.

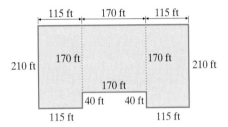

Now we know all the lengths that make up the building complex, so we can find its perimeter.

$$P = \underbrace{(115 + 210 + 115 + 40)}_{\text{Left side}} + \underbrace{(170 + 170)}_{\text{Center}}$$

$$+ \underbrace{(115 + 210 + 115 + 40)}_{\text{Right side}}$$

$$= 1{,}300$$

Note that the dashed lines are not part of the perimeter because they are not part of the outside of the figure. So 1,300 ft of drainage pipe are needed to go around the building complex.

PRACTICE 7

An interior decorator plans to place a wallpaper border at the top of the walls of a room whose ceiling is shown in the diagram. The ceiling consists of two rectangles. How many feet of the wallpaper border are needed?   82 ft

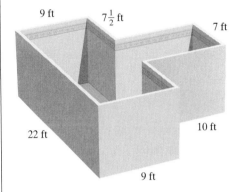

Exercises 11.2

Find the perimeter or circumference of each figure. Use π ≈ 3.14 when needed.

1. 2 in.   17 in.
6 in.
3 in.
1 in.
5 in.

2.   2 cm   24 cm
3 cm   3 cm
4 cm   4 cm
3 cm   3 cm
2 cm

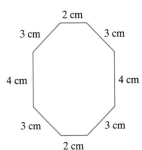

3.   25 m   60 m
5 m

4. 6.5 ft   26 ft
6.5 ft

5. $10\frac{1}{2}$ yd
$3\frac{1}{2}$ yd $3\frac{1}{2}$ yd
$3\frac{1}{2}$ yd

6.   13 m   30 m
5 m
12 m

7.   Approximately 62.8 m
10 m

8.   Approximately
75.36 m
24 cm

9.   Approximately
21.98 ft
7 ft

10.   Approximately
9.42 cm
1.5 cm

Find the perimeter of each composite geometric figure. Use π ≈ 3.14 when needed.

11. 6 ft   6 ft   42 ft
10 ft
10 ft

12.   26 cm
2 cm
8 cm   2 cm
3 cm

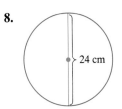

13. 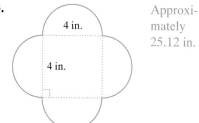 Approximately 25.12 in.

14. 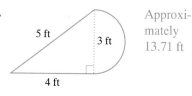 Approximately 13.71 ft

15. 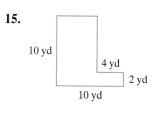 40 yd

16. 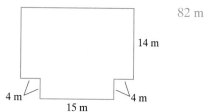 82 m

Find the perimeter or circumference. Use $\pi \approx 3.14$ when needed.

17. A square with side $5\frac{1}{4}$ yd long   21 yd

18. An equilateral triangle with side 8.5 km long   25.5 km

19. A rectangle of length $5\frac{3}{4}$ ft and width $3\frac{1}{4}$ ft   18 ft

20. A triangle whose sides measure 2 in., $1\frac{1}{2}$ in., and $\frac{7}{8}$ in. $4\frac{3}{8}$ in.

21. A circle whose radius measures 5 ft
Approximately 31.4 ft

22. A circle whose diameter measures 42 in.   Approximately 131.9 in.

23. An isosceles triangle whose equal sides measure $7\frac{1}{2}$ cm and whose third side measures 4 cm   19 cm

24. A rectangle with length 8 m and width $4\frac{1}{2}$ m   25 m

25. A circle whose diameter is 3.54 m long   Approximately 11.12 m

26. A polygon whose sides measure 22.75 ft, 25.73 ft, 15.94 ft, 18.23 ft, 21.65 ft, and 34.98 ft   139.28 ft

Applications

Solve.

27. Find the perimeter of the doubles tennis court shown below.   228 ft

28. If you drive from Atlanta to New York City to Chicago and back to Atlanta, what is the total mileage?   2,317 mi

29. As the following diagram shows, bicycle wheels come in different diameters.

In one wheel rotation, how much farther to the nearest inch does the 27-in. bicycle wheel go than the 25-in. bicycle wheel?   6 in.

30. Find the circumference of the circle shown.   Approximately 18.8 cm

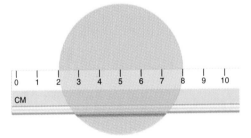

31. A garden 50 m wide and 100 m long is to be enclosed with a fence. If fence posts are placed every 10 m, how many posts are needed?   30 posts

32. If rug binding costs $1.95/ft, what is the cost of binding a rectangular rug that is 21 ft long and 12 ft wide? $128.70

33. Find the length of the belt needed for the two pulleys.   Approximately 69.98 ft

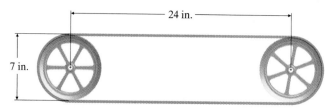

24 in.

7 in.

34. A carpenter plans to lay a wood molding in the room depicted. If the room has three doors, each 3 ft wide, what is the total length of floor molding required?   55 ft

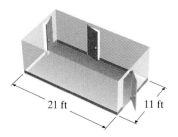

21 ft   11 ft

▦ *Use a calculator to solve the following problems, giving (a) the operation(s) carried out in your solution, (b) the rounded answer, and (c) an estimate of the answer.*

35. The radius of Earth is about 6,400 km. If a satellite is orbiting 400 km above Earth, find the distance to the nearest hundred kilometers that the satellite travels in one orbit.

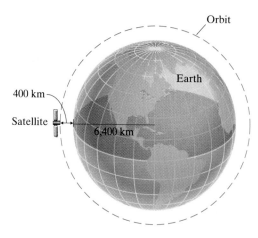

Orbit

Earth

400 km

Satellite

6,400 km

(a) Multiplication, (b) 42,700 km, (c) 42,000 km

36. If a crater on the moon is circular with a circumference of about 214.66 mi, what is the radius of the crater to the nearest mile?

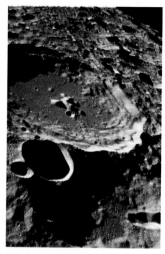

(a) Division, (b) 34 mi, (c) 33 mi

■ *Check your answers on page A-15.*

MINDSTRETCHERS

INVESTIGATION

1. Draw three triangles. Label the sides of each triangle a, b, and c. Measure each side, writing the measurements in the following table. Compare the sum of any two sides of a triangle with its third side.   Answers may vary.

	a	b	c	$a + b$	$a + c$	$b + c$
Triangle 1						
Triangle 2						
Triangle 3						

How does the length of the side of a triangle compare to the sum of the lengths of the other two sides?   The sum of the lengths of the other two sides is greater.

MATHEMATICAL REASONING

2. Consider the cart pictured. Which wheel do you think will wear out more quickly? Justify your answer.

Possible answer: The smaller wheel will wear out first because it will rotate more frequently than the large wheel in covering the same distance.

GROUPWORK

3. Explain how you can approximate the circumference of a circular room with a ruler. Compare your method with those of other members of the group.   Possible answer:

Measure the diameter and multiply by 3.14.

HISTORICAL NOTE

These mural decorations, called *litema,* from Southern Africa, show how simple geometric shapes can be combined to create diverse patterns. The twentieth-century pure mathematician G.H. Hardy wrote, "A mathematician, like a painter or a poet, is a maker of patterns."

Paulus Gerdes, *Women, Art and Geometry in Southern Africa* (Trenton, NJ: Africa World Press, Inc.), pp. 117 and 147.

11.3 Area

The Area of a Polygon and a Circle

OBJECTIVES

■ *To find the area of a polygon or a circle*
■ *To find the area of a composite figure*
■ *To solve word problems involving area*

Area is a measure of the size of a plane geometric figure. The floor space in a room, the size of a lake, and the sweep of a beam of light are all examples of areas.

To find the area of the rectangle shown, we split it into little squares, each representing 1 in^2.

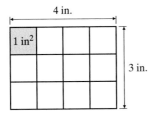

Then we count the number of square inches within the rectangle, which is 12 in^2.

Each row of the rectangle contains 4 in^2, and the rectangle has 3 rows. So a shortcut to counting the total number of square inches is to multiply 3×4, getting 12 in^2 in all. Note that areas are measured in square units, such as square inches (sq in. or in^2), square miles (sq mi or mi^2), or square meters (m^2).

Definition

Area is the number of square units that a figure contains.

In this section, we focus on finding the area of polygons and circles. We can compute these areas by using the following formulas. First, we consider polygons.

Figure	Formula	Example
Rectangle	$A = lw$ Area equals the length times the width. *l*, *w*	8 ft, 5 ft $A = lw$ $= 8 \cdot 5 = 40$ $= 40 \text{ ft}^2$
Square	$A = s \cdot s = s^2$ Area equals the square of a side. *s*, *s*	4 ft $A = s^2$ $= 4 \cdot 4 = 16$ $= 16 \text{ ft}^2$

continued

517

Triangle

$A = \dfrac{1}{2}bh$

Area equals one-half the base times the height.

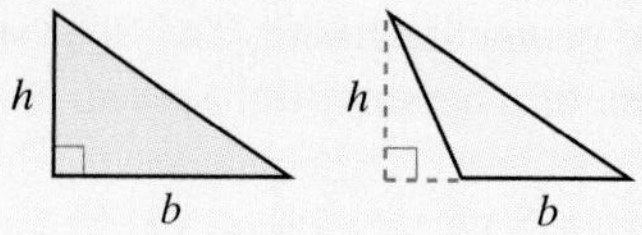

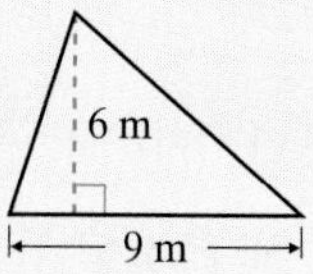

$A = \dfrac{1}{2}bh$

$= \dfrac{1}{2} \cdot 9 \cdot 8$

$= \dfrac{1}{2} \cdot 72 = 36$

$= 36 \text{ m}^2$

Parallelogram

$A = bh$

Area equals the base times the height.

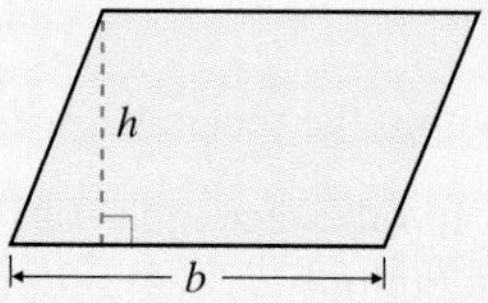

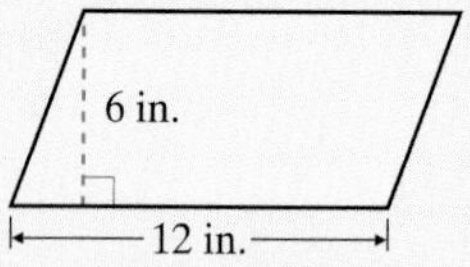

$A = bh$

$= 12 \cdot 6 = 72$

$= 72 \text{ in}^2$

Trapezoid

$A = \dfrac{1}{2}h(b + B)$

Area equals one-half the height times the sum of the bases.

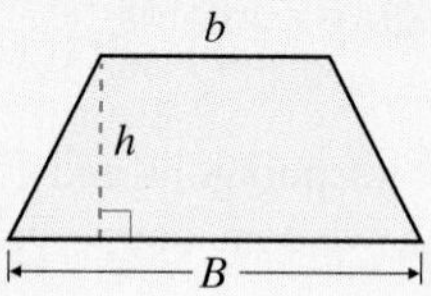

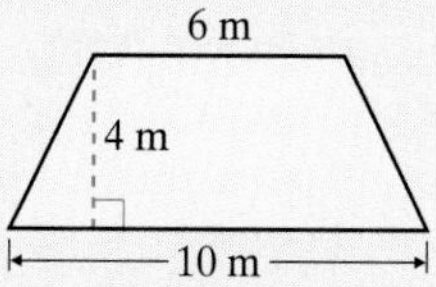

$A = \dfrac{1}{2}h(b + B)$

$= \dfrac{1}{2} \cdot 4(6 + 10)$

$= \dfrac{1}{2} \cdot 4 \cdot 16 = 32$

$= 32 \text{ m}^2$

Now, let's consider the area of a circle. As in the case of the circumference, the area of a circle is expressed in terms of π. Recall that π is approximately 3.14, or $\dfrac{22}{7}$.

Figure	Formula	Example
Circle	$A = \pi r^2$ Area equals π times the square of the radius. 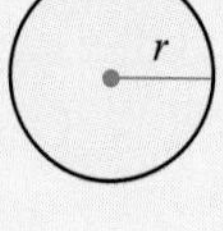	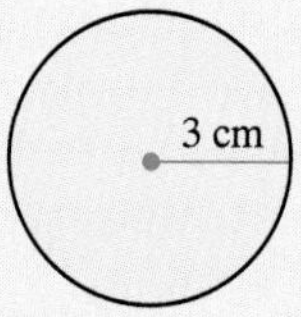 $A = \pi r^2$ $\approx 3.14(3)^2 \approx 3.14(9)$ ≈ 28.26 $\approx 28.26 \text{ cm}^2$

EXAMPLE 1

Find the area of a rectangle whose length is 5.5 ft and width is 3.5 ft.

Solution First, we draw a diagram to visualize the problem.

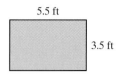

5.5 ft

3.5 ft

Then, we use the formula for the area of a rectangle.

$$A = lw$$
$$= (5.5)(3.5)$$
$$= 19.25$$

The area of the rectangle is 19.25 ft².

PRACTICE 1

A rectangle has length 6.3 cm and width 2.1 cm. Find its area.   13.23 cm²

EXAMPLE 2

Find the area of the square.

$4\frac{1}{2}$ in.

$4\frac{1}{2}$ in.

Solution We use the formula for the area of a square.

$$A = s^2$$
$$= \left(4\frac{1}{2}\right)\left(4\frac{1}{2}\right)$$
$$= \frac{9}{2} \cdot \frac{9}{2}$$
$$= \frac{81}{4}$$

The area of the square is $20\frac{1}{4}$ in².

PRACTICE 2

What is the area of a square with side 3.6 cm?   12.96 cm²

EXAMPLE 3

Find the area of a triangle with base 8 cm and height 5.9 cm.

Solution First, we draw a diagram.

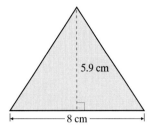

5.9 cm

8 cm

Next, we use the formula for finding the area of a triangle.

$$A = \frac{1}{2}bh$$

$$= \frac{1}{2}\overset{4}{(\cancel{8})}(5.9)$$

$$= 23.6$$

The area of the triangle is 23.6 cm².

PRACTICE 3

A triangle has a height of 3 in. and a base of 5 in. What is its area?   7.5 in²

EXAMPLE 4

What is the area of a parallelogram with base $6\frac{1}{2}$ m and height 3 m?

Solution We draw a diagram and then use the formula for the area of a parallelogram.

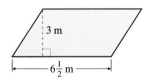

3 m

$6\frac{1}{2}$ m

$$A = bh$$

$$= \left(6\frac{1}{2}\right)(3)$$

$$= \frac{13}{2} \times \frac{3}{1}$$

$$= 19\frac{1}{2}$$

The area of the parallelogram is $19\frac{1}{2}$ m².

PRACTICE 4

Find the area of a parallelogram whose base is 5 ft and height is $2\frac{1}{2}$ ft. $12\frac{1}{2}$ ft²

EXAMPLE 5

What is the area of the trapezoid shown?

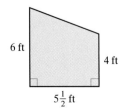

6 ft

4 ft

$5\frac{1}{2}$ ft

PRACTICE 5

Find the area of the following trapezoid.

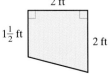

2 ft

$1\frac{1}{2}$ ft

2 ft

$3\frac{1}{2}$ sq ft

continued

Solution This polygon is a trapezoid, so we use the following formula to find its area.

$$A = \frac{1}{2}h(b + B)$$

$$= \frac{1}{2} \cdot 5\frac{1}{2}(6 + 4) \quad \text{Substitute } 5\frac{1}{2} \text{ for } h, \\ 6 \text{ for } b, \text{ and } 4 \text{ for } B.$$

$$= \frac{1}{2} \cdot \frac{11}{2} \cdot \overset{5}{\cancel{10}}$$
$$\phantom{= \frac{1}{2} \cdot} {}_{1}$$

$$= \frac{55}{2}$$

The area of the trapezoid is $27\frac{1}{2}$ ft^2.

EXAMPLE 6

What is the area of a circle whose diameter is 8 m?

Solution First, we draw a diagram.

We know that the radius is one half of 8 m, or 4 m. Then we use the formula for the area of a circle.

$$A = \pi r^2$$

$$\approx 3.14(4)^2$$

$$\approx 3.14(16)$$

$$\approx 50.24$$

The area of the circle is approximately 50.24 m^2.

PRACTICE 6

Find the area of a circle whose radius is 5 yd. 25π yd$^2 \approx 78.5$ yd^2

EXAMPLE 7

You want to buy an ad in a magazine that charges $1,000/in^2 for advertising space. How much will this ad cost?

2.5 in.

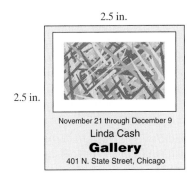

2.5 in.

November 21 through December 9
Linda Cash
Gallery
401 N. State Street, Chicago

Solution First, we need to find the area of the ad, which is square.

continued

PRACTICE 7

The following diagram shows the region in which the beam of light from a lighthouse can be seen in any direction in the fog. To the nearest square mile, what is the area of the region in which the light is visible? 79 mi^2

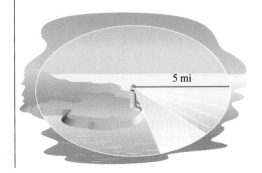

5 mi

$$A = s^2$$

$$= (2.5)(2.5)$$

$$= 6.25, \quad \text{or} \quad 6.25 \text{ in}^2$$

At $1,000/in^2, the ad will cost $6.25 \times 1{,}000$, or \$6,250.

Composite Figures

Recall that a composite figure comprises two or more simple figures. Let's consider finding areas of such figures.

EXAMPLE 8	PRACTICE 8

EXAMPLE 8

Find the area of the shaded portion of the figure.

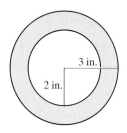

Solution To find the area of the shaded portion, we subtract the area of the small (inner) circle from the area of the large (outer) circle.

Shaded Area = Area of large circle − Area of small circle

$$\approx 3.14(3)^2 - 3.14(2)^2$$

$$\approx 3.14(9) - 3.14(4)$$

$$\approx 28.26 - 12.56$$

$$\approx 15.70$$

The area of the shaded figure is approximately 15.7 in^2.

PRACTICE 8

Find the shaded area.   8.75 m^2

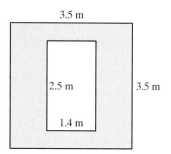

Be sure that students understand that the area of the large circle includes the area of the small circle.

Show students that the Distributive Property can simplify the solution.

EXAMPLE 9

At $19/ft^2, how much will it cost to carpet the ballroom pictured?

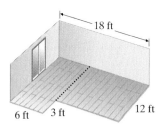

Solution First, we must find the area of the room. Note that the room consists of a rectangle, 15 ft × 6 ft, and a square 12 ft on a side.

PRACTICE 9

A coating of polyurethane is applied to the central circle on the gymnasium floor shown below. What is the area of the part of the floor that still needs coating?   4,421.5 sq ft

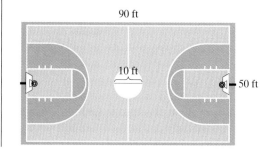

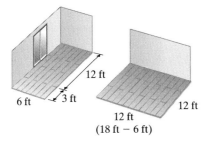

12 ft

6 ft   3 ft

12 ft
(18 ft − 6 ft)

12 ft

Total area = Area of rectangle + Area of square

$= l \cdot w + s^2$

$= 15 \cdot 6 + 12^2$

$= 90 + 144$

$= 234,$ or $234 \ ft^2$

The total area of the room is 234 ft². The carpet costs $19 per square foot, so we calculate the total cost as follows.

$$234 \ ft^2 \times \frac{\$19}{ft^2} = \$4,446$$

Carpeting the ballroom costs $4,446.

Exercises 11.3

Find the area of each figure. Use $\pi \approx 3.14$ when needed.

1.
25 m   125 m²
5 m

2. 10 in.   100 in²
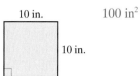
10 in.

3. 30 ft²
5 ft
12 ft

4. 90 cm²
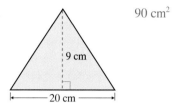
9 cm
20 cm

5. 290 yd²
10 yd
29 yd

6.

25 in.
55 in.
1,375 in²

7. Approximately 706.5 cm²
15 cm

8. Approximately 200.96 ft²
16 ft

9. 32 in²
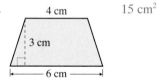
7 m
4 m
9 m

10. 15 cm²
4 cm
3 cm
6 cm

11. A parallelogram with base 4 m and height 3.9 m   15.6 m²

12. A parallelogram with base 6.5 in. and height 4 in.   26 in²

13. A circle with diameter 20 m   Approximately 314 in²

14. A circle with radius 100 ft   Approximately 31,400 ft²

15. A triangle with height 2.5 ft and base 5 ft   6.25 ft²

16. A triangle with base 8 in. and height $6\frac{1}{2}$ in.   26 in²

17. A trapezoid with height 4.2 yd and bases 7 yd and 14 yd.   44.1 yd²

18. A trapezoid with height 3.5 m and bases 4 m and 6.5 m   18.375 m²

19. A rectangle with length 2.6 m and width 1.4 m   3.64 m²

20. A rectangle with length $\frac{1}{2}$ ft and width $\frac{2}{3}$ ft $\frac{1}{3}$ ft²

21. A square with side $\frac{1}{4}$ yd long $\frac{1}{16}$ yd²

22. A square with side 15.5 cm long   240.25 cm²

23. 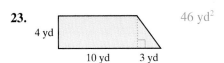   46 yd²

4 yd

10 yd   3 yd

24. 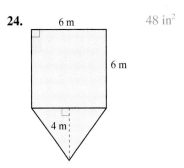   48 in²

6 m

6 m

4 m

25. 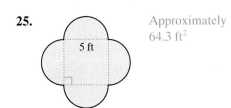   Approximately 64.3 ft²

5 ft

26. Approximately 4,314 yd²

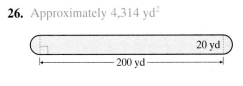

20 yd

200 yd

Find the shaded area.

🖩 **27.**   51.75 ft²

9 ft

1.5 ft

1.5 ft   6 ft

🖩 **28.**   Approximately 37.1 in²

2.5 in.

4.25 in.

Applications

Solve. Use π ≈ 3.14 when needed.

29. You want to tile a room that measures 9 ft × 12 ft with 1 ft² tiles that sell for $4.99 apiece. Will $500 be enough to pay for the tiles?   No. You need 108 tiles at a total cost of $538.92.

30. The base of the United Nations Secretariat Building is a rectangle with length 88 m and width 22 m. Tower 1 of the World Trade Center has a square base with side 121 m. Which building has the base with the larger area?   Tower 1 (14,641 m²) has a larger base than the UN building (1,936 m²).

31. A microscope allows you to see a circular region that is $\dfrac{1}{100}$ in. in diameter. What is the area of this region?   Approximately 0.00008 in²

32. A radar station can identify an enemy bomber within 10 mi of the station in any direction. What area does the station cover?   314 mi²

33. Suppose that you have an L-shaped house located on the rectangular lot shown. How much yard space do you have?   5,000 sq ft

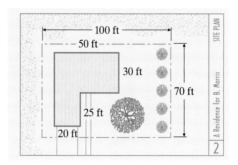

34. A walkway 2 yd wide is built around the entire building below. Find the area of the walkway.   284 sq yd

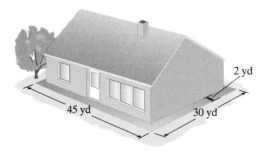

▦ *Use a calculator to solve the following problems, giving (a) the operation(s) carried out in the solution, (b) the answer, and (c) an estimate of the answer.*

35. Even though an LP record is larger than a CD, the CD holds twice as much music.

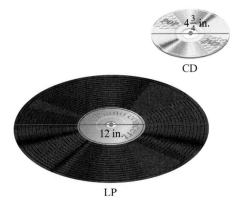

How much larger in area is an LP record than a CD? Round the answer to the nearest 10 in². (a) Multiplication and subtraction, (b) 100 in², (c) Possible estimate: 100 in²

36. Consider the two tables pictured. How much larger in area to the nearest square meter is the semicircular table than the rectangular table?

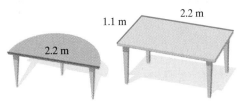

(a) Multiplication and subtraction, (b) 1 m², (c) Possible estimate: 1 m²

■ *Check your answers on page A-15.*

MINDSTRETCHERS

INVESTIGATION

1. ■ Draw a square.
   ■ Measure its side lengths.
   ■ Find its area.
   ■ Double the side length of the square.
   ■ Find the area of the new square.
   ■ Start with another square and repeat this process several times.
   ■ How does doubling the side length of a square affect its area?
   Doubling the side length quadruples the area.

GROUPWORK

2. In the following diagram, each small square represents 1 in². Working with a partner, estimate the area of the oval.   Possible answer: 16 in²

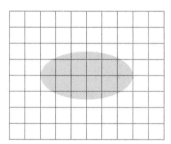

MATHEMATICAL REASONING

3. In the diagram below, $\overline{AC}$ is parallel to $\overline{ED}$.

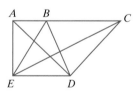

What relationship do you see between the areas of $\triangle EAD$, $\triangle EBD$, and $\triangle ECD$?   The areas are all equal.

11.4 Volume

OBJECTIVES
- To find the volume of a geometric solid
- To find the volume of a composite geometric solid
- To solve word problems involving volume

The Volume of a Geometric Solid

Volume is a measure of the amount of space inside a three-dimensional figure. The amount of water in an aquarium, the amount of juice in a can, or the amount of grain in a bin are all examples of volumes.

To find the volume of the box shown, we can split it into 24 cubes. Each edge of each cube is 1 in. long. Thus the volume of each cube is 1 cu in. (or in³), and the volume of the box is 24 in³. Note that we can also determine this volume by finding the product of the length, the width, and the height of the solid: $24 = 2 \cdot 3 \cdot 4$. Volumes are always expressed in cubic units.

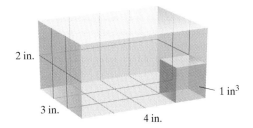

2 in.

3 in.

4 in.

1 in³

Definition

Volume is the number of cubic units required to fill a three-dimensional figure.

In this section, we consider three-dimensional objects in space and find their volume by using the following formulas.

Definition	Formula	Example
A **rectangular solid** is a solid in which all six faces are rectangles.	$V = lwh$ Volume equals length times width times height. h, l, w	3 ft, 4 ft, 6 ft $V = lwh$ $= 6 \cdot 4 \cdot 3$ $= 72$, or 72 ft^3
A **cube** is a solid in which all six faces are squares.	$V = e^3$ Volume equals the cube of the edge. e, e, e	3 cm, 3 cm, 3 cm $V = e^3$ $= (3)^3$ $= 27$, or 27 cm^3

continued

A **cylinder** is a solid in which the bases are circles and are perpendicular to the height.

$V = \pi r^2 h$

Volume equals π times the square of the radius times the height.

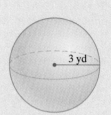

$V = \pi r^2 h$
$\approx 3.14(2)^2(5)$
$\approx 3.14(4)(5)$
$\approx 62.8,$ or 62.8 in^3

A **sphere** is a three-dimensional figure made up of all points a given distance from the center.

$V = \dfrac{4}{3}\pi r^3$

Volume equals $\dfrac{4}{3}$ times π times the cube of the radius.

$V = \dfrac{4}{3}\pi r^3$

$\approx \dfrac{4}{3}(3.14)(3)^3$

$\approx \dfrac{4}{3}(3.14)(27)$

$\approx \dfrac{339.12}{3}$

$\approx 113.04,$ or 113.04 yd^3

EXAMPLE 1

Find the volume of the rectangular solid shown.

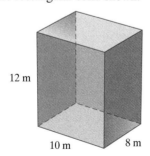

Solution To find the volume of a rectangular solid, use the formula $V = lwh$.

$$V = lwh$$
$$= 8 \cdot 10 \cdot 12$$
$$= 960$$

The volume of the rectangular solid is 960 m^3.

PRACTICE 1

What is the volume of this box?   72 ft^3

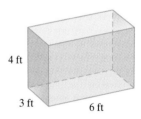

Solution 1
the cylinde

Volume

The volume

EXAMPLE 2

Find the volume of a cube whose edge measures 11 in.

Solution Use the formula for the volume of a cube, $V = e^3$.

$$V = e^3$$

$$= (11)^3$$

$$= 11 \cdot 11 \cdot 11$$

$$= 1,331$$

The cube's volume is therefore 1,331 in^3.

PRACTICE 2

What is the volume of a cube with an edge of 15 cm?   3,375 cm^3

EXAMPLE 3

What is the volume of the cylinder shown?

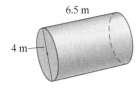

6.5 m

4 m

Solution In this cylinder the radius of the base is 2 m and the height is 6.5 m. We substitute these quantities into the formula for the volume of a cylinder.

$$V = \pi r^2 h$$

$$\approx 3.14(2)^2(6.5)$$

$$\approx 3.14(4)(6.5)$$

$$\approx 81.64$$

The cylinder's volume is approximately 81.64 m^3.

PRACTICE 3

Find the volume of this can.   Approximately 10.99 in^3

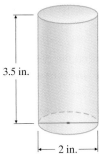

3.5 in.

2 in.

EXAMPLE 4

What is the volume of the sphere shown? Use $\pi \approx \dfrac{22}{7}$.

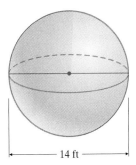

14 ft

PRACTICE 4

What is the volume of a ball whose diameter is 6 in.?   Approximately $113\dfrac{1}{7}$ in^3

continued

Solution W

and approxi

is 7 ft.

The volume

Exercises 11.4

Find the volume of each solid. Use $\pi \approx 3.14$ when needed.

1.   216 in³

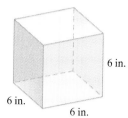

6 in.
6 in.
6 in.
6 in.

2.   1 cm³

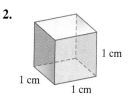

1 cm
1 cm
1 cm

3.   2,560 m³

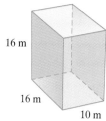

16 m
16 m
10 m

4.   8,000 in³

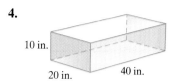

10 in.
20 in.
40 in.

EXAMPLE 5

A shipping
depth of 1.8

Solution T
mula $V = l$

5.

5 ft
2 ft

Approximately 62.8 ft³

6.

← 20 m →
10 m

Approximately 3,140 m³

The van car

EXAMPLE 6

Find the vo
deleted fro

7.

16 in.

Approximately 2,143.6 in³

8.

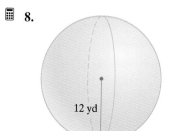

12 yd

Approximately 7,234.6 yd³

9. A rectangular solid with length 3.5 ft, width 5.5 ft, and height 6.5 ft.
125.125 ft³

10. A cube with length 45 m on a side.
91,125 m³

11.

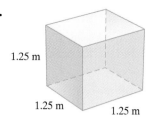

1.25 m

1.25 m   1.25 m

Approximately 1.95 m³

12.

← 5 in. →

15 in.

Approximately 294.4 in³

13.

6 ft

← 2 ft →

Approximately 92.1 ft³

14.

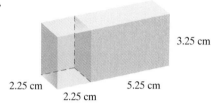

3.25 cm

2.25 cm   5.25 cm

2.25 cm

54.84 cm³

Applications

Use a calculator to solve the following problems, giving (a) the operation(s) carried out in the solution, (b) the exact answer, and (c) an estimate of the answer. Use $\pi \approx 3.14$ when needed.

15. In a chemistry course, you learn that a substance's density is found by dividing its weight by its volume. If a 300-g block of wood measures 7 cm by 5 cm by 3 cm, find the density of the wood. Round to the nearest 0.1 g/cm³.
(a) Multiplication and division,
(b) 2.9 g/cm³, (c) Possible estimate:
3 g/cm³

16. A cylindrical wheat silo has a height of 35 yd and a diameter of 5 yd. What is the maximum amount of wheat that can be stored in this silo? Round to the nearest $\frac{1}{10}$ yd³.   (a) Division and multiplication, (b) Approximately 686.9 yd³, (c) Possible estimate: 810 yd³

▦ *Solve.*

17. A rectangular gold bar is 20 cm long, 10 cm wide, and 6 cm high. If the bar weighs 13 kg, what is the weight of the bar per cubic centimeter? Round to the nearest 0.01 kg/cm^3. 0.01 kg/cm^3

18. A fish tank is 12 in. wide, 18 in. long, and 16 in. high. How much water, to the nearest 0.1 gal, will fill the tank if 1 gal equals 231 in^3?   15.0 gal

19. The crew of a spaceship began a search for a missing shuttlecraft 1,000 mi in every direction. How many cubic miles did the crew search? Round to the nearest $1,000 \text{ mi}^3$.
$4,186,667,000 \text{ mi}^3$

20. What is the volume of a hot air balloon, in the shape of a sphere, with a diameter of 30 ft? Round to the nearest 100 ft^3. $14,100 \text{ ft}^3$

21. The displacement of a car engine's cylinder is the volume of gas and air forced up by the piston. To calculate this volume, we multiply the distance that the top of the piston travels by the area of the base of the cylinder. Find the displacement of the cylinder shown, rounded to the nearest cubic inch.
46 in^3

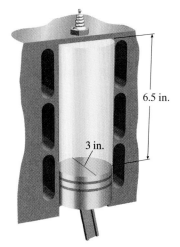

6.5 in.

3 in.

22. A board foot is a special measure of volume used in the lumber industry. If a board foot contains 144 in^3 of wood, how many board feet rounded to the nearest tenth are there in the board shown?   9.7 board ft

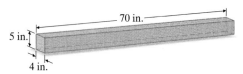

70 in.

5 in.

4 in.

23. A large box is placed in a delivery truck, as pictured. What is the volume of the space remaining in the truck? 15.19 m³

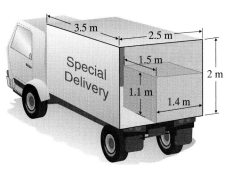

24. The metal machine part shown is cylindrical in shape, with a smaller cylinder drilled out of its center. How many cubic millimeters of metal does the machine part contain? Approximately 176.6 mm³

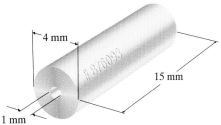

■ *Check your answers on page A-15.*

MINDSTRETCHERS

GROUPWORK

1. Working with a partner, decide which three-dimensional figure can be formed by folding each pattern. a. Cube; b. Cylinder

a.

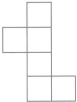

b.

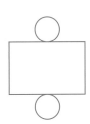

INVESTIGATION

2. ■ Draw a cube.
   ■ Measure its side lengths.
   ■ Find its volume.
   ■ Double the side length of the cube.
   ■ Find the volume of the new cube.
   ■ Start with another cube and repeat this process several times.
   ■ How does doubling the side length of a cube affect its volume?
   Doubling the side length multiplies the volume by 8.

WRITING

3. Explain in what way a sphere and a circle are alike. In what way are they different?
   Possible answer: Both are round; a sphere has volume, whereas a circle does not.

11.5 Similar Triangles

Identifying Corresponding Sides of Similar Triangles

OBJECTIVES
- *To identify corresponding sides of similar triangles*
- *To find the missing sides of similar triangles*
- *To solve word problems involving similar triangles*

When discussing ratio and proportion earlier, we looked at figures that have the same shape but different size. For example, on page 280, we noted that everything in the enlargement of a rectangular photo is the same shape as in the original — only larger. In this section we focus on triangles that have this relationship, which are called *similar triangles*.

Before they start this lesson, have students review Section 5.2.

Definition

Similar triangles are triangles that have the same shape but not necessarily the same size.

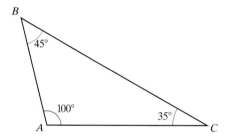

In the diagram above, $\triangle ABC$ is similar to $\triangle DEF$. We write $\triangle ABC \sim \triangle DEF$ (read "triangle *ABC* is similar to triangle *DEF*"). In the diagram each angle of $\triangle ABC$ corresponds to an angle of $\triangle DEF$, as follows:

$$\angle A = \angle D$$

$$\angle B = \angle E$$

$$\angle C = \angle F$$

Also, each side of $\triangle ABC$ corresponds to a side of $\triangle DEF$.

$\overline{AB}$ corresponds to $\overline{DE}$.

$\overline{BC}$ corresponds to $\overline{EF}$.

$\overline{AC}$ corresponds to $\overline{DF}$.

Definition

In similar triangles, **corresponding sides** are the sides opposite the equal angles.

Note that, when we write that two triangles are similar, we name them so that the order of corresponding angles in both triangles is the same. For example,

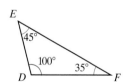

EXAMPLE 1

△RST ~ △XYZ. Name the corresponding sides of these triangles.

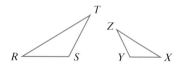

Solution Because △RST ~ △XYZ, ∠R = ∠X, ∠S = ∠Y, and ∠T = ∠Z. We know that the corresponding sides are opposite congruent angles. So we write the following.

Because ∠R = ∠X, $\overline{ST}$ corresponds to $\overline{YZ}$. $\overline{ST}$ is opposite ∠R, and $\overline{YZ}$ is opposite ∠X.

Because ∠S = ∠Y, $\overline{RT}$ corresponds to $\overline{XZ}$. $\overline{RT}$ is opposite ∠S, and $\overline{XZ}$ is opposite ∠Y.

Because ∠T = ∠Z, $\overline{RS}$ corresponds to $\overline{XY}$. $\overline{RS}$ is opposite ∠T, and $\overline{XY}$ is opposite ∠Z.

PRACTICE 1

△ABC ~ △GHI. List the corresponding sides of these triangles.

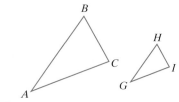

$\overline{AB}$ corresponds to $\overline{GH}$
$\overline{AC}$ corresponds to $\overline{GI}$
$\overline{BC}$ corresponds to $\overline{HI}$

Finding the Missing Sides of Similar Triangles

In similar triangles, corresponding sides are in proportion; that is, the ratios of their lengths are equal. Looking to the right at an example in which △ABC ~ △DEF, we have the following.

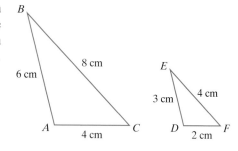

$$\frac{AB}{DE} = \frac{BC}{EF} = \frac{AC}{DF}$$

$$\frac{6}{3} = \frac{8}{4} = \frac{4}{2} = \frac{2}{1}$$

In a pair of similar triangles, such proportions can be used to find the length of a missing side.

> **To find a missing side of similar triangles**
>
> ■ write the ratios of the lengths of the corresponding sides,
> ■ write a proportion using a ratio with known terms and a ratio with an unknown term, and
> ■ solve the proportion for the unknown term.

EXAMPLE 2

In the following diagram, △TAP ~ △RUN. Find x.

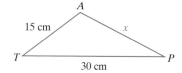

Solution Because △TAP ~ △RUN, we write the ratios of the lengths of the corresponding sides.

PRACTICE 2

In the following diagram, △DOT ~ △PAN. Find y.   y = 24

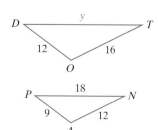

$$\frac{TA}{RU} = \frac{AP}{UN} = \frac{PT}{NR}$$

$$\frac{15}{9} = \frac{x}{12} = \frac{30}{18}$$

To solve for x, we can consider either the proportion $\frac{15}{9} = \frac{x}{12}$ or the

proportion $\frac{x}{12} = \frac{30}{18}$. Note that each proportion contains the unknown

term x. If we choose the first proportion and solve for x, we get the following.

$$\frac{15}{9} = \frac{x}{12}$$

$$9x = 180$$

$$x = 20$$

So x is 20 cm.

> Have students verify that using
> $\frac{x}{12} = \frac{30}{18}$ gives the same answer.

Similar triangles are useful in finding lengths that cannot be measured directly, as Example 3 shows.

EXAMPLE 3

A surveyor took the measurements shown. If $\triangle ABC \sim \triangle EFC$, find d, the distance across the river.

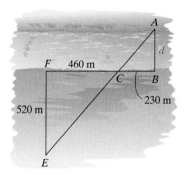

Solution Here, $\triangle ABC \sim \triangle EFC$, so we write the proportion $\frac{BC}{FC} = \frac{AB}{EF}$.

Then we substitute the values given in the diagram.

$$\frac{230}{460} = \frac{d}{520}$$

$$460d = (230)(520) \qquad \text{Cross multiply.}$$

$$\frac{\overset{1}{\cancel{460}}d}{\underset{1}{460}} = \frac{\overset{1}{\cancel{(230)}}(520)}{\underset{2}{460}} \qquad \text{Divide both sides by 460.}$$

$$d = 260$$

The distance across the river is 260 m.

PRACTICE 3

Suppose that your height and shadow form a triangle similar to that formed by a nearby tree and its shadow. What is the height of the tree?

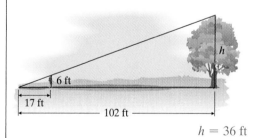

$h = 36$ ft

9.

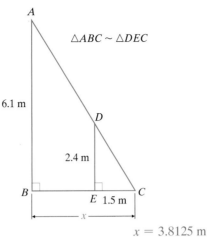

$\triangle ABC \sim \triangle DEC$

6.1 m

2.4 m

1.5 m

x

$x = 3.8125$ m

10.

$\triangle SRT \sim \triangle DEF$

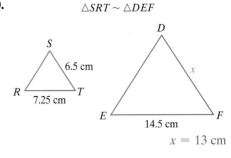

6.5 cm

7.25 cm

14.5 cm

x

$x = 13$ cm

Applications

Solve.

11. The tallest land animal is the giraffe. How tall is the giraffe pictured here? 576 cm

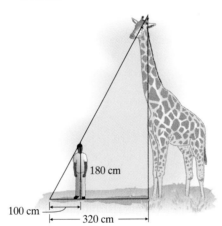

180 cm

100 cm

320 cm

12. Light from a flashlight shines through a dragon puppet onto a screen behind it, as shown below. Find the height of the puppet's image. 5 ft

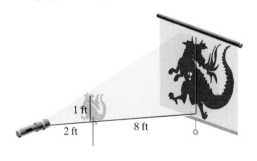

1 ft

2 ft

8 ft

13. A Coast Guard observer sees a ship out on the ocean and wants to know how far it is from the shore. Use the diagram to find that distance. 200 m

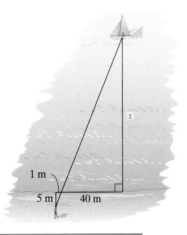

x

1 m

5 m   40 m

14. To measure the height of a building, you can position a mirror on the ground so that you see the top of the building reflected. As in all reflections of this type, $\triangle ECD \sim \triangle ACB$. Find the height of the building. 48 ft

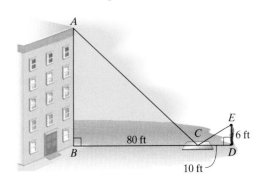

A

B 80 ft C

E

6 ft

D

10 ft

■ *Check your answers on page A-16.*

MINDSTRETCHERS

PATTERNS

1. List 10 pairs of similar triangles in the following square.

Some possible answers: $\triangle ABE \sim \triangle CDE$, $\triangle ABE \sim \triangle DBC$, $\triangle ABC \sim \triangle DCB$

WRITING

2. When you enlarge a photograph, the new image is similar to the old one. Give some other everyday examples of similarity.   <u>Answers may vary.</u>

GROUPWORK

3. Working with a partner, decide whether this statement is true: If two quadrilaterals have corresponding angles equal, the two quadrilaterals must have the same shape. Draw a diagram to support your answer.   Yes, it is true.

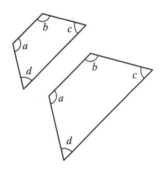

11.6 Square Roots and the Pythagorean Theorem

OBJECTIVES

- *To find the square root of a number*
- *To find the unknown side of a right triangle, using the Pythagorean theorem*
- *To solve word problems involving a square root or the Pythagorean theorem*

In Section 11.3, we found the area of a square by squaring the length of one of its sides. In this section, we look at the opposite of squaring a number, that is, finding its *square root*.

Finding the Square Root of a Number

Consider the following problem. Suppose that a square has an area of 25 ft². What is the length, s, of each side?

Recall that, for the area of a square, $A = s^2$. So for $25 = s^2$, we need to determine what whole number when multiplied by itself equals 25. Because $25 = 5 \cdot 5$, the whole number must be 5. So a square with an area of 25 ft² has sides of length 5 ft.

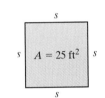

> You may want to remind students that $(-5)(-5) = 25$, so that $\sqrt{25} = \pm 5$, and to introduce the term *principal square root* for the +5.

Squaring the whole number 5 gives 25, so we say that the number 25 is a *perfect square* and that 5 is the *square root* of 25.

Definitions

A **perfect square** is a number that is the square of a whole number.

The **square root** of a number n, written $\sqrt{n}$, is the number whose square is n.

The square root of a perfect square is a whole number. Note that the columns are reversed in the following tables.

n	n^2
1	$1 \cdot 1 = 1$
2	$2 \cdot 2 = 4$
3	$3 \cdot 3 = 9$
4	$4 \cdot 4 = 16$
5	$5 \cdot 5 = 25$
6	$6 \cdot 6 = 36$

n	$\sqrt{n}$
1	$\sqrt{1} = 1$
4	$\sqrt{4} = 2$
9	$\sqrt{9} = 3$
16	$\sqrt{16} = 4$
25	$\sqrt{25} = 5$
36	$\sqrt{36} = 6$

EXAMPLE 1

Find the square root.

a. $\sqrt{36}$ **b.** $\sqrt{100}$ **c.** $\sqrt{81}$

Solution In each case, we need to find the whole number that when squared gives us the number under the square root sign.

a. $\sqrt{36} = 6$ because $6 \cdot 6 = 36$

b. $\sqrt{100} = 10$ because $10 \cdot 10 = 100$

c. $\sqrt{81} = 9$ because $9 \cdot 9 = 81$

PRACTICE 1

What is the square root of each perfect square?

a. $\sqrt{49}$ 7

b. $\sqrt{144}$ 12

c. $\sqrt{225}$ 15

Many numbers are not perfect squares. One such number is 28 because there is no whole number that when multiplied by itself equals 28.

If a number is not a perfect square, we can either estimate or use a calculator to find its square root.

EXAMPLE 2	PRACTICE 2

$\sqrt{28}$ lies between which two consecutive whole numbers?

Between which two consecutive whole numbers does $\sqrt{47}$ lie?   Between 6 and 7

Solution To begin, let's find the two consecutive perfect squares that 28 lies between. Examining the table of perfect squares, we see that the consecutive squares are 25 and 36.

n	$\sqrt{n}$
25	5
28	?
36	6

Because 28 lies between 25 and 36, $\sqrt{28}$ lies between 5 and 6.

EXAMPLE 3	PRACTICE 3

Using a calculator, find each square root to the nearest tenth.

a. $\sqrt{75}$ **b.** $\sqrt{21}$

Using a calculator, find each square root to the nearest hundredth.

a. $\sqrt{56}$ 7.48

Solution

a.

Input	75	$\sqrt{}$
Display	75.	8.660254038

So $\sqrt{75} \approx 8.7$, rounded to the nearest tenth.

b. $\sqrt{12}$ 3.46

b.

Input	21	$\sqrt{}$
Display	21.	4.582575695

Point out that different calculators may display a different number of decimal places.

Therefore $\sqrt{21} \approx 4.6$, rounded to the nearest tenth.

The Pythagorean Theorem

Recall that a right triangle is a triangle that has one 90° angle. In a right triangle, the side opposite the right angle is called the **hypotenuse**. The other two sides are called **legs**.

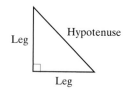

The following figure shows the relationship between the hypotenuse and the legs of a right triangle.

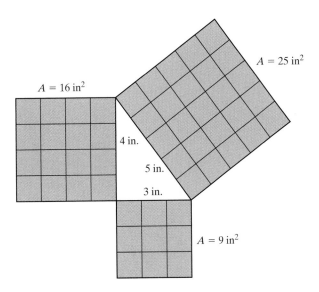

Note that the areas of the squares are related in the following way.

Area of the square on one leg + Area of the square on the other leg =
Area of the square on the hypotenuse.

So here we have the following, after substituting.

$$9 \text{ in}^2 + 16 \text{ in}^2 = 25 \text{ in}^2$$

If we let a and b represent the lengths of the legs and c represent the length of the hypotenuse, we get $a^2 + b^2 = c^2$.

Point out to students that equality is symmetric, so the equations $c^2 = a^2 + b^2$ and $a^2 + b^2 = c^2$ are equivalent.

This relationship is called the *Pythagorean theorem.*

The Pythagorean Theorem

For every right triangle, the sum of the squares of the two legs equals the square of the hypotenuse.

We can use the Pythagorean theorem to find the third side of a right triangle if we know the other two sides.

EXAMPLE 4

Find the length of the hypotenuse.

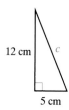

12 cm

c

5 cm

Solution To find the hypotenuse, we use the Pythagorean theorem.

$$a^2 + b^2 = c^2$$

$$5^2 + 12^2 = c^2$$

$$25 + 144 = c^2$$

$$169 = c^2$$

$$\sqrt{169} = c \quad \text{Take the square root of 169.}$$

$$13 = c, \quad \text{or} \quad c = 13$$

The hypotenuse is 13 cm long.

PRACTICE 4

Find the length of the unknown side in $\triangle ABC$.   10 in.

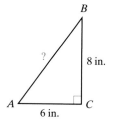

B

?

8 in.

A

6 in.

C

EXAMPLE 5

If $b = 1$ cm and $c = 2$ cm, what is the length of a in a right triangle where a and b are legs and c is the hypotenuse? Round the answer to the nearest tenth of a centimeter.

Solution First, we draw a diagram.

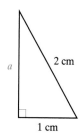

a

2 cm

1 cm

Then, we use the Pythagorean theorem and substitute the given values to obtain the following.

$$a^2 + b^2 = c^2$$

$$a^2 + 1^2 = 2^2$$

$$a^2 + 1 = 4$$

$$a^2 + 1 - 1 = 4 - 1$$

$$a^2 = 3$$

$$a = \sqrt{3}$$

Finally, we use a calculator and round to the nearest tenth to find that $\sqrt{3}$ is approximately 1.7. So $a \approx 1.7$ cm.

PRACTICE 5

In a right triangle, one leg equals 2 ft and the hypotenuse equals 4 ft. Find the length of the missing leg. Round your answer to the nearest tenth of a foot.   3.5 ft

EXAMPLE 6

Each rectangular section of a fence is reinforced with a piece of wood nailed across the diagonal of the rectangle. The height of the fence section is 6 ft. The length of the diagonal is 10 ft. What is the length of the fence section?

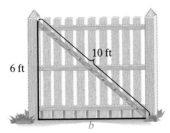

Solution The length and height of each fence section along with the diagonal form a right triangle. Because the height and diagonal are known, we can use the Pythagorean theorem to find the length.

$$a^2 + b^2 = c^2$$

$$6^2 + b^2 = 10^2$$

$$36 + b^2 = 100$$

$$36 - 36 + b^2 = 100 - 36$$

$$b^2 = 64$$

$$b = \sqrt{64}$$

$$b = 8$$

The length of each fence section is 8 ft.

PRACTICE 6

You use wire to secure a flag pole. Find the length W of the wire shown.   5 m

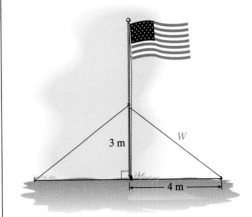

Exercises 11.6

Find each square root.

1. $\sqrt{9}$
3

2. $\sqrt{4}$
2

3. $\sqrt{36}$
6

4. $\sqrt{49}$
7

5. $\sqrt{81}$
9

6. $\sqrt{64}$
8

7. $\sqrt{144}$
12

8. $\sqrt{100}$
10

9. $\sqrt{169}$
13

10. $\sqrt{121}$
11

11. $\sqrt{400}$
20

12. $\sqrt{225}$
15

Determine between which two consecutive whole numbers each square root lies.

13. $\sqrt{2}$
1 and 2

14. $\sqrt{7}$
2 and 3

15. $\sqrt{14}$
3 and 4

16. $\sqrt{10}$
3 and 4

17. $\sqrt{39}$
6 and 7

18. $\sqrt{50}$
7 and 8

19. $\sqrt{80}$
8 and 9

20. $\sqrt{105}$
10 and 11

Find each square root to the nearest tenth.

21. $\sqrt{5}$
2.2

22. $\sqrt{11}$
3.3

23. $\sqrt{37}$
6.1

24. $\sqrt{74}$
8.6

25. $\sqrt{139}$
11.8

26. $\sqrt{165}$
12.8

27. $\sqrt{9,801}$
99.0

28. $\sqrt{8,649}$
93.0

Find each missing length to the nearest tenth of the indicated unit.

29. 30 cm / 34 cm / a
$a = 16$ cm

30.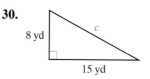
$c = 17$ yd

31. 1 m / c / 3 m
$c = 3.2$ m

32.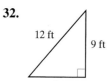
$b = 7.9$ ft

⊞ *Given a right triangle with legs a and b, and hypotenuse c, find the missing side to the nearest tenth of the indicated unit.*

	a	b	c
33.	24 m	7 m	25 m
34.	5 in.	12 in.	13 in.
35.	6 ft	8 ft	10 ft
36.	3 cm	4 cm	5 cm
37.	12 m	16 m	20 m
38.	12 in.	9 in.	15 in.
39.	7 cm	9 cm	11.4 cm
40.	2 yd	5 yd	5.4 yd
41.	8.7 ft	18 ft	20 ft
42.	2 in.	2 in.	2.8 in.

Solve. Where appropriate, use a calculator.

43. The following drawing shows a girl flying a kite, which is directly over a tree. How far is the girl from the base of the tree?   20 ft

44. While scuba diving, you swim away from the boat and then dive, as shown. How far, to the nearest foot, from the boat will you be?   28 ft

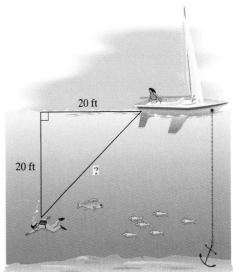

45. What is the length l of the rectangular plot shown? $l = 240$ ft

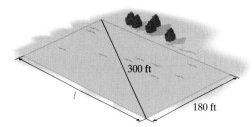

46. When Ronald Reagan was president, his helicopter and a private plane were involved in a "near miss" incident, as pictured. How close were the president's helicopter and the plane?   250 ft

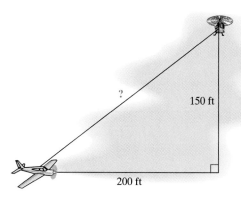

47. You are constructing a roof of wooden beams. According to the diagram, what is the length l of the sloping beam?

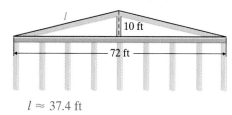

$l \approx 37.4$ ft

48. A college is constructing an access ramp for the handicapped to a door in one of its buildings, as shown. Find the length of the ramp.

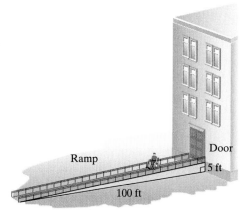

Approximately 100.1 ft

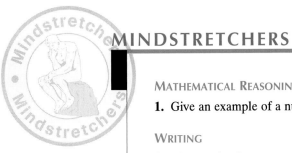

MINDSTRETCHERS

MATHEMATICAL REASONING

1. Give an example of a number that is smaller than its square root.   Possible answer: 0.25

WRITING

2. Thousands of years ago, the ancient Egyptians had a clever way of creating a right angle for their construction projects. They would use a rope, tying it in a circle with 12 equally spaced knots, as shown.

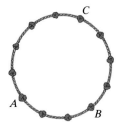

They would then pull on the knots labeled *A*, *B*, and *C*. Explain why doing so would create a right triangle. (*Source:* Peter Tompkins, *Secrets of the Great Pyramids*)

The triangle formed would have side lengths 3, 4, and 5; $3^2 + 4^2 = 5^2$

INVESTIGATION

3. Choose a whole number. Use a calculator to determine whether it is a perfect square.

Check whether the square root has decimal places. If not, the number is a perfect square.

KEY CONCEPTS and SKILLS

☐ = CONCEPT ☐ = SKILL

CONCEPT/SKILL	DESCRIPTION	EXAMPLE
[11.1] **Point**	An exact location in space, with no dimension.	•
[11.1] **Line**	A collection of points along a straight path, that extends endlessly in both directions.	
[11.1] **Line segment**	A part of a line having two endpoints. It has one dimension — its length.	
[11.1] **Ray**	A part of a line having only one endpoint.	
[11.1] **Angle**	Two rays that have a common endpoint called the *vertex* of the angle.	
[11.1] **Plane**	A flat surface that extends endlessly in all directions.	
[11.1] **Straight angle**	An angle that measures 180°.	180°
[11.1] **Right angle**	An angle that measures 90°.	
[11.1] **Acute angle**	An angle that measures less than 90°.	65°
[11.1] **Obtuse angle**	An angle that measures more than 90° and less than 180°.	120°
[11.1] **Complementary angles**	Two angles whose sum is 90°.	25° 65°
[11.1] **Supplementary angles**	Two angles whose sum is 180°.	40° 140°

CONCEPT/SKILL	DESCRIPTION	EXAMPLE
[11.1] Intersecting lines	Two lines that meet.	
[11.1] Parallel lines	Two lines on the same plane that do not intersect.	
[11.1] Perpendicular lines	Two lines that intersect to form right angles.	
[11.1] Vertical angles	Two equal angles that are formed by two intersecting lines.	
[11.1] Polygon	A closed plane figure made up of line segments.	
[11.1] Triangle	A polygon with three sides.	
[11.1] Quadrilateral	A polygon with four sides.	
[11.1] Equilateral triangle	A triangle with three sides equal in length.	
[11.1] Isosceles triangle	A triangle with two sides equal in length.	
[11.1] Scalene triangle	A triangle with no sides equal in length.	

continued

CONCEPT/SKILL	DESCRIPTION	EXAMPLE
[11.1] Acute triangle	A triangle with three acute angles.	
[11.1] Right triangle	A triangle with one right angle.	
[11.1] Obtuse triangle	A triangle with one obtuse angle.	
[11.1] Trapezoid	A quadrilateral with only one pair of opposite sides parallel.	
[11.1] Parallelogram	A quadrilateral with both pairs of opposite sides parallel. Opposite sides are equal in length, and opposite angles have equal measures.	
[11.1] Rectangle	A parallelogram with four right angles.	
[11.1] Square	A rectangle with four sides equal in length.	
[11.1] Circle	A closed plane figure made up of points that are all the same distance from a fixed point called the center.	
[11.1] Diameter	A line segment that passes through the center of a circle and has both endpoints on the circle.	
[11.1] Radius	A line segment with one endpoint on the circle and the other at the center.	

CONCEPT/SKILL	DESCRIPTION	EXAMPLE
[11.2] **Perimeter**	The distance around a polygon.	
[11.2] **Circumference**	The distance around a circle.	
[11.3] **Area**	The number of square units that a figure contains.	
[11.4] **Volume**	The number of cubic units required to fill a three-dimensional figure.	
[11.5] **Similar triangles**	Triangles that have the same shape but not necessarily the same size.	
[11.5] **Corresponding sides**	The sides opposite the equal angles in similar triangles.	In the similar triangles pictured, $\overline{AB}$ corresponds to $\overline{DE}$, $\overline{BC}$ corresponds to $\overline{EF}$, and $\overline{AC}$ corresponds to $\overline{DF}$.
[11.5] **To find the missing sides of similar triangles**	■ Write the ratios of the measures of the corresponding sides. ■ Write a proportion using a ratio with known terms and a ratio with an unknown term. ■ Solve the proportion for the unknown term.	$\triangle TRS \sim \triangle XYW$ Find a. $\dfrac{ST}{WX} = \dfrac{TR}{XY}$ $\dfrac{4}{6} = \dfrac{8}{a}$ $4a = 48$ $a = 12,$ or 12 in.

continued

CONCEPT/SKILL	DESCRIPTION	EXAMPLE
[11.6] **Perfect square**	A number that is the square of a whole number.	49 and 144
[11.6] **Square root of** n	The number, written $\sqrt{n}$, whose square is n.	$\sqrt{36}$ and $\sqrt{8}$
[11.6] **Pythagorean theorem**	For every right triangle, the sum of the squares of the two legs equals the square of the hypotenuse. $$a^2 + b^2 = c^2$$ a and b are legs, and c is the hypotenuse.	(right triangle: legs a, 24 yd, hypotenuse 25 yd) Find a. $a^2 + b^2 = c^2$ $a^2 + (24)^2 = (25)^2$ $a^2 + 576 = 625$ $a^2 + 576 - 576 = 625 - 576$ $a^2 = 49$ $a = \sqrt{49}$ $= 7,$ or 7 yd

KEY FORMULAS

FIGURE	FORMULA	EXAMPLE
[11.2]–[11.3] **Triangle**	Perimeter $$P = a + b + c$$ a, b, and c are the sides. Area $$A = \frac{1}{2}bh$$ b is the base and h is the height.	(triangle: sides 6 m, 8 m, base 10 m, height 4.8 m) $P = a + b + c$ $= 6 + 10 + 8$ $= 24,$ or 24 m $A = \frac{1}{2}bh$ $= \frac{1}{2}\overset{5}{10} \cdot 4.8$ $= 24,$ or 24 m^2
[11.2]–[11.3] **Rectangle**	Perimeter $$P = 2l + 2w$$ Area $$A = lw$$ l is the length and w is the width.	(rectangle: 7 in. by 3 in.) $P = 2l + 2w$ $= 2(7) + 2(3)$ $= 14 + 6$ $= 20,$ or 20 in. $A = lw$ $= 7 \cdot 3$ $= 21,$ or 21 in^2

FIGURE	FORMULA	EXAMPLE
[11.2]–[11.3] **Square**	Perimeter $$P = 4s$$ Area $$A = s^2$$ s is the side.	$\frac{1}{2}$ in. $P = 4s$ $= 4 \cdot \frac{1}{2}$ $= 2, \text{ or } 2 \text{ in.}$ $A = s^2$ $= \left(\frac{1}{2}\right)^2$ $= \frac{1}{4}, \text{ or } \frac{1}{4} \text{ in}^2$
[11.3] **Parallelogram**	Area $$A = bh$$ b is the base and h is the height.	3 ft, 6 ft $A = bh$ $= 6 \cdot 3$ $= 18, \text{ or } 18 \text{ ft}^2$
[11.3] **Trapezoid**	Area $$A = \frac{1}{2}h(b + B)$$ h is the height, b is the upper base, and B is the lower base.	3 in., 4 in., 5 in. $A = \frac{1}{2}h(b + B)$ $= \frac{1}{2}4(3 + 5)$ $= \frac{1}{2} \cdot \overset{2}{4} \cdot 8$ $= 16, \text{ or } 16 \text{ in}^2$
[11.2]–[11.3] **Circle**	Circumference $$C = \pi d \text{ or } C = 2\pi r$$ Area $$A = \pi r^2$$ d is the diameter, r is the radius, and π is approximately 3.14, or $\frac{22}{7}$.	8 cm $C = \pi d$ $\approx 3.14(8)$ $\approx 25.12, \text{ or } 25.12 \text{ cm}$ $A = \pi r^2$ $\approx 3.14(4)^2$ $\approx 3.14(16)$ $\approx 50.24, \text{ or } 50.24 \text{ cm}^2$ Note: $d = 8$ cm, so $r = 4$ cm.

continued

FIGURE	FORMULA	EXAMPLE
[11.4] Rectangular solid	Volume $$V = lwh$$ l is the length, w is the width, and h is the height.	5 cm · 7 cm · 15 cm $$V = lwh$$ $$= 15 \cdot 7 \cdot 5$$ $$= 525, \quad \text{or} \quad 525 \text{ cm}^3$$
[11.4] Cube	Volume $$V = e^3$$ e is the edge.	2 in. $$V = e^3$$ $$= (2)^3$$ $$= 2 \cdot 2 \cdot 2$$ $$= 8, \quad \text{or} \quad 8 \text{ in}^3$$
[11.4] Cylinder	Volume $$V = \pi r^2 h$$ r is the radius and h is the height.	12 m, 4 m $$V = \pi r^2 h$$ $$\approx 3.14(4)^2(12)$$ $$\approx 3.14(16)(12)$$ $$\approx 602.88, \quad \text{or} \quad 602.88 \text{ m}^3$$
[11.4] Sphere	Volume $$V = \frac{4}{3}\pi r^3$$ r is the radius.	2 ft $$V = \frac{4}{3}\pi r^3$$ $$\approx \frac{4}{3}(3.14)(2)^3$$ $$\approx \frac{4}{3}(3.14)(8)$$ $$\approx \frac{100.48}{3}$$ $$\approx 33.5, \quad \text{or} \quad 33.5 \text{ ft}^3$$ to the nearest tenth of a cubic foot

Chapter 11 Review Exercises

To help you review this chapter, solve these problems.

[11.1]

Sketch and label an example of each of the following. Where appropriate, use symbols to express your answer.

1. Line segment

2. Angle

3. Parallel lines

4. Isosceles triangle

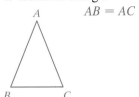

$AB = AC$

Find each missing angle.

5.

$x = 75°$

6.

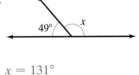

$x = 131°$

7.

$x = 51°$

8.

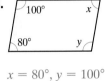

$x = 80°, y = 100°$

[11.2]

Find each perimeter or circumference. Use $\pi \approx 3.14$ when needed.

9. An equilateral triangle with side 1.8 m.   5.4 m

10. A circle with radius 10 in.   62.8 in.

11. A rectangle with length 6 cm and width $3\frac{1}{2}$ cm.
   19 cm

12. A polygon whose sides measure 4.5 ft, 9 ft, 7.5 ft, 3 ft, 3 ft, and 6 ft.   33 ft

[11.3]

Find the area of each figure. Use π ≈ 3.14 when needed.

13.
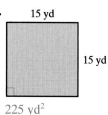
15 yd

15 yd

225 yd²

14.
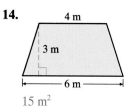
4 m

3 m

6 m

15 m²

15.
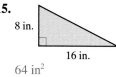
8 in.

16 in.

64 in²

16.
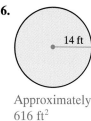
14 ft

Approximately
616 ft²

[11.4]

Find the volume of each figure. Use π ≈ 3.14 when needed. Round where appropriate.

17. A cylinder with radius 10 in. and height 4.2 in.
Approximately 1,318.8 in³

18. A rectangular solid with length 16 ft, width $4\frac{1}{2}$ ft, and
height 3 ft.   216 ft³

19. A cube with edge 1.25 m.   Approximately 1.95 m³

20. A sphere with diameter 2.5 cm.   Approximately
8.18 cm³

[11.2]–[11.4]

21. Find the perimeter of the figure shown, which is made
up of a semicircle and a trapezoid. Use π ≈ 3.14.

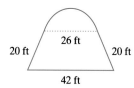
26 ft

20 ft   20 ft

42 ft

Approximately 122.82 ft

22. Find the area of the shaded portion of the figure. Use
π ≈ 3.14.

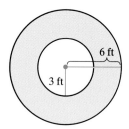
6 ft

3 ft

Approximately 84.78 ft²

[11.5]

Find each missing length.

23.
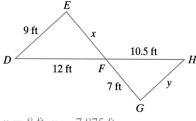
E

9 ft   x

10.5 ft

D   H

12 ft   F

7 ft   y

G

x = 8 ft, y = 7.875 ft

24.
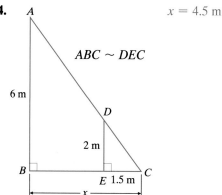
A

x = 4.5 m

ABC ~ DEC

6 m

D

2 m

B

E 1.5 m   C

x

[11.6]

Find the square root.

25. $\sqrt{9}$
3

26. $\sqrt{64}$
8

27. $\sqrt{121}$
11

28. $\sqrt{900}$
30

Determine between which two consecutive whole numbers each square root lies.

29. $\sqrt{3}$
1 and 2

30. $\sqrt{10}$
3 and 4

31. $\sqrt{40}$
6 and 7

32. $\sqrt{84}$
9 and 10

Find the square root. Round to the nearest hundredth.

33. $\sqrt{8}$
2.83

34. $\sqrt{29}$
5.39

35. $\sqrt{195}$
13.96

36. $\sqrt{1,225}$
35

For a given triangle with legs a and b and hypotenuse c, find the missing side to the nearest tenth of the indicated unit.

	a	b	c
37.	9 ft	12 ft	15 ft
38.	10 in.	24 in.	26 in.
39.	8 yd	5 yd	9.4 yd
40.	2 ft	2 ft	2.8 ft

Solve.

41. A roll of aluminum foil is 12 in. wide and 2,400 in. long. Find the area of the roll of aluminum foil.
28,800 in²

42. Following an explosion, poisonous gases spread in all directions for 2 mi. How big was the affected area, to the nearest square mile?   13 mi²

43. Suppose that your air conditioner can cool a room up to 3,000 ft³ in volume. Based on the floor plan of your living room and a ceiling height of 10 ft, can the air conditioner cool the room?

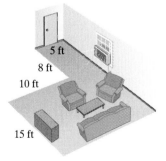

Yes, it can; the volume of the room is 2,650 ft³.

44. One television set has a screen with dimensions approximately 12 in. by 16 in., and another has dimensions 15 in. by 20 in. Is the area of the larger screen more than 50% greater than the area of the smaller screen?
Yes, the larger screen is $56\frac{1}{4}\%$ greater than the smaller screen.

45. A pilot flies 12 mi west from city A to city B. Then he flies 5 mi south from city B to city C. What is the straight-line distance from city A to city C?   13 mi

46. How high up on a wall will a 12-ft ladder reach if the bottom of the ladder is placed 6 ft from the wall? Round to the nearest foot.   10 ft

47. On the pool table shown, you hit the ball at point *E*. It ricochets off point *C* and winds up in the pocket at point *A*. If △*ABC* ~ △*EDC*, find *CD*.   *CD* = $2\frac{1}{4}$ ft

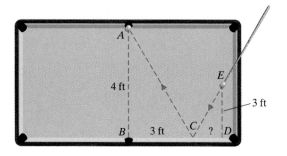

48. To measure *AB*, the distance across a certain lake, distances *BC, CD,* and *DE* were "staked out," as shown in the diagram below. If △*ABC* ~ △*EDC*, how wide is the lake?   *x* = 300 m

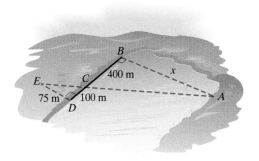

49. From the following drawing, find the total length of the building's walls.   160 ft

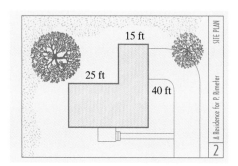

50. According to home economists, the distances between the refrigerator, stove, and sink usually form a work triangle. To be efficient, the perimeter of a work triangle must be no more than 22 ft. Determine whether the model kitchen shown is efficient.

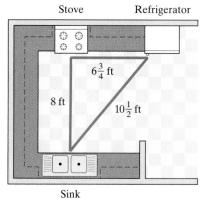

It is not efficient because the perimeter of the work triangle is $25\frac{1}{4}$ ft.

51. The coffee in this cylindrical can weighs 13 oz.

What is the weight of a cubic inch of coffee, to the nearest tenth of an ounce?   5.3 oz

52. How much soil is needed to fill the flower box shown to 1 cm from the top? Round to the nearest cubic centimeter.

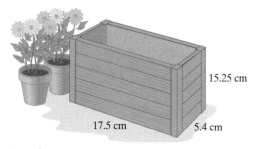

1,347 cm³ of soil

53. Find the area of the picture matting shown. $476\frac{1}{4}$ in^2

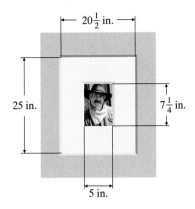

54. The game of racquetball is played with a small, hollow rubber ball, as depicted.

How much rubber, to the nearest tenth of a cubic inch, does a racquetball contain?   2.6 in^3

Chapter 11 Posttest

To see if you have mastered the topics in this chapter, take this test.

1. Sketch and label an example of each.

  a. An acute angle

  b. A ray

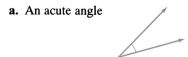

2. Find the square root of each.

  a. $\sqrt{49}$ 7

  b. $\sqrt{225}$ 15

3. Find the complement of 25°. 65°

4. What is the measure of an angle that is supplementary to 91°? 89°

Find each perimeter or circumference. Use $\pi \approx 3.14$ when needed.

5. A square with side $3\frac{1}{2}$ ft 14 ft

6. An equilateral triangle with side 1.5 m 4.5 m

7.

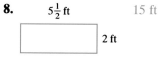

Approximately 25.12 cm

8 cm

8. $5\frac{1}{2}$ ft 15 ft

2 ft

Find each area. Use $\pi \approx 3.14$ when needed.

9. 6 ft 54 ft²

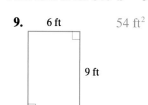

9 ft

10. 8 cm 110 cm²

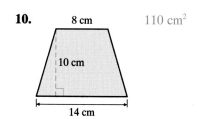

10 cm

14 cm

Find each volume. Use π ≈ 3.14 when needed.

11.

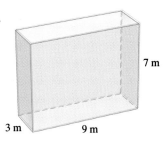

7 m

3 m   9 m

189 m³

12.

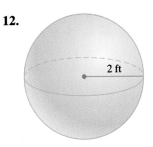

2 ft

Approximately 33.5 ft³

Find each unknown measure.

13.

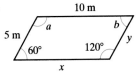

10 m

5 m   *a*   *b*

60°   120°   *y*

x

$a = 120°, b = 60°, y = 5$ m, $x = 10$ m

14.

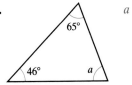

65°

46°   *a*

$a = 69°$

15.

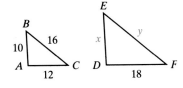

100°

x

$x = 80°$

16.

12 yd

5 yd   *y*

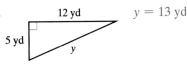

$y = 13$ yd

17. In the following diagram, $\triangle ABC \sim \triangle DEF$. Find *x* and *y*. $x = 15, y = 24$

E

B

10   16   *x*   *y*

A   C   D   F

12   18

Solve.

18. In constructing the foundation for a house, a contractor digs a hole 6 ft deep, 54 ft long, and 25 ft wide. How many cubic feet of earth are removed?   8,100 ft³

19. Suppose that you have a chain on your dog that is 5 m long. To the nearest square meter, what is the area of the dog's "run"? Use π ≈ 3.14.   79 m²

20. On level ground, a tree casts a shadow 36 m long. At the same time, a pole 9 m high casts a shadow 12 m long. What is the height of the tree?   27 m

■ *Check your answers on page A-16.*

Cumulative Review Exercises

To help you review, solve the following.

1. Estimate: $23{,}802 \div 396$ 60

2. Simplify: $\dfrac{24}{30}$ $\dfrac{4}{5}$

3. Round to the nearest tenth: 3.061 3.1

4. 40% of what number is 20?   50

5. Evaluate: $(-4)(-2) + 7$ 15

6. Solve for y: $\dfrac{y}{12} = \dfrac{2}{3}$ $y = 8$

7. A race track consists of two straightaways and two semicircles. What is the total length of the track?   Approximately 514 ft

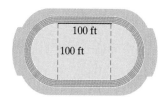

8. Find the width of the river pictured.   1 mi

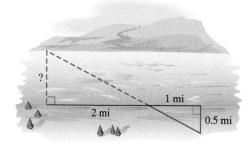

9. A common liquid insecticide for house plants requires mixing 50 parts of water with every part of insecticide. What fraction of the mixture is insecticide? $\frac{1}{51}$

10. This morning the temperature was 10° below 0°F. A few hours later, the temperature rose by 15°. Later in the day, it dropped by 5°. What was the final temperature?   0°F

■ *Check your answers on page A-16.*

Appendix

The Times Table

×	0	1	2	3	4	5	6	7	8	9
0	0	0	0	0	0	0	0	0	0	0
1	0	1	2	3	4	5	6	7	8	9
2	0	2	4	6	8	10	12	14	16	18
3	0	3	6	9	12	15	18	21	24	27
4	0	4	8	12	16	20	24	28	32	36
5	0	5	10	15	20	25	30	35	40	45
6	0	6	12	18	24	30	36	42	48	54
7	0	7	14	21	28	35	42	49	56	63
8	0	8	16	24	32	40	48	56	64	72
9	0	9	18	27	36	45	54	63	72	81

A Table of Common Fraction, Decimal, and Percent Equivalents

Fraction	Decimal	Percent
$\frac{1}{2}$	0.5	50%
$\frac{1}{3}$	0.33...	$33\frac{1}{3}\%$
$\frac{2}{3}$	0.66...	$66\frac{2}{3}\%$
$\frac{1}{4}$	0.25	25%
$\frac{3}{4}$	0.75	75%
$\frac{1}{5}$	0.2	20%
$\frac{2}{5}$	0.4	40%
$\frac{3}{5}$	0.6	60%
$\frac{4}{5}$	0.8	80%
$\frac{1}{6}$	0.16...	$16\frac{2}{3}\%$
$\frac{5}{6}$	0.83...	$83\frac{1}{3}\%$
$\frac{1}{8}$	0.125	$12\frac{1}{2}\%$
$\frac{3}{8}$	0.375	$37\frac{1}{2}\%$
$\frac{5}{8}$	0.625	$62\frac{1}{2}\%$
$\frac{7}{8}$	0.875	$87\frac{1}{2}\%$

A Table of Squares and Square Roots

n	n^2	$\sqrt{n}$
1	1	1
2	4	1.414
3	9	1.732
4	16	2
5	25	2.236
6	36	2.449
7	49	2.646
8	64	2.828
9	81	3
10	100	3.162
11	121	3.317
12	144	3.464
13	169	3.606
14	196	3.742
15	225	3.873
16	256	4
17	289	4.123
18	324	4.243
19	361	4.359
20	400	4.472

Introduction to Calculators

This introduction to calculators covers the basic functions of a scientific calculator that will be used in this text. The keystrokes that are covered in this introduction apply to most scientific calculators, although there are slight differences from one model to the next. Refer to your manual for specific instructions about your particular model.

Topics

- Basic Keystrokes
- Order of Operations
- Decimals and Rounding
- Memory Function

- Signed Numbers
- Exponents and Radicals
- Scientific Notation
- Other Calculator Functions

This secondary function takes the square root of number displayed.

Squares number displayed.

Activates secondary functions printed above certain keys. Also denoted (INV) or (2nd) .

Used when entering numbers in scientific notation. Also denoted (EXP) .

Finds reciprocal of number displayed.

Used to raise any base to a power. Also denoted (yˣ) or (∧).

Stores number displayed in memory. Also denoted (MIN) or (M) .

Recalls number stored in memory. Also denoted (MR) .

Used as an approximation for pi.

Clears all preceding numbers and operations. Also used to turn calculator on.

Used to perform indicated operation.

Used to control order in which certain operations are performed.

Used to change sign of number displayed.

Clears last number displayed but not preceding operations.

Used when entering decimal notation.

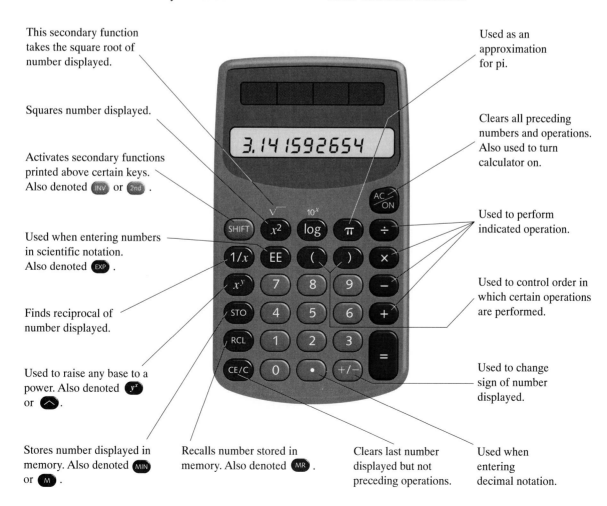

Basic Keystrokes

Most scientific calculators allow you to enter problems in the same order they appear on paper. For example, the problem $1{,}226 + 6{,}321 - 412$ is entered as follows:

Input	1226	+	6321	–	412	=
Display	1226	1226	6321	7547	412	7135

Note that numbers do not include commas when you use a calculator.

Press the CE/C key if the most recent entry needs to be changed. In the last example, if the number 6312 were accidentally entered instead of 6321, it can be corrected as follows:

Input	1226	+	6312	CE/C	6321	–	412	=
Display	1226	1226	6312	0	6321	7547	412	7135

The AC/ON key clears everything in the calculator and allows you to perform a new calculation.

Order of Operations

Most scientific calculators follow the accepted order of operations. Check by evaluating $3 + 5 \times 4$ as follows:

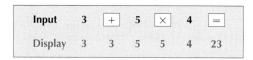

Input	3	+	5	×	4	=
Display	3	3	5	5	4	23

The correct answer is 23. If your calculator returns the answer 32, then it *does not* follow the order of operations. Do not assume that order of operations is always followed. For example, the expression $\dfrac{10 - 8}{4 \div 2}$ must be entered using parentheses, regardless of the type of calculator used. This expression is evaluated as follows:

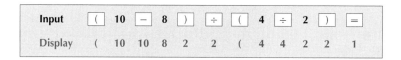

Input	(	10	–	8	)	÷	(	4	÷	2	)	=
Display	(	10	10	8	2	2	(	4	4	2	2	1

Decimals and Rounding

When entering decimal numbers, press the $\boxed{\cdot}$ key. The calculator will automatically add a decimal point if an answer is a decimal. Most scientific calculators can display between 8 and 10 digits. Answers will be rounded automatically to the place value of the last digit displayed.

Memory Function

Memory functions vary considerably depending on the calculator model. Most have a STO (or M) key that, when pressed, stores the number on the display in memory. The RCL (or MR) key recalls a stored number to the screen, and the MC key clears the contents of memory.

- The M– key evaluates the difference of a stored number and the number currently displayed on the screen. The new result is stored in memory.

- The M+ key evaluates the sum of a stored number and the number currently displayed on the screen. The new result is stored in memory.

Signed Numbers

Signed numbers are entered using the $\boxed{+/-}$ key. (Do not confuse this key with the $\boxed{-}$ key, which is used for subtraction.) The following keystrokes are used to evaluate $325 + (-428)$:

Input	325	$\boxed{+}$	428	$\boxed{+/-}$	$\boxed{=}$
Display	325	325	428	−428	−103

Basically, the $\boxed{+/-}$ key changes the sign of the number displayed.

Exponents and Radicals

Different keys can be used to evaluate a number raised to a power. Use the $\boxed{x^2}$ key when squaring a number. Use the $\boxed{x^y}$ (or $\boxed{\wedge}$) key when evaluating a number raised to any power. To evaluate 6^2, press 6 and the $\boxed{x^2}$ key. To evaluate 6^4, use the following keystrokes:

Input	6	$\boxed{x^y}$	4	$\boxed{=}$
Display	6	6	4	1296

To find the square root of a number, enter the number, and then press the $\boxed{\sqrt{}}$ key. For example, $\sqrt{144}$ can be found as follows:

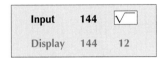

Input	144	$\boxed{\sqrt{}}$
Display	144	12

Scientific Notation

Use scientific notation to express numbers as the product of a number from 1 to 9 and 10 raised to a power. For example, to express 8,000,000 as 8×10^6 press 8, the $\boxed{EE}$ (or $\boxed{EXP}$) key, and then 6. Some scientific calculators will automatically put answers in scientific notation if they are too large or too small to fit on the screen. This can be checked by multiplying $50,000,000 \times 50,000,000 = 2.5 \times 10^{15}$. The calculator should display 2.5 E15.

Other Calculator Functions

- The $\boxed{\pi}$ key displays an approximation for the number pi to as many digits allowed on the screen.
- The $\boxed{1/x}$ key displays the reciprocal of a number.
- The $\boxed{SHIFT}$ (or $\boxed{2nd}$) key accesses any secondary functions located above certain keys.

Answers

Chapter 1

Pretest: Chapter 1, *p. 2*
1. Two hundred five thousand seven **2.** 1,235,000
3. Hundred thousands **4.** 8,100 **5.** 8,226 **6.** 4,714
7. 185 **8.** 29,124 **9.** 260 **10.** 308 R6 or 308 $\frac{3}{22}$
11. 2^3 **12.** 36 **13.** 5 **14.** 43 **15.** 75 years old
16. $675 **17.** 10^9 **18.** 10 sec **19.** $65 **20.** Room C
which measures 126 ft^2

Practices: Section 1.1, *pp. 3−8*
1, *p. 4:* a. Thousands **b.** Hundred thousands **c.** Ten
millions **2, *p. 4:*** Eight billion, three hundred seventy-six
thousand, fifty-two **3, *p. 4:*** $7,372,050; Seven million,
three hundred seventy-two thousand, fifty dollars **4, *p. 5:***
$95,000,003 **5, *p. 5:*** $375,000 **6, *p. 6:*** 2 ten thousands
+ 7 thousands + 0 hundreds + 1 ten + 3 ones = 20,000 +
7,000 + 10 + 3 **7, *p. 6:*** 1 million + 2 hundred thousands
+ 7 ten thousands + 0 thousands + 0 hundreds + 9 tens +
3 ones = 1,000,000 + 200,000 + 70,000 + 90 + 3
8, *p. 7:* a. 59,000 **b.** 60,000 **9, *p. 8:*** $100,000

Exercises 1.1, *p. 9*
1. 1,000,000,000; one billion **3.** 2,350,000; two million,
three hundred fifty thousand **5.** 975,135,000; Nine
hundred seventy-five million, one hundred thirty-five
thousand **7.** 487,500; Four hundred eighty-seven thousand,
five hundred **9.** 2,000,000,352; two billion, three hundred
fifty-two **11.** 10,120 **13.** 150,856 **15.** 6,000,055
17. 50,600,195 **19.** 400,072 **21.** 4,867 **23.** 316
25. 28,461,013 **27.** Hundred thousands **29.** Thousands
31. Billions **33.** 3 ones = 3 **35.** 8 hundreds + 5 tens +
8 ones = 800 + 50 + 8 **37.** 2 millions + 5 hundred
thousands + 4 ones = 2,000,000 + 500,000 + 4 **39.** 670
41. 7,100 **43.** 30,000 **45.** 700,000 **47.** 30,000
49.

To the nearest	135,800	816,533
Hundred	135,800	816,500
Thousand	136,000	817,000
Ten thousand	140,000	820,000
Hundred thousand	100,000	800,000

51. Seven hundred thousand **53.** Three hundred billion
55. One million, four hundred thousand, eighty-five
57. 100,000,000,000 **59.** $85,061,058 **61.** 1,400,000,000
63. 150 ft **65.** 20,000 mi **67.** 400 g

Practices: Section 1.2, *pp. 13−21*
1, *p. 14:* 235 **2, *p. 15:*** 1,936 **3, *p. 16:*** 16 mi
4, *p. 17:* 651 **5, *p. 18:*** 4,747 **6, *p. 18:*** 378
7, *p. 19:* Exact sum: 27,349; possible estimate: 27,300
8, *p. 29:* Exact sum: 2,802 **9, *p. 20:*** Exact: 2,791

Calculator Practices, *p. 20:* 49,532; 31,899; 2,499 ft

Exercises 1.2, *p. 22*
1. 177,778 **3.** 9,673 **5.** 5,034 **7.** 14,710 **9.** 10,467
11. 11,514 **13.** 14,002 **15.** 56,188 **17.** 6,978
19. 4,820 **21.** 413 **23.** 15,509 m **25.** 82 hr
27. $2,755 **29.** $104,831 **31.** 1,676 **33.** 29,602
35. $12,724 **37.** 31,200 tons
39.

+	400	200	1,200	300	Total
300	700	500	1,500	600	3,300
800	1,200	1,000	2,000	1,100	5,300
Total	1,900	1,500	3,500	1,700	8,600

41.

+	389	172	1,155	324	Total
255	644	427	1,410	579	3,060
799	1,188	971	1,954	1,123	5,236
Total	1,832	1,398	3,364	1,702	8,296

43. 13,296,657 **45.** 22,912,891 **47. a 49. a**
51. 217 **53.** 90 **55.** 421 **57.** 7,003 **59.** 362
61. 68,241 **63.** 175 **65.** 1,339 **67.** 1,999
69. 24,950 **71.** 5,186 **73.** 4,400 **75.** 120,310
77. 3,920 **79.** 368 **81.** 4,996 **83.** 982 **85.** 1,995 mi
87. $669 **89.** $3,609 **91.** 273 books **93.** 209 m
95. 2,001,000 **97.** 813,429 **99. c 101. a**
103. 75,000,000 people **105.** $500 **107. a.** Huskies: 8;

Ravens: 9 **b.** Ravens **109.** Possible answer: 43 years old
111. No, the elevator is not overloaded. The total weight
of passengers is 963 lb. **113.** 180 °F **115.** 31 m
117. $3,002 **119.** 126,000,000 sq mi **121.** Give the
bonus; rise = 11,301 **123. a.** Addition **b.** yes, $1,563
c. possible estimate: $1,600

Practices: Section 1.3, *pp. 29–35*
1, *p. 31:* 608 **2,** *p. 31:* 4,230 **3,** *p. 32:* 480,000
4, *p. 32:* 205,296 **5,** *p. 33:* 107 sq ft **6,** *p. 33:* No;
possible estimate = 20,000 **7,** *p. 34:* 112,840

Calculator Practices, *p. 34:* $1,026,015; 345,546; Possible
answer 1.219326311 $\times 10^{18}$

Exercises 1.3, *p. 36*
1. 400 **3.** 7,100 **5.** 850 **7.** 7,000,000 **9.** 1,194
11. 1,928 **13.** 12,700 **15.** 418 **17.** 10,840,000
19. 3,248,000 **21.** 31,220 **23.** 67,456 **25.** 817
27. 34,032 **29.** 3,003 **31.** 3,612 **33.** 57,019
35. 243,456 **37.** 220,025 **39.** 201,670 **41.** 121,706
43. 144,500 **45.** 123,830 **47.** 3,312 **49.** 2,106
51. 40,000 **53.** 23,085 **55.** 3,274,780 **57.** 54,998,850
59. c. Possible estimate: 480,000 **61.** b. Possible
estimate: 80,000 **63.** 3,300 yr **65.** 3,000,000 **67.** Yes;
you could go 576 mi without refilling. **69.** $8,700
71. 1,750 mi **73.** $2,793 **75. a.** Multiplication
b. Colorado; area = 106,700 sq mi **c.** Possible estimate:
120,000 sq mi

Practices: Section 1.4, *pp. 41–47*
1, *p. 43:* 807 **2,** *p. 43:* 7,002 **3,** *p. 44:* 5,291 R1
4, *p. 45:* 79 **5,** *p. 45:* 94 R10 **6,** *p. 46:* 607 R3
7, *p. 46:* 200 **8,** *p. 47:* 967

Calculator Practice, *p. 47:* 603

Exercises 1.4, *p. 48*
1. 400 **3.** 3,000,000 **5.** 50 **7.** 121 **9.** 560
11. 301 **13.** 3,003 **15.** 5,490 **17.** 202 **19.** 219
21. 500 **23.** 841 **25.** 30 **27.** 14 **29.** 160 **31.** 50
33. 651 R2 **35.** 11 R7 **37.** 117 **39.** 808 **41.** 159
43. 5,353 **45.** 1,002 **47.** 6,944 **49.** 723 R19
51. 428 R8 **53.** 1,001 **55.** 3,050 **57.** 721 **59.** 155
61. a. Possible estimate: 7,000 **63.** a. Possible estimate:
400 **65.** $89 **67.** 2 times **69.** 20 mpg **71.** 3
73. $2,496 **75. (a)** Division **(b)** Exact answer: 58 min
(c) Possible estimate: 50 min

Practices: Section 1.5, *pp. 53–58*
1, *p. 53:* 5^5 **2,** *p. 54:* 1 **3,** *p. 54:* 1,331
4, *p. 54:* 784 **5,** *p. 54:* 10^9 **6,** *p. 55:* 28
7, *p. 56:* 146 **8,** *p. 56:* 4 **9,** *p. 57:* $40
10, *p. 57:* No; the average this year is $100.

Calculator Practices, *p. 57:* 140,625; 131

Exercises 1.5, *p. 59*
1.

n	0	2	4	6	8	10	12	14	16	18	20
n^2	0	4	16	36	64	100	144	196	256	324	400

3.

n	0	2	4	6	8	10
n^3	0	8	64	216	512	1,000

5. 10^2 **7.** 10^4 **9.** 10^6 **11.** $2^2 \cdot 3^2$ **13.** $2^1 \cdot 3^3$
15. $2^2 \cdot 3^1$ **17.** 18 **19.** 2 **21.** 35 **23.** 343 **25.** 250
27. 36 **29.** 8 **31.** 92 **33.** 60 **35.** 28 **37.** 9
39. 2 **41.** 99 **43.** 99 **45.** 9 **47.** 419 **49.** 137,088
51. $\boxed{4} \cdot 3 + \boxed{6} \cdot 5 + \boxed{8} \cdot 7 = 98$ **53.** $(\boxed{8})(3 + \boxed{4}) -$
$2 \cdot \boxed{6} = 44$ **55.** $\boxed{8} + 10 \times \boxed{4} - \boxed{6} \div 2 = 45$
57. $(5 + 2) \cdot 4^2 = 112$ **59.** $(5 + 2 \cdot 4)^2 = 169$
61. $(8 - 4) \div 2^2 = 1$ **63.** 242 cm² **65.** 3,120 in.²
67.

Input	Output
0	$21 + 3 \times 0 = 21$
1	$21 + 3 \times 1 = 24$
2	$21 + 3 \times 2 = 27$

69. 25 **71.** 40 **73.** 4 **75.** 2,412 mi **77.** 8 **79.** 154
81. $1,100 **83.** Men spent more time; men's average was
3 hr, while women's was 2 hr. **85.** 625 sq ft **87.** $5^2 +$
$12^2 = 13^2$; 25 + 144 = 169 **89. (a)** Addition, division,
and subtraction **(b)** Newspaper B was more popular by 1,553
(c) The numbers are too close to estimate which is larger.

Practices: Section 1.6, *pp. 65–68*
1, *p. 66:* 10,670 **2,** *p. 67:* 2 yr **3,** *p. 67:* 1,551
4, *p. 68:* 180 lb

Exercises 1.6, *p. 69*
1. $2,150 **3.** 27 mi **5.** 12 shelves **7.** $1,827
9. 30,000,000 lb **11.** $1 less expensive **13.** 8 extra pens
15. 1952 was closer by 31 votes. **17. (a)** Subtraction and
division **(b)** $983 **(c)** Possible estimate: $1,000

Review Exercises: Chapter 1, *p. 73*
1. Ones **2.** Ten thousands **3.** Hundred millions
4. Ten billions **5.** Four hundred ninety-seven **6.** Two
thousand fifty **7.** Three million, seven **8.** Eighty-five
billion **9.** 251 **10.** 9,002 **11.** 14,000,025
12. 3,000,003,000 **13.** 3 hundreds + 8 ones
14. 2 millions + 5 hundred thousands **15.** 4 ten
thousands + 2 thousands + 7 hundreds + 7 tens
16. 3 ten millions + 1 ten + 2 ones **17.** 600
18. 1,000 **19.** 380,000 **20.** 70,000 **21.** 9,486
22. 65,692 **23.** 173,543 **24.** 150,895 **25.** 1,059,613
26. $223,067 **27.** 445 **28.** 300,000 **29.** 11,109
30. 5,600 **31.** 11,042,223 **32.** $2,062,852 **33.** 432
34. 1,200 **35.** 149,073 **36.** 12,000,000 **37.** 1,397,508
38. 188,221,590 **39.** 94 **40.** 39 **41.** 37 R10
42. 800 **43.** 25,625 **44.** 957 **45.** 343 **46.** 1
47. 72 **48.** 3,125 **49.** 5 **50.** 169 **51.** 5 **52.** 19
53. 10,833,312 **54.** 2,694 **55.** $7^2 \cdot 5^2$ **56.** $2^2 \cdot 5^3$
57. 39 **58.** 7 **59.** 6 **60.** 5 **61.** Two million, four
hundred thousand **62.** 150,000,000 **63.** $307 per week
64. 1758 **65.** 8,000,000 acres **66.** 308 ft/week

67. 9 **68.** 272 legs **69.** Possible answers: For a 28-day month: 2,800,000; for a 30-day month: 3,000,000; for a 31-day month: 3,100,000 **70.** 80,000 books **71.** 246,000 **72.** Gross margin: $255,000; net profit: $120,000 **73.** 81 sq ft **74.** 20 **75.** 1960 to 1972 (15,379,754 votes) **76.** 4,341 **77.** $59 **78.** 240,000,000,000,000 mi **79.** 29 mi^2 **80.** 162 cm

Posttest: Chapter 1, *p. 77*
1. 225,067 **2.** 1,768,405 **3.** One million, two hundred five thousand, seven **4.** 200,000 **5.** 1,894 **6.** 607 **7.** 147 **8.** 297,496 **9.** 70 **10.** 509 **11.** 125 **12.** $2^2 \cdot 3^3$ **13.** 84 **14.** 2 **15.** 5,700 sq mi **16.** $2,975 **17.** 10^6 transactions **18.** 87 **19.** 10 more yellow roses than pink roses **20.** 12 mg more

Chapter 2

Pretest: Chapter 2, *p. 80*
1. 1, 2, 4, 5, 10, 20 **2.** $2 \times 2 \times 2 \times 3 \times 3$ **3.** $\frac{2}{5}$
4. $\frac{61}{3}$ **5.** $1\frac{1}{30}$ **6.** $\frac{3}{4}$ **7.** 20 **8.** $\frac{1}{8}$ **9.** $1\frac{1}{5}$ **10.** $12\frac{5}{6}$
11. $2\frac{1}{4}$ **12.** $4\frac{5}{8}$ **13.** $3\frac{1}{2}$ **14.** 60 **15.** $\frac{2}{3}$ **16.** $2\frac{2}{35}$
17. $\frac{4}{9}$ **18.** 6 students **19.** $20\frac{7}{8}$ ft **20.** $\frac{1}{3}$ acre

Practices: Section 2.1, *pp. 81–87*
1, *p. 81:* 1, 7 **2,** *p. 82:* 1, 3, 5, 15, 25, 75 **3,** *p. 83:* 1, 2, 3, 5, 6, 9, 10, 15, 18, 30, 45, 90 **4,** *p. 83:* Yes (24 is a multiple of 3.) **5,** *p. 84:* **a.** Prime **b.** Composite **c.** Prime **d.** Composite **e.** Prime **6,** *p. 85:* $2^3 \times 7$
7, *p. 85:* $3^2 \times 5$ **8,** *p. 86:* 18 **9,** *p. 87:* 66 **10,** *p. 87:* 12 **11,** *p. 87:* 6 yr

Exercises 2.1, *p. 88*
1. 1, 3, 7, 21 **3.** 1, 17 **5.** 1, 2, 3, 4, 6, 12 **7.** 1, 31
9. 1, 2, 3, 4, 6, 9, 12, 18, 36 **11.** 1, 29 **13.** 1, 2, 4, 5, 10, 20, 25, 50, 100 **15.** 1, 2, 4, 7, 14, 28 **17.** Prime
19. Composite (2, 4, 8) **21.** Composite (7) **23.** Prime
25. Composite (3, 9, 27) **27.** 2^3 **29.** $2^2 \times 3$ **31.** $2^3 \times 3$ **33.** 2×5^2 **35.** 7×11 **37.** 3×17 **39.** 5^2
41. 2^5 **43.** 3×7 **45.** $2^3 \times 13$ **47.** 11^2 **49.** 2×71
51. $2^2 \times 5^2$ **53.** 5^3 **55.** $3^3 \times 5$ **57.** 15 **59.** 40
61. 90 **63.** 110 **65.** 72 **67.** 360 **69.** 300 **71.** 30
73. 105 **75.** 60 **77.** There was not a census in 1985 because 1985 is not a multiple of 10; there was a census in 1990 because 1990 is a multiple of 10. **79.** 24 in. × 24 in.

Practices: Section 2.2, *pp. 91–100*
1, *p. 92:* $\frac{5}{8}$ **2,** *p. 92:* $\frac{1,000}{2,451}$ **3,** *p. 92:* $\frac{125}{164}$
4, *p. 93:* 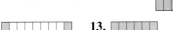 **5,** *p. 93:* $\frac{16}{3}$; $\frac{102}{5}$

6, *p. 94:* **a.** 2 **b.** $5\frac{5}{9}$ **c.** $2\frac{2}{3}$ **7,** *p. 96:* Possible answer: $\frac{4}{10}$, $\frac{6}{15}$, $\frac{8}{20}$ **8,** *p. 96:* $\frac{45}{72}$ **9,** *p. 97:* $\frac{2}{3}$ **10,** *p. 97:* $\frac{5}{2}$
11, *p. 98:* $\frac{15}{49}$ **12,** *p. 99:* $\frac{11}{16}$ **13,** *p. 100:* $\frac{8}{15}, \frac{23}{30}, \frac{9}{10}$
14, *p. 100:* Working: $\frac{1}{3} > \frac{7}{24}$

Exercises 2.2, *p. 101*
1. $\frac{1}{3}$ **3.** $\frac{3}{6}$ **5.** $1\frac{1}{4}$ **7.** $3\frac{2}{4}$ **9.**

11.   **13.**

15. **17.**
19. **21.** Proper **23.** Improper

25. Mixed **27.** Improper **29.** Proper **31.** Mixed
33. $\frac{13}{5}$ **35.** $\frac{55}{9}$ **37.** $\frac{57}{5}$ **39.** $\frac{5}{1}$ **41.** $\frac{59}{8}$ **43.** $\frac{88}{9}$ **45.** $\frac{38}{3}$
47. $\frac{98}{5}$ **49.** $\frac{14}{1}$ **51.** $\frac{54}{11}$ **53.** $\frac{115}{14}$ **55.** $\frac{202}{25}$ **57.** $1\frac{1}{3}$
59. $1\frac{1}{9}$ **61.** 3 **63.** 1 **65.** $19\frac{4}{5}$ **67.** $9\frac{1}{9}$ **69.** 1
71. $8\frac{2}{9}$ **73.** $13\frac{1}{2}$ **75.** $11\frac{1}{9}$ **77.** 27 **79.** 8
81. Possible answer: $\frac{2}{16}, \frac{3}{24}$ **83.** Possible answer: $\frac{4}{22}, \frac{6}{33}$
85. Possible answer: $\frac{6}{8}, \frac{9}{12}$ **87.** Possible answer: $\frac{2}{14}, \frac{3}{21}$
89. 9 **91.** 15 **93.** 40 **95.** 36 **97.** 40 **99.** 54
101. 36 **103.** 42 **105.** 6 **107.** 49 **109.** 32
111. 30 **113.** $\frac{1}{2}$ **115.** $\frac{2}{3}$ **117.** 1 **119.** $\frac{1}{3}$
121. $\frac{21}{5}$ or $4\frac{1}{5}$ **123.** $\frac{2}{3}$ **125.** $\frac{1}{4}$ **127.** $\frac{1}{8}$ **129.** $\frac{5}{4}$ or $1\frac{1}{4}$
131. $\frac{33}{16}$ or $2\frac{1}{16}$ **133.** $\frac{9}{16}$ **135.** $\frac{1}{20}$ **137.** $\frac{9}{7}$ or $1\frac{2}{7}$ **139.** $\frac{19}{51}$
141. 3 **143.** $\frac{1}{7}$ **145.** $\frac{3}{4}$ **147.** $\frac{3}{8}$ **149.** $4\frac{1}{2}$ **151.** $5\frac{2}{3}$
153. $7\frac{2}{5}$ **155.** 3 **157.** $\frac{7}{20} < \frac{11}{20}$ **159.** $\frac{1}{8} > \frac{1}{9}$
161. $\frac{2}{3} = \frac{6}{9}$ **163.** $2\frac{1}{3} < 2\frac{9}{15}$ **165.** $\frac{1}{3}$ **167.** $\frac{2}{3}$ **169.** $\frac{2}{3}$
171. a. $\frac{1}{50}$ **b.** $\frac{49}{50}$ **173.** $\frac{2}{5}$ **175.** No; $\frac{1}{4} = \frac{25}{100}$ which is greater than $\frac{23}{100}$. **177. a.** $\frac{3}{8}$ in. **b.** No **179.** $1\frac{1}{4}$ in.

Practices: Section 2.3, *pp. 109–122*
1, *p. 109:* $\frac{2}{3}$ **2,** *p. 110:* $1\frac{7}{40}$ **3,** *p. 110:* $\frac{2}{5}$ **4,** *p. 110:* Neither; they are both 1 mi. **5,** *p. 111:* $\frac{1}{2}$ yd
6, *p. 112:* $1\frac{2}{3}$ **7,** *p. 112:* $\frac{3}{10}$ **8,** *p. 112:* $\frac{71}{72}$
9, *p. 113:* 2 mi **10,** *p. 115:* $35\frac{1}{5}$ **11,** *p. 115:* $7\frac{1}{2}$
12, *p. 115:* $4\frac{2}{5}$ **13,** *p. 116:* 4 lengths **14,** *p. 116:* $2\frac{1}{8}$ in.
15, *p. 117:* $7\frac{5}{8}$ **16,** *p. 117:* $4\frac{1}{2}$ **17,** *p. 118:* $11\frac{5}{24}$
18, *p. 118:* $6\frac{1}{4}$ mi **19,** *p. 119:* $1\frac{2}{7}$ **20,** *p. 120:* $5\frac{1}{6}$
21, *p. 121:* $12\frac{21}{40}$ **22,** *p. 121:* No, there will be only $3\frac{5}{8}$ yd left. **23,** *p. 122:* $6\frac{7}{20}$; $8 - 4 + 2 = 6$

Exercises 2.3, *p. 123*
1. $1\frac{1}{3}$ **3.** $1\frac{1}{5}$ **5.** $1\frac{1}{4}$ **7.** $1\frac{1}{2}$ **9.** $\frac{4}{5}$ **11.** $\frac{3}{5}$ **13.** $1\frac{1}{6}$
15. $\frac{7}{8}$ **17.** $\frac{77}{100}$ **19.** $\frac{37}{40}$ **21.** $\frac{61}{72}$ **23.** $\frac{13}{24}$ **25.** $1\frac{5}{18}$
27. $1\frac{17}{100}$ **29.** $\frac{3}{4}$ hr **31.** $\frac{5}{12}$ mi **33.** $1\frac{1}{12}$ **35.** $1\frac{13}{40}$
37. $3\frac{2}{3}$; $1 + 2 = 3$ **39.** $15\frac{2}{5}$; $8 + 7 = 15$ **41.** $14\frac{1}{5}$; $7 + 7 = 14$ **43.** 15; $5 + 10 = 15$ **45.** 8; $4 + 4 = 8$
47. $5\frac{2}{5}$; $3 + 3 = 6$ **49.** $42\frac{5}{6}$; $38 + 5 = 43$
51. $14\frac{11}{12}$; $8 + 7 = 15$ **53.** $10\frac{5}{12}$; $5 + 5 = 10$
55. $3\frac{11}{15}$; $3 + 0 = 3$ **57.** $13\frac{13}{15}$; $8 + 6 = 14$ **59.** $6\frac{19}{24}$; $1 + 6 = 7$ **61.** $20\frac{1}{4}$; $10 + 11 = 21$ **63.** $96\frac{27}{100}$; $66 + 31 = 97$
65. $10\frac{3}{100}$; $6 + 4 = 10$ **67.** $45\frac{1}{2}$; $23 + 23 = 46$
69. $19\frac{1}{12}$; $11 + 8 = 19$ **71.** $36\frac{3}{50}$; $30 + 6 = 36$
73. $91\frac{7}{12}$; $80 + 1 + 11 = 92$ **75.** $15\frac{1}{4}$ hr; $6 + 6 + 4 = 16$ ft
77. $6\frac{3}{4}$ ft; $2 + 2 + 2 = 6$ ft **79.** $10\frac{33}{40}$ in.; $7 + 3 + 1 = 11$ in.
81. $11\frac{1}{24}$ min; $1 + 6 + 5 = 12$ min **83.** $22\frac{11}{16}$; $20 + 1 + 2 = 23 **85.** $\frac{1}{5}$ **87.** $\frac{2}{5}$ **89.** $\frac{4}{25}$ **91.** 1 **93.** $\frac{1}{2}$ **95.** 2
97. $\frac{1}{3}$ day **99.** $\frac{1}{12}$ **101.** $\frac{5}{18}$ **103.** $\frac{87}{100}$ **105.** $\frac{11}{100}$ **107.** $\frac{1}{14}$
109. $\frac{8}{45}$ **111.** $\frac{1}{2}$ **113.** $\frac{5}{72}$ **115.** $\frac{3}{20}$ **117.** $\frac{1}{66}$ **119.** $\frac{1}{6}$
121. $4\frac{2}{7}$ **123.** $7\frac{2}{5}$ **125.** $1\frac{3}{4}$ **127.** 20 **129.** $4\frac{1}{10}$
131. $9\frac{1}{2}$ **133.** $\frac{1}{3}$ **135.** $4\frac{5}{8}$ **137.** $3\frac{1}{3}$ **139.** $3\frac{3}{10}$
141. $6\frac{1}{3}$ **143.** $5\frac{1}{2}$ **145.** $4\frac{1}{2}$ **147.** $3\frac{1}{4}$ **149.** $8\frac{4}{5}$

151. $6\frac{2}{3}$ **153.** $7\frac{5}{6}$ **155.** $3\frac{13}{24}$ **157.** $15\frac{7}{18}$ **159.** $2\frac{29}{30}$
161. $\frac{1}{4}$ mi   **163.** $5\frac{5}{12}$ ft   **165.** $13\frac{39}{40}$ acres   **167.** $\frac{3}{4}$ pint
169. $1\frac{11}{40}$ **171.** $4\frac{1}{2}$ **173.** $16\frac{23}{30}$ **175.** $3\frac{1}{4}$ **177.** $10\frac{1}{15}$
179. $\frac{7}{8}$ in.   **181.** 5 hr   **183.** $35\frac{3}{8}$ mi   **185.** $\frac{1}{10}$ **187.** 1 lb

Practices: Section 2.4, pp. 131–140
1, p. 132: $\frac{15}{28}$ **2, p. 132:** 20   **3, p. 132:** $\frac{7}{22}$ **4, p. 133:** $\frac{2}{9}$
5, p. 133: $5\frac{1}{4}$ hr   **6, p. 133:** \$38,000   **7, p. 134:** $7\frac{7}{8}$
8, p. 134: 28   **9, p. 135:** $\frac{3}{16}$ lb   **10, p. 135:** $25\frac{1}{2}$ sq in.
11, p. 137: 6   **12, p. 137:** 8   **13, p. 138:** $2\frac{2}{3}$ yr
14, p. 138: $1\frac{3}{5}$ **15, p. 139:** $2\frac{2}{7}$ **16, p. 139:** 6 lb
17, p. 140: 6

Exercises 2.4, p. 141
1. $\frac{2}{15}$ **3.** $\frac{5}{12}$ **5.** $\frac{9}{16}$ **7.** $\frac{8}{25}$ **9.** $\frac{35}{32}=1\frac{3}{32}$ **11.** $\frac{45}{16}=2\frac{13}{16}$
13. $\frac{2}{9}$ **15.** $\frac{7}{12}$ **17.** $\frac{3}{40}$ **19.** $\frac{31}{30}=1\frac{1}{30}$ **21.** $\frac{5}{3}=1\frac{2}{3}$
23. $\frac{14}{5}=2\frac{4}{5}$ **25.** $\frac{40}{3}=13\frac{1}{3}$ **27.** $\frac{40}{3}=13\frac{1}{3}$ **29.** 4
31. $\frac{35}{4}=8\frac{3}{4}$ **33.** 4   **35.** 4   **37.** $\frac{8}{15}$ **39.** $\frac{4}{9}$ **41.** $1\frac{5}{16}$
43. $2\frac{1}{8}$ **45.** $\frac{25}{27}$ **47.** $2\frac{2}{3}$ **49.** 1   **51.** $\frac{7}{8}$ **53.** $1\frac{13}{35}$
55. $4\frac{41}{100}$ **57.** $7\frac{4}{5}$ **59.** 375   **61.** 8   **63.** 3   **65.** $41\frac{2}{3}$
67. $113\frac{1}{3}$ **69.** $1\frac{1}{6}$ **71.** $\frac{7}{12}$ **73.** $\frac{77}{100}$ **75.** $3\frac{3}{8}$ **77.** $\frac{9}{10}$
79. $\frac{32}{35}$ **81.** $3\frac{1}{2}$ **83.** $4\frac{4}{9}$ **85.** $1\frac{1}{2}$ **87.** $2\frac{1}{3}$ **89.** $1\frac{1}{5}$
91. $\frac{1}{4}$ **93.** $\frac{2}{21}$ **95.** $\frac{1}{30}$ **97.** $\frac{1}{10}$ **99.** $\frac{1}{9}$ **101.** 40
103. $16\frac{1}{3}$ **105.** $13\frac{1}{3}$ **107.** 7   **109.** $6\frac{11}{18}$ **111.** $1\frac{2}{3}$
113. $9\frac{22}{27}$ **115.** $100\frac{1}{2}$ **117.** $\frac{7}{90}$ **119.** $\frac{5}{26}$ **121.** $3\frac{1}{5}$
123. $\frac{21}{200}$ **125.** $\frac{35}{44}$ **127.** $1\frac{47}{115}$ **129.** $\frac{14}{27}$ **131.** $2\frac{1}{4}$
133. $1\frac{7}{18}$ **135.** $4\frac{13}{15}$ **137.** $\frac{87}{160}$ **139.** $1\frac{5}{27}$ **141.** $4\frac{1}{5}$
143. $3\frac{1}{8}$ **145.** $\frac{3}{8}$ mi   **147.** \$8,000   **149.** $6\frac{1}{4}$
151. $\$191\frac{1}{4}$ **153.** \$750 stocks; \$875 bonds   **155.** 7 times
157. 6 rolls

Review Exercises: Chapter 2, p. 150
1. 1, 2, 3, 5, 6, 10, 15, 25, 30, 50, 75, 150   **2.** 1, 2, 3, 4, 5,
6, 9, 10, 12, 15, 18, 20, 30, 36, 45, 60, 90, 180   **3.** 1, 3,
19, 57   **4.** 1, 2, 3, 4, 6, 8, 9, 12, 18, 24, 36, 72   **5.** Prime
6. Composite   **7.** Composite   **8.** Prime   **9.** $2^2 \times 3^2$
10. 3×5^2 **11.** $2^3 \times 5^2$ **12.** 2×3^3 **13.** 24   **14.** 10
15. 72   **16.** 60   **17.** $\frac{1}{3}$ **18.** $\frac{2}{4}=\frac{1}{2}$ **19.** $\frac{3}{6}=\frac{1}{2}$
20. $\frac{6}{12}=\frac{1}{2}$ **21.** $\frac{23}{3}$ **22.** $\frac{9}{5}$ **23.** $\frac{91}{10}$ **24.** $\frac{59}{7}$ **25.** $6\frac{1}{2}$
26. $4\frac{2}{3}$ **27.** $2\frac{3}{4}$ **28.** $4\frac{3}{5}$ **29.** $n=12$ **30.** $n=4$
31. $n=5$ **32.** $n=27$ **33.** $\frac{1}{2}$ **34.** $\frac{5}{7}$ **35.** $\frac{2}{3}$ **36.** $\frac{3}{4}$
37. $5\frac{1}{2}$ **38.** $8\frac{2}{3}$ **39.** $6\frac{2}{7}$ **40.** $8\frac{5}{7}$ **41.** $\frac{5}{8}>\frac{3}{8}$
42. $\frac{5}{6}>\frac{1}{6}$ **43.** $\frac{2}{3}<\frac{4}{5}$ **44.** $\frac{9}{10}>\frac{7}{8}$ **45.** $\frac{3}{4}>\frac{5}{8}$
46. $\frac{7}{10}>\frac{5}{9}$ **47.** $3\frac{3}{5}>1\frac{9}{10}$ **48.** $5\frac{1}{8}>5\frac{1}{9}$ **49.** $\frac{2}{7},\frac{3}{8},\frac{1}{2}$
50. $\frac{2}{15},\frac{1}{5},\frac{1}{3}$ **51.** $\frac{3}{4},\frac{4}{5},\frac{9}{10}$ **52.** $\frac{13}{18},\frac{7}{9},\frac{7}{8}$ **53.** $\frac{5}{9}$ **54.** $\frac{10}{8}=1\frac{1}{4}$
55. $\frac{11}{10}=1\frac{1}{10}$ **56.** $\frac{10}{6}=1\frac{2}{3}$ **57.** $\frac{6}{5}=1\frac{1}{5}$ **58.** $\frac{3}{4}$
59. $\frac{15}{8}=1\frac{7}{8}$ **60.** $\frac{3}{5}$ **61.** $\frac{11}{15}$ **62.** $1\frac{17}{24}$ **63.** $1\frac{2}{5}$
64. $1\frac{7}{40}$ **65.** 6   **66.** $9\frac{1}{2}$ **67.** $10\frac{2}{5}$ **68.** 8   **69.** $12\frac{1}{3}$
70. $4\frac{3}{10}$ **71.** $6\frac{7}{10}$ **72.** $17\frac{13}{24}$ **73.** $23\frac{5}{12}$ **74.** $46\frac{3}{8}$
75. $20\frac{3}{4}$ **76.** $56\frac{1}{24}$ **77.** $\frac{1}{4}$ **78.** $\frac{2}{3}$ **79.** 1   **80.** 0
81. $\frac{1}{4}$ **82.** $\frac{3}{8}$ **83.** $\frac{7}{20}$ **84.** $\frac{7}{30}$ **85.** $7\frac{1}{2}$ **86.** $2\frac{3}{10}$
87. $3\frac{3}{4}$ **88.** $18\frac{1}{2}$ **89.** $6\frac{1}{2}$ **90.** $1\frac{7}{10}$ **91.** $2\frac{1}{3}$ **92.** $\frac{1}{5}$
93. $1\frac{4}{5}$ **94.** $\frac{3}{4}$ **95.** $2\frac{1}{2}$ **96.** $3\frac{1}{3}$ **97.** $\frac{3}{10}$ **98.** $2\frac{7}{8}$
99. $\frac{7}{12}$ **100.** $3\frac{8}{9}$ **101.** $\frac{2}{15}$ **102.** $\frac{7}{16}$ **103.** $\frac{5}{8}$ **104.** $\frac{1}{6}$
105. 75   **106.** 160   **107.** $5\frac{1}{3}$ **108.** $\frac{7}{10}$ **109.** $\frac{1}{125}$

110. $\frac{8}{27}$ **111.** $\frac{1}{4}$ **112.** $\frac{7}{120}$ **113.** $\frac{24}{25}$ **114.** $1\frac{5}{9}$ **115.** $2\frac{2}{3}$
116. $\frac{2}{3}$ **117.** 6   **118.** $18\frac{5}{12}$ **119.** $8\frac{7}{16}$ **120.** $21\frac{1}{4}$
121. $\frac{9}{20}$ **122.** $1\frac{9}{16}$ **123.** $37\frac{1}{27}$ **124.** $3\frac{3}{8}$ **125.** $\frac{3}{2}$
126. $\frac{2}{3}$ **127.** $\frac{1}{8}$ **128.** 4   **129.** $\frac{7}{40}$ **130.** $\frac{5}{81}$ **131.** $\frac{2}{15}$
132. $\frac{1}{200}$ **133.** $\frac{3}{4}$ **134.** $1\frac{1}{3}$ **135.** 30   **136.** $8\frac{3}{4}$
137. $1\frac{1}{6}$ **138.** $1\frac{4}{5}$ **139.** 2   **140.** 4   **141.** $1\frac{3}{4}$ **142.** $\frac{4}{7}$
143. $1\frac{7}{12}$ **144.** $\frac{12}{19}$ **145.** $5\frac{1}{2}$ **146.** $2\frac{11}{20}$ **147.** 2
148. 3   **149.** No   **150.** 50¢   **151.** $\frac{1}{4}$ **152.** $\frac{2}{9}$ **153.** $\frac{1}{7}$
154. More   **155.** The Filmworks camera   **156.** Less
157. You got back more than $\frac{1}{3}$ because $\frac{275}{700}=\frac{11}{28}=\frac{33}{84}$ which
is greater than $\frac{1}{3}=\frac{28}{84}$.   **158.** Yes it should because $\frac{23}{32}$ is
greater than $\frac{2}{3}$. $\frac{23}{32}=\frac{69}{96}$ whereas $\frac{2}{3}=\frac{64}{96}$ **159. a.** $\frac{12}{23}$ **b.** $\frac{3}{4}$
160. a. Alisa Gregory **b.** Monica Yates
161.

Employee	Saturday	Sunday	Total
1	$7\frac{1}{2}$	$4\frac{1}{4}$	$11\frac{3}{4}$
2	$5\frac{3}{4}$	$6\frac{1}{2}$	$12\frac{1}{4}$
Total	$13\frac{1}{4}$	$10\frac{3}{4}$	24

162.

Worker	Hours per Day	Days Worked	Total Hours	Wage per Hour	Gross Pay
A	5	3	15	\$7	\$105.00
B	$7\frac{1}{4}$	4	29	\$10	\$290.00
C	$4\frac{1}{2}$	$5\frac{1}{2}$	$24\frac{3}{4}$	\$9	\$222.75

163. $1\frac{1}{16}$ **164.** $11\frac{3}{4}$ mi   **165.** 7 lb   **166.** \$18
167. 1,500 fps   **168.** 2,100   **169.** $7\frac{1}{2}$ in.
170. 500 lb/sq in.   **171.** 2 times   **172.** 19   **173.** $10\frac{10}{11}$ lb
174. $\frac{11}{12}$ oz

Posttest: Chapter 2, p. 158
1. 1, 3, 5, 15, 25, 75   **2.** $2^2 \times 3 \times 5$ **3.** $\frac{4}{9}$ **4.** $\frac{12}{1}$
5. $10\frac{1}{4}$ **6.** $\frac{7}{8}$ **7.** $\frac{5}{10}$ **8.** 24   **9.** $1\frac{13}{24}$ **10.** $8\frac{7}{40}$
11. $4\frac{2}{7}$ **12.** $5\frac{23}{30}$ **13.** $\frac{1}{81}$ **14.** 12   **15.** $\frac{7}{9}$ **16.** $1\frac{1}{2}$
17. $\frac{1}{3}$ **18.** $\frac{1}{6}$ of the pie   **19.** $90\frac{2}{3}$ sq ft   **20.** $3\frac{1}{3}$ in.

Cumulative Review: Chapter 2, p. 160
1. Five million, three hundred, fifteen   **2.** 581,400
3. 908   **4.** $\frac{3}{4}$ **5.** $6\frac{2}{5}$ **6.** $6\frac{1}{6}$ **7.** $\frac{3}{8}$ **8.** 1 million times
9. 549   **10.** Candle B; $\frac{1}{2}>\frac{1}{3}$

Chapter 3

Pretest: Chapter 3, p. 162
1. Hundredths   **2.** Four and twelve thousandths   **3.** 3.1
4. 0.0029   **5.** 21.52   **6.** 7.3738   **7.** 11.69   **8.** 9.81
9. 8,300   **10.** 18.423   **11.** 0.0144   **12.** 7.1   **13.** 0.00605
14. 32.7   **15.** 0.875   **16.** 2.83   **17.** One with a pH
balance of 2.95   **18.** \$8 million   **19.** 3 times   **20.** \$3.74

Practices: Section 3.1, *pp. 163–170*
1, *p. 164:* **a.** 0.3<u>6</u> **b.** 0.4<u>72</u> **c.** 0.0<u>2</u>51 **d.** 897.4<u>3</u>
e. 1,912.6<u>43</u>   **2, *p. 165:*** $2\frac{3}{100}$ **3, *p. 165:*** $\frac{7}{8}$ **4, *p. 165:***
a. $5\frac{3}{5}$ **b.** $5\frac{3}{5}$ **5, *p. 166:*** **a.** $7\frac{3}{500}$ **b.** $7\frac{3}{5}$ **6, *p. 166:***
a. Sixty-one hundredths **b.** Four and nine hundred twenty-
three thousandths **c.** Seven and five hundredths   **7, *p. 166:***
a. 0.043 **b.** 10.26   **8, *p. 167:*** 3.14   **9, *p. 167:*** 0.83 is
larger   **10, *p. 167:*** Uranus   **11, *p. 168:*** 3.8, 3.5, 2.9
12, *p. 168:* The one with the rating of 8.1, because 9 >
8.2 > 8.1.   **13, *p. 169:*** **a.** 748.1 **b.** 748.08 **c.** 748.077
d. 748 **e.** 700   **14, *p. 170:*** 7.30   **15, *p. 170:*** 163.0

Exercises 3.1, *p. 171*
1. Sixty-one hundredths   **3.** Three hundred five
thousandths   **5.** Six tenths   **7.** Five and seventy-two
hundredths   **9.** Twenty-four and two thousandths   **11.** $\frac{3}{5}$
13. $\frac{39}{100}$ **15.** $1\frac{1}{2}$ **17.** 8   **19.** $5\frac{3}{250}$ **21.** 0.8   **23.** 1.041
25. 60.01   **27.** 4.107   **29.** 3.2 m   **31.** One and four
hundred sixty-seven thousandths   **33.** Eighteen and seven
tenths to eighteen and eight tenths   **35.** Asia: Three
hundred one and three tenths; Africa: Fifty-five and nine
tenths; Europe: Two hundred sixty-eight and two tenths; North
America: Forty-six and six tenths; South America: Forty-three
and six tenths   **37.** One hundred thousandth; eight hundred
thousandths   **39.** 1.2 acres   **41.** 74.59 mph   **43.** 14.7 lb
45. $0.005   **47.** 352.1 kWh   **49.** Tenths place
51. Ones place   **53.** Tenths place   **55.** Hundredths place
57. Tenths   **59.** Hundredths   **61.** Thousandths
63. Ones   **65.** 3.21 > 2.5   **67.** 0.71 < 0.8   **69.** 9.123 >
9.11   **71.** 4 = 4.000   **73.** 8.125 ft < 8.2 ft   **75.** 7, 7.07,
7.1   **77.** 4.9, 5.001, 5.2   **79.** 9.1 mi, 9.38 mi, 9.6 mi
81. Less   **83.** Last winter   **85.** Yes   **87.** No   **89.** 17.4
91. 3.591   **93.** 37.1   **95.** 0.40   **97.** 7.06   **99.** 9 mi
101.

To the nearest	8.0714	0.9916
Tenth	8.1	1.0
Hundredth	8.07	0.99
Ten	10	0

103. $57.03   **105.** 0.001   **107.** 1.8

Practices: Section 3.2, *pp. 179–183*
1, *p. 179:* 10.387   **2, *p. 179:*** $39.30   **3, *p. 180:***
102.1 °F   **4, *p. 180:*** 46.21   **5, *p. 181:*** $485.43
6, *p. 181:* 13.5 mi   **7, *p. 181:*** 2.1 mi   **8, *p. 182:*** 0.863
9, *p. 182:* $4 million   **10, *p. 182:*** 0.079
11, *p. 182:* $480.00

Calculator Practices, *p. 183*: 79.23; 0.00002

Exercises 3.2, *p. 184*
1. 9.33   **3.** 0.9   **5.** 8.13   **7.** 21.45   **9.** 7.67
11. $77.21   **13.** 24.16   **15.** 44.422   **17.** 20.32 mm
19. 16.682 kg   **21.** 23.30595   **23.** 0.7   **25.** 16.8
27. 18.41   **29.** 75.63   **31.** 22.324   **33.** 0.17   **35.** 6.2
37. 15.37   **39.** 5.9   **41.** 6.21   **43.** 1.85 lb   **45.** 4.9°F
47. 41.40896   **49.** c   **51.** b   **53.** $1.03   **55.** 5,680 yr

or 56.8 centuries   **57.** $1.7 million   **59.** 6.84 in.
61. Yes; 2.8 + 2.9 + 2.6 + 1.6 = 9.9.   **63.** 102.5°F
65. **(a)** Addition, subtraction **(b)** Total 13.2 mg of iron; you
need 4.8 mg more **(c)** Possible estimate: 1 + 2 + 0 + 2 +
1 + 1 + 1 + 1 + 2 + 1 + 0 = 13

Practices: Section 3.3, *pp. 189–193*
1, *p. 189:* 9.835   **2, *p. 190:*** 1.4   **3, *p. 190:*** 0.01
4, *p. 190:* 0.024   **5, *p. 190:*** 9.91   **6, *p. 191:*** 325
7, *p. 191:* 327,000   **8, *p. 191:*** **a.** 18.01 **b.** 18   **9, *p. 192:***
Estimate: 0.004 ¥ 0.09 = 0.00036   **10, *p. 192:*** Possible
answer: 1,140 mi

Calculator Practices, *p. 193*: 815.6; 9.261

Exercises 3.3, *p. 194*
1. 2.99212   **3.** 0.0000969   **5.** 204.36   **7.** 2,492.0
9. 2,870.00   **11.** $0.73525   **13.** 0.18   **15.** 0.40
17. 0.02   **19.** 0.0008   **21.** 0.765   **23.** 2.016
25. 7.602   **27.** 0.5   **29.** 5.7   **31.** 3.735   **33.** 151.14
35. 0.084   **37.** 8,312.7   **39.** 23   **41.** 70   **43.** 0.09
45. 0.09   **47.** 0.000000001   **49.** 25.75   **51.** 42.5 ft
53. 1.4 mi   **55.** 3.29025   **57.** 272,593.75
59.

Input	Output
1	3.8 × **1** − 0.2 = 3.6
2	3.8 × **2** − 0.2 = 7.4
3	3.8 × **3** − 0.2 = 11.2
4	3.8 × **4** − 0.2 = 15

61. a: 50 × 1 = 50   **63.** b: 0.02 × 0.7 = 0.014
65. 2,900 fps   **67.** $2,600,000,000.00   **69.** 19.6 ft²
71. 1.25 mg
73. a.

Purchase	Quantity	Unit Price	Price
Belt	1	$11.99	$11.99
Shirt	3	$16.95	$50.85
Total Price			$62.84

b. $17.16   **75. (a)** Multiplication and addition **(b)** 88.81 in.
(c) 90 in.

Practices: Section 3.4, *pp. 199–205*
1, *p. 200:* 0.375   **2, *p. 200:*** 7.625   **3, *p. 200:*** 83.3
4, *p. 201:* 0.8   **5, *p. 202:*** 18.04; 18.04 × 0.15 = 2.706
6, *p. 202:* 2,050; 2,050 × 0.004 = 8.200 = 8.2
7, *p. 203:* 73.4   **8, *p. 203:*** 0.0341   **9, *p. 203:*** 0.00086
10, *p. 204:* 4 tablets   **11, *p. 204:*** 6.4   **12, *p. 205:*** 21.1;
possible estimate: 20

Calculator Practices, *p. 205*: 4.29; 0.2

Exercises 3.4, *p. 206*
1. 5.5   **3.** 5.25   **5.** 3.7   **7.** 1.625   **9.** 2.875   **11.** 21.03
13. 4.25   **15.** 4.2   **17.** 1.375   **19.** 8.5   **21.** 0.67   **23.** 0.78

25. 3.11   **27.** 5.06   **29.** 3.3   **31.** 0.3   **33.** 303.7   **35.** 6.6
37. 58.82   **39.** 6.9   **41.** 0.93   **43.** 2.8875   **45.** 27.17
47. 0.286   **49.** 4.3   **51.** 0.0015   **53.** 1.73   **55.** 2.875
57. 0.4   **59.** 0.704   **61.** 9.4   **63.** 12.5   **65.** 0.3   **67.** 0.2
69. 0.952   **71.** 8.16   **73.** 0.0027   **75.** 0.495   **77.** 384.0
79. 180.0   **81.** 0.0   **83.** 9,802.0   **85.** 0.3   **87.** 57.1
89. 9,666.7   **91.** 11.9   **93.** 5.4   **95.** 0.2   **97.** 325.2
99. 457.7   **101.** 8.6   **103.** 4.1   **105.** 67.4   **107.** 41.6
109.

Input	Output
1	$15 \div 1 - 0.2 = 14.8$
2	$15 \div 2 - 0.2 = 7.3$
3	$15 \div 3 - 0.2 = 4.8$
4	$15 \div 4 - 0.2 = 3.55$

111. c: $32 \div 0.6 \approx 53.3$ **113.** b: $0.6 \div 3 = 0.2$
115. 0.0037 in./yr   **117. a.** $\frac{21}{35}$; 0.6 **b.** $\frac{22}{40}$; 0.55 **c.** The
women's team has a better record. Its winning percentage is
60% and the men's team's is 55%.
119. a.

Car	Distance Driven (mi)	Gasoline Used (gal)	Miles/ Gallon
A	18.6	1.6	11.625
B	7.8	0.6	13
C	23.4	1.2	19.5

b. Car C   **121.** 2,000 shares   **123.** 1,000 times
125. 0.5 sec   **127. (a)** Division **(b)** 0.367 **(c)** 0.4

Review Exercises: Chapter 3, *p. 214*
1. Hundredths   **2.** Tenths   **3.** Tenths   **4.** Ten thousandths
5. $\frac{7}{20}$ **6.** $8\frac{1}{5}$ **7.** $4\frac{7}{1,000}$ **8.** 10   **9.** Seventy-two
hundredths   **10.** Five and six tenths   **11.** Three and nine
ten thousandths   **12.** Five hundred ten and thirty-six
thousandths   **13.** 0.007   **14.** 2.1   **15.** 0.03   **16.** 7.041
17. 0.04   **18.** 2.031   **19.** 5.12   **20.** 2   **21.** 0.72
22. 0.00057   **23.** 7.3   **24.** 0.039   **25.** 4.39   **26.** $899
27. 12.11   **28.** 52.75   **29.** $24.13   **30.** 12 m   **31.** 0.3
32. 28.78   **33.** 87.752   **34.** 1.62   **35.** 98.2033
36. $90,948.80   **37.** 2.912   **38.** 1,008   **39.** 0.00001
40. 13.69   **41.** 2,710   **42.** 0.034   **43.** 5.75   **44.** 13.5
45. 1,569.36846   **46.** 5,398.835596   **47.** 0.17   **48.** 0.63
49. 0.29   **50.** 0.22   **51.** 4.06   **52.** 90.13   **53.** 8.33
54. 11.83   **55.** 7.6   **56.** 0.7   **57.** 1.6   **58.** 0.2   **59.** 0.3
60. 0.0   **61.** 0.3   **62.** 0.0   **63.** 5.0   **64.** 750.0
65. 23.7   **66.** 16,358.3   **67.** 3.0   **68.** 0.3   **69.** Four ten
millionths   **70.** 95.92 points   **71.** 54.49 secs   **72.** 1.647 in.
73. No; $333.7 < 333.7095 < 333.75$ **74.** $44.00   **75.** No,
it would have travelled 0.585 mi, which is less than 0.75 mi.
76. 4.35 times   **77.** $0.06   **78.** Possible estimate: 15 in.
79. 7.19 g   **80.** 10.4 times   **81.** 36,162.45   **82.** Company

A, because it posted greater earnings over the four quarters.
(Company A: 5.085 v. Company B: 5.059)

Posttest: Chapter 3, *p. 218*
1. 0   **2.** Five and one hundred-two thousandths
3. 320.15   **4.** 0.00028   **5.** $3\frac{1}{25}$ **6.** .004   **7.** 4.354
8. $5.66   **9.** 20.9   **10.** 5.72   **11.** 0.001   **12.** 3.36
13. 0.0029   **14.** 32.7   **15.** 0.375   **16.** 4.17
17. 0.01 lb   **18.** 2.6 ft   **19.** Belmont Stakes
20. $2,733.00

Cumulative Review: Chapter 3, *p. 219*
1. 1,000,000   **2.** 32   **3.** 0.035   **4.** Hundredths   **5.** 2.1
6. 0.09   **7.** 0.58   **8.** 325   **9.** 1.6 in.
10. 33,016.2 mi

Chapter 4

Pretest: Chapter 4, *p. 222*
1. Possible answer: six less than y **2.** Possible answer:
quotient of x and 2   **3.** $m + 8$ **4.** $2n$ **5.** 4   **6.** $1\frac{1}{2}$
7. $x + 3 = 5$ **8.** $4y = 12$ **9.** $x = 6$ **10.** $y = 10$
11. $n = 13$ **12.** $a = 12$ **13.** $m = 6.1$ **14.** $n = 30$
15. $m = 0.48$ **16.** $n = 15$ **17.** $325.75   **18.** $12\frac{1}{12}$ ft
19. 72,000 sq mi   **20.** $2.68

Practices: Section 4.1, *pp. 223–227*
1, *p. 224*: a. One-half of p **b.** x less than 5 **c.** y divided
by 4 **d.** 3 more than n **e.** $\frac{3}{5}$ of b **2, *p. 225*: a.** $x + 9$
b. $10y$ **c.** $n - 7$ **d.** Possible answer: $n \div 5$ **e.** Possible
answer: $\frac{2}{5}n$ **3, *p. 225*:** $\frac{h}{4}$ hr   **4, *p. 226*: a.** 25 **b.** 0.38
c. 4.8 **d.** 26.6   **5, *p. 226*:** $55m$ words in m min and 1,650
words in 30 min   **6, *p. 227*:** You pay $15.45 + t$ dollars and
$18.45 for $t = $3.

Exercises 4.1, *p. 228*
1. 3 more than t: t plus 3   **3.** c minus 4: 4 subtracted
from c **5.** c divided by 3: The quotient of c and 3
7. 10 times s: The product of 10 and s **9.** y minus 10:
10 less than y **11.** 7 times a: The product of 7 and a
13. x divided by 4: The quotient of x and 4   **15.** x minus $\frac{1}{2}$:
$\frac{1}{2}$ less than x **17.** $\frac{1}{2}$ times w: $\frac{1}{2}$ of w **19.** x minus 2: The
difference between x and 2   **21.** 1 increased by x: x added
to 1   **23.** 3 times p: The product of 3 and p **25.** n
decreased by 1.1: n minus 1.1   **27.** y divided by 0.9: The
quotient of y and 0.9   **29.** $x + 10$ **31.** $n - 5$ **33.** $y + 5$
35. Possible answer: $t \div 6$ **37.** $10y$ **39.** $w - 5$ **41.** n
$+ 100$ **43.** Possible answer: $z \div 3$ **45.** $\frac{2}{5}x$
47. $k - 6$ **49.** $n + 5$ **51.** $n - 5\frac{1}{2}$ **53.** 26   **55.** 2.5
57. 15   **59.** $1\frac{1}{6}$ **61.** 1.1   **63.** 1
65.

x	$x + 8$
1	9
2	10
3	11
4	12

67.

n	$n - 0.2$
1	0.8
2	1.8
3	2.8
4	3.8

69.

x	$\frac{3}{4}x$
4	3
8	6
12	9
16	12

71.

z	$\frac{z}{2}$
2	1
4	2
6	3
8	4

73. $c + 2$ **75.** $30° + 90° + d°$, or $120° + d°$
77. 150 mi **79. a.** $2s$ dollars **b.** \$28

Practices: Section 4.2, *pp. 233–238*
1, *p. 234:* **a.** $n - 5.1 = 9$ **b.** $y + 5 = 12$ **c.** $n - 4 = 12$
d. $n + 5 = 7\frac{3}{4}$ **2,** *p. 234:* $p - 6 = 49.95$, where p is the
regular price. **3,** *p. 235:* $x = 9$; $9 + 5 \overset{?}{=} 14$, $14 \overset{\checkmark}{=} 14$
4, *p. 236:* $t = 2.7$; $2.7 - 0.9 \overset{?}{=} 1.8$, $1.8 \overset{\checkmark}{=} 1.8$ **5,** *p. 236:*
$m = 5\frac{1}{4}$; $5\frac{1}{4} + \frac{1}{4} \overset{?}{=} 5\frac{1}{2}$, $5\frac{2}{4} \overset{?}{=} 5\frac{1}{2}$, $5\frac{1}{2} \overset{\checkmark}{=} 5\frac{1}{2}$ **6,** *p. 237:*
a. $11 = m - 4$; $m = 15$ **b.** $12 + n = 21$; $n = 9$
7, *p. 237:* $809.46 = 144.95 + x$; $x = \$664.51$ **8,** *p. 238:*
$26\frac{1}{4} = x - 1\frac{5}{8}$; $x = \$27\frac{7}{8}$

Exercises 4.2, *p. 239*
1. $z - 9 = 25$ **3.** $7 + x = 25$ **5.** $t - 3.1 = 4$ **7.** $\frac{3}{2} + y = \frac{9}{2}$ **9.** $n - 3\frac{1}{2} = 7$ **11. a.** Yes **b.** No **c.** Yes **d.** No
13. Subtract 4 **15.** Add 11 **17.** Add 7 **19.** Subtract 2
21. $a = 31$ **23.** $y = 2$ **25.** $x = 12$ **27.** $n = 4$
29. $c = 47$ **31.** $x = 13$ **33.** $z = 2.9$ **35.** $n = 8.9$
37. $y = 0.9$ **39.** $x = 6\frac{2}{3}$ **41.** $m = 5\frac{1}{3}$ **43.** $x = 3\frac{3}{4}$
45. $m = 2$ **47.** $y = 90$ **49.** $y = 6\frac{1}{4}$ **51.** $n = \frac{1}{2}$
53. $x = 8.2$ **55.** $y = 19.91$ **57.** $x = 4.557$ **59.** $y = 10.251$ **61.** $n + 3 = 11$; $n = 8$ **63.** $y - 6 = 7$; $y = 13$
65. $n + 10 = 19$; $n = 9$ **67.** $x + 3.6 = 9$; $x = 5.4$
69. $n - 4\frac{1}{3} = 2\frac{2}{3}$; $n = 7$ **71.** Equation c
73. Equation d **75.** $621{,}000 = x - 13{,}000$; $x = \$634{,}000$
77. $40 + {-B} = 90°$; $-B = 50°$

Practices: Section 4.3, *pp. 243–248*
1, *p. 243:* **a.** $2x = 14$ **b.** $\frac{a}{6} = 1.5$ **c.** $\frac{n}{0.3} = 1$ **d.** $10 = \frac{1}{2}n$
2, *p. 244:* $15 = 3w$ **3,** *p. 245:* $x = 5$ **4,** *p. 245:*
$a = 6$ **5,** *p. 245:* $x = 4$ **6,** *p. 246:* $a = 2.88$
7, *p. 246:* $x = 16$ **8,** *p. 247:* **a.** $12 = \frac{z}{6}$, $z = 72$; $12 \overset{?}{=} \frac{72}{6}$,
$12 \overset{\checkmark}{=} 12$ **b.** $16 = 2x$, $8 = x$; $16 \overset{?}{=} 2(8)$, $16 \overset{\checkmark}{=} 16$
9, *p. 247:* $1.6 = 5x$; $x = 0.32$ km **10,** *p. 248:* $\frac{1}{5}x = 725$;
$x = \$3{,}625$

Exercises 4.3, *p. 249*
1. $\frac{3}{4}y = 12$ **3.** $\frac{x}{7} = \frac{7}{2}$ **5.** $\frac{1}{3}x = 2$ **7.** $\frac{x}{3} = \frac{1}{3}$ **9.** $9z = 27$
11. Divide by 3 **13.** Multiply by 2 **15.** Divide by $\frac{3}{4}$ or
multiply by $\frac{4}{3}$ **17.** Divide by 1.5 **19. a.** Yes **b.** No
c. No **d.** No **21.** $x = 6$ **23.** $x = 18$ **25.** $n = 4$
27. $x = 91$ **29.** $y = 4$ **31.** $b = 20$ **33.** $m = 157.5$
35. $t = 0.4$ **37.** $x = \frac{3}{2}$, or $1\frac{1}{2}$ **39.** $x = 36$ **41.** $t = 3$
43. $y = \frac{2}{5}$ **45.** $n = 700$ **47.** $x = 12.5$ **49.** $x = \frac{1}{2}$
51. $m = 6$ **53.** $x = 6.81$ **55.** $x = 4.949$ **57.** $8n = 56$;
$n = 7$ **59.** $\frac{3}{4}y = 18$; $y = 24$ **61.** $\frac{x}{5} = 11$; $x = 55$
63. $2x = 36$; $x = 18$ **65.** $\frac{1}{3}x = 4$; $x = 12$ **67.** $\frac{n}{5} = 1\frac{3}{5}$;
$n = 8$ **69.** $\frac{y}{2.5} = 10$; $y = 25$ **71.** Equation d
73. Equation a **75.** $4s = 60$; $s = 15$ **77.** $\frac{1}{6}x = 750$;
$x = 4{,}500$ ft **79. (a)** Multiplication **(b)** \$23 million
(c) Possible estimate: \$21 million

Review Exercises: Chapter 4, *p. 254*
1. $x + 1$ **2.** 4 more than y **3.** w minus 1 **4.** 3 less
than s **5.** c over 7 **6.** The quotient of a and 10 **7.** 2
times x **8.** The product of 6 and y **9.** y divided by 0.1
10. The quotient of n and 1.5 **11.** $\frac{1}{3}$ of x **12.** $\frac{1}{10}$ of w
13. $m + 16$ **14.** $b + \frac{1}{2}$ **15.** $y - 1.4$ **16.** $z - 8$
17. $\frac{6}{z}$ **18.** $n \div 2.5$ **19.** $3n$ **20.** $12n$ **21.** 12 **22.** 19
23. 0 **24.** 6 **25.** 0.3 **26.** 6.5 **27.** $1\frac{1}{2}$ **28.** $\frac{5}{12}$
29. 0.4 **30.** $4\frac{1}{2}$ **31.** $1\frac{1}{3}$ **32.** 9 **33.** $x = 9$
34. $y = 9$ **35.** $n = 26$ **36.** $b = 20$ **37.** $a = 3.5$
38. $c = 7.5$ **39.** $x = 11$ **40.** $y = 2$ **41.** $w = 1\frac{1}{2}$
42. $s = \frac{1}{3}$ **43.** $c = 6\frac{1}{4}$ **44.** $p = 11\frac{2}{3}$ **45.** $m = 5$
46. $n = 0$ **47.** $c = 78$ **48.** $y = 90$ **49.** $n = 11$
50. $x = 25$ **51.** $x = 31.0485$ **52.** $m = 26.6225$
53. $n - 19 = 35$ **54.** $t - 37 = 234$ **55.** $9 + x = 5\frac{1}{2}$
56. $s + 26 = 30\frac{1}{3}$ **57.** $2y = 16$ **58.** $25t = 175$
59. $\frac{n}{19} = 34$ **60.** $\frac{z}{13} = 17$ **61.** $\frac{1}{3}n = 27$ **62.** $\frac{2}{5}n = 4$
63. a. No **b.** No **c.** No **d.** No **64. a.** Yes **b.** No
c. No **d.** No **65.** $x = 5$ **66.** $t = 2$ **67.** $a = 105$
68. $n = 54$ **69.** $y = 9$ **70.** $r = 10$ **71.** $w = 90$
72. $x = 100$ **73.** $y = 20$ **74.** $a = 120$ **75.** $n = 32$
76. $b = 32$ **77.** $m = 3.15$ **78.** $z = 0.57$ **79.** $x = \frac{2}{5}$
80. $y = \frac{1}{2}$ **81.** $m = 1.2$ **82.** $b = 9.8$ **83.** $x = 12.5$
84. $x = 1.4847$ **85.** $12.99x$; \$103.92 **86.** $\frac{d}{40}$ dollars/hr;
\$9.55/hr **87.** $89p$ cents; \$2.67 **88.** $3{,}000 + d$ dollars;
\$3,225 **89.** $9.59 + x = 19.14$; $x = 9.55$
90. $\frac{1}{4}x = 500{,}000$; $x = 2{,}000{,}000$ **91.** $10{,}000t = 1{,}000$;
$t = 0.1$ mm **92.** $225 = x + 50$; $x = 175$ **93.** $2\frac{2}{5}w = 286$; $w = 110$ lb **94.** $1.8x = 6{,}696$; $x = 3{,}720$ km
95. $98.6 + x = 101$; $x = 2.4$ °F **96.** $y - 8 = 42$;
$y = \$50$

Posttest: Chapter 4, *p. 258*
1. Possible answer: x plus $\frac{1}{2}$ **2.** Possible answer: the
quotient of a and 3 **3.** $n + 8$ **4.** $\frac{1}{8}p$ **5.** 0 **6.** $\frac{1}{4}$
7. $x - 6 = 4\frac{1}{2}$ **8.** $\frac{y}{8} = 3.2$ **9.** $x = 0$ **10.** $y = 12$
11. $n = 27$ **12.** $a = 738$ **13.** $m = 7.8$ **14.** $n = 50$
15. $x = 5.4$ **16.** $n = 760$ **17.** $1\frac{3}{4} + x = 2\frac{1}{4}$; $x = \frac{1}{2}$ lb
18. $\frac{1}{3}x = 30{,}000$; $x = 90{,}000$ elephants **19.** $7.25x = 145$;
$x = 20$ hr **20.** $x - 14\frac{1}{2} = 156\frac{1}{2}$; $x = 171$ lb

Cumulative Review: Chapter 4, *p. 259*
1. $5\frac{3}{8}$ **2.** 0.0075 **3.** Yes **4.** $y = 3$ **5.** $w = 1\frac{2}{3}$
6. $n = 7.8$ **7.** $x = 32$ **8.** 7,200 **9.** You got back $\frac{2}{7}$ of
your money, which is less than $\frac{1}{3}$. **10.** 25.05 lb

Chapter 5

Pretest: Chapter 5, *p. 262*
1. $\frac{3}{4}$ **2.** $\frac{7}{9}$ **3.** $\frac{5}{3}$ **4.** $\frac{19}{51}$ **5.** $\frac{2 \text{ dental assistants}}{1 \text{ dentist}}$ **6.** $\frac{16 \text{ gal}}{5 \text{ min}}$
7. $\frac{1 \text{ basket}}{3 \text{ attempts}}$ **8.** $\frac{5 \text{ mg}}{3 \text{ hr}}$ **9.** $\frac{\$230}{1 \text{ box}}$ **10.** $\frac{1 \text{ page}}{5 \text{ min}}$ **11.** True
12. False **13.** $x = 9$ **14.** $x = 31\frac{1}{2}$ **15.** $x = 16$
16. $x = 160$ **17.** $\frac{4}{5}$ **18.** \$2/lb **19.** 204 **20.** 87 mi

Practices: Section 5.1, *pp. 263–267*

1, *p. 263:* $\frac{2}{3}$ **2,** *p. 264:* $\frac{9}{5}$ **3,** *p. 264:* $\frac{5}{19}$ **4,** *p. 265:*
a. $\frac{3\ nurses}{10\ patients}$ **b.** $\frac{1\ lab\ station}{4\ students}$ **5,** *p. 265:* **a.** 48 fps **b.** 0.375 hit
per times at bat **6,** *p. 266:* 75 wpm **7,** *p. 266:*
a. \$174/flight **b.** \$895/fax machine **8,** *p. 266:* The 32-oz
can at 2.875¢/oz

Exercises 5.1, *p. 268*

1. $\frac{2}{3}$ **3.** $\frac{2}{3}$ **5.** $\frac{11}{7}$ **7.** $\frac{3}{2}$ **9.** $\frac{1}{4}$ **11.** $\frac{4}{3}$ **13.** $\frac{1}{1}$ **15.** $\frac{5}{3}$
17. $\frac{7}{24}$ **19.** $\frac{20}{1}$ **21.** $\frac{8}{7}$ **23.** $\frac{4}{5}$ **25.** $\frac{5\ calls}{2\ days}$ **27.** $\frac{10\ children}{3\ families}$
29. $\frac{75\ students}{4\ faculty}$ **31.** $\frac{17\ baskets}{30\ attempts}$ **33.** $\frac{1\ car}{1\ person}$ **35.** $\frac{200\ mi}{9\ gal}$ **37.** $\frac{16\ males}{3\ females}$
39. $\frac{8\ Democrats}{7\ Republicans}$ **41.** $\frac{1\ lb}{8\ servings}$ **43.** $\frac{5\ pages}{6\ min}$ **45.** $\frac{1\ lb}{200\ ft^2}$
47. 55 mph **49.** 8 gal/day **51.** 0.3 tank/acre **53.** 1.6 yd/
dress **55.** 2 hr/day **57.** 0.25 km/min **59.** 18 mpg
61. \$0.45/bar **63.** \$2.95/roll **65.** \$66.67/plant
67. \$99/night **69.** a **71.** b **73.** b **75.** $\frac{114}{91}$ **77.** $\frac{2}{3}$
79. 170 Cal/oz **81.** 25 times/min **83.** 57.1 cm³/g
85. $\frac{1}{2}$ **87.** **(a)** Division **(b)** 0.51 **(c)** Possible answer: 0.5

Practices: Section 5.2, *pp. 273–277*

1, *p. 273:* Yes **2,** *p. 273:* Not a true proportion
3, *p. 274:* Different rates; Bob typed 60 wpm, and Tamika
typed $58\frac{1}{3}$ wpm. **4,** *p. 275:* $x = 8$ **5,** *p. 275:* $x = 12$
6, *p. 276:* 64,000 **7,** *p. 276:* 15 gal **8,** *p. 277:* 1.5 ft

Exercises 5.2, *p. 278*

1. True **3.** False **5.** True **7.** False **9.** True
11. True **13.** $x = 20$ **15.** $x = 38$ **17.** $x = 4$
19. $x = 13$ **21.** $x = 8$ **23.** $x = 4$ **25.** $x = 20$
27. $x = 15$ **29.** $x = 21$ **31.** $x = 13\frac{1}{3}$ **33.** $x = 20$
35. $x = 1.8$ **37.** $x = 21$ **39.** $x = 280$ **41.** $x = 100$
43. $x = 100$ **45.** $x = 280$ **47.** $x = 2\frac{8}{9}$ **49.** $x = 0.005$
51. Not the same **53.** $1\frac{7}{8}$ gal **55.** 54.5 g **57.** $1\frac{1}{8}$ lb
59. $41\frac{2}{3}$ in. **61.** 90 mg and 50 mg **63.** 0.25 ft
65. \$600 **67.** 12,000 fish **69.** 280 times
71. **(a)** Multiplication, division **(b)** 835,312.5 gal
(c) Possible answer: 880,000 gal

Review Exercises: Chapter 5, *p. 284*

1. $\frac{2}{3}$ **2.** $\frac{1}{2}$ **3.** $\frac{3}{4}$ **4.** $\frac{25}{8}$ **5.** $\frac{8}{5}$ **6.** $\frac{3}{4}$ **7.** $\frac{44\ ft}{5\ sec}$
8. $\frac{16\ applicants}{45\ positions}$ **9.** 0.0025 lb/sq ft **10.** 500,000,000 calls/day
11. 200 mph **12.** 1.5 VCRs/household
13. 10,500,000 vehicles/yr **14.** 76,000 commuters/day
15. \$118.75/night **16.** \$17/ticket **17.** \$2,500/station
18. \$93.64/share **19.** a **20.** b **21.** b **22.** a
23. True **24.** False **25.** False **26.** True
27. $x = 6$ **28.** $x = 3$ **29.** $x = 32$ **30.** $x = 30$
31. $x = 2$ **32.** $x = 8$ **33.** $x = 51\frac{4}{5}$ **34.** $x = 1\frac{1}{5}$
35. $x = 67\frac{1}{2}$ **36.** $x = 45$ **37.** $x = 28$ **38.** $x = 0.14$
39. $\frac{1}{15}$ **40.** $\frac{23}{45}$ **41.** \$90/day **42.** 0.125 in./mo **43.** $\frac{2}{7}$
44. 50,000 books **45.** No **46.** 55 cc **47.** 2 hr
48. 5 in. **49.** 0.68 g/cc **50.** 7.55 yd

Posttest: Chapter 5, *p. 287*

1. $\frac{2}{3}$ **2.** $\frac{5}{14}$ **3.** $\frac{55}{31}$ **4.** $\frac{12}{1}$ **5.** $\frac{13\ revolutions}{12\ sec}$ **6.** $\frac{1\ cm}{25\ km}$
7. 68 mph **8.** \$136/day **9.** $6\frac{2}{3}$ m/sec **10.** $17\frac{13}{17}$ mpg
11. No **12.** Yes **13.** $x = 25$ **14.** $x = 6$ **15.** $x = 28$

16. $x = 1$ **17.** 5 million for \$600 **18.** $\frac{3}{19}$ **19.** 25 ft
20. 48 beats/min

Cumulative Review: Chapter 5, *p. 288*

1. $\frac{2}{5}$ **2.** 8,200 **3.** $x = 10$ **4.** $\frac{1}{4}$ **5.** \$4/yd
6. Possible answer: 3 **7.** \$11/16D **8.** 54.54 sec
9. 7.5 in. **10.** \$180

Chapter 6

Pretest: Chapter 6, *p. 290*

1. $\frac{1}{20}$ **2.** $\frac{3}{8}$ **3.** 2.5 **4.** 0.03 **5.** 0.7% **6.** 800%
7. 67% **8.** 110% **9.** $37\frac{1}{2}$ ft **10.** 55 **11.** Possible
answer: \$48 **12.** 250 **13.** 40% **14.** 250% **15.** \$14
16. $33\frac{1}{3}$% **17.** $\frac{6}{25}$ **18.** 25% **19.** \$2,500 **20.** \$10,000

Practices: Section 6.1, *pp. 291–297*

1, *p. 292:* $\frac{21}{100}$ **2,** *p. 292:* 2 **3,** *p. 293:* $\frac{1}{8}$ **4,** *p. 293:* $\frac{9}{10}$
5, *p. 294:* 3 **6,** *p. 294:* 0.05 **7,** *p. 294:* 0.482
8, *p. 294:* 0.6225 **9,** *p. 295:* 0.11 **10,** *p. 295:* 2.5%
11, *p. 295:* 1% **12,** *p. 295:* 70% **13,** *p. 296:* 300%
14, *p. 296:* 72% **15,** *p. 296:* nitrogen; 78% > 0.93%, or
0.78 > 0.0093. **16.** p. 297: 16% **17.** p. 297: True. $\frac{2}{3} \approx$
67% > 60% **18.** 297: 22%

Exercises 6.1, *p. 298*

1. $\frac{1}{20}$ **3.** $2\frac{1}{2}$ **5.** $\frac{33}{100}$ **7.** $\frac{9}{50}$ **9.** $\frac{7}{50}$ **11.** $\frac{13}{20}$ **13.** $\frac{3}{400}$
15. $\frac{3}{1,000}$ **17.** $\frac{1}{8}$ **19.** $\frac{1}{3}$ **21.** 0.06 **23.** 0.72 **25.** 0.001
27. 1.02 **29.** 0.875 **31.** 0.182 **33.** 0.069 **35.** 0.035
37. 0.009 **39.** 0.0075 **41.** 31% **43.** 87.5% **45.** 30%
47. 4% **49.** 18% **51.** 129% **53.** 290% **55.** 287%
57. 101.6% **59.** 500% **61.** 75% **63.** 10%
65. 300% **67.** 80% **69.** 90% **71.** 12.5% **73.** $55\frac{5}{9}$%
75. $66\frac{2}{3}$% **77.** 162.5% **79.** $216\frac{1}{6}$%
81.

Fraction	Decimal	Percent
$\frac{1}{2}$	0.5	50%
$\frac{1}{4}$	0.25	25%
$\frac{3}{4}$	0.75	75%
$\frac{1}{5}$	0.2	20%
$\frac{2}{5}$	0.4	40%
$\frac{3}{5}$	0.6	60%
$\frac{4}{5}$	0.8	80%

83. 10% **85.** $\frac{6}{25}$ **87.** 970% **89.** $1\frac{7}{20}$ **91.** 0.8
93. The condition is more common among men. $\frac{1}{3} = 33\frac{1}{3}$%
< 40% **95.** **(a)** Subtraction, division, multiplication
(b) Yes **(c)** Possible estimate: 100%

Practices: Section 6.2, *pp. 303–312*

1, *p. 304:* **a.** $x \cdot 40 = 20$ **b.** $0.5 \cdot x = 10$ **c.** $x = 0.7 \cdot 80$
2, *p. 304:* 8 **3,** *p. 305:* 12 **4,** *p. 305:* 51 workers

5, *p. 305:* 200   **6,** *p. 306:* 57 students   **7,** *p. 306:* 50
8, *p. 306:* 2,500,000 sq ft   **9,** *p. 307:* 60 passes
10, *p. 307:* $83\frac{1}{3}\%$ **11,** *p. 308:* 120%   **12,** *p. 308:* 43%
13, *p. 309:* 270   **14,** *p. 309:* 1,080   **15,** *p. 310:* $233\frac{1}{3}\%$
16, *p. 310:* 11,190 employees   **17,** *p. 310:* $340,000
18, *p. 311:* 87%

Calculator Practices, *pp. 311–312:* $5.97; $254.96; 48.43;
22%

Exercises 6.2, *p. 313*
1. 6   **3.** 23   **5.** 2.87   **7.** $140   **9.** 0.65 L   **11.** 0.62
13. 0.1   **15.** $40   **17.** 4   **19.** $18.32   **21.** $6,000
23. 24.5 hr   **25.** 54 tables   **27.** 16 women   **29.** 32
31. $120   **33.** 75   **35.** 200 lb   **37.** 2.5   **39.** $200
41. $250   **43.** 1.75   **45.** 4,600   **47.** 2,800
49. 32 employees   **51.** 18,750,000 people
53. 1,500 tons   **55.** 50%   **57.** 75%   **59.** $83\frac{1}{3}\%$
61. 40%   **63.** 25%   **65.** 150%   **67.** 62.5%   **69.** $3\frac{1}{3}\%$
71. 10%   **73.** 15%   **75.** 25%   **77.** 20%   **79.** 60%

Practices: Section 6.3, *pp. 317–323*
1, *p. 317:* 300%   **2,** *p. 318:* 1929   **3,** *p. 319:* $57.25
4, *p. 319:* $412.50   **5,** *p. 320:* $69.60   **6,** *p. 320:* 100%
7, *p. 321:* $1,736   **8,** *p. 321:* $3,600   **9,** *p. 323:*
$2,524.95

Exercises 6.3, *p. 324*
1.

Original Value	New Value	Percent Increase or Decrease
10	12	20% increase
10	8	20% decrease
6	18	200% increase
35	70	100% increase
14	21	50% increase
10	1	90% decrease
$8	$6.50	$18\frac{3}{4}\%$ decrease
$6	$5.25	$12\frac{1}{2}\%$ decrease

3.

Selling Price	Rate of Sales Tax	Sales Tax
$30.00	5%	$1.50
$24.88	3%	$0.75
$51.00	$7\frac{1}{2}\%$	$3.83
$196.23	4.5%	$8.83

5.

Sales	Rate of Commission	Commission
$700	10%	$70.00
$450	2%	$9.00
$870	$4\frac{1}{2}\%$	$39.15
$922	7.5%	$69.15

7.

Original Price	Rate of Discount	Discount	Sale Price
$700.00	25%	$175.00	$525.00
$18.00	10%	$1.80	$16.20
$43.50	20%	$8.70	$34.80
$16.99	5%	$0.85	$16.14

9.

Selling Price	Rate of Markup	Markup	Original Price
$10.00	50%	$5.00	$5.00
$23.00	70%	$16.10	$6.90
$18.40	10%	$1.84	$16.56
$13.55	60%	$8.13	$5.42

11.

Principal	Interest Rate	Time (in years)	Interest	Final Balance
$300	4%	2	$24.00	$324.00
$600	4%	2	$48.00	$648.00
$300	8%	2	$48.00	$348.00
$300	4%	4	$48.00	$348.00
$375	10%	3	$112.50	$487.50
$70,000	6%	30	$126,000.00	$196,000.00

13.

Principal	Interest Rate	Time (in years)	Final Balance
$500	4%	2	$540.80
$6,200	3%	5	$7,187.50
$300	5%	8	$443.24
$20,000	4%	2	$21,632.00
$145	3.8%	3	$162.17
$810	2.9%	10	$1,078.05

15. 40%   **17.** 550%   **19.** 300%   **21.** $6.40
23. 7.6%, to the nearest tenth of a percent   **25.** $500
27. $9   **29.** $5 and 20%   **31.** $53\frac{1}{3}$%   **33.** $259.35
35. $150   **37.** $250   **39.** $3,370.80   **41.** 5,856

Review Exercises: Chapter 6, *p. 331*
1.

Fraction	Decimal	Percent
$\frac{1}{4}$	0.25	25%
$\frac{7}{10}$	0.7	70%
$\frac{3}{400}$	0.0075	$\frac{3}{4}$%
$\frac{5}{8}$	0.625	62.5%
$1\frac{1}{5}$	1.2	120%
$1\frac{1}{100}$	1.01	101%
$2\frac{3}{5}$	2.6	260%
$3\frac{3}{10}$	3.3	330%
$\frac{3}{25}$	0.12	12%
$\frac{7}{8}$	0.875	$87\frac{1}{2}$%
$\frac{1}{6}$	$0.16\frac{2}{3}$	$16\frac{2}{3}$%

2.

Fraction	Decimal	Percent
$\frac{3}{8}$	0.375	37.5%
$\frac{2}{5}$	0.4	40%
$\frac{1}{1,000}$	0.001	$\frac{1}{10}$%
$1\frac{1}{2}$	1.5	150%
$\frac{7}{8}$	0.875	87.5%
$\frac{5}{6}$	$.83\frac{1}{3}$	$83\frac{1}{3}$%
$2\frac{3}{4}$	2.75	275%
$1\frac{1}{5}$	1.2	120%
$\frac{3}{4}$	0.75	75%
$\frac{1}{10}$	0.1	10%
$\frac{1}{3}$	$0.33\frac{1}{3}$	$33\frac{1}{3}$%

3. 15%   **4.** 3.125%   **5.** 12   **6.** 120%   **7.** 50%   **8.** 20
9. 43.75%   **10.** 5.5   **11.** $6   **12.** 20%   **13.** 0.3   **14.** 16
15. $70   **16.** 1,000   **17.** 2,000%   **18.** 25%   **19.** $200
20. $44\frac{4}{9}$%   **21.** $12   **22.** $1,600   **23.** 17%   **24.** 42.42
25.

Original Value	New Value	Percent Decrease
24	16	$33\frac{1}{3}$%

26.

Selling Price	Rate of Sales Tax	Sales Tax
$50	6%	$3.00

27.

Sales	Rate of Commission	Commission
$600	4%	$24

28.

Original Price	Rate of Discount	Discount	Sale Price
$200	15%	$30	$170

29.

Selling Price	Rate of Markup	Markup	Original Price
$51	50%	$25.50	$25.50

30.

Principal	Interest Rate	Time (in years)	Simple Interest	Final Balance
$200	4%	2	$16	$216

31. $\frac{17}{20}$ **32.** 48% **33.** The agent that charges 11%.
34. 60% **35.** 20% **36.** Network XYZ **37.** $\frac{3}{4}$
38. 0.0625 **39.** 40% **40.** 10% **41.** 63% **42.** 7%
43. 18% **44.** Possible estimate: 80 in. **45.** 63 first serves
46. Yes **47.** $100,000 **48.** 20% **49.** No
50. 57,770,000 cats **51.** $165 **52.** 320 tons
53. 70,000 mi **54.** 9,460 hr **55.** $1,000 **56.** $30,000
57. $7,939.58 **58.** 122%

59.

Quarter	Income	Percent of Total Income
1	$375,129	27%
2	289,402	21%
3	318,225	23%
4	402,077	29%
Total	$1,384,833	100%

60. Individual income taxes: $526,050,000,000; social security taxes: $444,220,000,000; corporate income taxes: $105,210,000,000; excise taxes: $46,760,000,000; other: $46,760,000,000

Posttest: Chapter 6, *p. 335*
1. $\frac{1}{25}$ **2.** $\frac{5}{8}$ **3.** 1.5 **4.** 0.08 **5.** 0.9% **6.** 300%
7. 83% **8.** 220% **9.** 7.5 mi **10.** 48 **11.** Possible estimate: $6 **12.** 200 **13.** 60% **14.** 250% **15.** $30
16. 400% **17.** 90% **18.** 4 pt **19.** $15 **20.** $3\frac{1}{3}$%

Cumulative Review: Chapter 6, *p. 336*
1. 109 **2.** 0.83 **3.** 0.7 **4.** $5\frac{7}{10}$ **5.** 7.5
6. $1,000 **7.** $\frac{1}{8}$ **8.** 4 lb **9.**
10. 160 thousand

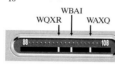

Chapter 7

Pretest: Chapter 7, *p. 338*
1. +7 **2.** −5 **3.** −17 **4.** 0 **5.** −7 **6.** 0
7. −75 **8.** −$\frac{1}{2}$ **9.** 64 **10.** $\frac{1}{4}$ **11.** 2 **12.** −$\frac{1}{8}$
13. 8 **14.** 8 **15.** 36 **16.** −19 **17.** −7 °F
18. −1 °F **19.** −$8,000 (a loss) **20.** −3

Practices: Section 7.1, *pp. 339–343*
1, *p. 340:*

2, *p. 341:* a. −9 **b.** $4\frac{9}{10}$ **c.** 2.9 **d.** −31 **3, *p. 341:* a.** 9
b. $1\frac{3}{4}$ **c.** 4.1 **d.** 5 **4, *p. 342:* a.** Sign: −; absolute value: 4 **b.** Sign: +; absolute value: $6\frac{1}{2}$ **5, *p. 342:* $\frac{1}{2}$

6, *p. 342:* −5 **7, *p. 343:*** 2 **8, *p. 343:*** +2
9, *p. 343:* Mums

Exercises 7.1, *p. 344*
1. −8 **3.** −10.2 **5.** 5 **7.** −$2\frac{1}{3}$ **9.** 4.1 **11.** 1.2
13, 15.   **17.** −$150
19. +14.5 °C **21.** Sign: +; absolute value: 9
23. Sign: −; absolute value: 4.3 **25.** Sign: −; absolute value: 7 **27.** Sign: +; absolute value: $\frac{1}{5}$ **29.** 6 **31.** $\frac{4}{5}$
33. 2 **35.** 0.6 **37.** Two; −5 and 5 **39.** No; absolute value is always positive, or 0. **41.** −4 **43.** 12 **45.** 2
47. −$2\frac{1}{3}$ **49.** −2 **51.** 9 **53.** −2 **55.** $7\frac{1}{4}$ **57.** T
59. T **61.** T **63.** T **65.** F **67.** T **69.** −3, 0, 3
71. −9, −4.5, 9 **73.** Better off today **75.** Gain 1 lb
77. Team A **79.** −1 °F **81.** Liquid chlorine

Practices: Section 7.2, *pp. 349–353*
1, *p. 350:* −25 **2, *p. 350:*** −$4\frac{1}{2}$ **3, *p. 350:*** 7 **4, *p. 351:*** 0 **5, *p. 351:*** $2\frac{3}{5}$ **6, *p. 351:*** 0 **7, *p. 352:*** 44 B.C.

Calculator Practices, *p. 353:* −12; 6; −15.592

Exercises 7.2, *p. 354*
1. 1 **3.** 3 **5.** 7 **7.** 0 **9.** 0 **11.** −5 **13.** 200
15. 6 **17.** −30 **19.** −5 **21.** −7 **23.** 0 **25.** 4.9
27. 0.1 **29.** 59.5 **31.** −5.9 **33.** −14.5 **35.** −6
37. −$\frac{3}{5}$ **39.** $1\frac{3}{5}$ **41.** −$1\frac{1}{10}$ **43.** −102 **45.** 0
47. −2 **49.** −6.9 **51.** 7.58 **53.** −4.914 **55.** 4.409
57. +7° **59.** $177 **61.** −1 **63.** −6 (6 years ago)
65. −2 (loss of 2 yards) **67.** Yes; you will have $1,228.60

Practices: Section 7.3, *pp. 359–361*
1, *p. 360:* −2 **2, *p. 360:*** 18 **3, *p. 360:*** −21.1
4, *p. 360:* −10 **5, *p. 361:*** 214 °F

Exercises 7.3, *p. 362*
1. 15 **3.** −41 **5.** −14 **7.** 44 **9.** −25 **11.** 21
13. 6 **15.** −38 **17.** −26 **19.** 26 **21.** −1
23. 1,000 **25.** −1.52 **27.** 9.7 **29.** 0 **31.** 10.5
33. −0.5 **35.** −5 **37.** $7\frac{3}{4}$ **39.** −$7\frac{1}{4}$ **41.** $7\frac{1}{4}$
43. 7 **45.** −5 **47.** −14 **49.** −3.842 **51.** −16.495
53. 1,000 mi **55.** 2,789 yr **57.** $359,819 **59.** −$3/8 (dropped)

Practices: Section 7.4, *pp. 365–367*
1, *p. 365:* 32 **2, *p. 366:*** −10 **3, *p. 366:*** 1 **4, *p. 366:*** −1 **5, *p. 366:*** −48 **6, *p. 367:*** 0.84 **7, *p. 367:*** 8
8, *p. 367:* No; the submarine will plunge 200 m in 4 min.

Exercises 7.4, *p. 368*
1. −10 **3.** −7,200 **5.** 25 **7.** 306 **9.** −16
11. −8,163 **13.** −40 **15.** −176 **17.** 800 **19.** 12
21. −10 **23.** −5 **25.** −10 **27.** 5.52 **29.** −8
31. −$\frac{5}{27}$ **33.** −$\frac{5}{6}$ **35.** −25 **37.** 10,000 **39.** 0.25
41. $\frac{9}{16}$ **43.** −1 **45.** −216 **47.** −60 **49.** 0 **51.** 5
53. −$\frac{32}{135}$ **55.** 0.094864 **57.** 0.14652 **59.** 5
61. −26 **63.** 0 **65.** −28 **67.** 1.25 **69.** −12
71. 9 **73. a.** 5 **b.** 2 **c.** −1 **d.** −4 **e.** −7 **75.** It fell $1\frac{1}{2}$ (−$1\frac{1}{2}$). **77.** Yes; the actual temperature was 31 °F.

79. -64 ft; 64 ft below the point of release   **81.** Your team scored 4 more points than its opponents ($+4$ points).

Practices: Section 7.5, pp. 373–375
1, p. 373: 12   **2, p. 374:** $-\frac{3}{5}$ **3, p. 374:** -0.28
4, p. 374: $-\frac{1}{6}$ **5, p. 375:** Each owes $5,493.33
($-\$5,493.33$)   **6, p. 375:** 0

Exercises 7.5, p. 376
1. 5   **3.** 0   **5.** -5 **7.** -2 **9.** 25   **11.** -25 **13.** 7
15. -2 **17.** 17   **19.** -5 **21.** $-\frac{5}{6}$ **23.** -21
25. -16 **27.** -0.3 **29.** -20 **31.** -16 **33.** 6.172
35. -3.5 **37.** $-\frac{1}{5}$ **39.** 1   **41.** $-\frac{2}{5}$ **43.** $5\frac{1}{2}$ **45.** $4\frac{1}{4}$
47. $\frac{3}{4}$ **49.** -2 **51.** -4 **53.** 5   **55.** 1   **57.** -16
59. -3 **61.** -0.06 **63.** -4 **65.** 2   **67.** 11
69. A decrease of 6,098.9/yr ($-6,098.9$)   **71.** $5,000/mo in expenses ($-\$5,000$)   **73.** A loss of $2\frac{1}{2}$ points per day ($-2\frac{1}{2}$ points per day)   **75.** Yes.

Review Exercises: Chapter 7, p. 381
1. -5 **2.** 4   **3.** $5\frac{1}{2}$ **4.** -10.1 **5.** 8   **6.** 2.5   **7.** $1\frac{1}{5}$
8. 12   **9.** $-2\frac{1}{4}$ **10.** 9   **11.** -8 **12.** 2   **13.** -8, -3.5, 8   **14.** -9.7, -6, 9   **15.** -2.9, $-2\frac{1}{2}$, 0   **16.** -4, $-1\frac{1}{4}$, 0   **17.** -20 **18.** -2 **19.** $6\frac{1}{2}$ **20.** -1
21. -4.1 **22.** -2 **23.** $-6\frac{1}{2}$ **24.** $-\frac{1}{4}$ **25.** 0
26. 28   **27.** -10 **28.** -11 **29.** 3   **30.** $-4\frac{1}{8}$
31. 100   **32.** -45 **33.** $-\frac{20}{33}$ **34.** -7.35 **35.** $\frac{1}{16}$
36. 9.61   **37.** 1   **38.** -5 **39.** -0.25 **40.** -2.5
41. $\frac{1}{32}$ **42.** -50 **43.** 15   **44.** 114   **45.** -100 **46.** 1
47. 2.73   **48.** -2.62 **49.** 95   **50.** -11 **51.** Yes
52. Yes   **53.** A loss of $250 ($-\$250$)   **54.** A loss of 7 lb (-7 lb)   **55.** 85 °F warmer   **56.** $200 per installment
57. 6.5 min   **58.** 2,010 years apart   **59.** An average profit of $10,000   **60.** 90 feet below the surface (-90 ft)   **61.** 5 below par   **62.** 100 ft above the point of release ($+100$)

Posttest: Chapter 7, p. 384
1. -10 **2.** $-\frac{1}{2}$ **3.** 0   **4.** -0.5 **5.** -49 **6.** 0
7. -207 **8.** -0.1 **9.** 144   **10.** $\frac{1}{16}$ **11.** 2   **12.** -8
13. 4   **14.** 21   **15.** 40   **16.** -18 **17.** -2 °C
18. You lost $5\frac{9}{10}$ lb/mo ($-5\frac{9}{10}$ lb/mo)   **19.** $+\$315$
20. 66.38 °C

Cumulative Review: Chapter 7, p. 385
1. 3,000   **2.** Possible answer: 45   **3.** 10   **4.** 14   **5.** $\frac{1}{6}$
6. 20%   **7.** $n = 65$ **8.** City University spent $25,000 more.   **9.** 2.75 hr   **10.** 25%

Chapter 8

Pretest: Chapter 8, p. 388
1. 9   **2.** $23,000   **3.** 1.6   **4.** 31   **5.** 3.1   **6.** 5.5 yr longer   **7.** Possible answer: 173 ft   **8.** There are more females (100) than males (approximately 97).   **9.** Stock B
10. Earnings account for a larger percent of their income (30%) than pensions (16%).

Practices: Section 8.1, pp. 391–395
1, p. 391: Yes, the cost was $2.87 above average.
2, p. 392: No, the average was 84.   **3, p. 393: a.** 7 **b.** 4
4, p. 394: a. $63.12 **b.** Above the median

5, p. 394: a. 2 **b.** 4 and 9   **6, p. 395:** 16   **7, p. 395:** 7
8, p. 395: The greater spread was at the previous party; the range at the later party was 6.

Exercises 8.1, p. 396
1.

	Numbers	Mean	Median	Mode(s)	Range
a.	8, 2, 9, 4, 8	6.2	8	8	7
b.	3, 0, 0, 3, 10	3.2	3	0 and 3	10
c.	6.5, 9, 8.5, 6.5, 8.1	7.7	8.1	6.5	2.5
d.	$3\frac{1}{2}$, $3\frac{3}{4}$, 4, $3\frac{1}{2}$, $3\frac{1}{4}$	$3\frac{3}{5}$	$3\frac{1}{2}$	$3\frac{1}{2}$	$\frac{3}{4}$
e.	4, -2, -1, 0 -1	0	-1	-1	6

3. $12,266.67   **5.** No; your GPA was 3.25   **7.** The mean amount is $100,000. We can't compute the median amount because we don't know the actual amounts given to each heir.   **9.** The number of this state's representatives is above average; the average is 8.7.   **11. a.** 7 **b.** 12
13. a. 28,000 mi **b.** 8,000 mi **c.** 8,000 mi **d.** 88,000 mi
15. (a) Division; subtraction **(b)** 200.32 **(c)** Possible answer: 250

Practices: Section 8.2, pp. 401–408
1, p. 402: a. No, the weight room is not open at 3 P.M. on Fridays. **b.** Both the pool and weight room are open for less time on Fridays, but the gym is open for the same length of time on all days. **c.** The gym is open and the pool closed on Mondays from $2-4$ P.M. and $7:45-8$ P.M. **d.** Yes, the gym stays open without interruption.   **2, p. 403: a.** The S&H charges are $4.95. **b.** You must pay $5.95. **c.** The S&H charges total $5.45.   **3, p. 404: a.** The decade ending in 1990 **b.** 4 to 1 **c.** Possible answer: The number of fatalities as a result of hurricanes in the United States declined by 7,975 from 1910 to 1990.   **4, p. 405: a.** 200,000 takeoffs and landings **b.** Dallas/Ft. Worth had about 700,000 and Haneda had 200,000. Therefore Dallas/Ft. Worth had about 500,000 more takeoffs and landings than Haneda. **c.** Los Angeles had 600,000 and Chicago had 800,000. Therefore Los Angeles had about 75% as many takeoffs and landings as Chicago.   **5, p. 405: a.** Approximately 2,000 more trucks and buses were sold in March than in February. **b.** About 5,000 more passenger cars than trucks and buses were sold in January.   **6, p. 406: a.** About $32.50 **b.** The value declined by about $1.50. **c.** The value increased by about $1.50.   **7, p. 407: a.** About 30% supported

Cardoso. **b.** September 1 **c.** During the period from June 1 to July 1 support shifted from da Silva to Cardoso. **8, *p. 408*: a.** $364,000,000,000 **b.** $86,000,000,000 **c.** $\frac{33}{23}$

Exercises 8.2, *p. 409*
1. a. More than; the actual area is 2,940,000 sq mi
b. Possible answer: 2.8 times **3. a.** 24.7 years old
b. 4 yr **c.** 0.4 yr **d.** Yes, they were. **5. a.** $60
b. $100 **c.** You will pay a lower commission if you sell 400 shares in a single deal; 400 shares in a single deal will cost you $70 or $90, whereas 2 deals of 200 shares will cost you $100 or $120. **7. a.** 8% **b.** $\frac{11}{100}$ **c.** 7
9. a. 24,000 households **b.** Possible answer: 30,000
c. 1980 **11. a.** Yes; there was between 300,000 and 350,000 gallons of water in the reservoir throughout January. **b.** September **c.** May **13. a.** Approximately 13 months **b.** Possible answer: 20 more words
15. a. 3¢ **b.** Yes; education and higher education accounted for 23¢ of every dollar (23%).

Review Exercises: Chapter 8, *p. 418*
1.

	List of numbers	Mean	Median	Mode	Range
a.	6, 7, 4, 10, 4, 5, 6, 8, 7, 4, 5	6	6	4	6
b.	1, 3, 4, 4, 2, 3, 1, 4, 5, 1	2.8	3	1 and 4	4

2. a. The husbands (83 yr) lived longer than the wives (70 yr). **b.** 13 yr **3. a.** Half of the people were younger than 32.9 and half were older. **b.** Answers will vary.
4. This machine is reliable because the range is 1.7 fl oz.
5. The average was 1,000; therefore the show was making a profit. **6. a.** 128 games **b.** Detroit **c.** Milwaukee
d. The total of losses at home and away (78) is not equal to the total losses for the season (80). **7. a.** 49,340,000 tons **b.** Valdez, Alaska (39,787 tons) **c.** 17,170,000 tons
d. All except Houston, Texas **8. a.** 21 theaters **b.** Every 10 yr the number of theaters has increased. Since 1970, the number has increased by 3 every 10 yr. **c.** 27
9. a. Possible answer: 65 launches **b.** February **c.** The number of launches peaked in February and then declined.
10. a. Run number 3 **b.** Approximately 3 min **c.** With practice, the rat ran through the maze more quickly.
11. a. 10% **b.** Yes; 37% of the immigrants came from Asia, which is more than $\frac{1}{3}$ (33.3%). **c.** Approximately 3.5 million immigrants **12. a.** 8 million **b.** Possible answer: $\frac{10}{7}$ **c.** Since 1940, the population of New York City had stayed almost constant (8 million) until the period between 1970 and 1980 when it fell by approximately 1 million and

then rebounded. Over the same time period (1940−1990) the population for the rest of the state had increased from 6 million to 10 million. **13. a.** Paper **b.** $\frac{7}{100}$ **c.** 17%

Posttest: Chapter 8, *p. 422*
1. 5 **2.** Yes it did, because the mean was 5.9 ppm.
3. $16,000 **4.** 5,500,000 **5.** 3.375 **6.** $7.04
7. Approximately 3,000 students **8.** Approximately 1,750,000; there were roughly 2,000,000 20-year olds and 250,000 30-year olds. **9.** $\frac{11}{20}$ **10.** 1961−1975; U.S. scientists received approximately 40 and U.K. scientists received approximately 20 Nobel Prizes.

Cumulative Review: Chapter 8, *p. 425*
1. $3,400 **2.** $\frac{4}{5}$ **3.** 40 **4.** 3,010 **5.** 2 **6.** $x = 17\frac{1}{4}$
7. $n = 21$ **8.** 5,000 yr **9.** 250% **10.** 64 more scientists

Chapter 9

Pretest: Chapter 9, *p. 428*
1. $y = -6$ **2.** $x = -2$ **3.** $x = -4$ **4.** $a = -5$
5. $x = 3$ **6.** $y = 1$ **7.** $c = -2$ **8.** $a = -10$
9. $w = -15$ **10.** $x = 2$ **11.** $x = -1$ **12.** $x = -4$
13. $l = \frac{1}{7}h$ **14.** $S = 91$ **15.** $45 = 4x + 5$, where x is your hourly rate; $x = \$10/hr$ **16.** $475 = 250 + 75x$, where $x =$ additional hours worked $x = 3$, so he worked a total of 6 hr **17.** $14x + 5 = 19$, where $x =$ cost of 1 book; $x = \$1$ **18.** $3(n + 7) = 30$; $n = 3$ **19. a.** $B = \frac{1}{12}twl$
b. $B = 13\frac{1}{3}$ board ft **20. a.** $s = r - d$ **b.** $s = \$301$

Practices: Section 9.1, *pp. 429–434*
1, *p. 429*: $x = -3$ **2, *p. 430*:** $m = -12$ **3, *p. 430*:** $y = -2$ **4, *p. 430*:** $y = 6$ **5, *p. 431*:** $2,500,000
6, *p. 431*: $x = -7$ **7, *p. 432*:** $k = 10$ **8, *p. 432*:** $d = 4$
9, *p. 433*: $x = -128$ **10, *p. 433*:** $2 + 0.2x = 10$, where x equals the number of oranges in the box; $x = 40$ oranges

Exercises 9.1, *p. 435*
1. $a = -14$ **3.** $b = -11$ **5.** $z = -7$ **7.** $x = -2$
9. $c = -19$ **11.** $z = -7.7$ **13.** $x = -3.6$ **15.** $y = -\frac{2}{3}$
17. $t = -6\frac{1}{4}$ **19.** $z = -12$ **21.** $x = -6$ **23.** $n = 4$
25. $m = -15$ **27.** $w = -240$ **29.** $x = 12$ **31.** $t = -30$
33. $y = -15$ **35.** $y = -\frac{2}{5}$ **37.** $n = -1.4$ **39.** $x = -14.8$
41. $n = 14$ **43.** $x = 2$ **45.** $k = -3$ **47.** $x = 0$
49. $h = -7$ **51.** $p = -5\frac{1}{4}$ **53.** $b = 2$ **55.** $a = -33$
57. $y = -36$ **59.** $x = 48$ **61.** $c = -21$ **63.** $x = 18$
65. $x = -19$ **67.** $r = -0.356603774$, or -0.4 **69.** $x = -0.0264$, or -0.0 **71.** $-3x = -21$, where x represents the number of months; $x = 7$ mo **73.** $3x = -6,000$, where x represents how much each one owes; $x = -2,000$; each partner owes $2,000. **75.** $15x - 25 = 155$, where x is the original price of each ticket; $x = \$12$ **77.** $8x + 3 = 43$, where x represents the number of people at the large tables; $x = 5$ people
79. **(a)** $251.20 = 8.95 + 4.75x$ **(b)** Exact answer: 51 hr
(c) Possible estimate: 48 hr

Practices: Section 9.2, *pp. 439–442*
1, *p. 439*: a. $12x$ **b.** $6y$ **c.** $-z + 6$ **2, *p. 440*:** $y = -8$
3, *p. 440*: $x = \frac{1}{2}$ **4, *p. 441*:** $x = -8$ **5, *p. 441*: a.** $x + 3x = x + 9$ **b.** $x = 3$ **6, *p. 442*:** 2.2 mi

Exercises 9.2, *p. 443*

1. $7x$ **3.** $3a$ **5.** $-3y$ **7.** $-n$ **9.** $3c + 12$ **11.** $8 - 6x$ **13.** $2y + 5$ **15.** $m = 4$ **17.** $y = 9$ **19.** $a = -7$ **21.** $b = -1\frac{3}{4}$ **23.** $n = 13$ **25.** $x = 3$ **27.** $n = 9$ **29.** $y = -5$ **31.** $r = -0.5$ **33.** $m = 12.4$ **35.** $x = 4$ **37.** $p = -2$ **39.** $x = -1$ **41.** $p = 1\frac{1}{2}$ **43.** $n = 12$ **45.** $t = 2$ **47.** $y = 1$ **49.** $x = 1\frac{1}{4}$ **51.** $x = 43{,}939.4$ **53.** $y = -1.1$ **55.** $n = 9$ **57.** $x = -2$ **59.** $n = 35$ **61.** $y = 15$ **63.** $y = 1\frac{1}{2}$ **65.** $n = 21$ **67.** $r = 2$ **69.** $x = -3$ **71.** $x = 2{,}500$ **73.** $y = 19.1$ **75.** $4x + 3x = 63; x = \$9/\text{hr}$ **77.** $140 + 2w = 200; w = 30$ ft **79.** $\frac{1}{5}x + \frac{1}{2}x = 1\frac{3}{4}; x = 2\frac{1}{2}$ hr **81.** $15n + 3{,}000 = 20n$; $n = 600$ books **83.** $180(n - 2) = 540; n = 5$ sides

Practices: Section 9.3, *pp. 447–448*

1, *p. 447:* $t = g - \frac{1}{200}a$ **2,** *p. 447:* $I = \$360$ **3,** *p. 448:* $p = \$302$ **4,** *p. 448:* $s = 64$ ft

Exercises 9.3, *p. 449*

1. $d = \frac{t}{3}$ **3.** $A = \frac{a + b}{2}$ **5.** $A = 6e^2$ **7.** $23°$ **9.** $\$225$ **11.** $0°$ **13.** 160 ft **15.** $v = \$13{,}500$ **17.** $C = 60$ mg **19.** **(a)** $T = 0.48A + 0.42A$, where T represents the total number of calves born. **(b)** Exact answer: 295 calves **(c)** Possible estimate: 270 calves

Review Exercises: Chapter 9, *p. 454*

1. $x = -6$ **2.** $y = -9$ **3.** $a = 0$ **4.** $c = -50$ **5.** $d = -9$ **6.** $w = -11$ **7.** $x = -9$ **8.** $y = -16$ **9.** $c = -18$ **10.** $p = 0$ **11.** $a = -2$ **12.** $b = -2$ **13.** $y = -6$ **14.** $d = -7$ **15.** $x = 0$ **16.** $w = 10$ **17.** $c = -8$ **18.** $p = 8$ **19.** $x = 4$ **20.** $a = 2$ **21.** $y = 5$ **22.** $w = 2$ **23.** $y = 3\frac{1}{2}$ **24.** $y = 2\frac{2}{3}$ **25.** $x = 5$ **26.** $a = 1$ **27.** $y = -5$ **28.** $w = -2$ **29.** $c = -12$ **30.** $a = -18$ **31.** $a = 24$ **32.** $w = 16$ **33.** $x = -25$ **34.** $b = -10$ **35.** $c = 0$ **36.** $d = 0$ **37.** $y = 9$ **38.** $b = 6$ **39.** $c = -2$ **40.** $x = 2\frac{1}{5}$ **41.** $y = 8$ **42.** $t = 3$ **43.** $x = 8$ **44.** $m = 3$ **45.** $n = \frac{1}{3}$ **46.** $a = -9$ **47.** $r = -8$ **48.** $y = -1$ **49.** $x = 9$ **50.** $y = -1$ **51.** $s = 2$ **52.** $x = -2$ **53.** $m = 2$ **54.** $x = 6$ **55.** $s = 11$ **56.** $z = \frac{7}{8}$ **57.** $w = 3$ **58.** $z = 0$ **59.** $x = 6$ **60.** $y = 10$ **61.** $y = -8$ **62.** $m = 2\frac{2}{3}$ **63.** $b = -1$ **64.** $d = -4$ **65.** $R = F + 460$ **66.** $C = \frac{W}{150} \cdot A$ **67.** $\frac{a + b + c + d + e}{5}$ **68.** $m = \frac{1}{2}(s + l)$ **69.** 50.4 **70.** 144 **71.** 880 **72.** -20 **73.** $p = \$5.38/\text{unit}$ **74.** 17 weeks **75.** $w = 9$ ft **76.** $s = \$271.05$ **77.** $D = 0.9$ g/ml **78.** $t = 37.2$ cm rounded to the nearest cm

Posttest: Chapter 9, *p. 457*

1. $x = -5$ **2.** $y = 0$ **3.** $a = -3$ **4.** $b = -10$ **5.** $x = \frac{1}{2}$ **6.** $y = -2\frac{1}{2}$ **7.** $w = -1\frac{2}{7}$ **8.** $y = -9$ **9.** $a = 0$ **10.** $x = -6$ **11.** $c = -2$ **12.** $x = 2$ **13.** $D = 4.9t^2$ **14.** $y = -2\frac{1}{2}$ **15.** $0.35 + 0.16x = 2.75$, where x represents the number of additional minutes; $x = 15$ min, so the total length of the call was 16 min. **16.** $100 + \frac{8}{9}P = 12{,}906$, where P is the suggested retail price; $P = \$14{,}406.75$. **17.** $(x + 300) + (x + 900) = 1{,}800; x = \$300$ **18.** $4(n + 3) = 5n; n = 12$ **19. a.** $d = 2.2r$ **b.** $d = 121$ ft **20. a.** $f = \frac{22}{15}m$ **b.** $f = 66$ fps

Cumulative Review: Chapter 9, *p. 458*

1. $3{,}800$ **2.** Three million, four hundred **3.** 0.001 **4.** $83\frac{1}{3}\%$ **5.** -2 **6.** $\frac{1}{17}$ **7.** $x = 6$ **8.** This change represented an increase because 0.1 is larger than 0.075 (by 0.025). **9.** Monday **10.** Possible answer: $i = 20$

Chapter 10

Pretest: Chapter 10, *p. 460*

1. 2 **2.** 10,000 **3.** 12 ft 6 in. **4.** b **5.** d **6.** 3,500 **7.** 2.1 **8.** 700 cm **9.** 500 ml **10.** 0.075 g

Practices: Section 10.1, *pp. 461–466*

1, *p. 463:* $2\frac{1}{2}$ **2,** *p. 463:* 7.5 gal **3,** *p. 464:* 86,400 sec **4,** *p. 464:* No; 3 pt = 1.5 qt **5,** *p. 465:* 5 lb 8 oz **6,** *p. 465:* 2 yr 9 mo **7,** *p. 465:* 3 qt **8,** *p. 466:* 6 sec

Exercises 10.1, *p. 467*

1. 4 **3.** 3 **5.** 3 **7.** 2 **9.** 3 **11.** 420 **13.** 512 **15.** $\frac{1}{2}$ **17.** 3,520 **19.** 35 **21.** $\frac{1}{48}$ **23.** 2 **25.** 3.5 **27.** 3,000 **29.** 310 **31.** 617,760 **33.** 1 lb 14 oz **35.** 29 lb 15 oz **37.** 9 yr 6 mo **39.** 2 gal 3 qt **41.** 4 qt **43.** 8 min 15 sec **45.** They both weigh the same. **47.** No; the refrigerator is too wide. **49.** 13 min 43 sec **51.** 1,044 mo **53.** 1 lb 5 oz **55.** 1 yr 10 mo **57. (a)** Division **(b)** 5.5 mi **(c)** Possible estimate: 6 mi

Practices: Section 10.2, *pp. 471–477*

1, *p. 475:* 3.1 **2,** *p. 475:* 25 **3,** *p. 476:* 5 km **4,** *p. 476:* Less than a liter **5,** *p. 477:* 38 L **b.** $x = 3$ **6,** *p. 477:* 608.1 mi

Exercises 10.2, *p. 478*

1. c **3.** b **5.** b **7.** b **9.** a **11.** 1 **13.** 0.75 **15.** 80 **17.** 3,500 **19.** 0.005 **21.** 4 **23.** 7 **25.** 4.13 **27.** 2,000 **29.** 3,250 m **31.** 98,025.6 g **33.** 840 **35.** 4 **37.** 1.2 **39.** 6 g **41.** Yes **43.** 3 in. **45.** 0.3 kg **47.** 750 cm **49.** 3.2 L **51.** 38,000,000 L **53.** 10.1 kg

Review Exercises: Chapter 10, *p. 484*

1. 15 **2.** $1\frac{2}{3}$ **3.** $\frac{1}{2}$ **4.** $3\frac{1}{3}$ **5.** 7,000 **6.** 136 **7.** 14 **8.** 32 **9.** 9 **10.** $2\frac{1}{2}$ **11.** 112 **12.** 2 **13.** 20 **14.** 15.2 **15.** 2 **16.** 125 **17.** 8 hr 10 min **18.** 18 ft 9 in. **19.** 6 min 9 sec **20.** 7 lb 2 oz **21.** 1 gal 3 qt **22.** 1 pt 2 fl oz **23.** c **24.** a **25.** b **26.** c **27.** b **28.** a **29.** a **30.** c **31.** 1.2 hr **32.** 12,800 pt **33.** *Frankenstein* **34.** 1 L **35.** 2 g **36.** 0.1 g **37.** 1.6 g/L **38.** 750,000 g **39.** 6 hr 30 min **40.** 10 in. **41.** 24 in. **42.** 5.5 L **43.** 4:21 P.M. **44.** 12 portions

Posttest: Chapter 10, *p. 487*

1. 2 **2.** 21 **3.** 3,600 sec **4.** b **5.** a **6.** 4 **7.** 0.5 **8.** 2 km **9.** 6,000,000 m **10.** 9 in.

Cumulative Review: Chapter 10, *p. 488*

1. 729 **2.** $5\frac{1}{2}$ **3.** 14 **4.** 36 **5.** 14 **6.** 8 **7.** 84 **8.** 22 qt **9.** $\frac{4}{5}$ **10.** According to an agricultural study, 5,000 g of grain are required to produce 1,000 g of steak.

Chapter 11

Pretest: Chapter 11, *p. 490*

1. a.  **b.**   **2. a.** 6 **b.** 11   **3.** 80°

4. 54°   **5.** 14 in.   **6.** 20 ft   **7.** Approximately 12.56 in.
8. 7.8 m   **9.** 30 sq in.   **10.** 50.24 sq ft   **11.** 125 cm³
12. Approximately 254.34 m³   **13.** $a = 75°$
14. $x = \sqrt{34}$ m   **15.** $a = b = 90°$; $x = 7$ ft; $y = 5$ ft
16. $a = 104°$ **17.** $y = 15$ cm   **18.** 7,200 ft³ of water
19. 3 mi²   **20.** 225 cm²

Practices: Section 11.1, *pp. 491–500*

1, *p. 494*: $\angle ABC$ **2, *p. 494*:** 53°

3, *p. 495*: 165°   **4, *p. 495*:** $x = 82°$
5, *p. 495*: $a = b = 153°$
6, *p. 499*: 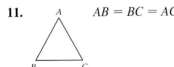 $\overline{AB} \parallel \overline{CD}$;
$\overline{AC} \parallel \overline{BD}$

7, *p. 499*: 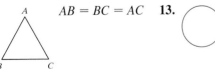 $\angle R = 60°$

8, *p. 500*: $\angle U = 60°$ **9, *p. 500*:** 1.34 ft

Exercises 11.1, *p. 501*

1. A **3.** $\overline{BC}$ **5.**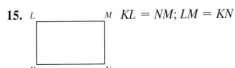

7.   **9.**

11. $AB = BC = AC$ **13.** ◯

15. $KL = NM$; $LM = KN$

17. $AB \neq BC \neq AC$ **19.**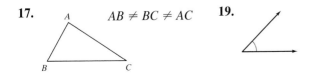

---

21. $x = 33°$ **23.** $\angle DEG = 53°$ **25.** 180°   **27.** 90°
29. 180°   **31.** 35°   **33.** $a = 53°$ **35.** $x = 90°$
37. 55°   **39.** 75°   **41.** 70°   **43.** 85°   **45.** 13 ft
47. $x = 97°$

Practices: Section 11.2, *pp. 505–510*

1, *p. 506*: 26 in.   **2, *p. 506*:** 14 m   **3, *p. 507*:** $70
4, *p. 508*: Approximately 132 in.   **5, *p. 508*:**
Approximately 113.0 ft   **6, *p. 509*:** Approximately
164.5 yd   **7, *p. 510*:** 82 ft

Exercises 11.2, *p. 511*

1. 17 in.   **3.** 60 m   **5.** $10\frac{1}{2}$ yd   **7.** Approximately
62.8 m   **9.** Approximately 21.98 ft   **11.** 42 ft
13. Approximately 25.12 in.   **15.** 40 yd   **17.** 21 yd
19. 18 ft   **21.** Approximately 31.4 ft   **23.** 19 cm
25. Approximately 11.12 m   **27.** 228 ft   **29.** 6 in.
31. 30 posts   **33.** 69.98 ft   **35. (a)** Multiplication
(b) 42,700 km   **(c)** 42,000 km

Practices: Section 11.3, *pp. 517–523*

1, *p. 519*: 13.23 cm²   **2, *p. 519*:** 12.96 cm²
3, *p. 520*: 7.5 in²   **4, *p. 520*:** $12\frac{1}{2}$ ft²   **5, *p. 520*:** $3\frac{1}{2}$ sq ft
6, *p. 521*: 25π yd² ≈ 78.5 yd²   **7, *p. 521*:** 79 mi²
8, *p. 522*: 8.75 sq m   **9, *p. 522*:** 4,421.5 sq ft

Exercises 11.3, *p. 524*

1. 125 m²   **3.** 30 ft²   **5.** 290 yd²   **7.** Approximately
706.5 cm²   **9.** 32 in²   **11.** 15.6 m²   **13.** Approximately
314 in²   **15.** 6.25 ft²   **17.** 44.1 yd²   **19.** 3.64 m²
21. $1\frac{1}{16}$ yd²   **23.** 46 yd²   **25.** Approximately 64.3 ft²
27. 51.75 ft²   **29.** No. You need 108 tiles at a total cost of
$538.92.   **31.** Approximately 0.00008 in²   **33.** 5,000 sq ft
35. (a) Multiplication and subtraction   **(b)** 100 in²
(c) Possible estimate: 100 in²

Practices: Section 11.4, *pp. 529–533*

1, *p. 530*: 72 ft³   **2, *p. 531*:** 3,375 cm³   **3, *p. 531*:**
Approximately 10.99 in³   **4, *p. 531*:** Approximately
$113\frac{1}{7}$ in³   **5, *p. 532*:** Approximately 4.19 m³
6, *p. 532*: The ball's volume is about 523.3 in³, whereas the
box's volume is 1,000 in³. Therefore the ball occupies more
than one-half the box's volume.

Exercises 11.4, *p. 534*

1. 216 in³   **3.** 2,560 m³   **5.** Approximately 62.8 ft³
7. Approximately 2,143.6 in³   **9.** 125.125 ft³
11. Approximately 1.95 m³   **13.** Approximately 92.1 ft³
15. (a) Multiplication and division   **(b)** 2.9 g/cm³
(c) Possible estimate: 3 g/cm³   **17.** 0.01 kg/cm³
19. 4,186,667 mi³   **21.** 46 in³   **23.** 15.19 m³

Practices: Section 11.5, *pp. 539–541*

1, *p. 540*: $\overline{AB}$ corresponds to $\overline{GH}$; $\overline{AC}$ corresponds to
$\overline{GI}$; $\overline{BC}$ corresponds to $\overline{HI}$ **2, *p. 540*:** $y = 24$
3, *p. 541*: $h = 36$ ft

Exercises 11.5, *p. 542*
1. $x = 10\frac{2}{3}$ in. **3.** $x = 24$ m **5.** $x = 15$ ft; $y = 20$ ft
7. $x = 24$ yd; $y = 12$ yd **9.** $x = 3.8125$ m **11.** 576 cm
13. 200 m

Practices: Section 11.6, *pp. 545–549*
1, *p. 545*: a. 7 **b.** 12 **c.** 15 **2, *p. 546*:** Between 6 and 7
3, *p. 547*: a. 7.48 **b.** 3.46 **4, *p. 548*:** 10 in.
5, *p. 548*: 3.5 ft **6, *p. 549*:** 5 m

Exercises 11.6, *p. 550*
1. 3 **3.** 6 **5.** 9 **7.** 12 **9.** 13 **11.** 20 **13.** 1 and 2
15. 3 and 4 **17.** 6 and 7 **19.** 8 and 9 **21.** 2.2
23. 6.1 **25.** 11.8 **27.** 99.0 **29.** $a = 16$ cm
31. $c = 3.2$ m **33.** 7 m **35.** 8 ft **37.** 20 m
39. 11.4 cm **41.** 8.7 ft **43.** 20 ft **45.** $l = 240$ ft
47. $l \approx 37.4$ ft

Review Exercises: Chapter 11, *p. 561*
1. 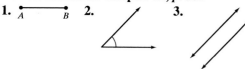 **2.** 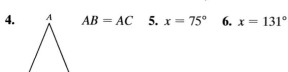 **3.**

4. $AB = AC$ **5.** $x = 75°$ **6.** $x = 131°$

7. $x = 51°$ **8.** $x = 80°$; $y = 100°$ **9.** 5.4 m
10. 62.8 in. **11.** 19 cm **12.** 33 ft **13.** 225 yd^2
14. 15 m^2 **15.** 64 in^2 **16.** Approximately 616 ft^2

17. Approximately 1,318.8 in^3 **18.** 216 ft^3
19. Approximately 1.95 m^3 **20.** Approximately 8.18 cm^3
21. Approximately 122.82 ft **22.** Approximately 84.78 ft^2
23. $x = 8$ ft; $y = 7.875$ ft **24.** $x = 4.5$ m **25.** 3
26. 8 **27.** 11 **28.** 30 **29.** 1 and 2 **30.** 3 and 4
31. 6 and 7 **32.** 9 and 10 **33.** 2.83 **34.** 5.39
35. 13.96 **36.** 35 **37.** 12 ft **38.** 10 in. **39.** 9.4 yd
40. 2.8 ft **41.** 28,800 in^2 **42.** 13 mi^2 **43.** Yes, it can;
the volume of the room is 2,650 ft^3 **44.** Yes, the larger
screen is $56\frac{1}{4}\%$ greater than the smaller screen. **45.** 13 mi
46. 10 ft **47.** $CD = 2\frac{1}{4}$ ft **48.** $x = 300$ m **49.** 160 ft
50. It is not efficient because the perimeter of the work
triangle is $25\frac{1}{4}$ ft **51.** 5.3 oz **52.** 1,347 cm^3 of soil
53. $476\frac{1}{4}$ in^3 **54.** 2.6 in^3

Posttest: Chapter 11, *p. 566*
1. a. **b.**  **2. a.** 7 **b.** 15 **3.** 65°

4. 89° **5.** 14 ft **6.** 4.5 m **7.** Approximately 25.12 cm
8. 15 ft **9.** 54 ft^2 **10.** 110 cm^2 **11.** 189 m^3
12. Approximately 33.5 ft^3 **13.** $a = 120°$; $b = 60°$;
$y = 5$ m; $x = 10$ m **14.** $a = 69°$ **15.** $x = 80°$
16. $y = 13$ yd **17.** $x = 15$; $y = 24$ **18.** 8,100 ft^3
19. 79 m^2 **20.** 27 m

Cumulative Review: Chapter 11, *p. 568*
1. 60 **2.** $\frac{4}{5}$ **3.** 3.1 **4.** 50 **5.** 15 **6.** $y = 8$
7. Approximately 514 ft **8.** 1 mi **9.** $\frac{1}{51}$ **10.** 0 °F

Glossary

absolute value [7.1] The absolute value of a number is its distance from zero on the number line.

acute angle [11.1] An acute angle is an angle that measures less than 90°.

acute triangle [11.1] An acute triangle is a triangle with three acute angles.

addends [1.2] In an addition problem, the numbers being added are called addends.

algebraic expression [4.1] An algebraic expression is an expression that contains one or more variables and may contain any number of constants.

amount (percent) [6.2] The amount is the result of taking the percent of the base.

angle [11.1] An angle consists of two rays that have a common endpoint.

area [11.3] The number of square units that a figure contains is called the area.

associative property of addition [1.2] The associative property of addition states that when adding three numbers, regrouping the addends gives the same sum.

associative property of multiplication [1.3] The associative property of multiplication states that when multiplying three numbers, regrouping the factors gives the same product.

average [1.5] An average is the sum of the numbers on a list, divided by however many numbers are on the list.

bar graph [8.2] A bar graph is a graph in which quantities are represented by thin, parallel rectangles called bars. The length of each bar is proportional to the quantity that it represents.

base (exponent) [1.5] The base is the number that is a repeated factor when written with an exponent.

base (percent) [6.2] The base is the number that we take the percent of. It always follows the word "of" in the statement of a percent problem.

circle [11.1] A circle is a closed plane figure made up of points that are all the same distance from a fixed point called the center.

circle graph [8.2] A circle graph is a graph that resembles a pie, representing the whole amount that has been cut into slices representing the parts of the whole.

circumference [11.2] The distance around a circle is called the circumference.

commission [6.3] Salespeople may work on commission instead of receiving a fixed salary. This means that the amount of money that they earn is a specified percent of the total sales for which they are responsible.

commutative property of addition [1.2] The commutative property of addition states that changing the order in which two numbers are added does not affect the sum.

commutative property of multiplication [1.3] The commutative property of multiplication states that changing the order in which two numbers are multiplied does not affect the product.

complementary angles [11.1] Complementary angles are two angles whose sum is 90°.

composite figure [11.2] A composite figure is the combination of two or more basic figures.

composite number [2.1] A composite number is a whole number that has more than two factors.

constant [4.1] A constant is a known number.

corresponding sides [11.5] The sides opposite the equal angles in similar triangles are called the corresponding sides.

cube [11.4] A cube is a solid in which all six faces are squares.

cylinder [11.4] A cylinder is a solid in which the bases are circles and are perpendicular to the height.

decimal [3.1] A decimal is a number written with three parts: a whole number, the decimal point, and a fraction whose denominator is a power of 10.

decimal place [3.1] The decimal place is the place to the right of the decimal point.

denominator [2.2] The number below the fraction line in a fraction is called the denominator. It stands for the number of parts into which the whole is divided.

diameter [11.1] A line segment that passes through the center of a circle and has both endpoints on the circle is called the diameter of the circle.

difference [1.2] The result of a subtraction problem is called the difference.

digits [1.1] Digits are the numbers 0, 1, 2, 3, 4, 5, 6, 7, 8, and 9.

discount [6.3] When buying or selling merchandise, the term "discount" means a reduction on the merchandise's original price.

distributive property [1.3] The distributive property states that multiplying a factor by the sum of two numbers gives us the same result as multiplying the factor by each of the two numbers and then adding.

dividend [1.4] In a division problem, the number into which another number is being divided is called the dividend.

divisor [1.4] In a division problem, the number that is being used to divide another number is called the divisor.

equation [4.2] An equation is a mathematical statement that two expressions are equal.

equilateral triangle [11.1] An equilateral triangle is a triangle with three sides equal in length.

equivalent fractions [2.2] Equivalent fractions are fractions that represent the same quantity.

evaluate [4.1] To evaluate an algebraic expression, replace each variable with the given number and carry out the computation.

exponent (or power) [1.5] An exponent (or power) is a number that indicates how many times another number is multiplied by itself.

exponential form [1.5] Exponential form is a shorthand way of representing a repeated multiplication of the same factor.

factors [1.3] In a multiplication problem, the numbers being multiplied are called the factors.

fraction [2.2] A fraction is any number that can be written in the form $\frac{a}{b}$, where a and b are whole numbers and b is not zero.

fraction line [2.2] The fraction line separates the numerator from the denominator, and stands for "out of" or "divided by."

graph [8.2] A graph is a picture or diagram of the data in a table.

hypotenuse [11.6] In a right triangle, the side opposite the right angle is called the hypotenuse.

identity property of addition [1.2] The identity property of addition states that the sum of a number and zero is the original number.

identity property of multiplication [1.3] The identity property of multiplication states that the product of any number and 1 is that number.

improper fraction [2.2] An improper fraction is a fraction greater than or equal to 1, that is, a fraction whose numerator is larger than or equal to its denominator.

integers [7.1] The integers are the numbers $\ldots, -4, -3, -2, -1, 0, +1, +2, +3, +4, \ldots$, continuing indefinitely in both directions.

intersecting lines [11.1] Two lines that meet are called intersecting lines.

isosceles triangle [11.1] An isosceles triangle is a triangle with two sides equal in length.

least common denominator (LCD) [2.2] The least common denominator (LCD) for any set of fractions is the least common multiple of the denominators.

least common multiple (LCM) [2.1] The least common multiple (LCM) of two or more numbers is the smallest nonzero number that is a multiple of each number.

legs [11.6] In a right triangle, the legs are the two sides which form the right angle.

like fractions [2.2] Like fractions are fractions which have the same denominator.

like terms [9.2] Like terms are terms that have exactly the same variables and exponents.

line [11.1] A line is a collection of points along a straight path, that extends endlessly in both directions

line graph (broken-line graph) [8.2] A line graph (broken-line graph) is a graph in which quantities are represented as points connected by straight line segments. At any point on a line, its height is read against the vertical axis.

line segment [11.1] A line segment is a portion of a line having two endpoints, and it only has one dimension — its length.

magic square [1.2] A magic square is a square array of numbers in which the sum of every row, column, and diagonal is the same number.

markup [6.3] The markup on an item is the difference between the selling price and the cost price.

mean (average) [8.1] The mean is the sum of a list of numbers, divided by however many numbers are on the list.

median [8.1] Given a list of numbers arranged in numerical order, the median is the number in the middle. If there are two numbers in the middle, the median is the average of the two numbers.

minuend [1.2] In a subtraction problem, the number that is being subtracted from is called the minuend.

mixed number [2.2] A mixed number is a number greater than 1 with a whole number part and a fractional part.

mode [8.1] The mode is the number (or numbers) occurring most frequently in a list of numbers.

multiplication property of 0 [1.3] The multiplication property of 0 states that the product of any number and 0 is 0.

negative number [7.1] A negative number is a number less than 0.

numerator [2.2] The number above the fraction line in a fraction is called the numerator. It tells us how many parts of the whole the fraction contains.

obtuse angle [11.1] An obtuse angle is an angle that measures more than 90° and less than 180°.

obtuse triangle [11.1] An obtuse triangle is a triangle with one obtuse angle.

opposites [7.1] Two numbers that are the same distance from 0 on the number line but on opposite sides of 0 are called opposites.

parallel lines [11.1] Two lines on the same plane that do not intersect are called parallel lines.

parallelogram [11.1] A quadrilateral with both pairs of opposite sides parallel is called a parallelogram. Also, opposite sides are equal in length, and opposite angles have equal measures.

percent [6.1] A percent is a ratio or fraction with denominator 100. A number written with the % sign means "divided by 100."

percent decrease [6.3] In a percent problem, if the quantity is decreasing, it is called a percent decrease.

percent increase [6.3] In a percent problem, if the quantity is increasing, it is called a percent increase.

perfect square [1.5, 11.6] A perfect square is a number that is the square of any whole number.

perimeter [11.2] The perimeter is the distance around a polygon.

periods [1.1] A period is a group of three numbers, which are determined by commas, when writing a large whole number in standard form.

perpendicular lines [11.1] Two lines that intersect to form right angles are called perpendicular lines.

pictograph [8.2] A pictograph is a variation of the bar graph in which images of people, books, coins, etc., represent the quantities.

place value [1.1] Each of the digits in a whole number in standard form has place value.

plane [11.1] A plane is a flat surface that extends endlessly in all directions.

point [11.1] An exact location in space, with no dimension, is called a point.

polygon [11.1] A closed plane figure made up of line segments is called a polygon.

positive number [7.1] A positive number is a number greater than 0.

prime factorization [2.1] Prime factorization is the process of writing a whole number as a product of its prime factors.

prime number [2.1] A prime number is a whole number that has exactly two different factors, itself and 1.

principal [6.3] The principal is the amount of money borrowed.

product [1.3] The result of a multiplication problem is called the product.

proper fraction [2.2] A proper fraction is a fraction less than 1, that is, a fraction whose numerator is smaller than its denominator.

proportion [5.2] A proportion is a statement that two ratios are equal.

Pythagorean Theorem [11.6] The Pythagorean Theorem states that for every right triangle, the sum of the squares of the two legs equals the square of the hypotenuse.

quadrilateral [11.1] A polygon with four sides is called a quadrilateral.

quotient [1.4] The result of a division problem is called the quotient.

radius [11.1] A line segment with one endpoint on the circle and the other at the center of a circle is called the radius of the circle.

range [8.1] The range of a list of numbers is the difference between the largest number and the smallest number on the list.

rate [5.1] A rate is a ratio of unlike quantities.

ratio [5.1] A ratio is a comparison in terms of a quotient of two numbers a and b, usually written as $\frac{a}{b}$.

ray [11.1] A part of a line having only one endpoint is called a ray.

reciprocal [2.4] The reciprocal of the fraction $\frac{a}{b}$ is $\frac{b}{a}$.

rectangle [11.1] A rectangle is a parallelogram with four right angles.

rectangular solid [11.4] A rectangular solid is a solid in which all six faces are rectangles.

reduced to lowest terms [2.2] A fraction is said to be reduced to lowest terms when the only common factor of its numerator and its denominator is 1.

right angle [11.1] A right angle is an angle that measures 90°.

right triangle [11.1] A right triangle is a triangle with one right angle.

rounding [1.1] Rounding is the process of approximating an exact answer by a number that ends in a given number of zeros.

scalene triangle [11.1] A triangle with no sides equal in length is called a scalene triangle.

signed number [7.1] A signed number is a number with a sign that is either positive or negative.

similar triangles [11.5] Similar triangles are triangles that have the same shape but not necessarily the same size.

simplest form [2.2] A fraction is said to be in simplest form when the only common factor of its numerator and its denominator is 1.

solution [9.1] A solution to an equation is a value of the variable that makes the equation a true statement.

solve [9.1] To solve an equation means to find all solutions to the equation.

sphere [11.4] A sphere is a three-dimensional figure made up of all points a given distance from the center.

square [11.1] A square is a rectangle with four sides equal in length.

square root of n [11.6] The square root of n is the number, written $\sqrt{n}$, whose square is n.

straight angle [11.1] A straight angle is an angle that measures 180°.

subtrahend [1.2] In a subtraction problem, the number that is being subtracted is called the subtrahend.

sum [1.2] The result of an addition problem is called the sum.

supplementary angles [11.1] Supplementary angles are angles whose sum is 180°.

table [8.2] A table is a rectangular display of data consisting of rows and columns.

trapezoid [11.1] A trapezoid is a quadrilateral with only one pair of opposite sides parallel.

triangle [11.1] A triangle is a polygon with three sides.

unit fraction [2.3] A fraction with 1 as the numerator is called a unit fraction.

unit price [5.1] The unit price is the price of one item.

unit rate [5.1] A unit rate is a rate in which the number in the denominator is 1.

unlike fractions [2.2] Unlike fractions are fractions which have different denominators.

unlike quantities [5.1] Unlike quantities are quantities that have different units.

unlike terms [9.2] Unlike terms are terms that do not have the same variable or the variables are not raised to the same powers.

variable [4.1] A variable is a letter that represents an unknown number.

vertex [11.1] The common endpoint of an angle is called the vertex.

vertical angles [11.1] Two angles that are formed by intersecting lines are called vertical angles.

volume [11.4] Volume is the number of cubic units required to fill a three-dimensional figure.

weighted average [8.1] A weighted average is a special kind of average (mean) used when some numbers on the list count more heavily than others.

Index

Index of Applications

Video Index